T0250646

Lecture Notes in Computer Science 626

Edited by G. Goos and J. Hartmanis

Advisory Board: W. Brauer D. Gries J. Stoer

E. Börger G. Jäger H. Kleine Büning
M. M. Richter (Eds.)

Computer Science Logic

5th Workshop, CSL '91
Berne, Switzerland, October 7-11, 1991
Proceedings

Springer-Verlag
Berlin Heidelberg New York
London Paris Tokyo
Hong Kong Barcelona
Budapest

Series Editors

Gerhard Goos
Universität Karlsruhe
Postfach 69 80
Vincenz-Priessnitz-Straße 1
W-7500 Karlsruhe, FRG

Juris Hartmanis
Department of Computer Science
Cornell University
5149 Upson Hall
Ithaca, NY 14853, USA

Volume Editors

Egon Börger
Dipartimento di Informatica, Università di Pisa
Corso Italia, 40, I-56100 Pisa, Italia

Gerhard Jäger
Universität Bern, Institut für Informatik und angewandte Mathematik
Länggassstraße 51, CH-3012 Bern, Switzerland

Hans Kleine Büning
FB 17, Mathematik/Informatik, Universität – GH Paderborn
Postfach 16 21, W-4790 Paderborn, FRG

Michael M. Richter
Fachbereich Informatik, Universität Kaiserslautern
Postfach 3049, W-6750 Kaiserslautern, FRG

CR Subject Classification (1991): D.3, F, G.2, H.2, I.1, I.2

ISBN 3-540-55789-X Springer-Verlag Berlin Heidelberg New York
ISBN 0-387-55789-X Springer-Verlag New York Berlin Heidelberg

This work is subject to copyright. All rights are reserved, whether the whole or part of the material is concerned, specifically the rights of translation, reprinting, re-use of illustrations, recitation, broadcasting, reproduction on microfilms or in any other way, and storage in data banks. Duplication of this publication or parts thereof is permitted only under the provisions of the German Copyright Law of September 9, 1965, in its current version, and permission for use must always be obtained from Springer-Verlag. Violations are liable for prosecution under the German Copyright Law.

© Springer-Verlag Berlin Heidelberg 1992
Printed in Germany

Typesetting: Camera ready by author/editor
45/3140-543210 - Printed on acid-free paper

Preface

The workshop CSL '91 (Computer Science Logic) was held at the University of Berne, Switzerland, October 7–11, 1991. It was the fifth in a series of workshops, following CSL '90 at the Max-Planck-Haus in Heidelberg, CSL '89 at the University of Kaiserslautern, CSL '88 at the University of Duisburg, and CSL '87 at the University of Karlsruhe. Forty-three talks were presented at the workshop, consisting of invited presentations and those selected from a hundred and one submissions.

In addition to the usual program of CSL meetings we had a panel discussion on the topic "Theoretische Informatik: Abstrakte Konzepte und praktische Anwendungen". It was our aim to bring together distinguished specialists from academia and industry to address this interesting and relevant issue in theoretical computer science. We thank the panelists K. Drangeid (IBM), E. Engeler (ETH Zürich), H. Lienhard (Landis & Gyr), H. Mey (Universität Bern, ASCOM Tech), P. Päppinghaus (Siemens) and M.M. Richter (Universität Kaiserslautern) for sharing their interesting views about the possible connections and interaction between theory and practice in the general field of computer science logic.

As was the case for CSL '87 (see Lecture Notes in Computer Science, Vol. 329), CSL '88 (see Lecture Notes in Computer Science, Vol. 385), CSL '89 (see Lecture Notes in Computer Science, Vol. 440), and CSL '90 (see Lecture Notes in Computer Science, Vol. 533), we collected the original contributions after their presentation at the workshop and began a review procedure which resulted in the selection of the papers in this volume. They appear here in final form.

We would like to thank the referees without whose help we would not have been able to accomplish the difficult task of selecting among the many valuable contributions. Several members of the Institut für Informatik und angewandte Mathematik of the University of Berne helped to organize the meeting and to prepare this volume, in particular Ursula Hadorn, Urs-Martin Künzi, Markus Marzetta, Robert Stärk, Thomas Strahm, Tyko Strassen and Susanne Thüler. We thank all of them for their commitment and continued support.

We gratefully acknowledge the financial sponsorship by the following institutions:

Max und Elsa Beer-Brawand-Fonds
Schweizerische Akademie der Naturwissenschaften
ASCOM Tech AG
Schweizerische Lebensversicherungs- und Rentenanstalt
Schweizerische Mobiliar Versicherungsgesellschaft
Sun Microsystems Schweiz AG

May 1992 E. Börger, G. Jäger, H. Kleine Büning, M. M. Richter

Table of Contents

This article was processed using the LaTeX macro package with LLNCS style

The Expressive Power of Transitive Closure and 2-way Multihead Automata

Yaniv Bargury* and Johann Makowsky*

Abstract

It is known, that over 1-dimensional strings, the expressive power of 2-way multihead (non) deterministic automata and (non) deterministic Transitive Closure formulas is (non) deterministic log space [Ib73, Im88]. However, the subset of formulas needed to simulate exactly k heads is unknown. It is also unknown if the automata and formulas have the same expressive power over more general structures such as multidimensional grids. We define a reduction from k-head automata to formulas of arity k, which works also for grids. The method used is a generalization of [Kl56], and the formulas obtained are a generalization of regular expressions to multihead automata and to grid languages. As simple applications, we use the reduction to show that the power of formulas of arity 1 over strings define (classical) regular languages, to give a simpler equivalent of the L=NL open problem, and to establish the equivalence of the automata and formulas over grids.

*Faculty of Computer Science, Technion—Israel Institute of Technology, Haifa, Israel. e-mail: yaniv@techunix.ac.il and janos@cs.technion.ac.il

1 Introduction

In this paper we aim at linking closer two characterizations of Log-Space computations both deterministic (L) and nondeterministic (NL). The first is the family of sets of words recognized by 2-way multihead automata. It is well known, see for example [Ib73], that this family is L for deterministic automata and NL for nondeterministic automata. The second is the family of sets of string models definable by Transitive Closure formulas. Immerman [Im88] proved that this family is L for deterministic Transitive Closure formulas and NL for nondeterministic Transitive Closure formulas.

Our target is finding the subset of formulas which capture exactly the same computational power as automata restricted to exactly k heads. To this end we generalize the Kleene [Kl56] reduction of classical automata to classical regular expressions to the context of many heads. In section 5 we reach this target for the nondeterministic case. A partial result that only embeds automata in formulas, applicable to the deterministic case, is obtained in section 3.

This reduction from automata to logic is intended to be a tool for transferring results between these two fields of research. Thus for example properties of some simpler automata can be expressed as properties of simpler formulas and investigated with model theoretical tools such as Ehrenfeucht-Fraïssé games. Such games for Transitive Closure logics where outlined in [Ca90] and are indeed simpler for the restricted formulas. As simple applications we obtained some new results:

(i) A characterization of regular sets in Transitive Closure formulas in section 4: Regular sets are exactly the sets accepted by Transitive Closure formulas of arity 1.

(ii) Using the fact that deterministic and nondeterministic space are the same iff any 3 head nondeterministic automata can be simulated in deterministic Log-Space we obtain a similar theorem in the context of formulas: $L = NL$ iff for every nondeterministic Transitive Closure formula of arity 3 there is an equivalent formula in deterministic Transitive Closure.

The formulas which capture exactly k heads are a generalization of regular expressions to multihead automata. We therefore call them k-regular formulas. They can be used to describe and investigate the languages accepted by k-head automata. We push this idea forward by expressing all the result in terms of multidimensional grids. Thus we also get a generalization of regular expressions to multidimensional grids, with strings and 2-D pictures being a special case. Little is known about the expressive power of automata on pictures, and we hope describing picture languages with k-regular expressions will prove helpful.

2 Background and definitions

2.1 Multidimensional grids

A structure w is an d-dimensional grid, or simply a grid, over the alphabet Σ, denoted $w \in \mathcal{W}(d, \Sigma)$ if

$$w = \langle W, \mathbf{min}, \mathbf{max}, s_1 \ldots s_d, L_1 \ldots L_{\lceil \log_2(|\Sigma|) \rceil} \rangle$$

Where the relations s_i are binary, and L_i are unary. This structure is actually a colored multi-graph, but we restrict ourself to structures which satisfy the following restrictions: Each $a \in \Sigma$ is coded as a binary number and the L_i predicates represent the binary digits in that number. Thus if $\Sigma = \{00, 01, 10\}$ and $w \in W$ has $L_1(w) =$ true and $L_2(w) =$ false we say that the cell w contains a 10. We sometime write $10(w)$ in formulas as short for $L_1(w) \wedge \neg L_2(w)$.

The s_i relations define the neighbors in an d-dimensional grid along the ith axis. Thus when $s_i(x, y) =$ true x is just below y on the ith axis. The s_i relations must satisfy certain conditions so that they will actually form a grid. A broader theory can be formed for the more general structures, which is beyond the scope of this discussion. All grid cells $a \in W$ for which there exists some i for which there is no x such that $s_i(x, a)$ or there is no x such that $s_i(a, x)$ are called edge cells. The L_i's always encodes for the edge cells special edge symbols, which are implicit elements of Σ. The edge cell s contains the edge symbol $E[e_1 \ldots e_d]$ where

$$e_i = \begin{cases} 1 & \text{if there is no } x \text{ s.t. } s_i(a, x) \\ -1 & \text{if there is no } x \text{ s.t. } s_i(x, a) \\ 0 & \text{otherwise} \end{cases}$$

The edge symbols are a technicality that will enable us to detect the edges of the grid. To avoid trivial cases we assume that the length of the grid in any axis is always at least 2.

For historical reasons we also have the two constants **min** and **max** in the grid. **min** is the grid cell for which the L_i's give $E[-1, -1, \ldots -1]$ and **max** is the cell for which the L_i's give $E[1, 1, \ldots 1]$.

2.2 Multihead automata

In this paper we concentrate on automata which can move their heads both ways in any dimension. They are usually called 2-way automata even though with d dimensions they can move in $2d$ ways.

M is a 2-way non-deterministic multihead finite automaton with k heads over d dimensional grids, or $M \in$ N2MFA(k, d) for short, if M is a tuple

$$M = \langle S, \Sigma, q_0, \delta, \tau, F \rangle$$

where S is a finite set of states, Σ is the alphabet i.e. the set of input symbols, $q_0 \in S$ is the initial state, $\delta : S \times \Sigma \rightarrow \mathcal{P}(S \times \{-1, 0, 1\})$ is the transition function,

$\tau : S \rightarrow \{1 \ldots k\} \times \{1 \ldots d\}$ is the selector function, and $F \subseteq S$ is the set of final states. We always assume without loss of generality that $S = \{1 \ldots |S|\}$.

A move of M consists of first consulting τ with the present state and getting the active head number h and the active dimension number d_c for this state. M now scans the symbol a under head h and consults δ which gives a set of pairs of next state and move direction m for the present state and a. M nondeterministically selects one element of the set. M now moves head h by m cells along the s_{d_c} relation if possible and switches to the next state. An impossible move is a move that positions a head off the grid, i.e. when $a = E[e_1 \ldots e_d]$ and $e_{d_c} = m$. In this case M halts and rejects the input. When $|\delta(s, a)| \leq 1$ for all s and a we say that M is deterministic and denote $M \in$ D2MFA(k, d).

M accepts an d-dimensional word, and we write $w \models M$, if when M is started in state q_0 with all heads pointing to **min** there is a sequence of moves that ends in a final state. We denote $L(M) = \{w \mid w \models M\}$. We also denote

$$\text{LN2MFA}(k, d) = \{L \mid L = L(M) \text{ for some } M \in \text{N2MFA}(k, d)\}$$

and similarly LD2MFA(k, d) for the deterministic case.

Recall that with usual 1-head finite automata over 1 dimensional strings we have:

Theorem 1 *Finite automata with one head over 1 dimensional strings which are either deterministic or nondeterministic, 1-way or 2-way, have the same expressive power.*

A proof of this theorem can be found in [HU79]. However, in the case of multihead automata the expressive power of 2-way automata is strictly larger than that of 1-way automata. Furthermore, many heads give much more computational power, as noted by [Ib73]:

Theorem 2

(i) $\bigcup_{k=1}^{\infty}$ *LD2MFA*$(k, 1) = $ *L. [Ib73]*

(ii) $\bigcup_{k=1}^{\infty}$ *LN2MFA*$(k, 1) = $ *NL. [Ib73]*

2.3 Transitive closure logics

We now recall the definition of transitive closure logics. To simplify the presentation we do not differentiate the semantic symbols in the grid structures from the syntactical symbols used in formulas. These logics where defined in [Im87] but we use a slightly different notation. First order logic (FOL) with non-deterministic transitive closure restricted to arity k, over an d dimensional vocabulary, or NTC(k, d) for short, is FOL expanded with NTC^k operators for some fixed $k \in \mathbb{N}$. k is called the *arity* of the operator. The syntax for the NTC^k operator is:

$$\psi = \mathrm{NTC}^k(\overline{xy}\ \varphi(\overline{zt}))$$

Where $\overline{x}$ and $\overline{y}$ are k-ary vectors of variables or constants and $\overline{z}$ and $\overline{t}$ are k-ary vectors of variables. $\overline{x}$ and $\overline{y}$ are free in ψ and in φ. $\overline{z}$ and $\overline{t}$ are free in φ but are bound in ψ by the NTC^k operator. The order of the variables indicated in $\varphi(\overline{zt})$ is relevant for the meaning of the formula. If instead of the symbolic φ we write an explicit formula the order of binding should be added to the syntax of the operator NTC^k. However, we omit the additional notational burden when the context makes the order clear. In this case we use the same variable names in $\overline{xy}$ and $\overline{zt}$ and write $\mathrm{NTC}^k(\overline{xy}\varphi(\overline{xy}))$ or just $\mathrm{NTC}^k(\overline{xy}\varphi)$. The intended meaning in both cases is $\mathrm{NTC}^k(\overline{xy}\varphi'(\overline{vw}))$ where $\overline{vw}$ are new variable names and φ' is φ with the $\overline{xy}$ variables substituted to the new names.

The meaning function for ψ gives "true" when there exists a series of k-ary vectors of elements in the world, $\overline{a}_1 \ldots \overline{a}_m$ for some m, such that $\overline{x} \approx \overline{a}_1$, $\overline{y} \approx \overline{a}_m$ and $\varphi(\overline{a}_i\overline{a}_{i+1})$ holds for all i. The notation $\varphi(\overline{a}_i\overline{a}_{i+1})$ stands for substituting elements of $\overline{zt}$ with the respective elements of $\overline{a}_i\overline{a}_{i+1}$.

We have a similar logic for deterministic transitive closure, which we call $\mathrm{DTC}(k,d)$. It uses the operator DTC^k instead of NTC^k, with the same syntax, but with the added restriction on the meaning function that if $\varphi(\overline{a}_i\overline{b})$ holds for some $\overline{b}$ and i then $\overline{b} \approx \overline{a}_{i+1}$.

We now denote $w \models \varphi$ if w is a grid which satisfies φ, $L(\varphi) = \{w \mid w \models \varphi\}$, $\mathrm{LNTC}(k,d) = \{L \mid L = L(\varphi) \text{ for some } \varphi \in \mathrm{NTC}(k,d)\}$ and similarly $\mathrm{LDTC}(k,d)$ for the deterministic case.

The unrestricted Transitive Closure logics originally defined by Immerman are $\bigcup_{k=1}^{\infty} \mathrm{NTC}(k,d)$ and $\bigcup_{k=1}^{\infty} \mathrm{DTC}(k,d)$. He also describes their expressive power over 1 dimensional strings (and everything that is 1st order reducible to strings):

Theorem 3 (Immerman)

(i) $\bigcup_{k=1}^{\infty} LDTC(k,1) = L$. *[Im88]*

(ii) $\bigcup_{k=1}^{\infty} LNTC(k,1) = NL$. *[Im88]*

We say that a $\mathrm{NTC}(k,d)$ ($\mathrm{DTC}(k,d)$) formula φ is in Immerman Normal form (INF) if it is of the form $\varphi = \mathrm{NTC}^k(\overline{\mathbf{min}\mathbf{max}}\ \psi(\overline{zt}))$ (respectively $\varphi = \mathrm{DTC}^k(\overline{\mathbf{min}\mathbf{max}}\ \psi(\overline{zt}))$) where ψ does not contain any quantifier or NTC^k (DTC^k) operator, i.e. ψ is a boolean combination of atomic formulas. Justification for the term normal form comes from the following lemma:

Lemma 1 (Immerman [Im87]) *For every $\varphi \in NTC(k,d)$ there is some k' and $\theta \in NTC(k',d)$ such that θ is in INF and $\theta \equiv \varphi$.*

And similarly For every $\varphi \in DTC(k,d)$ there is some k' and $\theta \in DTC(k',d)$ such that θ is in INF and $\theta \equiv \varphi$.

Comment: It is easy to simulate a Transitive Closure operator using a Transitive Closure operator of higher arity:

$$NTC^k(\overline{xy}\ \varphi(\overline{zt})) = NTC^{k+1}(\overline{xy}\ (\varphi \wedge x_{k+1} \approx y_{k+1})(\overline{zt})).$$

so clearly for every k and d we have

$$\mathrm{LNTC}(k,d) \subseteq \mathrm{LNTC}(k+1,d)$$

and

$$\mathrm{LDTC}(k,d) \subseteq \mathrm{LDTC}(k+1,d)$$

It is still open whether we can increase power forever or is there an arity k_0 for which any Transitive Closure operators can be encoded with formulas that only use operators of arity k_0.

3 From automata to logic

From theorems 2 and 3 we immediately get that over *strings*, each multihead automaton has an equivalent Transitive Closure formula and vice versa. This raises two questions: First, is this also true over more general structures such as multidimensional grids? Second, If we restrict ourself to a fixed number of heads, can we define a proper subset of the Transitive Closure formulas from which the equivalent formula can be selected (i.e. what is the parameter that corresponds to the number of heads in formulas)? Theorems 3 and 2 do not answer these questions. An affirmative answer for the first question is derived at the end of this section, and an answer to the second question can be found in this section and in section 5.

First we encode automata in formulas in the following lemma. Note that the lemma works for any dimensional grids, and that it gives an upper bound on the arity of Transitive Closure operators used in the formula which encodes the automaton.

Lemma 2 *For every k, where k is simultaneously the arity and number of heads, and for every dimension d and for every set A of d dimensional grids accepted by a (non)deterministic 2-way multihead finite automata with k heads, A is accepted by a (non)deterministic transitive closure formula of arity k. In our notations:*

$$LD2MFA(k,d) \subseteq LDTC(k,d)$$

and

$$LN2MFA(k,d) \subseteq LNTC(k,d)$$

Proof: The key observation here is that variables in a formula in NTC(k,d) (DTC(k,d)) can be viewed as pointers to cells in the grid. We therefore simulate the k heads of the automaton with k variables. A head move can be captured by the formula:

$$\text{mv}(p_{1h}, d_c, m, p_{2h}) = \begin{cases} s_{d_c}(p_{1h}, p_{2h}) & \text{if } m{=}1 \\ p_{1h} \approx p_{2h} & \text{if } m{=}0 \\ s_{d_c}(p_{2h}, p_{1h}) & \text{if } m{=}-1 \end{cases}$$

which is true iff a head h was in position p_{1h} and has moved m cells in dimension d_c to the new position p_{2h}. A transition of an automata can be captured by:

$$\begin{aligned} \text{tr}(a, h, d_c, m) = \quad & a(p_{1h}) \wedge \\ & \text{mv}(p_{1h}, d_c, m, p_{2h}) \wedge \\ & \bigwedge_{l \neq h} p_{1l} \approx p_{2l} \end{aligned}$$

which is true iff the same move was made while head h was scanning the symbol a, and all other heads stayed at the same place.

We now use the Kleene method for encoding a finite automaton with a regular expression. We first deal with the nondeterministic case. Given an automaton $M \in$N2MFA(k,d), we construct an equivalent formula $\chi \in$NTC(k,d).

Let $M = \langle S, \Sigma, q_0, \delta, \tau, F \rangle$. We build recursively the NTC(k,d) formulas $\chi^n_{ij}(\overline{p}_1\overline{p}_2)$, where $\overline{p}_i$ are vectors of variables of arity k. The formula $\chi^n_{ij}(\overline{p}_1\overline{p}_2)$ is true iff when M is started in state i with its head m positioned at p_{1m} for $m = 1 \ldots k$, M can move with 0 or more transitions to state j and head positions $p_{21} \ldots p_{2k}$, while not passing through states higher than n. Formally, we put:

$$\chi^0_{ij}(\overline{p}_1\overline{p}_2) = \bigvee_{\substack{a \in \Sigma \\ \langle j, m \rangle \in \delta(i, a) \\ \tau(i) = \langle h, d_c \rangle}} \text{tr}(a, h, d_c, m) \vee \begin{cases} \text{false} & i{\neq}j \\ \overline{p}_1 \approx \overline{p}_2 & i{=}j \end{cases}$$

Note that a move with 0 transitions, usually called an ϵ-move, is supplied by the formula $\overline{p}_1 \approx \overline{p}_2$ on the right. Now set

$$\begin{aligned} \chi^n_{ij}(\overline{p}_1\overline{p}_2) \;\; = \;\; & \exists \overline{p}_3\overline{p}_4 \left(\chi^{n-1}_{in}(\overline{p}_1\overline{p}_3) \wedge \text{NTC}^k(\overline{p}_3\overline{p}_4\; \chi^{n-1}_{nn}(\overline{p}_3\overline{p}_4)) \wedge \chi^{n-1}_{nj}(\overline{p}_4\overline{p}_2) \right) \vee \\ & \chi^{n-1}_{ij}(\overline{p}_1\overline{p}_2) \end{aligned}$$

Finally let

$$\chi = \exists \overline{p}_1 \bigvee_{i \in F} \chi^{|S|}_{q_0 i}(\overline{\textbf{min}}\,\overline{p}_1)$$

The formula χ says that if M is started with all heads pointing at $\overline{\textbf{min}}$ and in the initial state, then M can move to some head position $\overline{p_1}$ in a final state. This is exactly the condition for accepting the input, and therefore $L(M) = L(\chi)$.

We now cover the modifications needed in the deterministic case. We want to substitute the NTC^k operators in the χ_{ij}^n definition with DTC^k, but some details must be taken care of first. Each head position can lead through χ_{ij}^n both to a new head position in the same state, and to the same head position through an ϵ-move. Therefore we must dispose of the ϵ-moves as they would make all the DTC^k give false. We correct the induction step definition of χ_{ij}^n to compensate for the loss of the ϵ-moves.

This is not yet enough. Loops on a single state allows more than one $\overline{p}_2$ to satisfy $\chi_{nn}^{n-1}(\overline{p}_1, \overline{p}_2)$ for some fixed $\overline{p}_1$, and the DTC^k operator will give false in these cases. Suppose $\chi_{nn}^{n-1}(\overline{p}_1, \overline{x})$ and $\chi_{nn}^{n-1}(\overline{p}_1, \overline{y})$ where $\overline{x} \neq \overline{y}$. M is deterministic and has only one possible sequence of head positions, therefore $\chi_{nn}^{n-1}(\overline{y}, \overline{x})$ or $\chi_{nn}^{n-1}(\overline{x}, \overline{y})$ holds. Thus when M is started in state n with head positions $\overline{p}_1$, it either gets into an infinite loop and rejects, or else there must exist some unique head position $\overline{z}$ such that $\chi_{nn}^{n-1}(\overline{p}_1, \overline{z}) \wedge \neg\exists \overline{t}(\chi_{nn}^{n-1}(\overline{p}_1, \overline{t}) \wedge \chi_{nn}^{n-1}(\overline{z}, \overline{t}))$. Thus in the deterministic case we assume without loss of generality that there are no transitions from final states (to avoid infinite loops after reaching a final state) and change the definition of $\chi_{ij}^n(\overline{p}_1, \overline{p}_2)$ to

$$\chi_{ij}^0(\overline{p}_1\overline{p}_2) = \bigvee_{\substack{a \in \Sigma \\ \langle j, m\rangle \in \delta(i,a) \\ \tau(i) = \langle h, d_c\rangle}} \mathrm{tr}(a, h, d_c, m)$$

and

$$\begin{aligned} \chi_{ij}^n(\overline{p}_1\overline{p}_2) \;=\; & \exists \overline{p}_3\overline{p}_4 \\ & (\; \varphi_{in}^{n-1}(\overline{p}_1\overline{p}_3) \wedge \\ & \quad \mathrm{DTC}^k(\overline{p}_3\overline{p}_4\,(\chi_{nn}^{n-1}(\overline{p}_3\overline{p}_4) \wedge \neg\exists \overline{p}_5(\chi_{nn}^{n-1}(\overline{p}_3\overline{p}_5) \wedge \chi_{nn}^{n-1}(\overline{p}_4\overline{p}_5)))) \wedge \\ & \quad \varphi_{nj}^{n-1}(\overline{p}_4\overline{p}_2) \\ &) \;\vee \chi_{ij}^{n-1}(\overline{p}_1\overline{p}_2) \end{aligned}$$

where

$$\varphi_{ij}^n(\overline{p}_1\overline{p}_2) = \begin{cases} \chi_{ij}^n(\overline{p}_1\overline{p}_2) & i \neq j \\ \overline{p}_1 \approx \overline{p}_2 & i = j \end{cases}$$

and note that the proof still works. □

4 One head over strings

In this section we establish the fact that the expressive power of Transitive Closure logics restricted to arity 1 is exactly the regular languages. For this we need a powerful theorem that was proven by Büchi [Bu60], and independently by Trakhtenbrot [Tr61]:

Theorem 4 (Büchi, Trakhtenbrot) *Over strings, the set of languages definable by Monadic 2nd order formulas (LM2O) is equal to the set of regular languages (REG).*

For a proof the reader should rather consult [La77] or [Th80].

The next theorem is a complete map of the expressive power of the automata and logics we examine here, in the case of string languages and arity and number of heads 1. Our contribution to the map is the Transitive Closure logics.

Theorem 5 *The following define the same languages:*

REG *The set of regular languages (REG);*

LD2MFA(1,1) *Deterministic 2-way 1 head finite automata over strings;*

LN2MFA(1,1) *Non-deterministic 2-way 1 head finite automata over strings;*

LDTC(1,1) *Deterministic Transitive Closure logic of arity 1 over strings;*

LNTC(1,1) *Non-deterministic Transitive Closure logic of arity 1 over strings;*

LM2O *Monadic Second Order logic over strings.*

Proof: REG=LD2MFA$(1,1)$=LN2MFA$(1,1)$ is well known, see for example [HU79]. REG$(1,1)$=LM2O is theorem 4. LD2MFA$(1,1)\subseteq$LDTC$(1,1)$ is by lemma 2. LDTC$(1,1)\subseteq$LNTC$(1,1)$ is done by encoding the DTC^1 operator with NTC^1 thus:

$$\mathrm{DTC}^1(xy\ \varphi(zt)) = \mathrm{NTC}^1(xy\ (\varphi(zt) \wedge \forall s\ (\varphi(zs) \rightarrow s \approx t))(zt))$$

Finally LNTC$(1,1)\subseteq$LM2O is shown by encoding each NTC^1 operator in monadic 2nd order as follows. For each $\psi = \mathrm{NTC}^1(xy\ \varphi(zt))$ we define a formula $\theta(x)$ which says that the set A is the minimal set which contains x and is closed under φ:

$$\begin{aligned}\theta(x) \quad = \quad & (A(x) \wedge (\forall v \forall w\ A(v) \wedge \varphi(v,w) \rightarrow A(w))) \wedge \\ & (\forall B\ (B(x) \wedge (\forall v \forall w\ B(v) \wedge \varphi(v,w) \rightarrow B(w))) \rightarrow \\ & (\forall v\ A(v) \rightarrow B(v)))\end{aligned}$$

Now we substitute ψ with $\exists A(\theta \wedge A(y))$. □

This method of proof cannot be extended to bigger dimensions, as it heavily depend on Büchi's theorem which only works for strings. Obvious extensions to bigger arity and number of heads also fail to give productive result, as the encoding of the Transitive Closure operators require quantifying over relations with any arity. Recall that the expressive power of quantifying over general relations is the NP hierarchy.

5 k-Regular Formulas

We now define formulas which capture exactly the expressive power of k heads in the Non-deterministic case. Comments for the deterministic case are given at the end of this section.

Definition 1 ***k-regular formulas over d dimensional grids, or RTC(k,d) for short, are the minimal set such that:***

- *$tr(a,h,d_c,m) \in RTC(k,d)$, for all $a \in \Sigma$, $h \in \{1\dots k\}$, $d_c \in \{1\dots d\}$ and $m \in \{-1,0,1\}$.*
- *If $\varphi, \psi \in RTC(k,d)$ then $\varphi \circ \psi \in RTC(k,d)$*
- *If $\varphi \in RTC(k,d)$ then $\varphi^* \in RTC(k,d)$*
- *If $\varphi, \psi \in RTC(k,d)$ then $\varphi \vee \psi \in RTC(k,d)$*

We define a translation from k-regular formulas to NTC formulas $T : \mathrm{RTC}(k,d) \rightarrow \mathrm{NTC}(k,d)$ inductively as follows:

- **$T(tr(a,h,d_c,m))$ was defined in the proof of lemma 2.**
- **$T(\varphi \circ \psi) = \exists \overline{p}_3 \left(T(\varphi) \dfrac{\overline{p}_2}{\overline{p}_3} \wedge T(\psi) \dfrac{\overline{p}_1}{\overline{p}_3} \right)$ where $\theta\frac{\overline{x}}{\overline{y}}$ denote substituting all free occurrences of variables from $\overline{x}$ with the respective constant or variables from $\overline{y}$.**
- **$T(\varphi^*) = \mathrm{NTC}^k(\overline{p}_1\overline{p}_2\varphi)$**
- **$T(\varphi \vee \psi) = T(\varphi) \vee T(\psi)$**

Note that the names of the free variables in $tr(a,h,d_c,m)$ are $p_{11}\dots p_{1k}$ and $p_{21}\dots p_{2k}$. It is easily checked that the same is true for all k-regular formulas. It therefore makes sense to define the meaning for $\varphi \in \mathrm{RTC}(k,d)$ using the meaning in $\mathrm{NTC}(k,d)$ like this:

$$G \models \varphi \text{ iff } G \models \exists \overline{p}_2 T(\varphi) \frac{\overline{p}_1}{\mathbf{min}}$$

As usual we denote $\mathrm{LRTC}(k,d) = \{G \mid G \models \varphi \text{ and } \varphi \in \mathrm{RTC}(k,d)\}$. As promised, the following holds:

Theorem 6 *For every k and every d*

$$LN2MFA(k,d) = LRTC(k,d)$$

Proof: The proof that LN2MFA$(k,d) \subseteq$ LRTC(k,d) is almost identical to the proof of lemma 2. The inverse T translations of the χ_{ij}^n formulas are the k-regular formulas needed:

$$\chi_{ij}^0 = \bigvee_{\substack{a \in \Sigma \\ \langle j,m\rangle \in \delta(i,a) \\ \tau(i) = \langle h, d_c\rangle}} \mathrm{tr}(a,h,d_c,m) \vee \begin{cases} \text{false} & i \neq j \\ \bigvee_{a\in\Sigma} \mathrm{tr}(a,1,1,0) & i=j \end{cases}$$

$$\chi_{ij}^n = (\chi_{in}^{n-1} \circ (\chi_{nn}^{n-1})^* \circ \chi_{nj}^{n-1}) \vee \chi_{ij}^{n-1}$$

$$\chi = \bigvee_{i\in F} \chi_{q_0 i}^{|S|}$$

The proof that LN2MFA$(k,d) \supseteq$ LRTC(k,d) proceeds by the structure of the formulas. For each subformula we define an automaton which accepts the same language. The details are similar to the usual reduction of regular expressions to finite automata and we omit them. □

Note that the χ formulas in the previous proof are the *same* formulas used in the classical reduction of finite automata to regular expression. The only syntactical difference is the use of $\mathrm{tr}(a,h,d_c,m)$ instead elements of Σ. Classical finite automata have only one head, scan strings and only moves forward. Thus if we take a regular expression, and for each $a \in \Sigma$ substitute all a with $\mathrm{tr}(a,1,1,1)$, we get a k-regular formula with the same meaning. In this sense k-regular formulas generalize regular expressions, which justify their name.

Deterministic Transitive Closure formulas which capture the expressive power of the deterministic automata are supplied in lemma 2 but we can not give an elegant form for them as in k-regular formulas. The nondeterminism in k-regular formulas lies in the $\vee$ and $*$ operators. The modification of the k-regular formulas needed to take out the nondeterminism, will spoil their elegant form, and we do not pursue this further.

6 Many heads and many dimensions

First, we display the situation when we disregard the arity and number of heads.

Theorem 7 *For every dimension d*

$$\bigcup_{k=1}^{\infty} LNTC(k,d) = \bigcup_{k=1}^{\infty} LN2MFA(k,d) = \bigcup_{k=1}^{\infty} LRTC(k,d)$$

and

$$\bigcup_{k=1}^{\infty} LDTC(k,d) = \bigcup_{k=1}^{\infty} LD2MFA(k,d)$$

Proof:
$\bigcup_{k=1}^{\infty}$ LN2MFA(k,d) = $\bigcup_{k=1}^{\infty}$ LRTC(k,d) is from theorem 6.
$\bigcup_{k=1}^{\infty}$ LNTC(k,d) = $\bigcup_{k=1}^{\infty}$ LN2MFA(k,d) is clear when d=1, since each side is NL. With d>1, we have LN2MFA(k,d) $\subseteq$ LNTC(k,d) by lemma 2. To see the inverse inclusion, take any $\varphi \in$ NTC(k,d). By lemma 1 there exists some $\theta \in$ NTC(k',d) for some k' such that $\varphi \equiv \theta$ and θ is in INF. We can now simulate θ by a 2k+1 head automaton, but we omit the details for lack of space. A full 2 page proof can be found in [Ba91]. The deterministic case is proved similarly. □

Now lets look more closely at the arity, while fixing the dimension to 1. Two heads deterministic automaton can recognize non-regular languages such as $\{a^n b^n c^n \mid n \in \mathbb{N}\}$. This gives: LDTC(1,1)$\subsetneq$LDTC(2,1), LNTC(1,1)$\subsetneq$LNTC(2,1) and LRTC(1,1)$\subsetneq$LRTC(2,1). However for larger arity the following remains open:

Problem 1 ***Is it true that for for $k \geq 2$ LDTC(k,1)$\subsetneq$LDTC(k+1,1) and LNTC(k,1)$\subsetneq$LNTC(k+1,1)?***

For k-regular formulas we are more fortunate. Monien showed in [M76], that for all k, if LN2MFA(k,1) $\neq$ NL then LN2MFA(k,1)$\subsetneq$LN2MFA(k+1,1). This transforms through theorem 6 into

Theorem 8 ***For all k, if LRTC(k,1) $\neq$ NL then LRTC(k,1)$\subsetneq$LRTC(k+1,1).***

Another result by Monien [M76], which is is relevant here is as follows. If for some k LN2MFA(3,1) $\subset \bigcup_{k=1}^{\infty}$ LD2MFA(k,1) then L=NL. Clearly if L=NL then the inclusion holds, thus we immediately get:

Theorem 9 ***L=NL if and only if for some k LNTC(3,1) $\subset \bigcup_{k=1}^{\infty}$ LDTC(k,1).***

This problem can now be attacked with tools such as Ehrenfeucht-Fraïssé like games for Transitive Closure logics, defined in [Ca90] and [CM]. We now pursue in our research a suitable game for the k-regular formulas.

Another instance of the L=NL problem to be attacked by the games follows. 3DGAP is the set of 3 dimensional black and white grids in which min and max are connected by a path of black cells. Dahlhaus showed [Da84] that 3DGAP is NL-complete via first order reductions. Now, 3DGAP can be recognized by a 1 head nondeterministic automaton which guesses the path, thus we get:

Theorem 10 *L=NL if and only if for some d≥3 and k we have* $LRTC(1,d) \subset \bigcup_{k=1}^{\infty} LDTC(k,d)$.

Finally, lemma 2 and theorem 6 allow a new perspective of languages accepted by automata running on pictures. Basic problems are still open is this field. An extensive reference can be found in [IT88], and we give one simple example:

Problem 2 (2DGAP) *Is it possible for a deterministic 1 head automaton to recognize the set of black and white pictures on which* **min** *and* **max** *are connected by a path of black cells?*

Using Coy's method [Co77] it can be shown that this is impossible if the automata is restricted to go only on black cells. To the best of our knowledge the unrestricted problem is still open, and we propose a model theoretic approach to its equivalent:

Problem 3 *Is* **2DGAP** *in* $LDTC(1,2)$*?*

References

[Ba91] Yaniv Bargury. *The Hierarchy of Transitive Closure* Master thesis, Department of Computer Science, Technion—Israel Institute of Technology, Haifa, (In preparation).

[Bu60] J. R. Büchi. *Weak second-order arithmetic and finite automata* Z. Math. Logik Grundlagen Math **6**, 66–92, 1960.

[Ca90] Ariel Calò. *The Expressive Power of Transitive Closure* Master thesis, Department of Computer Science, Technion—Israel Institute of Technology, Haifa, 1990.

[CM] Ariel Calò and J.A. Makowsky, *The expressive power of transitive closure*, to appear in the proceedings of the Symposium on Logical Foundations of Computer Science 'Logic at Tver '92', Lecture Notes in Computer Science, 1992.

[Co77] W. Coy, *Automata in Labyrinths* Fundamentals of Computation Theory, LNCS **56**, 65–71, 1977.

[Da84] Elias Dahlhaus, *Reduction to NP-complete problems by interpretations* Logic and Machines Decision problems and Complexity, Lecture Notes in Computer Science, **171**, 357–365, 1984.

[HU79] J. E. Hopcroft and J. D. Ullman. *Introduction to Automata Theory, Languages, and Computation.* Addison Wesley, 1979.

[Ib73] Oscar. H. Ibarra. *On Two-Way Multihead Machines.* J. Comput. Sys. Sci. 7, 28–36, 1973.

[Im87] Neil Immerman. *Languages that Capture Complexity Classes.* SIAM J. Comput., Vol. 16, No. 4, August 1987.

[Im88] Neil Immerman. *Nondeterministic Space is Closed Under Complementation.* SIAM J. of Computation **17.5**, 935–938, 1988.

[IT88] K. Inoue and I. Takanami. *A Survey of Two-Dimensional Automata Theory* Machines, Languages and Complexity. Lecture Notes in Computer Science. **381**, 72–91, 1988.

[Kl56] S. C. Kleene. *Representation of events in nerve nets and finite automata* Automata studies, 3–42, Princeton Univ. Press, 1956

[La77] Richard E. Ladner. *Application of model theoretic games to discrete linear orders and finite automata.* Information and Control **33**, 281–303, 1977.

[M76] Burkhard Monien. *Transformational Methods and their Application to Complexity Problems* Acta Informatica 6, 95–108, 1976; Corrigenda, Acta Informatica **8**, 383–384, 1977.

[Th80] Wolfgang Thomas. *Classifying Regular Events in Symbolic Logic.* J. Comput. sys. sci. **25**, 360–376, 1982.

[Tr61] Boris A. Trakhtenbrot. *Finite automata and the logic of monadic predicates (in Russian.)* Dokl. Akad. Nauk CCCP **140**, 326–329, 1961.

Correctness proof for the WAM with types

Christoph Beierle
IBM Germany, Scientific Center
Inst. for Knowledge Based Systems
P.O. Box 80 08 80
D-7000 Stuttgart 80, Germany
beierle@dsØlilog.bitnet

Egon Börger
Dipartimento di Informatica
Università di Pisa
Corso Italia 40
I-56100 Pisa, Italia
boerger@dipisa.di.unipi.it

Abstract: We provide a mathematical specification of an extension of Warren's Abstract Machine for executing Prolog to type-constraint logic programming and prove its correctness. In this paper, we keep the notion of types and dynamic type constraints rather abstract to allow applications to different constraint formalisms like Prolog III or CLP(R). This generality permits us to introduce modular extensions of Börger's and Rosenzweig's formal derivation of the WAM. Starting from type-constraint Prolog algebras that are derived from Börger's standard Prolog algebras, the specification of the type-constraint WAM extension is given by a sequence of evolving algebras, each representing a refinement level. For each refinement step a correctness proof is given. Thus, we obtain the theorem that for every such abstract type-constraint logic programming system L and for every compiler satisfying the specified conditions, the WAM extension with an abstract notion of types is correct w.r.t. L. This is a first step towards our aim to provide a full specification and correctness proof of a concrete system, the PROTOS Abstract Machine (PAM), an extension of the WAM by polymorphic order-sorted unification as required by the logic programming language PROTOS-L.

1 Introduction

Recently, Gurevich's evolving algebra approach ([11]) has not only been used for the description of the (operational) semantics of various programming languages (Modula-2, Occam, Prolog, Prolog III, Smalltalk, Parlog, C; see [10]), but also for the description and analysis of implementation methods: Börger and Rosenzweig ([7,8]) provide a mathematical elaboration of Warren's Abstract Machine ([16], [1]) for executing Prolog. The description consists of several refinement levels together with correctness proofs, and a correctness proof w.r.t. Börger's phenomenological Prolog description ([5,6]).

In this paper we demonstrate how the evolving algebra approach naturally allows for modifications and extensions in the description of both the semantics of programming languages as well as in the description of implementation methods. Based on Börger and Rosenzweig's WAM description we provide a mathematical specification of a WAM extension to type-constraint logic programming and prove its correctness. Note that thereby our treatment covers also all extra-logical features (like the Prolog *cut*) whereas the WAM correctness proof of [12] deals merely with SLD resolution for Horn clauses.

The extension of logic programming by types requires in general not only static type checking, but types are also present at run time. For instance, if there are types and subtypes, restricting a variable to a subtype represents a constraint in the spirit of constraint logic programming. PROTOS-L ([2]) is a logic programming language that has a polymorphic, order-sorted type concept (derived from the slightly more general type concept of TEL [15], [14]) and a complete abstract machine implementation, called PAM ([13], [4]) that is an extension of the WAM by the required polymorphic order-sorted unification.

Our final aim is to provide a full specification and correctness proof of the concrete PAM system ([3]). In this paper, we keep the notion of types and dynamic type constraints rather abstract to allow applications to different constraint formalisms. Starting from type-constraint Prolog algebras that are derived from Börger's standard Prolog algebras, the specification of the type-constraint WAM extension is given by a sequence of evolving algebras, each representing a refinement level. We state precisely where no changes w.r.t. the WAM are needed (for instance the AND/OR structure can be taken essentially unchanged) and where - mostly orthogonal - extensions are required (in particular in the representation of terms). For each refinement step a correctness proof is given. As final result of this paper we obtain the main theorem: For every such abstract type-constraint logic programming system L and for every compiler satisfying the specified conditions the WAM extension with an abstract notion of types is correct w.r.t. L.

Although our description in this paper is oriented towards type constraints, it is modular in the sense that it should carry over to other constraint formalisms, like Prolog III or CLP(R), as well. Nevertheless, in order to avoid proliferation of different classes of evolving algebras, we will speak here in terms of PROTOS-L and PAM algebras (instead of type-constraint Prolog and type-constraint WAM algebras).

The paper is organized as follows: In Section 2 the derivation of PROTOS-L algebras from standard Prolog algebras is described. Section 3 defines PROTOS-L algebras with compiled AND/OR structure, and Section 4 introduces the representation of terms. The stack representation of environments and choicepoints is given in Section 5 which also contains the "Pure PROTOS-L" theorem stating the correctness of the PAM algebras developed so far w.r.t. the PROTOS-L algebras of Section 3. The notions of type constraint and constraint solving have been kept abstract through all refinement levels so far; thus, the development carried out in this paper applies to any type system satisfying the given abstract conditions. The introduction of PROTOS-L specific type constraint representation is carried out in [3].

To keep this paper within reasonable limits, we definitely suppose that the reader is familiar with [6] and [7,8] to which we will refer for many definitions and notations, including those concerning evolving algebras. The natural modularity of our approach will come out from presenting in this paper only those extensions and refinements which are needed in the presence of types. Due to severe space restrictions, examples could not be included in this paper; we refer the reader to [3].

2 PROTOS-L Algebras

2.1 Universes and Functions

The basic universes and functions in PROTOS-L algebras can be taken directly from the standard Prolog algebras ([5], [6]). In particular, we have the universes **TERM** and **SUBST** of terms and substitutions with a function

$$\texttt{subres:}\quad \textbf{TERM} \times \textbf{SUBST} \rightarrow \textbf{TERM}$$

yielding `subres(t,s)`, the result of applying `s` to `t`.

To be able to talk about (type constraints of) variables involved in substitutions we have to introduce a new universe

VARIABLE $\subseteq$ **TERM**

Since in PROTOS-L unification on terms is subject to type constraints on the involved variables, we have to distinguish between equating terms and satisfying type constraints for them. For this purpose we introduce a universe

EQUATION $\subseteq$ **TERM** $\times$ **TERM**

whose elements are written as $t_1 \doteq t_2$. Substitutions are then supposed to be finite sets of equations of the form $\{X_1 \doteq t_1, \ldots, X_n \doteq t_n\}$ with pairwise distinct variables X_i. The domain of such a substitution is the set of variables occurring on the left hand sides. (Note: If you want to have the logically correct notion of substitution - with occur check -, you should add the condition that no X_i occurs in any of the t_j.)

For a formalization of type constraints for terms - in the spirit of constraint logic programming - we introduce a new abstract universe **TYPETERM**, disjoint from **TERM** and containing all typeterms, that comes with functions

`TOP, BOTTOM:` **TYPETERM**
`inf:` **TYPETERM** $\times$ **TYPETERM** $\rightarrow$ **TYPETERM**

where `TOP` and `BOTTOM` can be thought of as 'maximal' and 'minimal' type terms, and `inf` yields the 'infimum' of two type terms, satisfying the integrity constraints

```
inf(TOP,tt) = inf(tt,TOP) = tt
inf(BOTTOM,tt) = inf(tt,BOTTOM) = BOTTOM
```

for any `tt` $\in$ **TYPETERM**. Type constraints are given by the universe

TYPECONS $\subseteq$ **TERM** $\times$ **TYPETERM**

whose elements are written as `t : tt`. A set `P` $\subseteq$ **TYPECONS** is called a *prefix* if it contains only type constraints of the form `X : tt` where `X` $\in$ **VARIABLE** and at most one such pair for every variable is contained in `P`. The *domain* of `P` is the set of all variables `X` such that `X : tt` is in `P` for some `tt`. We denote by **TYPEPREFIX** the universe of all type prefixes. Constraints are then defined as equations or type constraints, i.e.

CONSTRAINT $\subseteq$ **EQUATION** $\cup$ **TYPECONS**

Let **CSS** denote the set of all sets of constraints together with `nil` $\in$ **CSS** denoting an inconsistent constraint system.

The unifiability notion of ordinary Prolog is now replaced by a more general (for the moment abstract) constraint solving function:

`solvable:` **CSS** $\rightarrow$ **BOOL**

telling us whether the given constraint system is solvable or not. Every (solution of a) solvable constraint system can be represented by a pair consisting of a substitution and a type prefix. Thus, we introduce a function

`solution:` **CSS** $\rightarrow$ **SUBST** $\times$ **TYPEPREFIX** $\cup$ `{nil}`

where `solution(CS) = nil` iff `solvable(CS) = false`. For the trivially solvable empty constraint system we have `solution(∅) = (∅,∅)`, and the relationship between the three **TYPETERM** functions and constraint solving satisfies for any `t` $\in$ **TERM** and $tt_i \in$ **TYPETERM** the integrity constraints

```
solution({t : TOP}) = (∅,∅)
solution({t : BOTTOM}) = nil
solution({X : tt₁, X : tt₂}) = solution({X : inf(tt₁,tt₂)})
```

These are the only assumptions we make about the universe **TYPETERM** until we will introduce a special representation for it in Section 6. Thus, the complete development up to Section 5 applies to any concept of (type) constraints that exhibits the minimal requirements stated so far.

In the following, we will often abbreviate a Boolean condition of the form `p = true` by just writing p. We will also use the abbreviation

```
solution(CS) = (s, P)   ==   solvable(CS) & solution(CS) = (s, P).
```

The functions

subst-part: **CSS → SUBST**
prefix-part: **CSS → TYPEPREFIX**

are the two obvious projections of `solution`.

Having refined the notions of unifiability and substitution to constraint solvability and (solvable) constraint system, respectively, we can now also refine the related notion of substitution result to terms with type constrained variables. The latter involves three arguments

1. a term `t` to be instantiated,
2. type constraints for the variables of `t` given by a prefix P_t, and
3. a constraint system `CS` to be applied.

Since a `CS`-solution consists of an ordinary substitution s_{CS} together with variable type constraints P_{CS} via `solution(CS)` $= (s_{CS}, P_{CS})$, the result of the constraint application can be introduced by

conres(t, P_t, CS) = (t_1, P_1)

as a pair consisting of the instantiated term t_1 and type constraints P_1 for the variables of t_1. For this function

conres: **TERM** × **TYPEPREFIX** × **CSS** →
TERM × **TYPEPREFIX** ∪ {nil}

we impose the following integrity constraints:

$\forall$ t ∈ **TERM**, P_t ∈ **TYPEPREFIX**, CS ∈ **CSS** .
if solvable(P_t ∪ CS) then
conres(t, P_t, CS) = (t_1, P_1)
where:
t_1 = subres(t, subst-part(CS))
P_1 = prefix-part(P_t ∪ CS)$_{|var(t_1)}$
else
conres(t, P_t, CS) = nil

where $P'_{|var(t')}$ is obtained from P' by eliminating the type constraints for all variables not occurring in t'.

$P\backslash X$ will be an abbreviation for $P_{|domain(P)\backslash\{X\}}$, the prefix obtained from P by eliminating (if present) the constraint for X. - Thus, the condition that a constraint system CS "can be applied" to a term t with its variables constrained by P_t means that P_t is compatible with CS, i.e. solvable(CS ∪ P_t) = true.

Constraints in a program may refer to names defined in a separate definition part of the program (e.g. type definition, domain declaration, etc.) which will be element of a universe **DEFCONTEXT**. Thus, we assume that all constraint relevant universes and functions introduced above (like `solution`, `subst-part`, etc.) are parameterized by a `defcontext` ∈ **DEFCONTEXT**. For instance, strictly speaking `solvable` has therefore the signature

$$\texttt{solvable:}\quad \textbf{CSS} \times \textbf{DEFCONTEXT} \rightarrow \textbf{BOOL}$$

However, in order to ease our notation we will systematically drop this parameter, making it explicit only where needed. Just as we leave **TYPETERM** abstract for the moment, we also think of **DEFCONTEXT** as abstract, to be refined later on. For instance, in a particular PROTOS-L algebra the definition context will consist of all type and relation declarations occurring in the program.

In a PROTOS-L algebra a program is thus a pair consisting of a definition context and a sequence of clauses

$$\textbf{PROGRAM} \subseteq \textbf{DEFCONTEXT} \times \textbf{CLAUSE}^*$$

where a clause, depicted as

$$\{\texttt{P}\}\ \texttt{H <-- } G_1 \ \texttt{\&} \ \ldots \ \texttt{\&}\ G_n.$$

is an ordinary Prolog clause together with a set `P` of type constraints for (all and only) the variables occurring in the clause head and body. As in [9] we use three obvious projection functions

`clhead:`	**CLAUSE** → **TERM**
`clbody:`	**CLAUSE** → **TERM***
`clconstraint:`	**CLAUSE** → **TYPEPREFIX**

Every element of the universe **GOAL** comes with a type prefix for its variables and is written as

$$\{\texttt{P}\}\ G_1 \ \texttt{\&} \ \ldots \ \texttt{\&}\ G_n.$$

Every program (and every goal) must be well-typed; in particular this means that every clause must be well-typed under the type constraints for its variables given in its type prefix. Thus, at this point we assume that the compiler has done already static type checking and inferencing.

The `procdef` function of [6] now assumes the form

$$\texttt{procdef:}\quad \textbf{CLAUSE}^* \times \textbf{INDEX} \times \textbf{TERM} \times \textbf{SUBST} \times \textbf{TYPEPREFIX} \rightarrow \textbf{CLAUSE}^*$$

where `procdef(db, i, g, s, P)` is a list of clauses from `db`, renamed with variable index `i`, that might be relevant for solving the literal `g` under the substitution `s` and with the variables constrained to the types given in `P`. As in [6] this abstracts away from any indexing mechanism, but it does allow for a preselection of clauses both with regard to the given substitution as well as with regard to the variable types obtained so far in the deduction process.

An integrity constraint (to assure correctness) for `procdef` is that it does not leave out any relevant clauses, i.e. given `procdef(db, i, g, s, P) = l` we have:

$$\forall\, \texttt{cl} \in \texttt{db} \,.\, \texttt{cl} \notin \texttt{l} \;\Rightarrow$$
$$\texttt{solvable}(\{\texttt{g} \doteq \texttt{rename(clhead(cl), i)}\} \cup \texttt{s} \cup \texttt{P} \cup \texttt{rename(clconstraint(cl), i)}) = \texttt{false}$$

The basic part of Prolog algebras in [6] that holds the resolution state is a predecessor structure with universe **RESSTATE** and with functions

`currstate:`	**RESSTATE**
`nil:`	**RESSTATE**
`choicepoint:`	**RESSTATE** → **RESSTATE**
`declglseq:`	**RESSTATE** → **DECGOAL***

where **DECGOAL** ⊆ **GOAL** × **RESSTATE**. As in [9] we add a function

$$\texttt{constraint:}\quad \textbf{RESSTATE} \rightarrow \textbf{CSS}$$

to hold the constraint system (which includes its substitution part) accumulated so far. The abbreviations we use are

```
currdecglseq      ==  decglseq(currstate)
currdecgl         ==  head(currdecglseq)
curractivator     ==  first(head(currdecgl))
currcutpoint      ==  second(head(currdecgl))
currconstraint    ==  constraint(currstate))
currsubst         ==  subst-part(currconstraint)
currprefix        ==  prefix-part(currconstraint)
```

For an initially given goal {P} G_1 & ... & G_n the PROTOS-L algebra has as initial values

```
currdecglseq   = [<[SolveConstraint(P) & G1 & ... & Gn], nil>]
currconstraint = ∅
```

and all other initial values are as in [6].

2.2 PROTOS-L Transition Rules

Since the universe **TYPETERM** until now is abstract the constraints we considered so far can be treated as in [9]. However, whereas [9] deliberately takes unification and constraint solving as one abstract unit, for use in later refinements we distinguish here between adding an equation asking for unification and adding a clause's constraint. For this purpose, two literals representing the two different types of constraints are added in the selection rule to the decorated goal sequence as the first tasks to be solved for every newly created choicepoint.

```
if still working correctly                                              Selection
  & user-defined(curractivator) = true
  & procdef(database,varindex,curractivator,currsubst,currprefix) ≠ [ ]
then
  LET ClList = procdef(database,varindex,curractivator,currsubst,currprefix)
  LET CutPt  = choicepoint(currstate)
  LET Cont   = [tail(currdecgl) | tail(currdecglseq)]
  LET l      = length(ClList)
  EXTEND STATE by temp(1),...,temp(l)
  WHERE
    put temp(i) on top
    decglseq(temp(i)) :=
               [<[SolveConstraint(clconstraint(component(i,ClList))) &
                 curractivator = clhead(component(i,ClList))], .  >,
                <clbody(component(i,ClList)), CutPt> | Cont]
  ENDEXTEND
  varindex := succ(varindex)
```

where '.' is a dummy argument that is not used.

The unify rule of [6] is refined such that instead of instantiating the continuation sequence of goals with the most general unifier, the unification requirement is added to the current constraint system, provided it will still be solvable:

```
  if still working correctly                                          Unify
    & curractivator = (G1 = G2)
    & solvable(currconstraint                     | solvable(currconstraint
                 ∪ G1 = G2) = true                |       ∪ {G1 = G2}) = false
  then
    currconstraint := currconstraint              | backtrack
```

```
                    U {G1 = G2}                |
   currdecglseq := [tail(currdecgl)            |
                    | tail(currdecglseq)]      |
```

Similarly, if the current activator asks for solving a constraint system P, it will be added to the current constraint system, provided it remains solvable:

Solve Constraint

```
if still working correctly
   & curractivator = SolveConstraint(P)
   & solvable(currconstraint                | solvable(currconstraint
                    U P) = true             |                 U P) = false
then
   currconstraint := currconstraint         |  backtrack
                    U P                     |
   currdecglseq := [tail(currdecgl)         |
                    | tail(currdecglseq)]   |
```

Since we add a constraint to the current constraint system only if it remains solvable, we do not have to add a condition on its solvability to any of the other rules of [6], or to add a second backtracking rule, as in [9]: All other rules for 'pure' Prolog can be taken unchanged from [6]. Only for extra-logical built-in predicates (which, however, will not be discussed in this paper) that need the current activator instantiated - like `call`, `assert` or `retract` - we would have to do the instantiation in the rules: we just have to replace `curractivator` by `subres(curractivator, currsubst)`.

3 PROTOS-L Algebras with compiled AND / OR structure

Since the constraint feature is orthogonal to the conjunctive and disjunctive analysis of a logic program, one of the major design principles of the PAM was to adopt the AND/OR structure more or less directly from the WAM ([4]). For proving the correctness of the AND/OR structure in the PAM, we proceed along the lines of [7], presenting only the modifications and extensions needed for our PROTOS-L algebras, since the separation of the WAM specification development in [7] into several refinement levels enables us to locate the required modifications quite naturally with all correctness proofs obtained by straightforward adaptations.

Our compilation refines the compile function `compile`: **TERM** → **INSTR*** of [7] by introducing a new instruction `PutConstraint` for dealing with constraints. Formally, for clauses and goals we have:

```
compile({P} H <-- G1 & ... & Gn) =
     [PutConstraint(P), Unify(H), Call(G1),...,Call(Gn), Proceed]
compile({P} G1 & ... & Gn) =
     [PutConstraint(P), Call(G1),...,Call(Gn),Proceed]
```

Note that this treatment of constraints will be refined later by distributing the clause constraint over the individual literals such that each type constraint for a variable is only considered when the variable occurs for the first time (c.f. [4]). But we will postpone this until the explicit representation of terms is introduced.

The substitution component `s` from [7] becomes now `c`: **STATE** → **CSS**. If for a `state` ∈ **STATE**, `c(state)` is a solvable constraint system, it can be decomposed into a substitution part (as in the Prolog case) and a type prefix part, justifying the abbreviations

```
s == subst-part(c(currstate))
tp== prefix-part(c(currstate))
```

The logical structure of clauses is explicitly formalized by the structure of the new domain **CODEAREA**. Furthermore, at this level of refinement the information on the variable renaming index is formalized in the rule conditions. Therefore, following [7], the procdef function is now typed without **INDEX** as follows (c.f. Section 2):

$$\texttt{procdef: } \mathbf{PROGRAM} \times \mathbf{TERM} \times \mathbf{SUBST} \times \mathbf{TYPEPREFIX} \rightarrow \mathbf{CODEAREA}^*$$

As we will not discuss database updates in this paper, we will drop the insert and delete functions on **PROGRAM**, and all corresponding functions and rules without further mentioning.

Given a 'prototypical' goal as above, in the initial state of a PROTOS-L algebra, s = Id of initial Prolog algebras is replaced by c = ∅ where ∅ is the empty constraint system, and due to the refined compile function we have

p = load([PutConstraint(P),Call(G_1),...,Call(G_n),Proceed])

In the backtracking and selection rules we use procdef(db, G, s, tp), and c(temp(i)) := c replaces the substitution update. The modified rules for unification and the new rules for explicit constraint handling are straightforward transformations of the corresponding rules for our PROTOS-L algebras in Section 2:

```
if OK &                                                   Unify
   r_code = Unify(G) &
   solvable(c ∪ {g = G}) = true  |  ... = false
then
   c := c ∪ {g = G}              |  backtrack
   p := p+                       |
```

```
if OK &                                                   PutConstraint
   r_code = PutConstraint(P) &
   solvable(c ∪ P) = true        |  ... = false
then
   c := c ∪ P                    |  backtrack
   p := p+                       |
```

Correctness Theorem: The PROTOS-L algebras with compiled AND/OR structure as developed above are correct w.r.t. the PROTOS-L algebras of Section 2.

Proof: The proof is an easy adaption of the correctness proof of Section 2.1 in [7] by modifying the resolution state and decorated goal sequence recovering functions F and G (since our goal sequences in **RESSTATE** are not instantiated) and including the recovery of SolveConstraint(P) from PutConstraint(P):

F: **STATE** → **RESSTATE**
G: **ENV** → **DECGLSEQ**

```
F(nil) = nil
decglseq(F(St))      = [<goal_seq(p(St),vi(St),g(St)), F(ct(St))> |
                                   G(e(St))]

choicepoint(F(St)) = F(b(St))
constraint(F(St))    = c(St)

G(nil) = nil
G(E)    = [<goal_seq(p'(E),vi'(E), . ), F(ct'(E))> | G(ce(E))]
```

with the auxiliary function

$$\texttt{goal\_seq}: \quad \mathbf{CODEAREA} \times \mathbf{INDEX} \times \mathbf{GOAL} \to \mathbf{GOAL}$$

$$\texttt{goal\_seq(Ptr,Vi,G)} = \begin{cases} \texttt{true} & \text{if } \texttt{code(Ptr)=Proceed} \\ \texttt{SolveConstraint(P) \& goal\_seq(Ptr+,Vi,G)} & \text{if } \texttt{code(Ptr)=PutConstraint(P)} \\ \texttt{G = H \& goal\_seq(Ptr+,Vi,G)} & \text{if } \texttt{code(Ptr)=Unify(H)} \\ \texttt{decompile(Ptr,Vi)} & \text{otherwise} \end{cases}$$

This defines a partial function `F1` from PROTOS-L algebras with compiled AND/OR structure to the PROTOS-L algebras of Section 2. `F1` is extended to map the respective rules of the refining level to the corresponding rules of the refined level; for instance, `F1` maps the success part of **PutConstraint** to the success part of **SolveConstraint**. The proof now follows by showing that `F1` maps initial algebras to initial ones, and that for every algebra `A` with `F1(A)` being defined and every rule `R` that can be applied to `A` the diagram

```
                     F1(R)
      F1(A) ---------------> F1(A')
        ^                      ^
        |                      |
     F1 |                      | F1
        |          R           |
        A -------------------> A'
```

commutes, which is shown by straighforward case anaylsis. ■

All remaining parts of [7] can be adopted almost directly to our treatment of PROTOS-L algebras. For the explicit allocation and deallocation of environments and the last call optimization one just has to bear in mind that a PutConstraint instruction has to be generated for every clause. Also the compilation of the predicate structure carries over to our setting, as well as all corresponding correctness proofs.

4 Term representation

4.1 Universes and Functions

We adopt the notions of [8] and - adding a universe **SYMBOLTABLE** - use the pointer algebra

(DATAAREA, PO + MEMORY + SYMBOLTABLE;
`top, bottom; +, -; val)`

The functions `tag` and `ref` are defined on

`tag`: **PO → TAGS**
`ref`: **PO → DATAAREA + TYPETERM**

where, because of the type constraint treatment, a new tag `VAR` for indicating free variables is introduced into the universe

TAGS = {`REF, STRUC, VAR`}

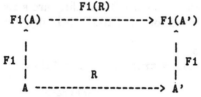

Special tags for representing constants, lists, built-in integers, etc. are also present in the PAM, but they are considered as optimizations that can be added later on without any difficulties. The tag `FUNC` from [8] has been removed since it is not needed.

The codomain of `ref` contains the universe **TYPETERM** since we will keep the type term representation abstract here; it will be refined later (see Section 6). **SYMBOLTABLE** is a universe (also abstract for the moment) that comes with functions (further functions will be added later)

`atom:` **SYMBOLTABLE** $\rightarrow$ **ATOM**
`arity:` **SYMBOLTABLE** $\rightarrow$ **NAT**
`entry:` **ATOM** $\times$ **NAT** $\rightarrow$ **SYMBOLTABLE**

of which we assume `entry(atom(s),arity(s)) = s` for any `s` $\in$ **SYMBOLTABLE** and `atom(entry(f,n)) = f`, `arity(entry(f,n)) = n` for any atom `f` with arity `n`.

Here are some new resp. modified abbreviations, integrity constraints, and functions which we add to the ones from [8]:

```
unbound(l)          ==   tag(l) = VAR
mk_unbound(l)       ==   mk_unbound(l,TOP)
mk_unbound(l,t)     ==   tag(l) := VAR
                         insert_type(l,t)
insert_type(l,t)    ==   ref(l) := t
```

Note that an unconstrained free variable gets the trivial type restriction `TOP`, representing no restriction at all (c.f. Section 2.1). As integrity constraints we have:

```
if unbound(l)              then  ref(l) : TYPETERM
if tag(l) = STRUC          then  val(ref(l)) : SYMBOLTABLE
if tag(l) : {REF, STRUC}   then  ref(l) : DATAAREA
                                 term(l) : TERM
                                 typeprefix(l) : TYPEPREFIX
```

`deref:` **DATAAREA** $\rightarrow$ **DATAAREA**

$$\texttt{deref(l)} = \begin{cases} \texttt{deref(ref(l))} & \text{if } \texttt{tag(l)} = \texttt{REF} \\ \texttt{l} & \text{otherwise} \end{cases}$$

`term:` **DATAAREA** $\rightarrow$ **TERM**

$$\texttt{term(l)} = \begin{cases} \texttt{mk-var(l)} & \text{if } \texttt{unbound(l)} \\ \texttt{term(deref(l))} & \text{if } \texttt{tag(l)} = \texttt{REF} \\ \texttt{f(a1,...,an)} & \text{if } \texttt{tag(l)} = \texttt{STRUC} \text{ and } \texttt{f = atom(val(ref(l)))} \\ & \qquad \texttt{n = arity(val(ref(l)))} \\ & \qquad \texttt{ai = term(ref(l)+i)} \end{cases}$$

`type-prefix:` **DATAAREA** $\rightarrow$ **TYPEPREFIX**

$$\texttt{type-prefix(l)} = \begin{cases} \texttt{mk-var(l) : ref(l)} & \text{if } \texttt{unbound(l)} \\ \texttt{type-prefix(deref(l))} & \text{if } \texttt{tag(l)} = \texttt{REF} \\ \texttt{P1} \cup \ldots \cup \texttt{Pn} & \text{if } \texttt{tag(l)} = \texttt{STRUC} \text{ and} \\ & \quad \texttt{n = arity(val(ref(l)))} \\ & \quad \texttt{Pi = type-prefix(ref(l)+i)} \end{cases}$$

4.2 Unification

Unification in the PAM can be carried out as in the WAM (see [1]) if we refine the bind operation into one that takes into account also the type constraints of the variables ([4]). The bind operation may thus also fail and initiate backtracking if the type constraints are not satisfied. Thus, we can use the treatment of unification as described in [8], while leaving the bind operation abstract for the moment, not only in order to postpone the discussion of occur check and trailing but also to stress the fact that the bind operation will take care of the type constraints for the variables. We impose the following modified

BINDING CONDITION: For any `l1, l2, l` ∈ **DATAARRA**, with `term`, `term'` values of `term(l)` and with `prefix`, `prefix'` values of `type-prefix(l)` before and after execution of `bind(l1, l2)`, we have if `unbound(l1)` holds:

```
LET CS = {mk-var(l1) = term(l2)} ∪ type-prefix(l1)
                                  ∪ type-prefix(l2)
if solvable(CS) = true
then (term', prefix') = conres(term, prefix, CS)
else backtrack update will be executed
```

With this generalized binding assumption we obtain the following modified

UNIFICATION LEMMA: If `pdl-- = nil`, `term(left)`, `term(right)` ∈ **TERM**, and `type-prefix(left)`, `type-prefix(right)` ∈ **TYPEPREFIX**, the effect of setting `what_to_do` to `Unify`, for any `l` ∈ **DATAAREA** such that `term(l)` ∈ **TERM** and `type-prefix(l)` ∈ **TYPEPREFIX** is as follows:

Let `term`, `term'` be the values of `term(l)` and `prefix`, `prefix'` be the values of `type-prefix(l)` when setting `what_to_do` to `Unify` and when `what_to_do` has been set back to `Run` again, respectively. Then we have:

```
LET CS = {term(left) = term(right)} ∪ type-prefix(left)
                                     ∪ type-prefix(right)
if solvable(CS) = true
then (term', prefix') = conres(term, prefix, CS)
else backtrack update will be executed
```

Proof: As in [8] the proof of the Unification Lemma is by induction, this time relying on our generalized Binding Condition. ∎

4.3 Putting of terms

In the code for generating terms on the heap we have to consider our slightly modified term representation: The value cell pointed to by a `STRUC` reference does not contain the top level atom `f` of the term and its arity `a`, but is a pointer to **SYMBOLTABLE**, from which we will get the atom, its arity, its target sort, etc. Thus we assume that in the generated code every pair `(f,a)` is replaced by `entry(f,a)`.

The code developed in Section 1.2 of [8] for constructing terms in body goals uses put instructions which assume that, for all variables `Yi` of the term `t` to be built on the heap, there is already a term denoting `yi` ∈ **DATAAREA** available. Since this means in particular that no variables are created during this process, we can use (with the obvious modification mentioned above) the same put instructions (i.e. `put_value`, `unify_value` in `Write` mode, `put_structure`). Furthermore, we may assume that for the variables `Yi` we have no type constraints to formalize here because they have already been associated to the corresponding location `yi` (i.e. the variable `term(yi)` which is - up to renaming - equal to `Yi`). This gives us the following

PUTTING LEMMA: If all variables occurring in a term `t` ∈ **TERM** are among {`Y1, ..., Yl`}, and if for `n` ∈ {`1, ..., l`}, `yn` ∈ **DATAAREA** with

    term(yn) ∈ TERM
    type-prefix(yn) ∈ TYPEPREFIX

and `Xi` is a fresh variable, and `CS` is the constraint system consisting of the substitution associating every `Yn` with `term(yn)` and of the union of the type constraints `type-prefix(yn)`, i.e.

    CS = ⋃ₙ {Yn ≐ term(yn)} ∪ type-prefix(yn)

then the effect of setting p to

    load(append(put_code(Xi = t), More))

with subsequent fresh indices generated by nfs being non-top level, is that the pair

    (term(xi), type-prefix(xi))

at the moment of passing to `More`, gets value of

    conres(t, ∅, CS)

Proof: The proof is a direct generalization of the proof of the Putting Lemma in [8], observing that no type related actions like variable creation or variable binding is involved here. ■

4.4 Getting of terms

Unlike putting of terms that does not involve unification the getting of terms does involve unification where parts of it are compiled into the getting instructions (like `get_structure` followed by a sequence of `unify` instructions) and the remaining unification tasks are handled by the lowlevel `unify` procedure.

The `get_value`, `unify_value`, and `unify_variable` (the latter both in `Read` and `Write` mode) are as in [8], where for the generation of a heap variable in `Write` mode of `unify_variable` we use

```
mk_heap_var(l) == mk_unbound(h)
                  bind(l, h)
                  h := h+
```

Note that we assume that the bind update will require several rule applications that will be triggered by the abbreviation `bind(l,h)` - this is important since the binding rules themselves must see the new value of `val(h)` (which would not be the case if the `bind` abbreviation consisted only of e.g. function updates that did not trigger any rules). This point had been left open in [8] where the `bind` abbreviation in `mk_heap_var` is not consistent with the use of it in the unify rules.

The first `get_structure` rule for PROTOS-L is as in the WAM case (where the structure pointer `s` is replaced by `nextarg`):

Get-Structure-1

```
if  RUN
  &
   code(p) = get_structure(f, xi)
  &
   tag(deref(xi)) = STRUC
  &
   val(ref(deref(xi))) = f        | val(ref(deref(xi))) ≠ f
then
  nextarg := ref(deref(xi))+      | backtrack
  mode := Read                    |
  succeed                         |
```

When in get_structure(entry(f,n),xi) xi is unbound, xi must be bound to a newly created term with top-level symbol f. Whereas in the WAM this will always succeed, in the PAM case the type constraint of xi must be taken into account. Indeed, what is happening here is the binding of a variable X with a type constraint, say tt, to a term t starting with f. In abstract terms this amounts to solving the constraint system

$$\{X \doteq t,\ X : tt\}$$

We still want to leave the details of variable binding abstract here; what is of interest for this special case occurring in get_structure is which type constraints stemming from tt and (the declaration of) f must be propagated onto the argument terms of t = f(...). Therefore, we introduce the function

$$\texttt{propagate\_list}: \ \mathbf{SYMBOLTABLE} \times \mathbf{TYPETERM} \rightarrow \mathbf{TYPETERM}^* \cup \{\texttt{nil}\}$$

yielding for arguments entry(f,n) and tt the list of type terms the arguments of f must satisfy. To be more precise, we have the following integrity constraint:

$$\texttt{propagate\_list(entry(f,n),tt)} = (tt_1,\dots,tt_n)$$
$$\text{iff}$$
$$\texttt{prefix-part}(\{f(X_1,\dots,X_n) : tt\}) = \{X_{i1} : tt_{i1},\dots,\ X_{ik} : tt_{ik}\}$$

where $\{i1,\dots,ik\} \subseteq \{1,\dots,n\}$, and for $j \in \{1,\dots,n\}\setminus\{i1,\dots,ik\}$ we have tt_j = TOP.

If the constraint system $\{f(X_1,\dots,X_n) : tt\}$ is not solvable, no propagation is possible, and if it reduces to the trivially solvable empty constraint system, propagate_list yields a list containing only TOP. Thus we introduce the abbreviations

$$\texttt{can\_propagate(entry(f,n),tt)} \ == \ \texttt{solution}(\{f(X_1,\dots,X_n) : tt\}) \neq \texttt{nil}$$
$$\texttt{trivially\_propagates(entry(f,n),tt)} \ == \ \texttt{solution}(\{f(X_1,\dots,X_n) : tt\}) = \emptyset$$

Get-Structure-2

```
if  RUN
  &
   code(p) = get_structure(f, xi)
  &
   unbound(deref(xi))
  &
   can_propagate(f,ref(deref(xi)))
     = true                                                      | = false
  &                                                              |
   trivially_propagates(f,ref(deref(xi)))                        |
     = true            | = false                                 |
then
  h  <- <STRUC,h+>                                               | backtrack
  bind(deref(xi),h)                                              |
  val(h+)   :=  f                                                |
  h    :=  h++                                                   |
  mode := Write        | nextarg := h++                          |
                       | mk_unbounds(h+,propagate_list(f,ref(deref(xi))) |
                       | mode := Read                            |
  succeed                                                        |
```

For l ∈ **DATAAREA** and tt1,...,ttn ∈ **TYPETERM**, the update

```
mk_unbounds(l,(tt1,...,ttn)) ==  FORALL i = 1,...,n DO
                                      mk_unbound(l+i, tti)
                                 ENDFORALL
```

puts n type restricted variables at the locations l+1,...,l+n on the heap. When this update is executed in the rule above the machine continues in **read** mode so that the subsequent n unify instructions take into the account these type restrictions.

GETTING LEMMA: If all variables occurring in a term t ∈ **TERM** are among {Y1, ..., Yl}, and if for n ∈ {1, ..., l}, yn ∈ **DATAAREA** with

unbound(yn)
ref(yn) ∈ **TYPETERM**

and Xi is a fresh variable with xi ∈ **DATAAREA** and

term(xi) ∈ **TERM**
type-prefix(xi) ∈ **TYPEREFIX**

and CS is the constraint system consisting of the equation $t \doteq$ term(xi) together with type-prefix(xi) and the union of the type constraints type-prefix(yn), i.e.

$$CS = \{t \doteq \mathtt{term(xi)}\} \cup \mathtt{type\text{-}prefix(xi)} \cup \bigcup_n \mathtt{type\text{-}prefix(yn)}$$

then the effect of setting p to

load(append(get_code(Xi = t), More))

for any l ∈**DATAAREA** with term = term(l) ∈ **TERM** and typeprefix = type-prefix(l) ∈ **TYPEPREFIX** being the values before execution, is as follows:

If solvable(CS) = true then p reaches More without backtracking and the pair

(term(l), type-prefix(l))

at the moment of passing to More, gets value of

conres(term, typeprefix, CS)

else backtracking will occur before p reaches More.

Proof: The proof is an easy adaption of the proof in [8]. Note that if CS is solvable, then conres(term, typeprefix, CS) ≠ nil because CS ∪ typeprefix is also solvable since the intersection between typeprefix and any type-prefix(yn) is already contained in CS. ∎

In order to uphold the HEAP VARIABLE CONSTRAINT the instruction unify_local_value in Write mode creates a new heap variable for a so-called local variable. In the PROTOS-L case the type restriction of the local variable must be taken into account. However, this is done by the binding update in our (modified) mk_heap_variable abbreviation. Thus, the HEAP VARIABLE CONSTRAINT (with tag(l) = VAR instead of REF) as well as the HEAP VARIABLES LEMMA carry over to the PROTOS-L case.

4.5 Putting of Constraints

In this section we will still keep the type constraint representation abstract, while specifying the conditions about the constraint handling code (for realization of PutConstraint of Section 3) in order to prove a theorem corrresponding to the Pure Prolog Theorem of [8] (see 5). The compile function will be refined using

```
put_constraint_seq({Y1:tt1,...,Yr:ttr}) = [put_constraint(y1,tt1),
                                             ...
                                           put_constraint(yr,ttr)]
```

for which we introduce a new instruction `put_constraint(yn,tt)` (where `tt` ∈ **TYPETERM**) and the following rule:

Put-Constraint

```
if  RUN
    &
     code(p) = put_constraint(l, tt)
then
    insert_type(l, tt)
    succeed
```

The update for inserting a type restriction has still the straightforward definition given in 4.1 (i.e. `ref(l) := tt`), but will be refined later when we introduce a representation of type terms. In any case it must satisfy the following

TYPE INSERTING CONDITION: For any `l1, l` ∈ **DATAARRA**, with `term`, `term'` values of `term(l)` and with `prefix`, `prefix'` values of `type-prefix(l)` before and after execution of `insert_type(l1,tt)` we have if `unbound(l1)` holds:

```
(term', prefix') = conres(term, prefix\mk-var(l1), {mk-var(l1) : tt})
```

For the definition given above the type inserting condition is obviously satisfied.

5 PAM Algebras

5.1 Environment and Choicepoint Representation

Both the stack of states and environments of PROTOS-L algebras with compiled AND/OR structure of 3 are represented by a subalgebra of **DATAAREA**. This is done as in [8] since the only type-related action is in the allocation of `n` free variable cells in the rule for `Allocate`: This situation is covered by our modified `mk_unbound` abbreviation that assigns the `TOP` type restriction to it.

5.2 Trailing

Since variables in PROTOS-L carry a type restriction represented in the `ref` value of a location - which is updated when binding the variable -, the type restriction must be saved upon binding and recovered upon backtracking. Strictly speaking, it would be sufficient to save only the `ref` value of a location; however, for use in a later refinement -when we will introduce different tags for free variables - we also trail the tag of a location. Therefore, we extend the codomain of the `ref''` function of the trail

(**TRAIL**, **DATAAREA** × **PO**; `tr, botr; +, -; ref''`)

to record also the old `val` decoration. The trail update, to be executed when binding a location `l`, is then:

```
trail(l)  ==   ref''(tr) := (l, val(l))
               tr := tr+
```

Note that this is a non-optimized version of the trailing operation; as in [8] we could have also used a conditional trailing governed by the condition `l : heap & l < hb OR l : stack & l < b` .

For `t` ∈ **TRAIL** with `ref''(t) = (l, v)` we use the following abbreviation for the two obvious projections on `ref''(t)`:

```
location(t)  ==  l                        value(t)  ==  v
```

Unwinding of the trail upon backtracking is defined by

```
unwind_trail  ==  FORALL  t = tr-,...,tr(b)  DO
                      location(t) <- value(t)
                  ENDFORALL
```

where `value(t)` retrieves the previous tag and type restriction of `location(t)`.

We still leave the binding update abstract, but pose the following
TRAILING CONDITION: Let `l1, l2, l` ∈ **DATAAREA**. If `val(l)` before execution of `bind(l1, l2)` is different from `val(l)` after successful execution of `bind(l1, l2)`, then the location `l` has been trailed.

Note that due to the update on the type restrictions of a variable the trailing of both locations `l1` and `l2` may be triggered by `bind(l1, l2)`; moreover, if e.g. `l2` denotes a polymorphic term containing variables these variables also have to be trailed if they get another type restriction in the binding process (see Section 5.4).

As in [8] all our rules given up to now satisfy the STACK VARIABLE PROPERTY. Instead of formulating a particular binding discipline already here - e.g. requiring always binding the higher location to the lower one - we impose therefore as condition on bind that it also satisfies this property.

5.3 Pure PROTOS-L theorem

PURE PROTOS-L THEOREM: The PAM algebras developed so far are correct w.r.t. the PROTOS-L algebras with compiled AND/OR structure of Section 3.
Proof: We will define a (partial) function `F` form PAM algebras and (sequences of) transitions rules to PROTOS-L algebras and (sequences of) transition rules such that

```
                 F(R)
      F(A) -----------------> F(A')
        ^                       ^
        |                       |
     F  |                       | F
        |          R            |
        A -------------------> A'
```

commutes for any `A, R` such that `F(A)` and `F(R)` is defined. This generalizes the proof of the PURE PROLOG THEOREM of Section 2.1.4 of [8] from which we take the notation and all auxiliary functions with the following modifications and extensions (Note that all functions on the PROTOS-L level thus have the suffix '1'):

```
compile ( {P} H <-- G1 & ... & Gn ) =
        splice( [Allocate(r), put_constraint_seq(P) | get_seq(H)],
                call_seq(G1),
                ...
                call_seq(Gn),
                [Deallocate, Proceed] )

compile1 ( {P} H <-- G1 & ... & Gn ) =
                [Allocate, PutConstraint(P), Unify(H),
                Call(G1),
                ...
                Call(Gn),
                Deallocate, Proceed]
```

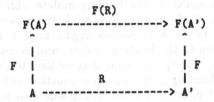

For the definition of the correspondence between the constraint representation we add the functions

$$\texttt{css}: \ \mathbf{TRAIL} \rightarrow \mathbf{CSS}$$

$$\texttt{typeprefix}: \ \mathbf{DATAAREA} \times \mathbf{TRAIL} \rightarrow \mathbf{TYPEPREFIX}$$

where css is defined by

$$\texttt{css(lt)} = \bigcup_{botr \leq l < tr} \{\texttt{mk\_var(location(l))} \doteq \texttt{term(location(l),lt)}\} \cup \texttt{typeprefix(location(l),lt)}$$

and `typeprefix(l,lt)` yields the value `type-prefix(l)` would take after having unwound the trail down to `lt`.

We find `subst` in `css(lt)` as the equation part, i.e.

```
subst(lt) = subst-part(css(lt))
```

and the PROTOS-L algebra constant holding the constraint system is defined by

```
cl := css(tr)
```

The correspondence between the rules is extended by associating the sequence of `put_constraint` instructions to the `PutConstraint` rule of the PROTOS-L algebras:

```
put_constraint_seq --> PutConstraint
```

The rest of the proof is a generalization of the proof in [8], relying on our (modified) conditions and lemmas. In particular, we observe the fact that `Allocate` allocates a new variable location (with `TOP` restriction) for every variable occurring in the clause. These locations are used by the `put_constraint` instructions, so that the preconditions for the TYPE INSERTING CONDITION hold. ∎

5.4 Binding

We are now ready for a first refinement of the binding update which will take into account the bind direction, occur check, and trailing, while the type constraints still remain abstract. We introduce two new constants `arg1`, `arg2` ∈ **DATAAREA** which will hold the locations given to the binding update, and extend the values of `what_to_do` by {`Bind_direction`, `Bind`} indicating that we have to choose the direction of the binding resp. do the binding itself. The new constant `return_from_bind` will take values of the domain of `what_to_do`, indicating where to return when the binding is finished. (Remember that the binding update is used in different places, e.g. in the unify update or in the creation of a new heap variable).

For `l1,l2` ∈ **DATAAREA** the binding update and some new abbreviations are defined by

```
bind(l1, l2)    ==  arg1 := l1
                    arg2 := l2
                    return_from_bind := what_to_do
                    what_to_do := Bind_direction
bind_success    ==  what_to_do := return_from_bind
BIND            ==  OK & what_to_do = Bind
trail(l1,l2)    ==  ref''(tr)  := (l1, val(l1))
                    ref''(tr+) := (l2, val(l2))
                    tr := tr++
```

In order to reset also the constant `what_to_do` upon backtracking, we refine the `backtrack` update `p := val(b-2)` of [8] which sets the next-instruction-pointer `p` to the next alternative of the current choicepoint `b` (which happens to be stored in location `b-2`), to

```
backtrack  ==  p := val(b-2)
               what_to_do := Run
```

For unbound(l1) there are two alternative conditions on the update occur_check(l1,l2), depending on whether the unification should perfom the occur check (which is required for being logically correct) or not (which is done in most Prolog implementations for efficiency reasons):

OCCUR CHECK CONDITION: If no occur check should take place then the update occur_check(l1,l2) is empty; otherwise it has the following effect: If mk_var(l1) is among the variables of term(l2) then the backtrack update will be executed.

We will leave the occur check update abstract, and all correctness proofs are thus implicitly parameterized by the decision whether it actually performs the occur check or not.

Bind-1 (Bind-Direction)

```
if  OK
   &
     what_to_do = Bind_direction
   &
    unbound(arg1)
   &
    (NOT (unbound(arg2))  |  unbound(arg2)
       or                 | &
     arg2 < arg1)         |  arg2 > arg1          |  arg1 = arg2
then
   what_to_do := Bind     | what_to_do := Bind  |  bind_success
                          | arg1 := arg2        |
                          | arg2 := arg1        |
```

When binding two unbound variables their type constraints must be 'joined' using the inf function (Section 2).

Bind-2 (Bind-Var-Var)

```
if  BIND
   &
    unbound(arg2)
   &
    inf(ref(arg1),ref(arg2))
         ≠ BOTTOM                                                  | = BOTTOM
   &                                                               |
    inf(ref(arg1),ref(arg2))                                       |
         ≠ ref(arg2)                            | = ref(arg2)      |
then
   trail(arg1,arg2)                             | trail(arg1)      | backtrack
   insert_type(arg2,inf(ref(arg1),ref(arg2)))   |                  |
   arg1 <- <REF, arg2>                                             |
   bind_success                                                    |
```

When binding an unbound variable to a non-variable term, the type restriction of the variable must be propagated to the variables occurring in the term. As a special case this situation already occured in get_structure(f,xi) when the dereferenced value of xi is a type-restricted variable. In that situation where the term was still to be built upon the heap, we ensured the propagation by writing arity(f) free value cells on the heap with appropriate type restrictions and continuing in read mode; the actual propagation was then achieved by the immediately following sequence of `unify` instructions. In the general case occurring in the binding rules, the arguments of the term are not just variables but arbitrary terms. However, as we will not go into the details of type constraint solving here, we assume an abstract propagate update satisfying the following:

PROPAGATION CONDITION: For any `l1, l2, l` ∈ **DATAARRA**, with `term, term'` values of `term(l)`, with `prefix, prefix'` values of `type-prefix(l)`, and with `val, val'` values of `val(l)`, before and after execution of `propagate(l1,l2)` we have if `unbound(l1)`, `ref(l1)` ∈ **TYPETERM**, `tag(l2) = STRUC`, and `term(l2)` ∈ **TERM**,

```
LET CS = {term(l2) : ref(l1)}
if solvable(CS) = true
then  (a) (term', prefix') = conres(term, prefix, CS)
      (b) if val ≠ val' then the location l will be trailed
else  backtrack update will be executed
```

With this update at hand the third binding rule is

Bind-3 (Bind-Var-Struc)

```
if  BIND
   &
    NOT (unbound(arg2))
then
   trail(arg1)
   arg1 <- <REF, arg2>
   occur_check(arg1,arg2)
   propagate(arg1,arg2)
```

BINDING LEMMA: The bind rules are a correct realization of the binding update of Section 4.2, i.e. the BINDING CONDITION, the TRAILING CONDITION as well as the STACK VARIABLES PROPERTY are preserved.

Proof: The proof for the update `bind(l1, l2)` is by case analysis and induction on the size of `term(l2)`, relying on the integrity conditions for the infimum function on type terms when binding one type-restricted variable to another one (Bind-2), resp. on the Propagation Condition when binding a variable to a non-variable term (Bind-3). ∎

6 Conclusions and Outlook

Putting everything together, we obtain the

Main Theorem: The PAM algebras developed so far are correct w.r.t. PROTOS-L algebras. Thus, since we kept the notion of types abstract, for every such type-constraint logic programming system L and for every compiler satisfying the specified conditions, the WAM extension with this abstract notion of types is correct w.r.t. L.

Thus, any type system satisfying the minimal preconditions on the `inf` and `solution` functions stated in Section 2 is covered by the development above. In [3] we continue the work reported here in several directions. Leaving the notion of types still abstract, we show how the WAM optimizations of environment trimming, last call optimization, and initialization of temporary and permanent variables carry over to this setting. The universe **TYPETERM** together with its functions is refined in several levels, finally leading to PROTOS-L specific type constraints. Furthermore, [3] also deals with some PAM specific optimizations, namely a switch on typed variables and a special representation of typed variables. The latter refinement leads to a situation where when only variables with trivial TOP restriction are used - this corresponds to the untyped case - the WAM comes out a special case [4].

References

[1] H. Aït-Kaci. *Warren's Abstract Machine: A Tutorial Reconstruction.* MIT Press, Cambridge, MA, 1991.

[2] C. Beierle. Types, modules and databases in the logic programming language PROTOS-L. In K. H. Bläsius, U. Hedtstück, and C.-R. Rollinger, editors, *Sorts and Types for Artificial Intelligence.* LNAI 418, Springer-Verlag, Berlin, Heidelberg, New York, 1990.

[3] C. Beierle and E. Börger. *A WAM extension for type-constraint logic programming: Specification and correctness proof.* IWBS Report 200, IBM Germany, Scientific Center, Inst. for Knowledge Based Systems, Stuttgart, 1991.

[4] C. Beierle, G. Meyer, and H. Semle. Extending the Warren Abstract Machine to polymorphic order-sorted resolution. In V. Saraswat and K. Ueda, editors, *Logic Programming: Proceedings of the 1991 International Symposium*, pages 272–286, MIT Press, Cambridge, MA, 1991.

[5] E. Börger. A logical operational semantics of full Prolog. Part I. Selection core and control. In E. Börger, H. Kleine Büning, and M. M. Richter, editors, *CSL'89 - 3rd Workshop on Computer Science Logic.* LNCS 440, pages 36–64, Springer-Verlag, Berlin, Heidelberg, New York, 1990.

[6] E. Börger. A logical operational semantics of full Prolog. Part II. Built-in predicates for database manipulations. In B. Rovan, editor, *MFCS'90 - Mathematical Foundations of Computer Science.* LNCS 452, pages 1–14, Springer-Verlag, Berlin, Heidelberg, New York, 1990.

[7] E. Börger and D. Rosenzweig. From Prolog algebras towards WAM - a mathematical study of implementation. In E. Börger, H. Kleine Büning, M. M. Richter, and W. Schönfeld, editors, *Computer Science Logic.* LNCS 533, pages 31–66, Springer-Verlag, Berlin, Heidelberg, New York, 1991.

[8] E. Börger and D. Rosenzweig. WAM algebras - a mathematical study of implementation, Part II. In *Russian Conference on Logic Programming '91.* LNCS , Springer-Verlag, 1992. (to appear). Preliminary version in: Technical Report CSE-TR-88-91, The University of Michigan, Department of Electrical Engineering and Computer Science, Ann Arbor, Michigan.

[9] E. Börger and P. H. Schmitt. A formal operational semantics for languages of type Prolog III. In E. Börger, H. Kleine Büning, M. M. Richter, and W. Schönfeld, editors, *Computer Science Logic.* LNCS 533, pages 67–79, Springer-Verlag, Berlin, Heidelberg, New York, 1991.

[10] Y. Gurevich. Evolving algebras. A tutorial introduction. *EATCS Bulletin*, 43, February 1991.

[11] Y. Gurevich. Logic and the challenge of computer science. In E. Börger, editor, *Trends in Theoretical Computer Science*, pages 1–57, Computer Science Press, 1988.

[12] D. M. Russinoff. *A Verified Prolog Compiler for the Warren Abstract Machine.* Technical Report ACT-ST-292-89, MCC, Austin, Texas, 1989. (To appear in *Journal of Logic Programming*).

[13] H. Semle. *Extension of an Abstract Machine for Order-Sorted Prolog to Polymorphism.* Diplomarbeit Nr. 583, Universität Stuttgart und IBM Deutschland GmbH, Stuttgart, April 1989. (in German).

[14] G. Smolka. *Logic Programming over Polymorphically Order-Sorted Types.* PhD thesis, FB Informatik, Univ. Kaiserslautern, 1989.

[15] G. Smolka. *TEL (Version 0.9), Report and User Manual.* SEKI-Report SR 87-17, FB Informatik, Universität Kaiserslautern, 1988.

[16] D. Warren. *An Abstract PROLOG Instruction Set.* Technical Report 309, SRI, 1983.

Model Checking of Persistent Petri Nets[1]

Eike Best and Javier Esparza
Institut für Informatik
Universität Hildesheim
Marienburger Platz 22
W-3200 Hildesheim, Germany
e-mail: {E.Best,esparza}@informatik.uni-hildesheim.de

Abstract

In this paper we develop a model checking algorithm which is fast in the size of the system. The class of system models we consider are safe persistent Petri nets; the logic is S_4, i.e. propositional logic with a 'some time' operator. Our algorithm does not require to construct any transition system: We reduce the model checking problem to the problem of computing certain Parikh vectors, and we show that for the class of safe marked graphs these vectors can be computed – from the structure of the Petri net – in polynomial time in the size of the system.

1 Introduction

Model checking - the algorithmic determination of truth or falsehood of a modal or temporal logic formula, given a model - faces, when applied to concurrent systems, the state explosion problem: the size of the transition system (when finite) can at best be assumed exponential in the size of the underlying system. Therefore, in order to be able to verify properties of non–toy systems, the algorithms have to be able to accept as input graphs containing millions of nodes.
Much work has been done on how to palliate this problem, following two approaches. The first is to improve the efficiency of existing general algorithms: explicit knowledge about concurrency can be used in order to obtain condensed transition systems [9,11,18]. These techniques have the advantage of being generally applicable; however, it is very difficult to know *a priori* if they will be really effective.
The second approach attemps to take advantage of special properties of the underlying model in order to speed up the model checking algorithm. This could lead to efficient, albeit special purpose methods. Two examples of this line of work are [5,15]. However, these papers also require the construction of transition systems. In this contribution, we are more radical: we investigate the possibility of obtaining model checkers which do not require at all to construct the associated transition system, i.e. model checkers that work directly on the syntax. We consider the modal logic S_4

[1] Partly supported by the Esprit Basic Research Action 3148 DEMON

[10] tailored for safe Petri nets (our syntax), and concentrate on a particular subclass of models, namely safe persistent Petri nets [13]. The logic can express properties such as reachability of a marking, liveness of a transition or mutual exclusion of a set of transitions. It also allows some "counting": properties such as "in order to reach a marking in which place s has one token, transition t has to occur 4 times" can be expressed as well.

(Safe) persistent nets are being currently used to model self-timed circuits [17]. In general, concurrent but deterministic systems (applications appear mainly in hardware design) can be modelled using persistent nets.

Given a safe persistent system Σ and a formula ϕ, we show how to reduce the model checking problem to a set of Linear Programming problems. For the subclass of safe T-systems, we prove that the model checker is polynomial in the size of Σ, although exponential in the length of ϕ. Since formulae are usually short, while systems can be very large, this is a very satisfactory result; moreover, as shown in the paper, a model checker polynomial in both the size of Σ and the length of ϕ can exist only if $P = NP$.

The paper is organised as follows. Section 2 introduces the logic. Section 3 discusses briefly the model checking problem. Section 4 introduces the models: persistent and strongly persistent systems. Section 5 presents some results on net processes; in particular, that strongly persistent systems have one single maximal process. The main theorem for the construction of the model checker is proved in Section 6. The model checker itself is described in Section 7. The particular case of T-systems is studied in Section 8. Some basic definitions are contained in an Appendix, although reading this paper is easier if the reader is familiar with the basic notions of Petri nets (otherwise, see [16]).

2 A Modal Logic for Safe Petri Nets

We define a simple modal logic over computations (more precisely occurrence sequences) of safe marked Petri nets, with the following basic propositions:

- Assertions of the form s, to be used with a model containing a place named s. The intended meaning is 'after the present computation, a token is on s'.
- Assertions of the form $t \leq 4$, to be used with a model containing a transition named t. The intended meaning is 'in the present computation t occurs no more than 4 times'.

Our logic is S_4 [10], i.e. , propositional logic augmented with the modal operator $\Diamond$, meaning 'it is possible that ... '.

Definition 2.1 *Syntax of formulae*

The formulae ϕ of our logic have the following form:

$$\begin{array}{lll}\phi ::= & \mathbf{true} & \text{(Truth)}\\ & s & \text{(Place assertion)}\\ & t \leq k \quad (k \in \mathrm{N} \cup \{-1\} \cup \{\omega\}) & \text{(Transition assertion)}\\ & \neg\phi & \text{(Negation)}\\ & \phi_1 \wedge \phi_2 & \text{(Conjunction)}\\ & \phi_1 \vee \phi_2 & \text{(Disjunction)}\\ & \Diamond\phi & \text{(Some time } \phi\text{).}\end{array}$$

A literal is a (possibly negated) place or transition assertion. A formula is called propositional iff it does not contain a modal operator. ■ 2.1

Derived formulae and operators are: $(\phi_1 \Rightarrow \phi_2) = \neg\phi_1 \vee \phi_2$; etc., and the modal operator $\Box\,\phi = \neg\Diamond\neg\phi$ ('always ϕ'). We also write $t > k$ for $\neg(t \leq k)$.
This logic can be interpreted on safe marked Petri nets $\Sigma = (S, T, F, M_0)$. We define what it means for Σ to satisfy a formula by defining inductively what it means that a formula is satisfied by an occurrence sequence σ. We fix some notations first: the set of all occurrence sequences from the initial marking M_0 is denoted by $\mathcal{L}(\Sigma)$, the language of Σ. For two occurrence sequences σ, τ, $\sigma \leq \tau$ if σ is a prefix of τ. Given a total ordering on the set T of transitions, the Parikh vector of an occurrence sequence σ, denoted by $\mathcal{P}(\sigma)$, is given by:

$$\mathcal{P}(\sigma)(t_i) = \text{ number of times } t_i \text{ appears in } \sigma.$$

Definition 2.2 *Satisfaction*

Let ϕ be a formula (of the above form), let $\Sigma = (S, T, F, M_0)$ be a safe marked Petri net and $\sigma \in \mathcal{L}(\Sigma)$. We define $\sigma \models \phi$ (σ satisfies ϕ) inductively:

$$\begin{array}{lll}\sigma \models \mathbf{true} & & \text{always.}\\ \sigma \models s & \text{iff} & M_0[\sigma\rangle M \wedge M(s) = 1.\\ \sigma \models t \leq k & \text{iff} & \mathcal{P}(\sigma)(t) \leq k.\\ \sigma \models \neg\phi & \text{iff} & \text{not } \sigma \models \phi.\\ \sigma \models \phi_1 \wedge \phi_2 & \text{iff} & \sigma \models \phi_1 \text{ and } \sigma \models \phi_2.\\ \sigma \models \phi_1 \vee \phi_2 & \text{iff} & \sigma \models \phi_1 \text{ or } \sigma \models \phi_2.\\ \sigma \models \Diamond\phi & \text{iff} & \exists \tau \geq \sigma : \tau \models \phi.\end{array}$$

Finally, $\Sigma \models \phi$ iff $\varepsilon \models \phi$ (where ε is the empty sequence). ■ 2.2

Notice that $(t \leq \omega) \Leftrightarrow$ **true** and $(t \leq -1) \Leftrightarrow$ **false**. These formulae are introduced just out of syntactic convenience in order to write formulae in a more compact form. The logic permits one to express safety properties such as:

- Reachability of a marking. The system of figure 1 satisfies $\Diamond(\neg s_1 \wedge s_2 \wedge \neg s_3 \wedge s_4)$ iff the marking $(s_1, s_2, s_3, s_4) = (0, 1, 0, 1)$ is reachable.

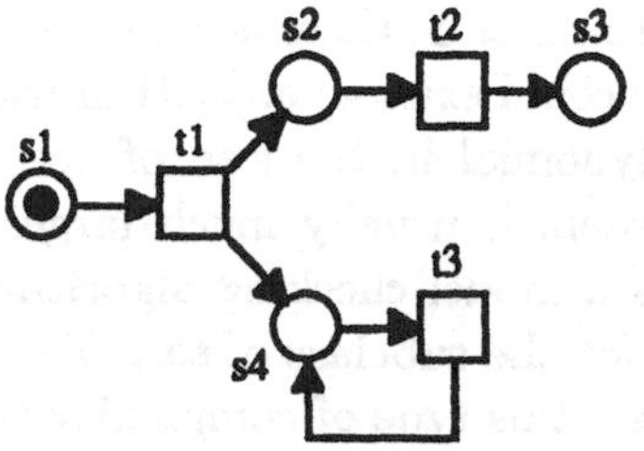

Figure 1: A safe system

- Concurrency of transitions. The system satisfies $\Diamond(s_1 \wedge s_2)$ iff transitions t_1 and t_2 are concurrently enabled at some reachable marking.
- Liveness of a transition. The system satisfies $\Box\Diamond s_4$ iff transition t_3 is live.

These properties involve place assertions only. Transition assertions permit to express properties as 'in order to reach a state with $M(s_3) = 1$, transition t_3 has to occur': $\Box(s_3 \Rightarrow t_3 > 0)$.

3 The Model Checking Problem

The Model Checking Problem (MCP) investigated in this paper is the problem of determining, given a formula ϕ in our logic and a system Σ, whether or not $\Sigma \models \phi$. It is easy to give a lower bound on the complexity of the MCP: since the reachability problem is known to be PSPACE–complete for safe Petri nets [12], and reachability is expressible in our logic, the MCP is PSPACE–hard. Moreover, even for the simplest concurrent systems the problem is still NP–hard.
Let $i \in \mathbf{N}$ and $\Sigma_i = (S_i, T_i, F_i, M_i)$ be the system given by:

$$\begin{aligned} S_i &= \{s_1, \ldots, s_i\} \\ T_i &= \{t_1, \ldots, t_i\} \\ F_i &= \{(s_1, t_1), \ldots, (s_i, t_i)\} \\ M_i(s_j) &= 1 \text{ for all } j, 1 \leq j \leq i \end{aligned}$$

Define $\mathcal{S} = \{\Sigma_i \mid i \in \mathbf{N}\}$.

Proposition 3.1

The MCP for $\mathcal{S}$ is NP–hard.

Proof: By reduction from SAT (satisfiability of propositional logic). Let K be a propositional formula on variables $x_1, \ldots, x_k$. We construct the formula $\Diamond\phi$, where ϕ is obtained from K by replacing x_j by s_j for all $j, 1 \leq j \leq k$. It is easy to show that K is satisfiable iff $\Sigma_k \models \Diamond\phi$ (see [3]). ■ 3.1

So even for such a simple class as $\mathcal{S}$, there is little hope to find a polynomial algorithm for MCP. However, there could exist an algorithm which is exponential on the length of the formula, but polynomial in the size of the system. Such a result carries interest, because the system is usually much larger than the formula. We design in the following sections a model checking algorithm for safe (strongly) persistent systems, and show that for the subclass of safe T–systems – of which $\mathcal{S}$ is, in turn, a trivially small subclass – this type of complexity (exponentiality in the size of the formula but polynomiality in the size of the system) can be obtained.

4 Persistent and strongly persistent net systems

We are interested in the class of nets without conflicts: when two transitions are enabled at a marking, then they are concurrent. We call these systems strongly persistent. Strongly persistent systems are a subclass of the slightly larger and well known class of persistent systems [13], in which when two transitions are enabled at a marking then they can occur in any order, but not always concurrently. The model checking problem for persistent systems reduces easily to the problem for strongly persistent systems.

Definition 4.1 *Persistence and strong persistence*

(i) A Petri net Σ is persistent iff for all $M \in [M_0\rangle$ and for all $t_1, t_2 \in T, t_1 \neq t_2$, if $M[t_1\rangle$ and $M[t_2\rangle$ then $M[t_1 t_2\rangle$ and $M[t_2 t_1\rangle$.

(ii) Σ is strongly persistent iff for all $M \in [M_0\rangle$ and for all $t_1, t_2 \in T, t_1 \neq t_2$, if $M[t_1\rangle$ and $M[t_2\rangle$ then $M[\{t_1, t_2\}\rangle$. ■ 4.1

Strong persistence implies persistence, but the converse is not true.

Persistent systems can be translated into strongly persistent systems, so that a formula is true of a persistent system if and only if it is true of its translation. The proof of this result can be found in [3].

Lemma 4.2

Let Σ be a persistent system. There exist a polynomial time algorithm to construct a system Σ', with the same set of transitions as Σ, enjoying the following properties:

(i) *Σ' is strongly persistent.*

(ii) $\mathcal{L}(\Sigma) = \mathcal{L}(\Sigma')$.

(iii) *There is a function f from the formulae of Σ in the formulae of Σ' such that*
$\sigma \models F$ *in* Σ *iff* $\sigma \models f(F)$ *in* Σ'. ■ 4.2

(Strongly) persistent systems find applications in the design of switching circuits [17] and when modelling deterministic concurrent systems by means of Petri nets.

5 The lattice of cuts. Processes of strongly persistent systems

In this section, we state some elementary properties of finite occurrence nets. Basic definitions are given in the Appendix. The proofs appear either in [8] or in [3].

Throughout the section, let $N = (B, E, F)$ be a finite occurrence net, let $\mathcal{C}$ denote the set of B-cuts of N, and let $(X, \prec) = (B \cup E, F')$ be the partial order associated with N [4]. Whenever we speak of cuts in the sequel, we shall always mean B-cuts.

Figure 2.(a) shows a strongly persistent system; its reachability graph is shown in Figure 2.(b). Finally, Figure 2.(c) shows one of its processes and some cuts. For $c_1, c_2 \in \mathcal{C}$, let $c_1 \sqsubseteq c_2$ iff

$$\forall b_1 \in c_1 \, \forall b_2 \in c_2 \colon \neg(b_2 \prec b_1).$$

We study some properties of this definition.

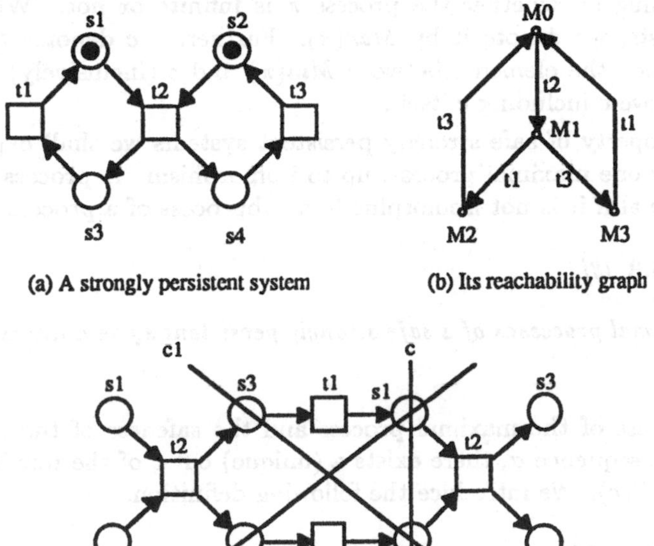

(a) A strongly persistent system (b) Its reachability graph

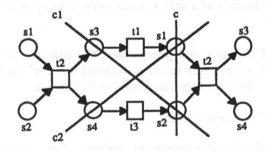

(c) One of its processes, and some cuts.

Figure 2: Illustration of the notion of process and cut

Lemma 5.1

$\sqsubseteq$ *is transitive and antisymmetric.* ■ 5.1

For $x \in B \cup E$ and $c \in \mathcal{C}$, let $x \prec c$ denote $\exists b' \in c \colon x \prec b'$.

Lemma 5.2

Let $c_1, c_2 \in C$. Then

(i) $c_1 \sqcup c_2 \;=\; (c_1 \cup c_2)\backslash(\{b_1 \in c_1 \mid b_1 \prec c_2\} \cup \{b_2 \in c_2 \mid b_2 \prec c_1\})$
is the lowest upper bound of c_1, c_2 with respect to $\sqsubseteq$.

(ii) $c_1 \sqcap c_2 \;=\; (c_1 \cup c_2)\backslash(\{b_1 \in c_1 \mid c_2 \prec b_1\} \cup \{b_2 \in c_2 \mid c_1 \prec b_2\})$
is the greatest lower bound of c_1, c_2 with respect to $\sqsubseteq$.

(iii) $c_1 \cap c_2 \;\subseteq\; (c_1 \sqcup c_2) \cap (c_1 \sqcap c_2)$

(iv) $c_1 \cup c_2 \;\supseteq\; (c_1 \sqcup c_2) \cup (c_1 \sqcap c_2)$

(v) $\sqcap$ *and* $\sqcup$ *are monotonic with respect to* $\sqsubseteq$. ■ 5.2

In the process of Figure 2.(c), $c = c_1 \sqcup c_2$ is the lowest upper bound of c_1 and c_2. As a consequence of parts (i) and (ii) of this lemma, $(C, \sqcup, \sqcap)$ is a lattice. This lattice has always a least element, denoted by $Min(\pi)$, but not always a maximal one, depending on whether the process π is infinite or not. When the maximal element exists, we denote it by $Max(\pi)$. Further, $\Downarrow c$ denotes the subprocess of π below c, i.e. the elements between $Min(\pi)$ and c (inclusively). $\Uparrow c$ denotes the elements above c including c itself.

The nice property of safe strongly persistent systems we shall exploit is that they have exactly one maximal process, up to isomorphism. A process of a system Σ is called maximal if it is not isomorphic to a subprocess of a process of Σ.

Theorem 5.3 *[3]*

All maximal processes of a safe strongly persistent system are isomorphic to each other.

The uniqueness of the maximal process and the safeness of the system guarantee that, given a sequence σ, there exists a (unique) cut c of the maximal process such that $\sigma \in Lin(\Downarrow c)$. We introduce the following definition.

Definition 5.4 *The mapping Cut*

Let Σ be a safe strongly persistent system. The mapping $Cut\colon \mathcal{L}(\Sigma) \rightarrow C$ is defined by

$Cut(\sigma) =$ the cut c of the maximal process of Σ such that $\sigma \in Lin(\Downarrow c)$.

■ 5.4

6 Conjunctive Propositional Formulae

In this section we show that strongly persistent nets, due to Theorem 5.3, have interesting properties with respect to our logic. We start, however, with an observation valid for all safe Petri nets: occurrence sequences with the same Parikh vector satisfy the same properties.

Lemma 6.1

Let σ, τ be two occurrence sequences of a system such that $\mathcal{P}(\sigma) = \mathcal{P}(\tau)$. Then, for every formula ϕ: $\sigma \models \phi \Leftrightarrow \tau \models \phi$.

Proof: Follows easily from the definition of $\models$ and the fact that both σ and τ lead to the same marking: $M_0[\sigma\rangle M$ and $M_0[\tau\rangle M$. ■ 6.1

For the rest of the section $\Sigma = (S, T, F, M_0)$ denotes a strongly persistent safe Petri net and $\pi = (B, E, F', p)$ its unique maximal process.

Lemma 6.2

Let σ_1, σ_2 be two occurrence sequences. Let $c_1 = Cut(\sigma_1)$ and $c_2 = Cut(\sigma_2)$. If $\sigma \in Lin(\Downarrow(c_1 \sqcup c_2))$, then $\mathcal{P}(\sigma) = \max\{\mathcal{P}(\sigma_1), \mathcal{P}(\sigma_2)\}$.

Proof: (i) $\mathcal{P}(\sigma) \geq \max\{\mathcal{P}(\sigma_1), \mathcal{P}(\sigma_2)\}$.

We have:

$$\begin{aligned}
\mathcal{P}(\sigma_1)(t) &= |\{e \in E \mid e \prec c_1 \wedge p(e) = t\}| \\
\mathcal{P}(\sigma_2)(t) &= |\{e \in E \mid e \prec c_2 \wedge p(e) = t\}| \\
\mathcal{P}(\sigma)(t) &= |\{e \in E \mid e \prec (c_1 \sqcup c_2) \wedge p(e) = t\}|
\end{aligned}$$

It follows easily that for every transition t, $\mathcal{P}(\sigma)(t) \geq \mathcal{P}(\sigma_1)(t)$ and $\mathcal{P}(\sigma)(t) \geq \mathcal{P}(\sigma_2)(t)$. Hence, (1) holds.

(ii) $\mathcal{P}(\sigma) \leq \max\{\mathcal{P}(\sigma_1), \mathcal{P}(\sigma_2)\}$.

We prove the following claim first.

Claim. If there exist $e_1, e_2 \in E$ such that $p(e_1) = p(e_2) = t$, then $\neg((c_2 \prec e_1 \prec c_1) \wedge (c_1 \prec e_2 \prec c_2))$.

Proof. Suppose, on the contrary, that $(c_2 \prec e_1 \prec c_1) \wedge (c_1 \prec e_2 \prec c_2)$. Then $e_1\, co\, e_2$ and $e_1 \neq e_2$; this is because if (for instance) $e_1 \preceq e_2$, then $b_2 \prec e_1 \preceq e_2 \prec b_2'$, for some $b_2, b_2' \in c_2$, contradicting the fact that c_2 is a cut. But since $e_1\, co\, e_2$, Theorem 3.19 of [2] shows that $t = p(e_1) = p(e_2)$ can be concurrently enabled, contradicting the safeness of Σ. *End of proof.*

The claim implies that we have either one of the following two cases:

$$\begin{array}{ll}
\text{(a)} & \{e \in E \mid e \prec (c_1 \sqcup c_2) \wedge p(e) = t\} \subseteq \{e \in E \mid e \prec c_1 \wedge p(e) = t\} \\
\text{(b)} & \{e \in E \mid e \prec (c_1 \sqcup c_2) \wedge p(e) = t\} \subseteq \{e \in E \mid e \prec c_2 \wedge p(e) = t\}.
\end{array}$$

In the first case, $\mathcal{P}(\sigma) \leq \mathcal{P}(\sigma_1)$. In the second, $\mathcal{P}(\sigma) \leq \mathcal{P}(\sigma_2)$. Hence, (2) holds. ■ 6.2

As a corollary of this lemma, we obtain the following result of [13]. In fact, the result was proved there for arbitrary (strongly) persistent nets, not just safe ones.

Corollary 6.3 [13]

Let σ_1, $\sigma_2 \in \mathcal{L}(\Sigma)$. There exist sequences τ_1, τ_2 such that $\sigma_1\tau_1$, $\sigma_2\tau_2$ are occurrence sequences and

$$\mathcal{P}(\sigma_1\tau_1) = \mathcal{P}(\sigma_2\tau_2) = \max\{\mathcal{P}(\sigma_1), \mathcal{P}(\sigma_2)\}.$$

Proof: Let $c_1 = Cut(\sigma_1)$ and $c_2 = Cut(\sigma_2)$.
Take $\tau_1 \in Lin(\Uparrow c_1 \cap \Downarrow(c_1 \sqcup c_2))$, $\tau_2 \in Lin(\Uparrow c_2 \cap \Downarrow(c_1 \sqcup c_2))$ (the intersection is defined componentwise, and can easily be shown to be a process).
Then $\sigma_1\tau_1$, $\sigma_2\tau_2 \in Lin(\Downarrow(c_1 \sqcup c_2))$, and the result follows from Lemma 6.2.
■ 6.3

We are now ready to prove the following theorem:

Theorem 6.4

Let χ a conjunction of literals. Let $\sigma_1, \sigma_2 \in \mathcal{L}(\Sigma)$.
If $\sigma_1 \models \chi$ and $\sigma_2 \models \chi$, then for every $\sigma \in \mathcal{L}(\Sigma)$ such that $\mathcal{P}(\sigma) = \max\{\mathcal{P}(\sigma_1), \mathcal{P}(\sigma_2)\}$

we have $\sigma \models \chi$.

Proof: By Lemma 6.1, it suffices to prove the property for a particular σ satisfying the condition on the Parikh vector. Let $c_1 = Cut(\sigma_1)$, $c_2 = Cut(\sigma_2)$. Using Lemma 6.2, we choose σ as one of the linearisations of $\Downarrow(c_1 \sqcup c_2)$. We prove the claim separately for the possible literals in χ.

(i) $(\sigma_1 \models t \leq k \wedge \sigma_2 \models t \leq k) \Rightarrow \sigma \models t \leq k$
We have $\mathcal{P}(\sigma_1)(t) \leq k$ and $\mathcal{P}(\sigma_2)(t) \leq k$.
Then $\mathcal{P}(\sigma)(t) = \max\{\mathcal{P}(\sigma_1)(t), \mathcal{P}(\sigma_2)(t)\} \leq k$.

(ii) $(\sigma_1 \models t > k \wedge \sigma_2 \models t > k) \Rightarrow \sigma \models t > k$
We have $\mathcal{P}(\sigma_1)(t) > k$ and $\mathcal{P}(\sigma_2)(t) > k$.
Then $\mathcal{P}(\sigma)(t) = \max\{\mathcal{P}(\sigma_1)(t), \mathcal{P}(\sigma_2)(t)\} > k$.

(iii) $(\sigma_1 \models s \wedge \sigma_2 \models s) \Rightarrow \sigma \models s$
Let $M_0[\sigma_1\rangle M_1$ and $M_0[\sigma_2\rangle M_2$.
Because $\sigma_1 \models s$ and $\sigma_2 \models s$, we have:

$$\exists b_1 \in c_1 : p(b_1) = s \quad \text{and} \quad \exists b_2 \in c_2 : p(b_2) = s$$

Case 1: $b_1 = b_2$.
Then $b_1 = b_2 \in c_1 \cap c_2$, and by Lemma 5.2(iii), also $b_1 = b_2 \in c_1 \sqcup c_2$.
Case 2: $b_1 \neq b_2$.
Then by the safeness of Σ (and Theorems 3.15, 3.17 and 3.19 of [2]), it cannot be the case that b_1 *co* b_2. Therefore, either $b_1 \prec b_2$ or $b_2 \prec b_1$.
In the former case, $b_2 \in c_1 \sqcup c_2$, in the latter case, $b_1 \in c_1 \sqcup c_2$.
In all cases, $\exists b \in c_1 \sqcup c_2 : p(b) = s$.

(iv) $(\sigma_1 \models \neg s \wedge \sigma_2 \models \neg s) \Rightarrow \sigma \models \neg s$
Let $M_0[\sigma_1\rangle M_1$ and $M_0[\sigma_2\rangle M_2$.
Since $\sigma_1 \models \neg s$ and $\sigma_2 \models \neg s$, we have $s \notin p(c_1)$ and $s \notin p(c_2)$. Because of $c_1 \sqcup c_2 \subseteq c_1 \cup c_2$ (Lemma 5.2(iv)), we also have $s \notin p(c_1 \sqcup c_2)$, and hence $\sigma \models \neg s$.

From (i)–(iv), it follows that if both σ_1 and σ_2 satisfy a conjunction of literals χ, then σ satisfies it as well. ■ 6.4

Remark 6.5

Theorem 6.4 is false if χ is allowed to be a disjunction. Consider the formula $\phi = s_3 \vee s_4$ in the example of Figure 2. We have $t_2 t_1 \models \phi$ and $t_2 t_3 \models \phi$, but $t_2 t_1 t_3 \not\models \phi$. ■ 6.5

We associate to a conjunction of literals χ a Parikh vector in the following way:

Definition 6.6 *The mapping $Last_\chi$*

Let χ be a conjunction of literals. The mapping $Last_\chi: T \rightarrow \mathbb{N} \cup \{-1\} \cup \{\omega\}$ (with $\omega > k$ for every $k \in \mathbb{N}$) is defined as follows:

$$Last_\chi(t) = \begin{cases} -1 & \text{if no occurrence sequence satisfies } \chi \\ \sup\{\mathcal{P}(\sigma)(t) \mid \sigma \models \chi\} & \text{otherwise} \end{cases}$$

■ 6.6

Remark 6.7

By Theorem 6.4, $Last_\chi = \sup\{\mathcal{P}(\sigma) \mid \sigma \models \chi\}$. ■ 6.7

The interest of this definition lies in the following result. Loosely speaking, $Last_\chi(t)$ indicates the maximum number of times (arbitrarily many if $Last_\chi(t) = \omega$) that transition t can occur without losing the possibility of extending the current occurrence sequence to one satisfying χ.

Lemma 6.8

Let $\sigma \in \mathcal{L}(\Sigma)$. σ can be extended to an occurrence sequence $\tau \geq \sigma$ with $\tau \models \chi$ iff $\mathcal{P}(\sigma) \leq Last_\chi$.

Proof: ($\Rightarrow$): Follows easily from the definition of $Last_\chi$.

($\Leftarrow$): If $\mathcal{P}(\sigma) = Last_\chi$, then take $\tau = \sigma$. If $\mathcal{P}(\sigma) \neq Last_\chi$ then we have:

$$
\begin{aligned}
&\mathcal{P}(\sigma) \neq Last_\chi \\
&\Rightarrow \{\text{ Remark 6.7 }\} \\
&\exists \sigma' \colon \sigma' \models \chi \wedge \mathcal{P}(\sigma) \leq \mathcal{P}(\sigma') \leq Last_\chi \\
&\Rightarrow \\
&\max\{\mathcal{P}(\sigma), \mathcal{P}(\sigma')\} = \mathcal{P}(\sigma') \\
&\Rightarrow \{\text{ Corollary 6.3 }\} \\
&\exists \sigma'' \colon\ \sigma\sigma'' \text{ is an occurrence sequence } \wedge \mathcal{P}(\sigma\sigma'') = \mathcal{P}(\sigma') \\
&\Rightarrow \{\sigma' \models \chi,\ \text{Lemma 6.1 }\} \\
&\exists \sigma'' \colon\ \sigma\sigma'' \text{ is an occurrence sequence } \wedge \sigma'' \models \chi.
\end{aligned}
$$

Taking $\tau = \sigma\sigma''$, the result follows. ■ 6.8

The following theorem is the kernel of our model checker: it shows how to replace a formula with one modality by a propositional formula.

Theorem 6.9

Let Σ be a safe persistent system. For every conjunction χ of literals and every occurrence sequence σ:

$$\sigma \models \Diamond\chi \Leftrightarrow \sigma \models \bigwedge_{t \in T} t \leq Last_\chi(t).$$

Proof: By definition of $\models$, $\sigma \models \Diamond\chi$ iff there exists $\tau \geq \sigma$ such that $\tau \models \chi$. By Lemma 6.8, this is the case iff $\mathcal{P}(\sigma) \leq Last_\chi$. By definition of $\models$ again, $\mathcal{P}(\sigma) \leq Last_\chi$ iff $\sigma \models \bigwedge_{t \in T} t \leq Last_\chi(t)$. ■ 6.9

7 The Model Checker

Before presenting formally the model checker, we need to introduce a standard form for the formulae of our logic.

Definition 7.1 *Standard form*

A formula $\Diamond\phi$ is a *first-degree formula* iff ϕ contains no modalities. A formula ϕ is in *standard form* iff:

- ϕ contains no derived operators, and
- for every first-degree subformula $\Diamond\phi'$ of ϕ, ϕ' is a conjunction of literals.

■ 7.1

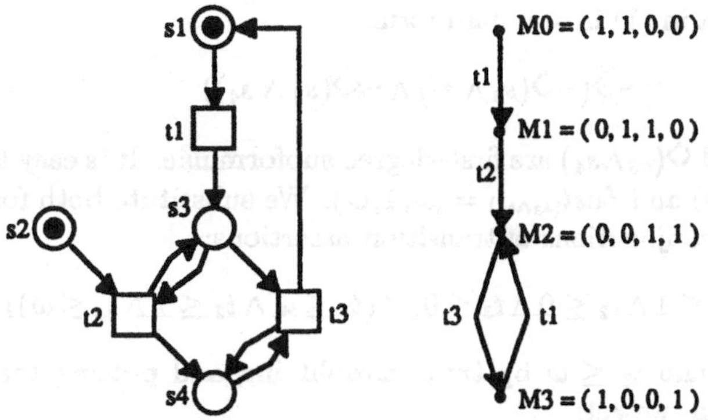

Figure 3: A system on which to use the model checker

Proposition 7.2

Every formula is equivalent to a (not necessarily unique) formula in standard form.

Proof: Easy, using propositional calculus and the schema $\Diamond(\phi_1 \vee \phi_2) = \Diamond\phi_1 \vee \Diamond\phi_2$, valid in our logic. ■ 7.2

Algorithm 7.3 *The model checker.*

Input: A safe strongly persistent system $\Sigma = (N, M_0)$ and a formula ϕ.
Output: $\Sigma \models \phi$ or $\Sigma \not\models \phi$.

begin
  while ϕ contains modalities **do**
    $\phi := \phi$ in standard form;
    for every first–degree subformula $\Diamond\chi$ of ϕ **do**
      compute $Last_\chi$;
      substitute $\Diamond\chi$ by $\bigwedge_{t \in T} t \leq Last_\chi(t)$
    endfor
  endwhile
  (⋆ Now ϕ contains no modalities ⋆)
  check if $\Sigma \models \phi$ using the definition and answer accordingly
end

■ 7.3

We apply the model checker to a small example. Consider the safe strongly persistent system Σ on the left of Figure 3.

We use the model checker to answer:

$$\Sigma \overset{?}{\models} \Box \Diamond(s_3 \wedge (s_2 \vee s_4)) \tag{1}$$

We put the formula in (1) in standard form:

$$\neg\Diamond(\neg\Diamond(s_3 \wedge s_2) \wedge \neg\Diamond(s_3 \wedge s_4)) \tag{2}$$

Both $\Diamond(s_3 \wedge s_2)$ and $\Diamond(s_3 \wedge s_4)$ are first–degree subformulae. It is easy to see that $Last_{(s_3 \wedge s_2)} = (1,0,0)$ and $Last_{(s_3 \wedge s_4)} = (\omega, 1, \omega)$. We substitute both formulae by the corresponding conjunctions of transition assertions:

$$\neg\Diamond\neg((t_1 \leq 1 \wedge t_2 \leq 0 \wedge t_3 \leq 0) \vee (t_1 \leq \omega \wedge t_2 \leq 1 \wedge t_3 \leq \omega)) \tag{3}$$

Replacing $t_1 \leq \omega$ and $t_3 \leq \omega$ by **true**, simplifying and putting the result in standard form again, we get:

$$\neg(\Diamond(t_1 > 1 \wedge t_2 > 1) \;\vee\; \Diamond(t_2 > 0 \wedge t_2 > 1) \;\vee\; \Diamond(t_3 > 0 \wedge t_2 > 1)) \tag{4}$$

There is no sequence satisfying $t_1 > 1 \wedge t_2 > 1$. Hence, the corresponding *Last* vector is $(-1,-1,-1)$, and similarly for the other two disjuncts of (4) :

$$\neg((t_1 \leq -1 \wedge t_2 \leq -1 \wedge t_3 \leq -1) \vee \ldots \vee (t_1 \leq -1 \wedge t_2 \leq -1 \wedge t_3 \leq -1)) \tag{5}$$

Replacing $t_i \leq -1$ by **false**, we get:

$$\neg(\textbf{false} \vee \textbf{false} \vee \textbf{false}) \tag{6}$$

which evaluates to **true**. Since $\varepsilon \models$ **true**, Σ satisfies the original formula.

Putting ϕ in standard form can make the size of ϕ grow exponentially (this may happen, for instance, if $\phi = \Diamond\phi'$, where ϕ' is a propositional formula in conjunctive normal form). Therefore, the number of *Last* vectors to be computed is in the worst case exponential in the length of ϕ (and independent of the size of Σ).

Our model checker is completely specified only after giving the description of a procedure for computing this vector. This is done in the next section for the class of safe T–systems (which can easily be shown to be persistent).

8 Computing Last$_\chi$ for safe T–systems

Definition 8.1

A net (S,T,F) is a T–net iff for every $s \in S$: $|{}^\bullet s| \leq 1$ and $|s^\bullet| \leq 1$.
(S,T,F,M_0) is a T–system iff (S,T,F) is a T–net. ■ 8.1

Throughout this section, $\Sigma = (S,T,F,M_0)$ is a safe T–system, and C the incidence matrix of (S,T,F), which we suppose to be weakly connected (this constraint is introduced to simplify the presentation; the results can be extended to non–connected T-systems by computing their connected components first, and considering them separately). We shall make use of some results on T–systems that are presented now. They are immediate consequences of results of [7,14].

Theorem 8.2

Let T_d be the set of transitions t of Σ that do not appear in any sequence of $\mathcal{L}(\Sigma)$. X is an integer solution of the system of linear (in)equalities

$$\begin{aligned} M_0 + C \cdot X &\geq 0 \\ \forall t \in T_d \colon X(t) &= 0 \\ X &\geq 0 \end{aligned}$$

iff there exists an occurrence sequence σ such that $\mathcal{P}(\sigma) = X$. ■ 8.2

The constraints on the transitions of T_d are necessary. If they are suppressed then, for instance, $X = (1)$ is a solution of the equation system corresponding to the T-net $(\{s\}, \{t\}, \{(s,t),(t,s)\})$ with marking $M_0(s) = 0$. However, there is no occurrence sequence with X as Parikh vector. Once the constraint $X(t) = 0$ is added, the only solution is $X = (0)$, which corresponds to the empty occurrence sequence.

The set T_d can be very easily computed in polynomial time in the size of Σ, as shown in [3].

The computation of Last_χ can now be reduced to the solution of a Linear Programming problem. Let us see first how to associate linear constraints to the basic propositions of our logic.

Let χ be a conjunction $\chi_1 \wedge \chi_2 \ldots \wedge \chi_n$ of literals. The system of inequalities S_χ is obtained by adding to the system of Theorem 8.2 a linear constraint for each literal χ_i in the following way:

(1) If $\chi_i = s$, then add $(M_0 + C \cdot X)(s) = 1$.

(2) If $\chi_i = \neg s$, then add $(M_0 + C \cdot X)(s) = 0$.

(3) If $\chi_i = t \leq k$ then add $X(t) \leq k$.

(4) If $\chi_i = t > k$ then add $X(t) > k$.

S_χ has the following properties:

Lemma 8.3

(i) If $\sigma \models \chi$, then $\mathcal{P}(\sigma)$ is solution of S_χ.

(ii) If X is solution of S_χ, then there exists σ such that $\mathcal{P}(\sigma) = X$ and $\sigma \models \chi$.

(iii) If S_χ has infinitely many solutions, then for every $k \in \mathbb{N}$ there exists a solution $X \geq \vec{k} = (k, k, \ldots, k)$.

Proof: (i) and (ii) follow from the definition of $\models$ and Theorem 8.2. (iii) follows from results of [7], taking into account that we only consider connected T–systems. ■ 8.3

We define the Linear Programming problem LP_χ

$$\begin{array}{ll} \text{maximise} & \sum_{t\in T} X(t) \\ \text{subject to} & S_\chi \end{array}$$

LP_χ has the following property:

Lemma 8.4

Optimal solutions of LP_χ are integer.

Proof: S_χ can be written in compact form in the following way:

$$\begin{array}{l} P_1 \leq M_0 + C \cdot X \leq P_2 \\ T_1 \leq X \leq T_2 \end{array}$$

for adequate vectors P_1, P_2, T_1, T_2. In particular, $\vec{0} \leq P_1 \leq P_2 \leq \vec{1}$; T_2 may have ω–components, meaning that there is no upper bound for the corresponding component of X.
We show that if X is a solution of S_χ, so is $\lceil X \rceil$. The lemma then follows from

$$\sum_{t\in T} X(t) \leq \sum_{t\in T} \lceil X(t) \rceil .$$

Let $M_1 = M_0 + C \cdot X$ and $M_2 = M_0 + C \cdot \lceil X \rceil$. Let s be a place. We have $|{}^\bullet s| \leq 1$ and $|s^\bullet| \leq 1$. Then, taking into account that P_1, P_2 are vectors over $\{0, 1\}$, we have

$$P_1(s) \leq \lfloor M_1(s) \rfloor \leq M_2(s) \leq \lceil M_1(s) \rceil \leq P_2(s).$$

Moreover, since T_1, T_2 are vectors on $\mathbf{N} \cup \{\omega\}$, we have $T_1 \leq X \leq \lceil X \rceil \leq T_2$. ■ 8.4

We can now give the following computational characterisation of $Last_\chi$.

Theorem 8.5

Let Σ be a safe T–system and χ as above.

(i) If LP_χ has no solution, then $Last_\chi = -\vec{1}$.

(ii) If LP_χ has solutions but no optimal solution, then $Last_\chi = \vec{\omega}$.

(iii) If LP_χ has an optimal solution X then $Last_\chi = X$.

Proof: (*i*) If LP_χ has no solution, then by Lemma 8.3(1) no occurrence sequence satisfies χ. By definition, $Last_\chi = -\vec{1}$.

(*ii*) In this case, S_χ has infinitely many solutions. By Lemma 8.3(3), there exists for every $k \in \mathbf{N}$ a solution σ with $\mathcal{P}(\sigma) \geq \vec{k}$. By Lemma 8.3(2), $Last_\chi > \vec{k}$ for every $k \in \mathbf{N}$. Hence, $Last_\chi = \vec{\omega}$.

(*iii*) By Lemma 8.4, X is integer. By Lemma 8.3(2), there exists σ such that $\mathcal{P}(\sigma) = X$ and $\sigma \models \chi$. By the optimality of X, there is no $\tau \geq \sigma$ such that $\tau \models \chi$. By Lemma 6.8, $Last_\chi = X$. ■ 8.5

Since Linear Programming is known to be polynomial in the size of the equation system, Theorem 8.5 implies that the computation of Last_X is polynomial in the size of the T–system. Therefore, our model checker is polynomial in the size of the system (although it has exponential worst-case complexity in the length of the formula).
It is well known that simplex seems to have better average complexity than the existing polynomial algorithms for Linear Programming, and therefore it is the algorithm that should be used in practice.

9 Conclusions

We have tailored the modal logic S_4 for safe Petri nets. The resulting logic is rather modest; for instance, it is not possible to express liveness properties (such as 'the system will eventually reach a state with one token in place s') but many useful safety properties (reachability, reversibility, liveness of transitions) can be expressed, and the logic has some counting power. We have proved that even for this simple logic (in fact, for any logic extending the propositional calculus) and the simplest classes of concurrent systems, the model checking problem is NP–hard. This shows that model checkers polynomial in both the size of the system and the length of the formula are very unlikely to exist: however, there can still be model checkers exponential in the length of the formula, but polynomial in the size of the system.

We have designed a model checking algorithm for safe strongly persistent systems; the model checking problem is reduced to obtaining a set of so called ***Last*** vectors, whose number can be exponential in the length of the formula. We have shown that for the class of safe T–systems, the ***Last*** vectors can be computed solving a Linear Programming problem, in polynomial time in the size of the system. Therefore, for this class we have obtained a model checker with the complexity mentioned above. This is the first time, to the best of our knowledge, that such a result is obtained for a non–trivial class of systems and a non–trivial logic.

Classical results (see, for instance, [7]) showed that particular properties such as liveness or reachability could be verified in polynomial time for T–systems. These results are subsumed by our paper (for the safe case); we have shown that the polynomiality result can be extended to the whole class of properties expressible by a logic.

Our results are also related to the the constrained expression formalism of [1]. In this framework, linear constrains on the behaviour of the system are obtained from its structure. Then, Integer Linear Programming is used as a decision algorithm to verify properties. In this paper, we have shown how this method can be improved for the particular class of T–systems. In particular, Integer Linear Programming, which is known to be NP-complete, can be replaced by Linear Programming. Linear Programming is also used in the work of [6] on semidecision algorithms of some first–order assertions.

Acknowledgement

The authors wish to thank Raymond Devillers for very careful reading of the text.

References

[1] G.S. Avrunin, U.A. Buy, J.C. Corbett, L.K. Dillon and J.C. Wileden: Automated Analysis of Concurrent Systems with the Constrained Expression Toolset. COINS Technical Report 90–116, University of Massachussets at Amherst (1990).

[2] E. Best and R. Devillers: Sequential and Concurrent Behaviour in Petri Net Theory. TCS Vol. 55, 87–136 (1987).

[3] E. Best and J. Esparza: Model Checking of Persistent Petri Nets. Hildesheimer Informatik Fachbericht 11/91 (1991).

[4] E. Best and C. Fernández: Nonsequential Processes – a Petri Net View. EATCS Monographs on Theoretical Computer Science Vol.13 (1988).

[5] E.M. Clarke, O. Grumberg, M.C. Browne: Reasoning about Networks with many identical finite–state processes. Proceedings of the Fifth Annual Symposium on Principles of Distributed Computing, 240–248 (1986).

[6] J.M. Colom: Structural Analysis Techniques Based on Linear Programming and Convex Geometry. Ph. D. Thesis, University of Zaragoza (1990) (in spanish).

[7] F. Commoner, A.W. Holt, S. Even and A. Pnueli: Marked Directed Graphs. Journal of Computer and System Science Vol.5, 511–523 (1971).

[8] C. Fernández, M. Nielsen and P.S. Thiagarajan: Notions of Realisable Non–Sequential Processes. Fundamenta Informaticae IX, 421–454 (1986).

[9] P. Godefroid: Using Partial Orders to Improve Automatic Verification Methods. Proc. of Computer-Aided Verification Workshop, Rutgers, New Jersey (1990).

[10] G.E. Hughes and M.J. Creswell: An Introduction to Modal Logic. Methuen and Co. (1968).

[11] R. Janicki and M. Koutny: Optimal Simulation for the Verification of Concurrent Systems. Technical report, McMaster University (1989).

[12] M. Jantzen: Complexity of Place/Transition Nets. Petri Nets: Central Models and Their Properties. W. Brauer, W. Reisig, G. Rozenberg (eds.), LNCS 254, 397–412 (1987).

[13] L.H. Landweber and E.L. Robertson: Properties of Conflict–free and Persistent Petri Nets. JACM Vol.25, No.3, 352–364 (1978).

[14] T. Murata: Petri Nets: Properties, Analysis and Applications. Proc. of the IEEE Vol. 77, No. 4, 541–580 (1989).

[15] H. Qin: Efficient Verification of Determinate Processes. CONCUR'91, J.C.M. Baeten, J.F. Groote (eds.), LNCS 527, 470–479 (1991).

[16] W. Reisig: Petri Nets – an Introduction. EATCS Monographs on Theoretical Computer Science, Vol. 4, Springer Verlag (1985).

[17] M. Tiusanen: Some Unsolved Problems in Modelling Self-timed Circuits Using Petri Nets. EATCS Bulletin, Vol. 36, 152–160 (1988).

[18] A. Valmari: Stubborn Sets for Reduced State Space Generation. Advances in Petri Nets 1990, G. Rozenberg (ed.), LNCS 483, 491–515 (1990).

Appendix: Basic Notions

An occurrence net $N = (B, E, F')$ is an acyclic net without branched places, i.e., $F'^{+} \cap (F'^{-1})^{+} = \emptyset$ (acyclicity) and $\forall b \in B: |^{\bullet}b| \leq 1 \wedge |b^{\bullet}| \leq 1$ (no branching of places). Elements of E are called events and elements of B are called conditions. An occurrence net may be infinite, or may contain isolated conditions (but no isolated events). To every occurrence net, a poset $(X, \preceq) = (B \cup E, F'^{*})$ can be associated.
We denote $li = \preceq \cup \succeq$ and $co = ((X \times X)\setminus li) \cup id|_X$. $c \subseteq X$ is called a co-set iff any two elements in c are unordered, i.e. in relation co.
A cut $c \subseteq X$ is a maximal co-set. A cut c is called B-cut iff $c \subseteq B$.

$Min(N)$ is defined as $\{x \in X \mid {}^{\bullet}x \cap X = \emptyset\}$; $Max(N)$ is defined similarly.

A process $\pi = (N, p) = (B, E, F', p)$ of a marked net $\Sigma = (S, T, F, M_0)$ consists of an occurrence net $N = (B, E, F')$ together with a labelling $p: B \cup E \rightarrow S \cup T$ which satisfy appropriate properties such that π can be interpreted as a concurrent run of Σ. To be a process of Σ, π must satisfy the following properties:

(i) $p(B) \subseteq S$ and $p(E) \subseteq T$.

(ii) $\forall x \in B \cup E: |\downarrow x| \in \mathbb{N}$ (this implies that $Min(N)$ is a cut).

(iii) $\forall e \in E: p({}^{\bullet}e) = {}^{\bullet}p(e), |p({}^{\bullet}e)| = |{}^{\bullet}p(e)|$ and $p(e^{\bullet}) = (p(e))^{\bullet}, |p(e^{\bullet})| = |(p(e))^{\bullet}|$
(transition environments are respected).

(iv) $\forall s \in S: M_0(s) = |p^{-1}(s) \cap Min(N)|$ (i.e., $Min(N)$ corresponds to the initial marking M_0).

$Lin(\pi)$ denotes the set of occurrence sequences which are linearisations of π. For a detailed explanation of these notions (and a proof that the set $Lin(\pi)$ is always nonempty), the reader is referred to [2]. An example is given in the paper (Figure 2(c)).

Provability in TBLL: A Decision Procedure

Jawahar Chirimar
and
James Lipton
Dept. of Mathematics
University of Pennsylvania

Abstract

We prove the decidability of the Tensor-Bang fragment of linear logic and establish an upper (doubly exponential) bound.

1 Introduction

Since Lincoln et al. ([11]) discovered in 1990 that propositional linear logic is undecidable, there has been a great deal of interest in determining the complexity of different fragments. Here we investigate one such fragment, using a direct analysis of deduction in the fragment somewhat in the spirit of formal language theory. The tensor-bang fragment of linear logic (TBLL) is comprised of the following symbols and rules of inference. The **language** consists of a (linearly ordered) set Σ of *propositional letters* $\{A_1, \ldots, A_n\}$. The *logical constants* are the binary connective *tensor* ($\otimes$) and the unary (prefix) connective *bang*[1] (!).

Rules of Inference: Sequents $\Gamma \vdash \theta$ consist of finite sequences Γ of propositions (separated by commas) and single propositions θ. In the rule called "axiom", only propositional letters occur.

$$A \vdash A \qquad \text{"axiom"}$$

$$\frac{\Gamma, \varphi, \psi, \Lambda \vdash \varphi}{\Gamma, \varphi \otimes \psi, \Lambda \vdash \varphi} \;\; \otimes\text{ intro-left}\,(\otimes L) \qquad \frac{\Gamma \vdash \varphi \qquad \Lambda \vdash \psi}{\Gamma, \Lambda \vdash \varphi \otimes \psi} \;\; \otimes\text{ intro-right}\,(\otimes R)$$

$$\frac{\Gamma, \theta \vdash \varphi}{\Gamma, !\theta \vdash \varphi} \;\; \text{Dereliction}\,(\mathbf{D}) \qquad \frac{\Gamma, !\theta, !\theta \vdash \varphi}{\Gamma, !\theta \vdash \varphi} \;\; \text{Contraction}\,(\mathbf{C})$$

$$\frac{\Gamma \vdash \varphi}{\Gamma, !\theta \vdash \varphi} \;\; \text{Weakening}\,(\mathbf{W}) \qquad \frac{!\Gamma \vdash \varphi}{!\Gamma \vdash !\varphi} \;\; !\text{-introduction}\;\;(!I)$$

The only structural rules are *exchange* ($\times$):

$$\frac{\Gamma, \varphi, \psi, \Delta \vdash \theta}{\Gamma, \psi, \varphi, \Delta \vdash \theta}$$

[1] also called "of course"

and cut

$$\frac{\Gamma \vdash \theta \qquad \Delta, \theta \vdash \varphi}{\Gamma, \Delta \vdash \varphi}.$$

The following result will simplify our work below.

Theorem 1.1 *TBLL admits cut-elimination.*

The proof is almost identical to that for the full propositional logic. See e.g. [6].

2 A decision algorithm for deducibility in TBLL

To begin with we will need a series of translations and technical lemmas to define our main algorithm.

Definition 2.1 *A formula α is in* **TLL** *(tensor-linear logic) if it is built up from propositional letters using only the connective $\otimes$.*

Definition 2.2 *A formula in TBLL is called banged if it is of the form $!\alpha$ where α is any TBLL- formula, or if it is of the form $\alpha \otimes \beta$ where α and β are banged[2]. For a multiset Γ, $\otimes\Gamma$ is the tensor product of all the formulas in Γ. A multiset Γ is banged if $\otimes\Gamma$ is.*

The following properties of banged formulas will be useful.

Lemma 2.3 *The weakening and contraction rules of inference remain valid for arbitrary banged formulas β in place of $!\theta$ above. We can also replace the premiss in the rule of !-introduction by any banged formula.*

The proof is immediate by structural induction.

We define a canonical form which will be repeatedly used in the sequel.

Lemma 2.4 (decomposition lemma) *Every TBLL formula α is either*

1. *in TLL or*

2. *banged or*

3. *equivalent (with respect to TBLL-deducibility) to a formula of the form $\alpha_0 \otimes \alpha_1$ where α_0 is in TLL and α_1 is banged.*

proof: If (1) and (2) do not hold, then α is of the form $\gamma \otimes \delta$ where not both γ and δ are in TLL. Inducting on length of formulas, and applying the tensoring rules of inference, there is a formula $\gamma_0 \otimes \gamma_1 \otimes \delta_0 \otimes \delta_1$, with γ_0 and δ_0 in TLL, γ_1 and δ_1 banged, which is equivalent to α (the cases γ in TLL, δ in TLL are left to the reader). Therefore $\alpha \equiv (\gamma_0 \otimes \delta_0) \otimes (\gamma_1 \otimes \delta_1)$. ∎

We will call case (3) in the statement of the previous lemma a **decomposition** of a TBLL formula α. The first formula will be called the *unbanged* or *pure tensor part* of the decomposition. Formulas in case (3) are said to be nontrivial, or to have a

[2] This agrees with Girard's recent definition of *positive polarity* of a formula

nontrivial decomposition.

We first tackle the problem of deciding when one can deduce sequents of the form $\Gamma \vdash \alpha$ with Γ in TBLL and α in TLL. The number of occurrences of each letter of Σ in α will help determine the possible replications (contractions) of banged formulas in the antecedent required in the associated deduction we are trying to reconstruct. In order to make this precise we define the notion of an **instance** of a TBLL formula, namely a certain expression obtained by "decorating" the ! symbols with natural numbers.

Definition 2.5 (instances) *Let γ be a TBLL formula.*

1. *If γ is an atom A, then A is the sole* **instance** *of γ.*

2. *$\frac{!}{\mathbf{0}}(\delta')$ is an instance of $!\delta$ if and only if (δ') is an instance of δ and every ! in the subexpression (δ') is instantiated to* $\mathbf{0}$.

3. *$\frac{!}{\mathbf{m}}(\delta')$ $(m > 0)$ is an* **instance** *of $!\delta$ if δ' is an instance of δ and m is a natural number.*

4. *$\delta' \otimes \gamma'$ is an* **instance** *of $\delta \otimes \gamma$ if δ' (resp. γ') is an instance of δ (resp. γ).*

It will also be convenient to define a symbolic version of an instance.

Definition 2.6 *A* **symbolic instance** *of a TBLL formula is an indexing of every ! with a fresh variable.*

Note that if we pick a standard list of variables, and demand that consecutive variables be used as we proceed along the formula from left to right, a symbolic instance of a formula is unique. We now define a translation from TBLL formulas to a "polynomial expression" in the propositional letters of the language.

Definition 2.7 **Σ-polynomial expressions** *are (ordered) expressions of the form*

$$A_1^{e_1} A_2^{e_2} \cdots A_n^{e_n}$$

where $A_i \in \Sigma$, $A_i < A_{i+1}$ in the Σ-order, and each e_i is a sum $\sum_{ij} e_{ij}$ of terms of the form k_{ij} or $k_{ij}x_{ij}$ where k_{ij} is a natural number and x_{ij} a variable. The product of two such expressions $A_1^{i_1} A_2^{i_2} \cdots A_n^{i_n}$ and $A_1^{j_1} A_2^{j_2} \cdots A_n^{j_n}$ is $A_1^{i_1+j_1} A_2^{i_2+j_2} \ldots A_n^{i_n+j_n}$.

Let θ' be a symbolic instance of a proposition of TBLL. We define the **polynomial form** *of θ' to be the following Σ-polynomial expression $p(\theta')$, by induction on the structure of θ':*

1. *$p(A) \stackrel{\text{def}}{\equiv} A$ for $A \in \Sigma$.*

2. *$p(\beta' \otimes \gamma') \stackrel{\text{def}}{\equiv} p(\beta')p(\gamma')$*

3. *$p(\frac{!}{\mathbf{x}}\gamma') \stackrel{\text{def}}{\equiv} p(\gamma')^{(x)}$ where we define the* **exponent** *$(\theta')^{(x)}$ as follows:*

- $A^{(x)} \stackrel{\text{def}}{\equiv} A^x$
- $(uv)^{(x)} \stackrel{\text{def}}{\equiv} u^{(x)}v^{(x)}$
- $(u^{(y)})^{(x)} \stackrel{\text{def}}{\equiv} u^{(y)}$

We will also call such an expression a polynomial form of the original TBLL-formula θ itself.

Note in the preceding definition that p and the exponent are defined via a series of rewrite rules. Strictly speaking $p(\theta)$ is the polynomial obtained by applying the above rewritings until a polynomial expression results (which is unique, once Σ is ordered, up to the names of the variables). Also note that the translation proceeds somewhat like an instantiation of regular expressions (with ! taking the place of $*$) except for the unusual iterated exponentiation "law" $(u^{(y)})^{(x)} \stackrel{\text{def}}{\equiv} u^{(y)}$. Note also that for α in TLL, $p(\alpha)$ has no variables, and is essentially a lexicographical re-ordering of α. When instantiated with natural numbers, the exponents of the polynomials just defined will correspond to certain choices of instantiations (copies) of banged formulas in the proof theory.

We are now in a position to define a *string test* which will be repeatedly used in our decision procedure, and which is, as we shall see below, sufficient to decide the special case $\Gamma \vdash \alpha$ when α is in TLL.

Definition 2.8 *Let Γ and α be propositions in TBLL and TLL respectively. Then the pair $\langle\Gamma, \alpha\rangle$ is said to satisfy the* **string test** *(in symbols $\Gamma \models_s \alpha$) if*

1. *There are natural number values for the variables in the polynomials $p(\otimes(\Gamma))$, $p(\alpha)$ such that the following equation has a solution:*
$$p(\otimes(\Gamma)) = p(\alpha) \tag{1}$$
and
2. *These numerical values for the variables in $p(\otimes(\Gamma))$ correspond to a legal instance (in the sense of definition 2.5) of the original formula (multiset) Γ when substituted for the same variables in the corresponding symbolic instance of θ. In particular any solution to 1 that results in the assignment of a nonzero decoration to a bang within the scope of a $\underset{0}{!}$ does not qualify.*

We remark that certain variables from the symbolic instance of θ may not occur in $p(\theta)$ because of the collapse of exponents. This is always the case when there is a subformula of θ of the form $!\alpha$ where α is banged. In this case, our notion of "checking the legality" of a solution to equation (1) requires comment. We first replace those variables which occur in the polynomial equation with their numerical values. Then all uninstantiated variables in the scope of a $\underset{0}{!}$ are set equal to 0. All the remaining ones are set equal to 1.

We will call an instantiation of the variables in $p(\theta)$ a *polynomial instance* for θ. If α is in TLL, Γ is in TBLL, and $\Gamma \models_s \alpha$, we call the corresponding instance of Γ the one "induced by the string test" $\Gamma \models_s \alpha$.

In the next definition and the lemma that follows it, it will be convenient to introduce the TLL- *unit* **1**. For the purposes of the lemma we add the axiom $\vdash \mathbf{1}$ as a legal proof rule to the rules of TBLL. We call the resulting fragments T(B)LL1.

Definition 2.9 *Every instance θ has an associated* **TLL-normal form**, *namely the TLL1 formula obtained as follows.*

1. $T(A) \stackrel{\text{def}}{\equiv} A$ *for* $A \in \Sigma$.
2. $T(\beta \otimes \gamma) \stackrel{\text{def}}{\equiv} T(\beta) \otimes T(\gamma)$
3. $T(\underset{\mathbf{0}}{!}\gamma) \stackrel{\text{def}}{\equiv} \mathbf{1}$
4. $T(\underset{\mathbf{m}}{!}\gamma)\ (0 < m) \stackrel{\text{def}}{\equiv} \underbrace{\alpha \otimes \cdots \otimes \alpha}_{\text{m times}} \otimes T(\beta)$ *where $\alpha \otimes \beta$ is the canonical decomposition of γ (i.e. α in TLL and β banged).*

We lift the definition of instances to multisets of formulas in the obvious way.

Lemma 2.10 *Let γ be in TBLL. Then for any instance γ' of γ. the following sequent is derivable in* TBLL1: $\gamma \vdash T(\gamma')$.

proof: The proof is by induction on the structure of γ

1. If γ is an atom A then its instance is A. The conclusion is immediate.
2. Suppose γ is $\phi \otimes \delta$. By definition we have that $T(\phi' \otimes \delta') = T(\phi') \otimes T(\delta')$. The result is thus obtained by induction hypothesis and $\otimes - R$.
3. Suppose γ is $!\phi$. By definition we have that $T(\underset{\mathbf{0}}{!}\phi') = \mathbf{1}$. But then we have the deduction

$$\frac{\vdash \mathbf{1}}{!\phi \vdash \mathbf{1}}\,\text{W}$$

If $m > 0$ we have

$$T(\underset{\mathbf{m}}{!}\phi') = \underbrace{\alpha \otimes \cdots \otimes \alpha}_{\text{m times}} \otimes T(\beta')$$

where $\alpha \otimes \beta$ is a canonical decomposition of ϕ, and β' is an instance of β. We take m to be 2 for ease of notation below. The induction hypothesis $\beta \vdash T(\beta')$ along with uses of $\otimes - R$ and the derived rule of weakening for banged formulas completes the proof shown below.

$$\cfrac{\cfrac{\cfrac{\alpha, \beta, \alpha, \beta \vdash \alpha \otimes \alpha \otimes T(\beta')}{\phi, \phi \vdash \alpha \otimes \alpha \otimes T(\beta')}\,(\otimes - L)}{!\phi, !\phi \vdash \alpha \otimes \alpha \otimes T(\beta')}\,(D)}{!\phi \vdash \alpha \otimes \alpha \otimes T(\beta')}\,(C)$$

∎

We give an example:

$$\begin{aligned} p(A \otimes !(B \otimes !A) \otimes (!(A \otimes !B \otimes B))) &= \\ p(A \otimes \underset{\mathbf{x}}{!}(B \otimes \underset{\mathbf{y}}{!}A) \otimes (\underset{\mathbf{z}}{!}(A \otimes \underset{\mathbf{u}}{!}B \otimes B))) &= A^{1+y+z} B^{x+u+z}. \end{aligned} \qquad (2)$$

If we take, e.g., $x = 2$, $y = 2$, $z = 0$ and $u = 0$ we obtain the *instance*

$$A \otimes \frac{!}{\mathbf{2}}(B \otimes \frac{!}{\mathbf{2}}A) \otimes (\frac{!}{\mathbf{0}}(A \otimes \frac{!}{\mathbf{0}}B \otimes B))$$

The corresponding polynomial instance is A^3B^2 and the associated tensor normal form is

$$A \otimes A \otimes A \otimes B \otimes B.$$

We consider a join operation on instances.

Definition 2.11 (join) *Let C', C'' be instances of the TBLL-formula C. We define the* **join** *$C' \bowtie C''$ as follows:*

- *If C is an atom A, $C \bowtie C = C$*
- *If C is of the form $!\theta$, with $C' = \frac{!}{\mathbf{m}}\theta'$ and $C'' = \frac{!}{\mathbf{n}}\theta''$ then*

$$\frac{!}{\mathbf{m}}\theta' \bowtie \frac{!}{\mathbf{n}}\theta'' = \frac{!}{\mathbf{m+n}}\theta' \bowtie \theta''$$

- *If C is of the form $\alpha \otimes \beta$, and $C' = \alpha' \otimes \beta'$ and $C'' = \alpha'' \otimes \beta''$ then*

$$(\alpha' \otimes \beta') \bowtie (\alpha'' \otimes \beta'') = (\alpha' \bowtie \alpha'') \otimes (\beta' \bowtie \beta'')$$

Observe that if α is in TLL then $\alpha \bowtie \alpha$ is α.

Lemma 2.12 (TLL-join lemma) *Let γ' and γ'' be instances of the same* **banged** *formula γ*

$$T(\gamma' \bowtie \gamma'') \vdash T(\gamma') \otimes T(\gamma'')$$

proof: Suppose $\gamma = !\varphi$. Then γ', γ'' are of the form $\frac{!}{\mathbf{m}}\varphi'$ and $\frac{!}{\mathbf{n}}\varphi''$ for some m and n. Then we have

$$T(\gamma' \bowtie \gamma'') = T(\frac{!}{\mathbf{m}}\varphi' \bowtie \frac{!}{\mathbf{n}}\varphi'') = T(\frac{!}{\mathbf{n+m}}\varphi' \bowtie \varphi'') \tag{3}$$

Let $\alpha \otimes \beta$ be the canonical decomposition of φ. Then $\varphi' \bowtie \varphi''$ must be equivalent to $\alpha \otimes (\beta' \bowtie \beta'')$ where β', β'' are the instances of β occurring in the instances φ', φ''. By definition of T, the expressions in (3) are equivalent to

$$\underbrace{\alpha \otimes \cdots \otimes \alpha}_{m+n \text{ times}} \otimes T(\beta' \bowtie \beta'').$$

This is easily shown equivalent to $T(\frac{!}{\mathbf{m}}\varphi') \otimes T(\frac{!}{\mathbf{n}}\varphi'')$ once we apply the definition of T and the induction hypothesis to β' and β''. The case $\theta = \delta \otimes \gamma$ is a straightforward induction and rearrangement of tensorands. The reader can easily check that in the $\frac{!}{\mathbf{0}}$ cases the appropriate TBLL1 sequent is provable. In particular if the left hand side is $\mathbf{1}$ or a tensor of $\mathbf{1}$'s, the right hand side must also be of this form. ∎

Theorem 2.13 *Let α be a proposition in TLL logic, and Γ a formula in TBLL. Then*

$$\Gamma \vdash \alpha \qquad \text{iff} \qquad \Gamma \models_s \alpha.$$

proof: We establish "soundness" ($\Gamma \vdash \alpha \Rightarrow \Gamma \models_s \alpha$) by induction on the length of proofs.

If $\Gamma \vdash \alpha$ is a one step proof by "axiom": $A \vdash A$ then the polynomials are identical.

Now suppose that $\Gamma \vdash \alpha$ is a proof with $n+1$ steps and inductively assume soundness holds for all proofs of shorter length. We consider the possible cases for the last step of the proof.

case $\otimes R$ The last step in the proof is an inference of the form

$$\frac{\Delta \vdash \beta \qquad \Lambda \vdash \gamma}{\Delta, \Lambda \vdash \beta \otimes \gamma}.$$

By the induction hypothesis and the definition of p the result is immediate.

case $\otimes - L$ Soundness for tensor left is built into the definition of the string-test.

case D The last step is an inference using *dereliction*:

$$\frac{\Delta, \theta \vdash \alpha}{\Delta, !\theta \vdash \alpha}.$$

we can obtain a solution to $p(\Delta \otimes !\theta) = p(\alpha)$ by using the instantiations obtained by inductive hypothesis, extended to the new premiss by taking the variable corresponding to the introduced ! to be 1 (assuming it is not eliminated due to collapse of exponents).

case C Suppose the last step in the proof is an instance of the rule of *contraction*:

$$\frac{\Gamma, !\theta, !\theta \vdash \alpha}{\Gamma, !\theta \vdash \alpha}$$

By induction there is a polynomial instance of the left hand side of the sequent $\Gamma, !\theta, !\theta \vdash \alpha$ agreeing with $p(\alpha)$ and satisfying the legality criterion of the string test. Note that the polynomial instances u and v corresponding to the first and second indicated occurrences of $!\theta$ may be different. Let $i_1 \ldots, i_n$ and $j_1 \ldots, j_n$ be the corresponding instantiations of exponents in u and v. Observe that the sequence of natural numbers $i_1 + j_1, \ldots, i_n + j_n$ instantiating the polynomial associated with the indicated occurrence of $!\theta$ in the sequent $\Gamma, !\theta \vdash \alpha$, satisfies

$$p(\Gamma, !\theta) = p(\alpha).$$

The instance is legal, since, if some $\frac{!}{i_r + j_r}$ occurs on the scope of a $\frac{!}{i_s + j_s}$ where $i_s + j_s = 0$ then $i_s = j_s = 0$ so by legality of the original solutions, $i_r = j_r = 0$.

case W Suppose the last rule used was *weakening*:

$$\frac{\Gamma, \vdash \alpha}{\Gamma, !\theta \vdash \alpha}$$

by inductive hypothesis we have a solution to

$$p(\Gamma) = p(\alpha)$$

By instantiating all the variables associated with the indicated $!\theta$ to 0 we get a solution for

$$p(\Gamma \otimes !\theta) = p(\alpha)$$

case $\times$ This is true by induction hypothesis and the fact that the polynomials are unaffected by exchange.

Now we prove the other direction: "completeness" ($\Gamma \models_s \alpha \Rightarrow \Gamma \vdash \alpha$), i.e., if there is a solution to $p(\Gamma) = p(\alpha)$ then $\Gamma \vdash \alpha$. Lemma 2.10 gives us that $\Gamma \vdash T(\Gamma')$, for any instance Γ' of Γ. It is easy to show that the instance obtained by the String Test is some permutation of α, thus Γ proves $T(\Gamma')$ and $T(\Gamma')$ proves α. Hence by Cut we are done. ∎

The string test algorithm

The string test problem can be solved in deterministic exponential time. (see [10] for a proof that the decidability of the TBLL-TLL -fragment is NP-Complete.)

We can describe a deterministic string test algorithm informally as follows. The input is an ordered pair $\langle \Gamma, \alpha \rangle$ with Γ in TBLL and α in TLL. We then compute the *polynomial forms* $p(\alpha)$ and $p(\otimes\Gamma)$ associated with these formulas (see definition 2.7). For α in TLL, $p(\alpha)$ is an expression of the form

$$A_1^{f_1} A_2^{f_2} \cdots A_n^{f_n}$$

and $p(\Gamma)$ is of the form

$$B_1^{e_1} B_2^{e_2} \cdots B_m^{e_m}$$

where the $f_i > 0$ are natural numbers and the e_j are sums of terms of the form k_{ij} or $k_{ij}x_{ij}$.

The string test algorithm checks that every A_i is among the B_j. If A_i matches B_j then we get an equation $f_i = e_j$ else it fails as $f_i = 0$ has no solution. If all letters in A_i's are found among the B_j's then we get a set of equations of the following kind.

$$f_1 = e_{i_1} \tag{4}$$

$$\vdots$$

$$f_n = e_{i_n}$$

$$\vdots$$

$$0 = e_{i_m} \tag{5}$$

Then it tries to solve these equations using bounded search, the bound being given by f_j for the variables in e_{i_j}. Moreover the algorithm checks that the solution corresponds to a legal instantiation of Γ (in the sense of definition 2.5).

The number of variables in e_i is bounded by b_Γ, the number of bangs in Γ and f_i are bounded by l_α, the length of α. Thus the search is bounded by $O(l_\alpha^{b_\Gamma})$.

An Example

Consider the candidate sequent "$A \otimes !(B \otimes !A) \otimes (!(A \otimes !B \otimes B)) \vdash A \otimes (B \otimes A)$." Applying the string test reduces to finding if there are $x, y, z, u \in \mathcal{N}$ for which

$$\begin{aligned} 2 &= 1 + y + z \\ 1 &= x + u + z \end{aligned}$$

$x = 1, y = 1, u = z = 0$ is a solution, and a legal instance, so the test **succeeds**, showing that the sequent is in fact derivable. Of course, the test can always be performed by a bounded search. Each equation generated is of the form

$$k = a_1 x_1 + a_2 x_2 + \cdots + a_m x_m$$

where each a_i on the right is a natural number, and each x_i on the right satisfies $0 \leq x_i \leq k$.

The following example suggests that there is no immediate generalization of the string test to deal with the general case of TBLL-sequents. Consider $\Gamma := !(A \otimes A \otimes A) \otimes !(A \otimes A)$ and $\theta := A \otimes A \otimes !A$.

The set of polynomial instances of Γ is $\{A^{3x+2y} : x, y \in \mathbf{N}\}$ which is precisely the same set $\{A^{x+2z} : z \in \mathbf{N}\}$ of instances associated with θ (Using a little algebra: the set $\{3x + 2y : x, y \in \mathbf{N}\}$ is the (semi)- ideal of $\mathbf{N}$ generated by gcd(3,2) intersected with the set of numbers greater than (3-1)(2-1) = 2). However $\Gamma \not\vdash \theta$ and $\theta \not\vdash \Gamma$, as can be checked using the algorithm described next.

Now we are ready to deal with general TBLL-sequents.

Definition 2.14 (reductions) *Let θ' be an instance of a TBLL-formula φ (i.e. a decoration of every ! in the original formula with a numerical subscript, as in definition 2.5). We abuse language and define a* **decomposition** *of the instance θ' to be the "decorated" decomposition of the original formula φ in which we retain the subscripts of the instance θ'. Let the* **reduced form** *$r(\theta')$ be the following associated TBLL-formula defined by induction on the structure of θ'. Primed formulas denote instances, and are dropped for TLL-formulas, (whose sole instances are themselves).*

1. $r(a) = a$ *(a an atom).*
2. $r(\varphi' \otimes \psi') = r(\varphi') \otimes r(\psi')$
3. $r(\underset{0}{!}\varphi') = !\varphi$.
4. $r(\underset{k}{!}\varphi')$ $(k > 0) = !\varphi \otimes \alpha^k \otimes r(\beta')$, *where $\alpha \otimes \beta'$ is a decomposition of φ'.*

We establish a few properties of the reduction r.

Lemma 2.15 *Let φ be any TBLL-formula, with instance φ'. Then*

$$\varphi \vdash r(\varphi')$$

proof: By structural induction. The atomic case is axiom. In the case that φ' is a tensor, the result is immediate from the induction hypothesis. Suppose φ' is $\frac{!}{m}\psi'$. Then use the induction hypothesis on ψ' together and the derived rules of weakening and contraction on banged formulas (2.3) to obtain the proof. ▮

Lemma 2.16 *Let $\alpha \otimes \beta$ be a decomposition of a nontrivial TBLL-formula. Then*

$$\Gamma \vdash \alpha \otimes \beta \quad \Rightarrow \quad \Gamma \models_{s} \alpha.$$

proof: It is clear that $\alpha \otimes \beta \vdash \alpha$, by weakening on β which is banged (2.3). By the cut rule $\Gamma \vdash \alpha$. Now by the Soundness of the String Test we get the result. ▮

Lemma 2.17 *Suppose α is in TBLL and it is banged. If $\Gamma \vdash \alpha$ then Γ is banged.*

proof: The lemma is shown by an easy induction on proofs. The only rule to check is that of $\otimes - R$, as in all other rules if the context Γ is banged prior to the application of the rule then it remains so after the application.

$$\frac{\Gamma_0 \vdash \beta \qquad \Gamma_1 \vdash \delta}{\Gamma_0, \Gamma_1 \vdash \beta \otimes \delta}.$$

then Γ_0 and Γ_1 are banged by induction hypothesis, hence so is $\Gamma \equiv \Gamma_0, \Gamma_1$. ▮

Definition 2.18 *Let γ' be an instance of the TBLL formula γ. The* **!-closure** $\widehat{\gamma'}$ *of γ' is a banged formula (or the empty multiset) defined inductively as follows.*

- *If a is an atom $\widehat{a}$ is the empty multiset.*
- $\widehat{\frac{!}{0}\theta'}$ *is* $!\theta$.
- $\widehat{\frac{!}{m}\theta'}$ *is* $!\theta \otimes \widehat{\theta'}$ $\quad (m > 0)$.
- $\widehat{\theta' \otimes \gamma'}$ *is* $\widehat{\theta'} \otimes \widehat{\gamma'}$

Lemma 2.19 (TBLL-join lemma) *Let γ' and γ'' be instances of the same* **banged** *formula γ*

$$\widehat{\gamma' \bowtie \gamma''} \vdash \widehat{\gamma'} \otimes \widehat{\gamma''}$$

The proof of this is a straight forward induction on the structure of the formula, and is quite similar to the proof of TLL-join lemma.

Theorem 2.20 (bang-closure theorem) *Let $\alpha \otimes \beta$ be a decomposition of a TBLL-formula, and suppose*

$$\Lambda \vdash \alpha \otimes \beta.$$

Then there is an instance Λ' of Λ such that

$$T(\Lambda') \vdash \alpha$$

and

$$\widehat{\Lambda'} \vdash \beta.$$

Conversely, suppose α is in TLL, β is a banged formula, and $\Lambda \models_s \alpha$, inducing an instance Λ'. Then

$$\widehat{\Lambda'} \vdash \beta \Rightarrow \Lambda \vdash \alpha \otimes \beta$$

proof: ($\Rightarrow$) Suppose $\Lambda \vdash \alpha \otimes \beta$. By the preceding lemma $\Lambda \models_s \alpha$, i.e., we have solution to $p(\Lambda) = p(\alpha)$ which yields an instance Λ'. Now by lemma 2.10 we have that $T(\Lambda') \vdash \alpha$.

Now we show $\widehat{\Lambda'} \vdash \beta$. By the cut elimination theorem, there is a proof of $\Lambda \vdash \alpha \otimes \beta$ whose last right-introduction is $(\otimes - R)$ resulting in the formation of $\alpha \otimes \beta$, i.e., there is a proof of the form:

$$\dfrac{\dfrac{\dfrac{\Delta_0 \vdash \alpha \qquad \Delta_1 \vdash \beta}{\Delta_0, \Delta_1 \vdash \alpha \otimes \beta}}{\vdots \quad \Big\} \; (*)}}{\Lambda \vdash \alpha \otimes \beta} \tag{6}$$

where the final steps are left introductions. Our proof that $\widehat{\Lambda'} \vdash \beta$ will be by induction on the length of the displayed proof of $\Lambda \vdash \alpha \otimes \beta$ starting from $\Delta_0, \Delta_1 \vdash \alpha \otimes \beta$, in particular on the length n of the "mid-section" (*). Our induction hypothesis is that if the proof above has length n then $\Lambda \models_s \alpha$ and for the instance Λ' induced by the string test, $\widehat{\Lambda'} \vdash \beta$.

Base case: If $n = 0$ then Λ is Δ_0, Δ_1 and we must show that $\widehat{\Delta_0'}, \widehat{\Delta_1'} \vdash \beta$. Since the bang-closure of a formula is banged, it suffices to notice that for banged formulas φ, $\widehat{\varphi'} \vdash \varphi$ (almost immediate from the definition). As Λ_1 is banged as β is, then we have $\widehat{\Delta_1'} \vdash \Delta_1$ and $\Delta_1 \vdash \beta$, which, by the cut rule gives $\widehat{\Delta_1'} \vdash \beta$, whence by the derived rule of weakening for banged formulas (see 2.3), we obtain $\widehat{\Delta_0'}, \widehat{\Delta_1'} \vdash \beta$.

Inductive case: suppose the induction hypothesis true for all shorter derivations than the one displayed above in (6). We consider all possible rules used in the last step.

D:: Suppose that the last step of (6) is a dereliction:

$$\dfrac{\dfrac{\vdots}{\Gamma, C \vdash \alpha \otimes \beta}}{\Gamma, !C \vdash \alpha \otimes \beta}.$$

By our induction hypothesis, $\Gamma, C \models_s \alpha$ and the instance Γ', C' induced by the string test satisfies $\widehat{\Gamma \otimes C'} \vdash \beta$. It is easy to see that $\Gamma, !C \models_s \alpha$ with the following instantiation

$$\Gamma', \underset{1}{!}C'$$

By decorating the outermost ! of $!C$ with a 1, we preserve precisely the same instance we had before. Now by the definition of !-closure

$$\widehat{\Gamma', \underset{1}{!}C'} = \widehat{\Gamma'}, \widehat{\underset{1}{!}C'} = \widehat{\Gamma'}, !C, \widehat{C'}.$$

But we have $\widehat{\Gamma'}, \widehat{C'} \vdash \beta$ by the induction hypothesis, hence, by weakening $\widehat{\Gamma'}, !C, \widehat{C'} \vdash \beta$.

$\underline{\otimes\text{-L}}$: This is immediate: the string test and the definition of instance and !-closure remain unchanged when a subformula of a premiss C, D is rewritten as $C \otimes D$.

$\underline{\text{W}}$: Suppose the last step of the proof (6) is weakening:

$$\frac{\begin{array}{c}\vdots\\ \Gamma \vdash \alpha \otimes \beta\end{array}}{\Gamma, !C \vdash \alpha \otimes \beta}.$$

We take the zero instance $\underset{0}{!}C$ for $!C$, and use the same instance of Γ. Then note that $\widehat{\underset{0}{!}C'} = !C$ By induction hypothesis $\widehat{\Gamma'} \vdash \beta$ and so $\widehat{\Gamma'}, !C \vdash \beta$ as well.

$\underline{\text{C}}$: We are left with the only slightly delicate case, namely where the last step in the proof (6) is contraction:

$$\frac{\begin{array}{c}\vdots\\ \Gamma, !C, !C \vdash \alpha \otimes \beta\end{array}}{\Gamma, !C \vdash \alpha \otimes \beta}.$$

and the induction hypothesis that there are instances $\Gamma', !C', !C''$ induced by the string-test, for which $\widehat{\Gamma'}, \widehat{!C'}, \widehat{!C''} \vdash \beta$. Now $!C' \bowtie !C''$ provides us an instantiation such that the instantiation of the conclusion is the same as that of the premiss. Now use TBLL-join lemma along with the Induction hypothesis to get $\widehat{\Gamma'}, \widehat{!C' \bowtie !C''} \vdash \beta$.

($\Leftarrow$) Suppose that $\Lambda \models_s \alpha$. Let Λ' be the *instance of Λ induced by the string test* as defined in definition (2.5). Now by (lemma 2.15) we get that

$$\Lambda \vdash r(\Lambda')$$

Notice that $r(\Lambda')$ is (equivalent to) $\widehat{\Lambda'}, T(\Lambda')$. This gives us the result. ∎

We are now ready to describe a decision procedure for TBLL.

The Algorithm: "decide_TBLL"

Input: An ordered pair $\langle \Gamma, \theta \rangle$ where Γ is a multiset of TBLL formulas, and θ is a formula in TBLL.

begin

1. if $\theta \in$ TLL then execute string-test for $\langle \Gamma, \theta \rangle$, else
2. if θ is *banged* then check Γ is *banged* and

   case 1: θ is $!\varphi$. Then apply `decide_TBLL` to $\langle \Gamma, \varphi \rangle$

   case 2: θ is $\sigma \otimes \tau$, (where σ and τ are *banged*): then apply `decide_TBLL` to $\langle \Gamma, \sigma \rangle$ and $\langle \Gamma, \tau \rangle$.

3. else let $\alpha \otimes \varphi$ be a nontrivial decomposition of θ (α in TLL, φ banged). Then do:

   3.1 apply string-test to $\langle \Gamma, \alpha \rangle$. If it fails then **fail**. If it succeeds, let $\Gamma' :=$ an instance induced by the string test not already used up. If all are used up then **fail**.

   3.2 let $\widehat{\Gamma'}$ be the *bang-closure* of Γ'

   3.3 apply `decide_TBLL` to $\langle \widehat{\Gamma'}, \varphi \rangle$. If it fails, then call this instance Γ' used up, and **goto 3.1**.

end

Remark on the algorithm Note that in step 3. we backtrack through all instantiations produced by the string test.

We briefly reconsider the example discussed following the string test algorithm, above.

We claimed that

$$\Gamma \not\vdash \theta \qquad \text{and} \qquad \theta \not\vdash \Gamma$$

where $\Gamma :=!(A \otimes A \otimes A) \otimes !(A \otimes A)$ and $\theta := A \otimes A \otimes !A$.

As our algorithm shows in two cycles, $\Gamma \vdash \theta$ is not a theorem of TBLL: $(A \otimes A) \otimes !A$ is a decomposition of θ, so we apply `decide_TBLL` to $\langle \Gamma, A \otimes A \rangle$ and $\langle \Gamma, !A \rangle$. The first input yields "yes" (by the string-test) but the second one yields the application of `decide_TBLL` to $\langle \Gamma, A \rangle$ which *fails* the string-test. The failure of $\theta \vdash \Gamma$ is immediate: θ is not banged.

2.1 Correctness and complexity of the algorithm

The correctness of the algorithm decide_TBLL is essentially the content of lemma 2.13 (which justifies step 1.), the decomposition lemma 2.4, lemma 2.4, lemma 2.17 on banged formulas, and the bang-closure theorem 2.20. The complexity is bounded by the complexity of the string test times number of calls to the string test on the pair $\langle\widehat{\Gamma}, \alpha\rangle$ where $\widehat{\Gamma}$ is the bang-closure of Γ. $\widehat{\Gamma}$ is bounded by the "full" bang-closure obtained by the taking the multiset of all banged subformulas of Γ whose length is bounded by $lth(\Gamma^2)$. The number of calls is bounded by the $2^{(\otimes\alpha)}$ where $(\otimes\alpha)$ = number of tensors in α, which is bounded by l. Letting $l = lth(\alpha) + lth(\Gamma), !!!l = lth(\widehat{\Gamma}) \leq l^2$, This gives a bound of $(!!l)^{l^2} \cdot 2^l$ steps. We state this bound as a theorem and leave the details of formalizing this argument to the reader. (We also recall that an NP lower bound for just the TBLL/TLL-fragment is established in [10]).

Theorem 2.21 *There is an $o(l^{2l^2})$ algorithm for deciding the derivability of sequents in TBLL.*

The authors would like to thank Andre Scedrov for comments and suggestions, and Dirk Roorda for pointing out errors in earlier drafts and for helpful suggestions. We are also indebted to Ramesh Subramanian and Jean Gallier for insightful discussions.

References

[1] Abramsky, S., [1991] Computational Interpretations of linear logic. TCS, 1991.

[2] Danos, V., and Reignier, L., [1989] The Structure of Multiplicatives, Arch. Math. Logic 28.

[3] Gallier, J. [1991] *Constructive Logics. Part II: Linear Logic and Proof Nets* PRL (Paris Research Laboratory) Research Report 9, Paris.

[4] Garey, Michael R. and Johnson, Davis S.[1979] *Computers and intractability : a guide to the theory of NP-completeness* San Francisco : W. H. Freeman.

[5] Girard, Lafont, Taylor [1989] *Proofs and Types*, Cambridge University Press (Cambridge Tracts in Theoretical Computer Science 7), Cambridge.

[6] J.-Y. Girard. Linear logic. *Theoretical Computer Science*, 50:1–102, 1987.

[7] J.-Y. Girard. Towards a geometry of interaction. In *Categories in Computer Science and Logic*, volume 92 of *Contemporary Mathematics*, pages 69–108, held June 1987, Boulder, Colorado, 1989.

[8] Girard, J-Y, [1991] "Quantifiers in Linear Logic II", Technical Report 19, Equipe de Logique Mathematique, Univ. Paris 7.

[9] Horowitz, E. and Sahni S. [1978] *Fundamentals of Computer Algorithms*, Computer Science Press.

[10] Chirimar, J. and Lipton, J. [1992] *Kripke Semantics, Tableaux and Decision Procedures for some fragments of Linear Logic*, Technical Report, University of Pennsylvania, to appear.

[11] Lincoln, P. Mitchell, J., Scedrov, A. and Shankar, N. [1990] "Decision Problems for Propositional Linear Logic", proc. 31st IEEE symp. on Foundations of Computer Science.

[12] Troelstra, A. S. [1991], *Lectures on Linear Logic* and *Lectures on Linear Logic: errata and Supplement*, lecture notes, Institute for Language, logic and information, Department of mathematics and Computer Science, University of Amsterdam. To appear as a book in the CSLI- Stanford series.

[13] [Troelstra-van Dalen]Troelstra, A. S. and D. van Dalen [1988], *Constructivism in Mathematics: An Introduction*, Vol. I, Studies in Logic and the Foundations of Mathematics, Vol. 121, North-Holland, Amsterdam.

J. Lipton was partially supported by NSF grant CCR91-02753.

How to Implement First Order Formulas in Local Memory Machine Models

Elias Dahlhaus

Department of Computer Science, University of Bonn, Römerstrasse 164, D-5300 Bonn 1 *

Abstract. The paper indicates how the validity of a fixed first order formula in any finite structure can be checked by realistic local memory parallel computers.

1 Introduction.

There are a sample of results on the connections between the expressive power of formulas of certain type and the computational complexity of the classes of their finite structures [9, 12, 14, 17]. Such characterizations of complexity classes are of interest as they give syntactic descriptions of complexity classes. The practical use of such considerations could be that we get a tool to develop programming languages covering computational complexity classes like P, NP, logspace, and nondeterministic logspace. Recently Immerman [13] proved that, with auxiliary predicates arising from the input order, first order formulas cover those classes of finite models which can be recognized by a certain parallel random access machine model, the CRAM (concurrent random access machine), in constant time with a polynomial number of processors. The CRAM is an extension of the concurrent read concurrent write parallel random access machine (CRCW-PRAM) with the shift of a binary number as an additional in constant time operation. The degree of the polynomial assigning the number of processors depends on the number of variables of the first order formula [13]. Here we consider the parallel complexity of the evaluation of first order formulas in finite structures on more realistic local memory parallel computation models like the hypercube, the butterfly, and the shuffle exchange network (see for example [2]).

There are a couple of general techniques to translate parallel random access machine algorithms into local memory machine programs [18, 16, 1]. All these techniques have one of the following disadvantages: they use probabilistic elements or they are based on routing techniques on sorting networks. All deterministic sorting networks of logarithmic depth and size $O(n \log n)$ are based on expanders. The constants for the size and depth of expander networks are very high. In so far expander networks are no realistic sorting networks. The most common sorting networks are

* present address: Basser Department of Computer Science, University of Sydney, NSW 2006, Australia

the odd-even merge network and the bitonic sorting network [5]. Both networks can be simulated on a butterfly network or on a hypercube but they have a depth of $O(\log^2 n)$. Recently Cypher and Plaxton [8] developed a deterministic sorting algorithm on the hypercube with a linear number of processors and a time bound of $O(\log n \log\log n)$. This algorithm is only of theoretical interest because the constant in the time bound is very high.

One possibility to implement first order formulas in local memory machines like the hypercube, the butterfly, or the shuffle exchange network is the simulation of the corresponding CRAM-program by some general techniques as mentioned above. Here we shall show that the direct implementation into any of the well known local memory networks as the hypercube or the butterfly network or the shuffle exchange network is more efficient and can be done without probabilistic elements. We shall prove that first order formulas without function symbols can be evaluated by a hypercube, a shuffle exchange network, and by a butterfly in logarithmic time using polynomially many processors. The degree of the polynomial coincides with the number of variables of the first order formula (up to a logarithmic factor). Note that, in general, the processor number is optimal (up to a logarithmic factor), because the sequential time to evaluate a a first order formula with k variables is $O(n^k)$. Moreover, also in case that parallel random access machines can be simulated in a logarithmic time loss by realistic local memory networks, the parallel techniques to evaluate first order formulas on finite models as they are presented in this paper may remain interesting, because they are quiet simple.

In section 2, the basic concepts and notions are presented. Section 3 introduces the parallel communication structures considered in this paper. In section 4, the main results are presented. Section 5 discusses the problem how to behave if the certain formula has function symbols. In the last section, we discuss some conclusions related to the results of this paper.

2 Basic Concepts

A *similarity type* or *signature* L consists of a set C of *constant symbols*, sets R_i of *relation symbols* of arity i, and sets F_i of *function symbols* of arity i.

A similarity type without function symbols is called a *relational similarity type*.

Without loss of generality, we assume that the *domain* of each finite structure is an initial segment $\{0, \ldots, n-1\}$ of the natural numbers. We denote the set of structures of the similarity type L with an initial segment of the natural numbers as domain by $Mod(L)$.

The first order language of the similarity type L is denoted by $FO(L)$.

The set of variables of a first order formula ϕ is denoted by $Var(\phi)$. Note that the certain variable x may appear in different places and is bounded by different quantifiers. Such variable is still counted only once.

Suppose the formula $\phi(x_1, \ldots, x_k) \in FO(L)$ has the free variables $x_1, \ldots, x_k$. Then, for each $M \in Mod(L)$, we set

$$F_\phi(M) = \{(u_1, \ldots, u_k) | M \text{ satisfies } \phi(u_1, \ldots, u_k)\}.$$

Suppose $\phi \in FO(L)$. Then

$$Mod(\phi) = \{M \in Mod(L) | M \text{ satisfies } \phi\}.$$

In this paper, a RAM-program is built up from basic operations and basic predicates, and the while-loop.

Basic operations are

1. increasing and decreasing a natural number by one,
2. multiplication and integer division by two,
3. checking a certain bit of a natural number in binary representation,
4. changing a certain bit of a natural number in binary representation.

Basic predicates are the comparison operations $<$, $>$, $\leq$, $\geq$, $=$.

The evaluation of basic operations and predicates need one time unit.

A parallel random access machine (PRAM) consists of infinitely many numbered processors which execute the same program on the same storage units. Different behaviors of processors is based on different processor numbers. For extended studies, we refer to [11] of [10].

Local memory machines consist of communicating RAMs. Each RAM has its own storage and fixed communication links to other RAMs. Also in the case of local memory machines, we assume that each RAM executes the same program. The different behavior is based on the fact that each processor has its own number. The number of processors of a local memory machine program is defined as the number of active processors.

3 A Survey on Local Memory Architectures

3.1 The Hypercube

For $m = 2^k$, the m-hypercube consists of processors with number $0, \ldots, m-1$. Two processors i and j have a communication link $(i - j)$ iff their numbers i and j differ in exactly one bit in their k-bit binary representation. That means the hypercube of size 2^k is the k-dimensional cube.

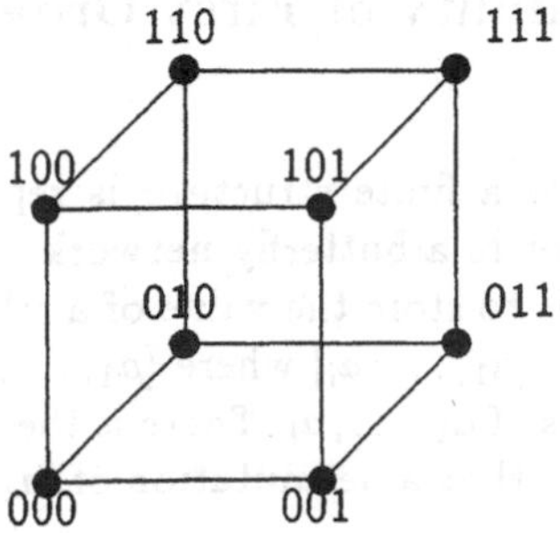

Figure 1: the hypercube

3.2 The Butterfly

The $2^k(k+1)$-*butterfly* consists of processors with labels (i,j) such that

$$i \in \{0,\ldots,2^k-1\}$$

and

$$j \in \{0,\ldots,k\}.$$

The communication structure of a butterfly network is defined as follows: Suppose i_1 and i_2 differ exactly in the $j-th$ bit, $j = 0,\ldots,k-1$. Then processors (i_1,j) and $(i_2,j+1)$ communicate and processors $(i_1,j+1)$ and (i_2,j) communicate.

Remark: If, in a butterfly, one contracts, for each i, the processors (i,j) to one processor with number i then one gets a hypercube.

3.3 The Shuffle-Exchange-Network

The 2^k-*shuffle-exchange-network* consists of processors numbered by $0,\ldots,2^k-1$.

The communication links of a shuffle-exchange-network consist of ***shuffle links*** and *exchange links*.

Let $i_0\ldots i_{k-1}$ be the binary representation of the label i. Then $i_0\ldots i_{k-1} - i_1\ldots i_{k-1}i_0$ is a shuffle link.

Processors i and j communicate via an exchange link iff they differ in exactly the last bit.

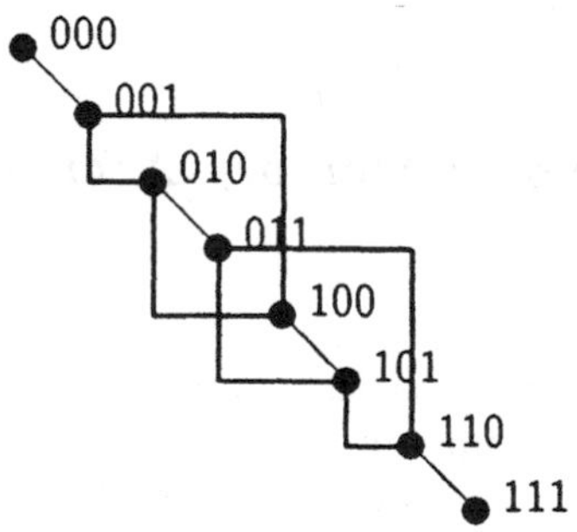

Figure 2: the shuffle-exchange network

4 Checking the Validity of First Order Formulas in Local Memory Machines

At first we have to fix how a finite structure is represented in a hypercube, in a shuffle exchange network, or in a butterfly network.

The most natural way is to store the value of a relation R for a tuple $(a_1,\ldots,a_l)$ in a processor with number $[a_1,\ldots,a_l]$ where $[a_1,\ldots,a_l]$ is the concatenation of the k-bit binary representations of $a_1,\ldots,a_l$. There is the problem that, in the evaluation of a formula, it is necessary that a permutation of $[a_1,\ldots,a_l]$ needs the value of the relation R in $(a_1,\ldots,a_l)$.

To surround this problem, we proceed as follows:

1. We store values of relations only in processors $[a_1, \dots, a_l]$ such that $a_1 \leq \dots \leq a_l$.
2. Let R be a p-ary relation. Let $b_1 \leq \dots \leq b_p < 2^k$. Then, for each permutation $\pi : \{1, \dots, p\} \rightarrow \{1, \dots, p\}$, the value of $R(b_{\pi(1)}, \dots, b_{pi(p)})$ is stored in a shuffle exchange network or a hypercube of size 2^{kl} in the processor numbered by $[0^{l-p}, b_1, \dots, b_p]$. In a hypercube of size $2^{kl}(kl+1)$ these values are stored in the processor labeled by $([0^{l-p}, b_1, \dots, b_p], 0)$.

We can express this proposal also as follows.

Call a relation R *faithful* iff, for all $a_1, \dots, a_p$ such that $R(a_1, \dots, a_p)$, we have $a_1 \leq \dots \leq a_p$.

We split the relation R in its *faithful components* R_π, say $R_\pi(a_1, \dots, a_p)$ iff $a_1 \leq \dots a_p$ and $R(a_{\pi(1)}, \dots, a_{\pi(p)})$.

Clearly each relation can be expressed by its faithful components.

We call a finite structure *faithful* iff all its relations are faithful and denote the set of faithful finite structures in $Mod(L)$ by $Mod_F(L)$.

W assume that all variables are of the form v_i, $i \in \omega$. We call a sequence $v_{i_1}, \dots, v_{i_l}$ of variables *ascending* iff $i_1 < \dots < i_l$. Suppose $x = v_i$, $y = v_j$ and $i < j$. Then we write also $x < y$.

We call a formula $\phi(x_1, \dots, x_q)$ with ascending free variables $x_1, \dots, x_q$ of a signature L faithful iff in each faithful finite model M of L the formula $\phi(x_1, \dots, x_q)$ defines a faithful relation. That means $\{(u_1, \dots, u_q) | M \text{ satisfies } \phi(u_1, \dots, u_q)\}$ is a faithful relation on the domain of M.

Theorem 1. : *Each faithful first order formula ϕ is equivalent to a first order formula ϕ' with the same number of variables such that each subformula is faithful.*

Proof. : Let $\phi(x_1, \dots, x_l)$ be a faithful formula and $\phi(y_1, \dots, y_l)$ a renaming of $\phi(x_1, \dots, x_l)$ such that the free variables have the largest index.

Lemma 2. *$\phi(x_1, \dots, x_l)$ can be expressed by $\exists$, $\wedge$, $\vee$, $=$, and $\phi(y_1, \dots, y_l)$ without renaming and without new variables.*

Proof. : Clearly the index of x_i is at most as large as the index of y_i.

Therefore

$$\phi(x_1, \dots, x_j, y_{j+1}, \dots, y_l)$$

and

$$\exists y_j (y_j = x_j \wedge \phi(x_1, \dots, x_{j-1}, y_j, \dots y_l))$$

are equivalent. By an easy induction on j, we can express each formula

$$\phi(x_1, \dots, x_j, y_{j+1}, \dots, y_l)$$

by the symbols $\exists$, $\wedge$, $\vee$, and $=$ without any new variable and without any renaming.

□(Lemma)

In the same way, we also can prove the following.

Lemma 3. : *$\phi(y_1, \dots, y_l)$ can be expressed by $\exists$, $\wedge$, $\vee$, $=$, and $\phi(x_1, \dots, x_l)$ without renaming and without new variables.*

We may assume that each faithful formula ϕ is of the form $\phi'(x_1, \ldots, x_l) \wedge x_1 \leq x_2, \ldots, x_l$. We call ϕ' the *core* of ϕ. We prove the theorem by induction on the depth of the core of ϕ.

Suppose $\phi' = \phi_1 \wedge \phi_2$. Then $\phi'(x_1, \ldots x_l) \wedge x_1 \leq \ldots x_l$ is equivalent to the conjunction of the formulas $\phi_i(x_1 \ldots, x_l) \wedge x_1 \leq \ldots \leq x_l$. The cores of these formulas are of smaller depth.

Suppose $\phi' = \neg\phi_1$. Then $\phi'(x_1, \ldots x_l) \wedge x_1 \leq \ldots \leq x_l$is equivalent to the formula $\neg(\phi_1 \cap x_1 \leq \ldots \leq x_l) \wedge x_1 \leq \ldots \leq x_l$.

Suppose $\phi'(x_1, \ldots, x_l) = \exists y \phi_1(y, x_1, \ldots, x_l)$. By Lemma 2 and 3, we may assume that $x_1, \ldots, x_l$ are of larger index than y. $\phi'(x_1, \ldots, x_l) \wedge x_1 \leq \ldots \leq x_l$ is equivalent to the disjunction of formulas $\psi_i = \exists y\, (x_1 \leq \ldots \leq x_l \wedge x_i \leq y \leq x_i + 1 \wedge \phi_1(y, x_1, \ldots, x_l))$ for $i = 0, \ldots, l$. Let $\psi_i' = x_1 \leq \ldots \leq x_l \wedge x_i \leq y \leq x_i + 1 \wedge \phi_1(y, x_1, \ldots, x_l)$. It remains to make ψ_i' faithful if $i > 0$. But if, in ψ_i', we replace x_1 by y, x_j by x_{j-1}, for $j = 2 \ldots, i$ and y by x_i, we get a faithful formula, say ψ_i''. The formula $\phi_i'' = \exists x_i \psi_i''$ is equivalent to the formula which arises from ψ_i by replacing x_1 by y, x_j by x_{j-1}, for $j = 2 \ldots, i$ and y by x_i. By lemma 3, we can express ψ_i by ψ_i'' without any renaming. Note that ψ_i' is a subformula of ψ_i'' with smaller depth.

It remains to consider the case that the core is a basic formula $r(x_{i_1}, \ldots, x_{i_l})$ but, for some j, $i_j > i_{j+1}$. Then $r(x_{i_1}, \ldots, x_{i_l}) \wedge x_1 \leq \ldots \leq x_l$ is equivalent to $r(x_{i_1}, \ldots, x_{i_{j-1}}, x_{i_{j+1}}, x_{i_j}, x_{i_{j+1}}, \ldots, x_{i_l}) \wedge x_1 \leq \ldots \leq x_{i_{j+1}} = \ldots = x_{i_j} \leq \ldots \leq x_l$. After a finite number of changes, r is in the faithful order.

□(Theorem)

4.1 Implementation of Faithful Formulas in a Hypercube and in a Shuffle Exchange network

Boolean operations can be executed in constant time and do not need any communication.

By previous considerations, it remains to realize the existential quantifier.

Let R be a k-ary relation. Then, to realize the existential quantifier

$$\exists x_i R(x_1, \ldots, x_k),$$

we have to inform each processor

$$[x_1, \ldots, x_{i-1}, x, x_{i+1}, \ldots, x_k]$$

on the existence of such an x_i.

Suppose, we work on a finite structure of size $n = 2^l$. Then, in a hypercube of size 2^{kl}, we proceed as follows:

Let $x_i^{l-1} \ldots x^0$ be the l-bit representation of x_i. Then the j-th neighbor of x_i, denoted by $N(x_i, j)$ is that x_i' which arises from x_i by changing exactly the bit x_i^j. Note that $[x_1, \ldots, x_{i-1}, x_i, x_{i+1}, \ldots, x_k]$ and $[x_1, \ldots, x_{i-1}, N(x_i, j), x_{i+1}, \ldots, x_k]$ are hypercube neighbors.

We proceed as follows:

1. At the beginning, each processor $[x_1, \ldots, x_k]$ with $R(x_1, \ldots, x_k)$ knows the existence of some x with $R(x_1, \ldots, x_{i-1}, x, x_{i+1}, \ldots, x_k)$.

2. For $j = 0, \ldots, l-1$, each processor $[x_1, \ldots, x_{i-1}, y, x_{i+1}, \ldots, x_k]$ knowing the existence of an x making $R(x_1, \ldots, x_{i-1}, x_{i+1}, \ldots, x_k)$ true informs the processor with the number $[x_1, \ldots, x_{i-1}, N(y,j), x_{i+1}, \ldots, x_k]$ on the existence of such x.

In the step j, all processors $[x_1, \ldots, x_{i-1}, y, x_{i+1}, \ldots, x_k]$, such that y and x coincide in the bits $j+1, \ldots, l$, are informed on the existence of an x which makes R true. Therefore in step $l-1$, all processors $[x_1, \ldots, x_{i-1}, y, x_{i+1}, \ldots, x_k]$ know the existence of such an x.

Therefore:

Theorem 4. *Suppose R is a k-ary relation of a finite structure of size $n = 2^l$. Then a hypercube of size n^k can compute the* projection

$$\{(x_1, \ldots, x_{i-1}, x_i, x_{i+1}, \ldots, x_k) | \exists x\, R(x_1, \ldots, x_{i-1}, x, x_{i+1}, \ldots, x_k)\}$$

in $O(\log n)$ time.

In a shuffle exchange network, we proceed as follows. For any binary string $u_1 \ldots u_q$ of length q, its i-th *cyclic permutation* $u_{i+1} \ldots u_q u_1 \ldots u_i$ is denoted by $per(u_1 \ldots u_p, i)$.

Again the size of the finite structure to be considered is $n = 2^l$.

Using the shuffle communication links, after $l \cdot (k-i)$ steps, processor

$$per([x_1, \ldots, x_k], il)$$

knows the value of $R(x_1, \ldots, x_k)$. We proceed in a similar way as in the hypercube:

For $j = 0, \ldots, l-1$, each processor $per([x_1, \ldots, x_k], il-j)$ knowing the existence of an x with $R(x_1, \ldots, x_{i-1}, x, x_{i+1}, \ldots, x_k)$ informs its exchange link neighbor processor $per([x_1, \ldots, x_{i-1}, N(x_i, j), \ldots, x_k], il-j)$ on the existence of such an x. After l steps, each processor $[x_i, \ldots, x_k, x_1, \ldots, x_{i-1}]$ knows the existence of such an x. By shuffeling back, we make each processor $[x_1, \ldots, x_k]$ aware of the existence of such an x.

The time we need to evaluate an existential quantifier is $O(\log n)$.

Therefore we get the following.

Theorem 5. : *Suppose R is a k-ary relation of a finite structure of size $n = 2^l$. Then a shuffle exchange network of size n^k can compute the* projection

$$\{(x_1, \ldots, x_{i-1}, x_i, x_{i+1}, \ldots, x_k) | \exists x\, R(x_1, \ldots, x_{i-1}, x, x_{i+1}, \ldots, x_k)\}$$

in $O(\log n)$ time.

4.2 Implementation of Faithful First order Formulas in Butterflies

We may assume that the value of $R(x_1, \ldots, x_k)$ is stored in processor $([x_1, \ldots, x_k], 0)$.

Again assume that the given finite structure is of size $n = 2^l$. That means $x_i \in \{0, \ldots, 2^l - 1\}$.

By the definition of the $2^{kl} \cdot kl + 1$-butterfly, for each $i = 1, \ldots, k$, and each $j = 0, \ldots, l-1$, $([x_1, \ldots, x_k], (i-1)l + j)$ and $([y_1, \ldots, y_k], (i-1)l + j + 1)$ are connected by a communication link iff

1. for $j \neq i$, $x_j = y_j$,
2. x_i and y_i differ in at most one bit.

Therefore:

Proposition 6. : *For each* $a_1, \ldots, a_{i-1}, a_{i+1}, \ldots, a_k \in \{0, \ldots, n-1\}$, *the set*

$$\{([a_1, \ldots, a_{i-1}, u, a_{i+1}, \ldots, a_k], (i-1)l + j) | j = 0, \ldots l,\ u = 0, \ldots, n-1\}$$

induces a subbutterfly of size $n(l+1)$.

We can proceed as follows:

Suppose $(a_1, \ldots, a_k) \in R$. Then all processors $([a_1, \ldots, a_k], j)$ are on this fact. This can be done in logarithmic time.

Then all processors of the subbutterfly $\{([a_1, \ldots, a_{i-1}, u, a_{i+1}, \ldots, a_k], (i-1)l + j) | j = 0, \ldots l,\ u = 0, \ldots, n-1\}$ are informed on the fact that there is an x such that $R(a_1, \ldots, a_{i-1}, x, a_{i+1}, \ldots, a_k)$. Since the number of neighbor processors is bounded by four and the diameter of a butterfly is $O(\log n)$, this step can be done in logarithmic time.

The last step is to inform all processors $([a_1, \ldots, a_{i-1}, u, a_{i+1}, \ldots, a_k], j)$ on the existence of an x making $R(a_1, \ldots, a_{i-1}, x, a_{i+1}, \ldots, a_k)$ true. This can be done in logarithmic time.

Therefore we get the following result.

Theorem 7. : *Suppose* R *is a* k*-ary relation of a finite structure of size* $n = 2^l$. *Then a butterfly of size* $O(n^k \log n)$ *can compute the* projection

$$\{(x_1, \ldots, x_{i-1}, x_i, x_{i+1}, \ldots, x_k) | \exists x\, R(x_1, \ldots, x_{i-1}, x, x_{i+1}, \ldots, x_k)\}$$

in $O(\log n)$ *time.*

4.3 Main Result

By Theorem 5 and 7 and the fact that all faithful formulas can be transformed in such a way that all subformulas are faithful (Theorem 4) and all basic formulas with relation symbols different from $=$ and $\leq$ use only variables of highest index, all faithful formulas with k variables can be checked in logarithmic time using $O(n^k)$ processors or $O(n^k \log n)$ processors on a hypercube, a shuffle-exchange network, and on a butterfly.

Since we decomposed all relations into faithful relations, we get the following.

Theorem 8. *(Main Theorem) For suitable implementation of the relations of any finite structure, the validity of a fixed first order formula with* k *variables can be checked for each finite structure with a domain of size* n *in* $O(\log n)$ *time using* $O(n^k)$ *processors on a hypercube and on a shuffle exchange network and in* $O(\log n)$ *time using* $O(n^k \log n)$ *processors on a butterfly network.*

5 Introducing Function Symbols

In case that function symbols are used, it might be necessary that any processor with processor number i has access to any processor j. If a term to evaluate has as many variables as the processor exponent then we are forced to use general simulation techniques of PRAMs based on sorting or on probabilistic elements. In many cases the arity of a term to be evaluated has less variables. We can generalize a simple technique of Awerbuch, Israeli, and Shiloach [3] to simulate a PRAM-step with n processors by a local memory machine with n^2 processors in logarithmic time.

As in the case of relational structures, we assume that, for each permutation π and for each increasing sequence $x_1, \ldots, x_k < n$, the value of $f(x_{\pi(1)}, \ldots, x_{\pi(k)})$ is stored in a processor $[0^{l-k}, x_1, \ldots, x_k]$ or in a processor $([0^{l-k}, x_1, \ldots, x_k], 0)$

Theorem 4: **Any term with k variables and maximal function arity k' can be evaluated by a hypercube and a shuffle-exchange network with $n^{k+k'}$ processors and by a butterfly with $O(n^{k+k'} \log n)$ processors in logarithmic time.**

Proof. We use the following auxiliary result.

Lemma 9. *: For any increasing sequence $y_1, \ldots, y_l$ containing $x_1, \ldots, x_k$, the value of $f(x_1, \ldots, x_k)$ can be sent to processor $[y_1, \ldots, y_l]$ in a hypercube and a shuffle-exchange network in $O(\log n)$ time and to processor $([y_1, \ldots, y_l], 0)$ in logarithmic time.*

Proof. In the same way as we realize the existential quantifier, we can propagate the value of $f(x_1, \ldots, x_k)$ to any processor

$$[y_1, \ldots, y_{l-k}, x_1, \ldots, x_k]$$

or

$$([y_1, \ldots, y_{l-k}, x_1, \ldots, x_k], 0).$$

Such a step needs logarithmic time.

For example also each processor

$$[y_1, \ldots, y_{i-1}, x_1, y_{i+1}, \ldots, y_{l-k}, x_1, \ldots, x_k]$$

or processor

$$([y_1, \ldots, y_{i-1}, x_1, y_{i+1}, \ldots, y_{l-k}, x_1, \ldots, x_k], 0)$$

knows the value of $f(x_1, \ldots, x_k)$. Again, in logarithmic time, each processor

$$[y_1, \ldots, y_{i-1}, x_1, y_{i+1}, \ldots, y_{l-k+1}, x_2, \ldots, x_k]$$

or

$$([y_1, \ldots, y_{i-1}, x_1, y_{i+1}, \ldots, y_{l-k+1}, x_2, \ldots, x_k], 0)$$

gets to know the value of $f(x_1, \ldots, x_k)$. Repeating this procedure, we can place $x_1, \ldots, x_k$ in anywhere in increasing order in a sequence $y_1, \ldots, y_l$ such that processor $[y-1, \ldots, y_l]$ or $([y_1, \ldots, y_l], 0)$ knows the value of $f(x_1, \ldots, x_k)$.

□(Lemma)

To continue the proof of the theorem, we consider some term $t = f(t_1, \ldots, t_p)$ with free variables $x_1, \ldots, x_k$. Without loss of generality, we evaluate t for an increasing sequence $a_1, \ldots, a_k$. Let $y_1, \ldots, y_{k+p}$ be the increasing sequence which contains all a_i and all values of $t_1, \ldots, t_k$ applied to $a_1, \ldots, a_k$. Without loss of generality, we may assume that $p = k'$. Then the processor numbered by $[y_1, \ldots, y_{k+p}]$ or by $([y_1, \ldots, y_{k+p}, 0)$ can evaluate t in one step. It remains to propagate the value of t in $a_1, \ldots, a_k$ to all processors $[y_1, \ldots, y_p, a_1, \ldots, a_k]$ or $([y_1, \ldots, y_p, a_1, \ldots, a_k], 0)$. Also this can be done in logarithmic time by applying lemma 9 backward.

□(Theorem)

6 Conclusions

One immediate consequence of the main results of this paper is that the graph connectivity algorithm of Chin, Lam, and Chen [6] can be implemented in a hypercube and a shuffle exchange network of size n^2 and a butterfly network of size $O(n^2 \log n)$. The time bound is $O(\log^2 n)$. Several papers [4, 15] discussed the implementation of the graph connectivity problem in realistic local memory machines. Their results can be seen as corollaries of the main results of this paper.

Immerman [12] suggested to to develop a programming language based on the logical characterizations of complexity classes. Extending Immerman's recommendations I would suggest to develop a parallel programming language based on the iterative application of first order formulas on finite structures. This paper gives a tool how to advise the compiler to reduce all to all communication.

References

1. F. Abolhassan, J. Keller, W. Paul: On physical realizations of the theoretical PRAM model, Technical Report # 21/90, Sonderforschungsbereich 124 - VLSI-Entwurfsmethoden und Parallelität, University of Saarbrücken.
2. , S. Akl: The design and analysis of parallel algorithms, Prentice Hall, Englewood Cliffs, New Jersey, 1989
3. B. Awerbuch, A. Israeli, Y. Shiloach: Efficient simulation of PRAM by ultracomputer, preprint, Technion- Israel Institute of Technology, 1983.
4. B. Awerbuch, Y. Shiloach: New connectivity and MSF algorithms for shuffle exchange network and PRAM, IEEE-Transactions on Computing C-36 (1987), pp. 1258-1263.
5. K. Batcher: Sorting networks and their applications, Proceedings of the AFIPS 1968, Atlanta City, New jersey, pp. 307-314.
6. Y. Chin, Y. Lam, I. Chen: Efficient parallel algorithms for some graph problems, Communication of the ACM 25 (1982), pp. 659-665.
7. S. Cook: A taxonomy of problems with fast parallel algorithms, Information and Control 64 (1985), pp. 2-22.
8. R. Cypher, G. Plaxton: Deterministic sorting in nearly logarithmic time on a hypercube and related computers, $2^n d$ ACM-STOC (1990), pp. 193-203.
9. R. Fagin: Generalized first order spectra and polynomial time recognizable sets, in "Complexity of Computations" (R. Karp ed.), SIAM-AMS-proceedings 7 (1974), pp.27-41.
10. S. Fortune, J. Wyllie: Parallelism in random access machines, 10th ACM-STOC (1978), pp. 114-118.

11. A. Gibbons, W. Rytter: Efficient Parallel Algorithms, Cambridge University Press, Cambridge, 1989.
12. N. Immerman: Languages which capture complexity classes, 15th ACM-STOC (1983) pp. 347-354.
13. N. Immerman: Expressibility and parallel complexity, SIAM-Journal on Computing 18 (1989), pp. 625-638.
14. N. Jones, A. Selman: Turing machines and spectra of first order formulas, Journal of Symbolic Logic 39 (1974), pp. 139-150.
15. R. Miller, Q. Stout: Graph and image processing algorithms for hypercube, Proceedings 1986 SIAM Conference on Hypercube Multiprocessors (1987), pp. 418-425.
16. A. Ranade: How to emulate shared memory, IEEE-FOCS 1987, pp.185-194.
17. D. Rödding, H. Schwichtenberg: Bemerkungen zum Spektralproblem, Zeitschrift für Mathematische Logik 18 (1972), pp. 1-12.
18. E. Upfal: An $O(\log N)$ deterministic packet routing scheme, 21st ACM-STOC (1989), pp. 241-250.

This article was processed using the LaTeX macro package with LLNCS style

A New Approach to Abstract Data Types II Computation on ADTs as Ordinary Computation

Solomon Feferman[1]

Department of Mathematics, Stanford University
Stanford, CA 94305, USA
sf@csli.stanford.edu

Abstract. A notion of abstract computational procedure is introduced here which meets the criteria for computation over ADTs, for the general theory of such presented in Part I of this paper. This is provided by a form of generalized recursion theory (g.r.t.) which uses schemata for explicit definition, conditional definition and least fixed point (LFP) recursion in partial functions and functionals of type level ≤ 2 over any appropriate structure. It is shown that each such procedure is preserved under isomorphism and thus determines an abstract procedure over ADTs. The main new feature of the g.r.t. developed here is that abstract computational procedures reduce to computational procedures in the ordinary sense when confined to data structures in a recursion-theoretic interpretation.

1 Introduction

The main purpose of this paper is to give a precise definition of *abstract computational procedure* on abstract data types which meets the requirements of §8.1 of Part I (=[1]). Such a procedure π is to associate with each structure $\mathcal{A}$ of a given (finite) signature Σ an object $\pi^{\mathcal{A}}$ of a specified arity (individual, function, or functional) over $\mathcal{A}$. Before explaining the requirements, we need to say a word about the form of such $\mathcal{A}$. For present purposes, we take

$$(1) \qquad \mathcal{A} = (A_0, A_1, \ldots, A_n, =_{A_0}, =_{A_1}, \ldots, =_{A_n}, F_0, \ldots, F_m),$$

where for $i = 0, \ldots, n$, $=_{A_i}$ is an equality relation on $\mathcal{A}$ (not necessarily the identity relation), A_0 is the Boolean type $B = \{tt, ff\}$, $=_{A_0}$ is the identity relation on $\mathbb{B}$, and for each $k = 1, \ldots, m$ each F_k is either a constant (0-ary function), partial function or partial functional over $\mathcal{A}$ of specified arity. The constants tt, ff are supposed to be among the constants of $\mathcal{A}$, and each function or functional among the F_k is supposed to preserve the given equality relations $=_{A_i}$; moreover those F_k which are functionals are supposed to be monotonic.

By way of comparison with [1], the Boolean type A_0 was only implicitly assumed there to be part of $\mathcal{A}$. Furthermore, $\mathcal{A}$ was permitted to contain additional relations R_j, as needed to specify an ADT for structures of signature Σ. These are left aside here, since only the F_k are accessible to a computation procedure on $\mathcal{A}$. However,

[1] Research supported by NSF grant #CCR-8917606. Invited paper for the conference CSL '91, Berne Oct. 7-11, 1991.

the equality relations are retained, since it must be verified that each $\pi^{\mathcal{A}}$ preserves these. The importance of including equality relations in structures for a sufficiently general theory of ADTs was argued in [1] and is taken for granted here.

The criteria that an abstract computational procedure π is to meet are as follows:

C1. π associates with each $\mathcal{A}$ an object $\pi^{\mathcal{A}}$ (individual, partial function, or partial functional) of specified arity over $\mathcal{A}$.
C2. $\pi^{\mathcal{A}}$ is determined by the (individual, function and functional) constants of $\mathcal{A}$.
C3. If $\mathcal{A} \cong_{(h,h')} \mathcal{A}'$ then $\pi^{\mathcal{A}}$ corresponds to $\pi^{\mathcal{A}'}$ under (h, h').
C4. $\pi^{\mathcal{A}}$ preserves the equality relations on $\mathcal{A}$.
C5. For $\mathcal{A}$ with domains contained in ω, $\pi^{\mathcal{A}}$ reduces to an ordinary computational procedure.

These requirements are met here by a form of generalized recursion theory (g.r.t.) which provides a notion of computability over arbitrary structures of the kind described above. In order to satisfy C3 we must insure that whenever an object is defined by recursion it is uniquely specified. For (partial) functions this will be as a least fixed point (LFP) of a suitable monotonic functional. There are two forms of g.r.t. available in the literature which feature LFP as a central scheme (along with explicit definition and conditional definition), namely those of Moschovakis [5], [6] and the earlier Platek [8]. The latter is formally simpler but uses objects of arbitrary finite type level over $\mathcal{A}$. The former uses only objects of finite type level ≤ 2, which is sufficient for practice, where the initial objects F_k all meet that condition. (And, it was proved in [8] that computability from $F_0, \ldots, F_m$ of type level ≤ 2 can be managed without use of higher type levels.) Neither g.r.t. considers structures with general equality relations and the resulting criterion C4, but they are readily adapted to this more general situation. Also Platek [8] makes stronger assumptions on the structures $\mathcal{A}$ than are necessary (in effect, building in the structure $\mathcal{N}$ of natural numbers—cf. the discussion in [2]).

The form of LFP g.r.t. that is adapted here is that of Moschovakis [5], [6], but it is presented in a rather different way. Moschovakis uses a formal system of terms involving a complicated-looking term builder for simultaneous LFP recursion. Our approach uses instead schemata much like those of Kleene's S1–S9 in his fundamental paper [4], but with S9 replaced by a form of simple LFP recursion.[2] The schemata for abstract computational procedures will be found in §7 below; these meet C1 and C2 by construction. The criteria C3 and C4 are verified in §8, and closure under simultaneous LFP recursion in §9.

The g.r.t. developed here has another dimension of generality which is novel. It applies to a wide variety of data universes V with relatively weak closure conditions on the classes of partial functions and functionals over V. There are two extremes of interpretation: (i) V is the full cumulative hierarchy and "all" functions and functionals are admitted. (ii) $V = \omega$ and we only admit partial recursive functions and functionals. The setting (i) is the usual one for g.r.t., while the setting (ii) serves

---

[2] Kleene makes a similar modification in [5], but his work in both [4] and [5] is confined to computability in finite types over the structure $\mathcal{N}$ of natural numbers.

for the precise formulation of the criterion C5; the statement and proof of that is the main new contribution of this paper (§11 below).

2 Data Universes

The notion of computability on ADTs introduced below is applicable to a variety of interpretations as to the kinds of data, operations on them and data types which may be admitted. These include—at the extremes—set-theoretic and recursion-theoretic interpretations. We thus provide a setting of some generality, as follows.

2.1 There is given an underlying non-empty *universe* V, out of which the data objects are considered to be drawn; the lower case Latin letters $a, b, c, \ldots, x, y, z$ range over V. It is assumed that V is closed under a *pairing operation* $x, y \mapsto (x, y)$; n-tupling $(x_1, \ldots, x_n)$ is then defined by iterated pairing. V is assumed to contain $\mathbb{B} = \{tt, ff\}$ and $\mathbb{N} = \{0, 1, 2, \ldots\}$ as subsets. The latter is not needed for the general notion of computability on ADTs, but figures in most examples.

2.2 A collection PFn of objects called *partial functions* on V is supposed to be given, along with an *application relation* $App \subseteq PFn \times V \times V$. We use Greek letters $\varphi, \psi, \theta, \ldots \xi, \eta, \zeta$ to range over PFn, and write $\varphi x \simeq y$ for $App(\varphi, x, y)$, where $\varphi x \simeq y_1 \wedge \varphi x \simeq y_2 \Rightarrow y_1 = y_2$. φx is said to be defined, and we write $\varphi x \downarrow$, when $\exists y(\varphi x \simeq y)$; $\text{dom}(\varphi) = \{x \mid \varphi x \downarrow\}$. We also write $\varphi(x)$ for φx (when defined), and $\varphi(x_1, \ldots, x_n)$ when $x = (x_1, \ldots, x_n)$. Put $\varphi \subseteq \psi$ when $\forall x[\varphi x \downarrow \Rightarrow \psi x \downarrow \wedge \varphi x = \psi x]$. The members of PFn are not required to be extensional, i.e., it is *not* assumed that $\varphi \subseteq \psi \wedge \psi \subseteq \varphi \Rightarrow \varphi = \psi$. PFn is supposed to be closed under explicit definition and conditional definition, as will be explained in precise terms below.

2.3 A collection DT of subsets of V, called *data types,* is supposed to be given; upper case Latin letters $A, B, C, \ldots, X, Y, Z$ are used to range over DT. It is assumed that $\mathbb{B}, \mathbb{N}$ and V are in DT, and that DT is closed under Cartesian product $A \times B$ in the sense of the pairing operation on V. $A_1 \times \ldots \times A_\nu$ and A^ν are defined in terms of this in the usual way for $\nu > 0$; in case $\nu = 0$ these are identified with a singleton, e.g. $\{0\}$.

2.4 For any φ, A, B, define

(i) $(\varphi: A \underset{\smile}{\rightarrow} B) \Leftrightarrow \forall x[x \in A \wedge \varphi x \downarrow \Rightarrow \varphi x \in B]$,

(ii) $(\varphi: A \rightarrow B) \Leftrightarrow \forall x[x \in A \Rightarrow \varphi x \downarrow \wedge \varphi x \in B]$.

Note that it is *not* assumed in either case that dom $(\varphi) \subseteq A$; thus we may consider the same $\varphi: A_1 \underset{\smile}{\rightarrow} B_1$, $\varphi: A_2 \underset{\smile}{\rightarrow} B_2$ for different A_i, B_i. Let $(A \underset{\smile}{\rightarrow} B) = \{\varphi \mid \varphi: A \underset{\smile}{\rightarrow} B\}$ and $(A \rightarrow B) = \{\varphi \mid \varphi: A \rightarrow B\}$. It is also *not* assumed that DT is closed under these operations; when it is, we call these *function types.*

In the next section, we also consider basic assumptions at one formal type level higher.

3 Functional Notions and Notation

Any given data type structure $\mathcal{A}$ will have one or more data types $A_0, A_1, \ldots, A_n$ as basic domains, where we always take $A_0 = \mathbb{B}$. The following notions are relative to an arbitrary choice of $A_1, \ldots, A_n$, simply indicated by '$\mathcal{A}$'. Let i, j, k range over $\{0, \ldots, n\}$ and $\bar{\imath}, \bar{\jmath}, \bar{k}$ range over finite sequences of such, possibly empty. For $\bar{\imath} = (i_1, \ldots, i_\nu)$, put $A_{\bar{\imath}} = A_{i_1} \times \ldots \times A_{i_\nu}$. By an *arity for a partial function on* $\mathcal{A}$ is meant a symbol $\bar{\imath} \underset{\cdot}{\rightarrow} j$; it is said to be *proper* if length$(\bar{\imath}) > 0$. In that case, φ is said to be of arity $\bar{\imath} \underset{\cdot}{\rightarrow} j$ on $\mathcal{A}$ if $\varphi: A_{\bar{\imath}} \underset{\smile}{\rightarrow} A_j$. The letters σ, τ are used to range over proper arities of partial functions on $\mathcal{A}$. For $\sigma = (\bar{\imath} \underset{\cdot}{\rightarrow} j)$, put $A_\sigma = (A_{\bar{\imath}} \underset{\smile}{\rightarrow} A_j)$. Then for $\bar{\sigma} = (\sigma_1, \ldots, \sigma_\mu)$ with $\mu \geq 0$, put $A_{\bar{\sigma}} = A_{\sigma_1} \times \ldots \times A_{\sigma_\mu}$. By an *arity for a partial functional on* $\mathcal{A}$ is meant a symbol $(\bar{\sigma}, \bar{\imath}) \underset{\cdot}{\rightarrow} j$; this is said to be *proper* if length$(\bar{\sigma}) > 0$, and a partial map $F: A_{\bar{\sigma}} \times A_{\bar{\imath}} \underset{\smile}{\rightarrow} A_j$ is said to be of arity $(\bar{\sigma}, \bar{\imath}) \underset{\cdot}{\rightarrow} j$ on $\mathcal{A}$. Given $\varphi = (\varphi_1, \ldots, \varphi_\mu)$, and $x = (x_1, \ldots, x_\nu)$, we write $F(\varphi, x)$ for $F(\varphi_1, \ldots, \varphi_\mu, x_1, \ldots, x_\nu)$ when defined, and $F(\varphi, x) \simeq y$ when the value of F at $(\varphi_1, \ldots, \varphi_\mu, x_1, \ldots, x_\nu)$ is y. When $\nu = 0$, F is just a functional of its partial function arguments, $F(\varphi) \simeq y$. For simplicity, we also consider the improper cases $\mu = 0, \nu > 0$, where F reduces to a partial function $F(x) \simeq y$. Finally, it is useful to allow the degenerate case $\mu = \nu = 0$, in which case F is identified with an element of A_j (a "constant" of sort j). The collection of all partial functions of proper arity on $\mathcal{A}$ is denoted $PFnl$; it is supposed to be given along with the notions prescribed in §2. We use $F, G, H, \ldots$ to range over $PFnl$ and, by extension, over PFn and the constants in $\mathcal{A}$. These are supposed to be closed under explicit definition and conditional definition, as explained below. In particular, we use the λ notation for the result of abstraction with respect to any particular arguments.

4 Set-Theoretic and Recursion-Theoretic Interpretations

4.1 The Full Set-Theoretic Interpretation

Let V be the class of all sets in the cumulative hierarchy; pairing is defined as usual in set theory. Then the partial functions are taken to be those sets of pairs in V which are functions. The types range over all the sets in V; these are closed under Cartesian product and (partial and total) function types. Functionals are just those partial functions in V of the form $F(\varphi, x)$ where the φ_k's are partial functions in V. The type $\mathbb{N}$ is identified with ω and $\mathbb{B}$ with $\{1, 0\}$.

There are obvious generalizations of the preceding, e.g., to the cumulative hierarchy over any set of urelements and/or to the hierarchy cut off at any limit level $> \omega$.

4.2 The Set-Theoretic Interpretation on Computational Data

For computational purposes, all data should be represented in finite symbolic form; without loss of generality, we can take the universe to be $V = \omega$. Pairing in this case is taken to be a primitive recursive function on ω. The partial functions here are arbitrary $\varphi: \omega \xrightarrow{\smile} \omega$, and the types are arbitrary subsets of ω. We thus have closure under Cartesian products but not under function types. We identify $\mathbb{N}$ with ω and $\mathbb{B}$ with $\{1,0\}$. Partial functionals $F(\varphi, x)$ in this interpretation take arbitrary partial function arguments φ on ω. Special interest attaches below to those F which are *partial recursive* (p.r.) or have a partial recursive extension. Note that p.r. functionals are not closed under abstraction when the remaining function arguments are not p.r.

4.3 The Recursion-Theoretic Interpretation, Extensional Form

Here again we take $V = \omega$ and pairing as in 4.2. The types in this case range over arbitrary subsets of ω (or, more generally, over any collection of subsets closed under arithmetical definability). The collection PFn is taken to be the p.r. functions on ω, and $PFnl$ the p.r. functionals of p.r. function arguments. Thus we have closure under abstraction in this case. Every (proper) p.r. functional F has a canonical p.r. extension F^* defined at arbitrary partial function arguments by continuity: for example for $\bar{\sigma} = (\sigma_1)$, $F^*(\varphi, x) \simeq y \Leftrightarrow \exists\psi[\psi \subseteq \varphi \wedge \psi \text{ finite } \wedge F(\psi, x) \simeq y]$.

4.4 The Recursion-Theoretic Interpretation, Intensional (Index) Form

$V(=\omega)$, pairing, and types are as in 4.3. Now PFn is taken to be the set of all Gödel numbers e of p.r. functions, with $ex \simeq \{e\}(x)$ (in Kleene's notation). In this case not only is $A \times B$ again a type for types A and B, but also $(A \xrightarrow{\smile} B) = \{z \mid \forall x[x \in A \wedge zx \downarrow \Rightarrow zx \in B]\}$ and $(A \to B) = \{z \mid \forall x[x \in A \Rightarrow zx \downarrow \wedge zx \in B]\}$ are types. $PFnl$ is taken to coincide with PFn in this interpretation, but special interest will attach below to those functionals which are extensional (or "effective" in the sense of Myhill/Shepherdson [6]).

5 Monotonic Functionals and Least-Fixed-Point Operators

We return to the general setting of sections 2, 3. Given $\mathcal{A} = (A_0, A_1, \ldots, A_n, \ldots)$ and φ, ψ of (proper) arity $\bar{\imath} \xrightarrow{\cdot} j$ on $\mathcal{A}$, define

(1) $$\varphi \subseteq_{\mathcal{A}} \psi \Leftrightarrow \forall x \in A_{\bar{\imath}}[\varphi(x) \downarrow \Rightarrow \psi(x) \downarrow \wedge \varphi(x) = \psi(x))].$$

Then for $\varphi = (\varphi_1, \ldots, \varphi_\mu)$, $\psi = (\psi_1, \ldots, \psi_\mu)$, take

(2) $$\varphi \subseteq_{\mathcal{A}} \psi \Leftrightarrow \varphi_k \subseteq_{\mathcal{A}} \psi_k \text{ for } k = 1, \ldots, \mu.$$

Next, given F of (proper) arity $(\bar{\sigma}, \bar{\imath}) \xrightarrow{\cdot} j$ on $\mathcal{A}$, we say that F is $\mathcal{A}$-*monotonic* if

(3) $$\forall \varphi, \psi \in A_{\bar{\sigma}} \forall x \in A_{\bar{\imath}}[F(\varphi, x) \downarrow \wedge \varphi \subseteq_{\mathcal{A}} \psi \Rightarrow F(\psi, x) \downarrow \wedge F(\psi, x) = F(\varphi, x)].$$

For F_1, F_2 of the same arity $(\bar{\sigma}, \bar{\imath}) \xrightarrow{\cdot} j$ on $\mathcal{A}$, put

(4) $$F_1 \subseteq_{\mathcal{A}} F_2 \Leftrightarrow \forall \varphi \in A_{\bar{\sigma}} \forall x \in A_{\bar{\imath}}[F_1(\varphi, x) \downarrow \Rightarrow F_2(\varphi, x) \downarrow \wedge F_1(\varphi, x) = F_2(\varphi, x)].$$

The following special case of $\mathcal{A}$-monotonic functionals will be particularly important. Suppose $\sigma = (\bar{\imath} \rightharpoondown j)$ and $G: A_\sigma \times A_{\bar{\imath}} \rightharpoondown A_j$. Define $\widehat{G}: A_\sigma \rightarrow A_\sigma$ by

$$(5) \qquad (\widehat{G}\varphi) = \lambda x.G(\varphi, x), \text{ i.e., } (\widehat{G}\varphi)x \simeq G(\varphi, x) \text{ for } \varphi \in A_\sigma, x \in A_{\bar{\imath}}.$$

If G is $\mathcal{A}$-monotonic then so is $\widehat{G}$ in the sense that

$$(6) \qquad \varphi \subseteq_{\mathcal{A}} \psi \Rightarrow \widehat{G}\varphi \subseteq_{\mathcal{A}} \widehat{G}\psi.$$

We shall want in such cases to attach a *least fixed point* LG to $\widehat{G}$; this is not necessarily assured in the general interpretation of §§2–3.

Definition. L is called an *LFP operator on $\mathcal{A}$* if for any $\sigma = (\bar{\imath} \rightharpoondown j)$ and $\mathcal{A}$-monotonic functional $G: A_\sigma \times A_{\bar{\imath}} \rightharpoondown A_j$, we have:
(i) $LG \in A_\sigma$ and $\widehat{G}(LG) \subseteq_{\mathcal{A}} LG$;
(ii) whenever $\psi \in A_\sigma$ and $\widehat{G}(\psi) \subseteq_{\mathcal{A}} \psi$ then $LG \subseteq_{\mathcal{A}} \psi$.
It is furthermore required that whenever G_1, G_2 are $\mathcal{A}$-monotonic functionals of the same arity,
(iii) $G_1 \subseteq_{\mathcal{A}} G_2$ implies $LG_1 \subseteq_{\mathcal{A}} LG_2$.
The question of existence of LFP operators in the various interpretations will be delayed until §10 below. Note that all of the notions in this section depend only on the domains $A_0, A_1, \dots A_n$ of $\mathcal{A}$.

6 Equality Preserving Functions and Functionals

We now assume that each A_i is equipped with an "equality" relation $=_{A_i}$, i.e., an equivalence relation on A_i. In some cases, this may be ordinary identity ($=$) as, e.g., with $A_0 (= \mathbb{B})$, or if $\mathbb{N}$ is one of the domains of $\mathcal{A}$. Given $x = (x_1, \dots, x_\nu)$ and $y = (y_1, \dots, y_\nu)$ in $A_{\bar{\imath}}$ where $\bar{\imath} = (i_1, \dots, i_\nu)$, put

$$(1) \qquad x =_{\mathcal{A}} y \Leftrightarrow x_k =_{A_{i_k}} y_k \text{ for } k = 1, \dots, \nu.$$

Then for $\varphi: A_{\bar{\imath}} \rightharpoondown A_j$, we say that φ *preserves $\mathcal{A}$-equalities* if

$$(2) \qquad \forall x, y \in A_{\bar{\imath}}[x =_{\mathcal{A}} y \wedge \varphi(x) \downarrow \Rightarrow \varphi(y) \downarrow \wedge \varphi(x) =_{A_j} \varphi(y)].$$

Next, for φ, $\psi \in A_\sigma$ with $\sigma = (\bar{\imath} \rightharpoondown j)$, define

$$(3) \qquad \varphi =_{\mathcal{A}} \psi \Leftrightarrow \forall x \in A_{\bar{\imath}}[\varphi(x) \downarrow \vee \psi(x) \downarrow \Rightarrow \varphi(x) \downarrow \wedge \psi(x) \downarrow \wedge \varphi(x) =_{A_j} \psi(x)]$$

Note that

$$(4) \qquad \varphi =_{\mathcal{A}} \psi \Leftrightarrow \varphi \subseteq_{\mathcal{A}} \psi \wedge \psi \subseteq_{\mathcal{A}} \varphi.$$

For $\varphi = (\varphi_1, \ldots, \varphi_\mu)$, $\psi = (\psi_1, \ldots, \psi_\mu)$, both in $A_{\bar{\sigma}}$, put

(5) $$\varphi =_{\mathcal{A}} \psi \Leftrightarrow \varphi_k =_{\mathcal{A}} \psi_k \text{ for each } k = 1, \ldots, \mu .$$

Finally, given $F: A_{\bar{\sigma}} \times A_{\bar{\imath}} \rightharpoonup A_j$, we say that F *preserves $\mathcal{A}$-equalities* if

(6)
$$\forall \varphi, \psi \in A_{\bar{\sigma}} \forall x, y \in A_{\bar{\imath}}[\varphi =_{\mathcal{A}} \psi \wedge x =_{\mathcal{A}} y \wedge F(\varphi, x) \downarrow \Rightarrow F(\psi, y) \downarrow \wedge F(\varphi, x) =_{A_j} F(\psi; y)]$$

We say that F is *strongly $\mathcal{A}$-monotonic* if it is $\mathcal{A}$-monotonic and preserves $\mathcal{A}$-equalities. This is equivalent to:

(7)
$$\forall \varphi, \psi \in A_{\bar{\sigma}} \forall x, y \in A_{\bar{\imath}}[\varphi \subseteq_{\mathcal{A}} \psi \wedge x =_{\mathcal{A}} y \wedge F(\varphi, x) \downarrow \Rightarrow F(\psi, y) \downarrow \wedge F(\varphi, x) =_{A_j} F(\psi, y)] .$$

7 Schemata for Abstract Computational Procedures

For present purposes, a *signature* Σ of functional structures is given by a specification of $sort(\Sigma) = \{0, 1, \ldots, n\}$ and $Fnl(\Sigma) = \{\mathbf{F}_0, \ldots, \mathbf{F}_m\}$ where each $\mathbf{F}_k (k = 0, \ldots, m)$ is a *functional symbol* of specified arity $\bar{\sigma}_k \times \bar{\imath}_k \rightharpoonup j_k$. When $\bar{\sigma}_k$ is empty, this reduces to a *function symbol* of arity $\bar{\imath}_k \rightharpoonup j_k$, and when both $\bar{\sigma}_k$ and $\bar{\imath}_k$ are empty, it reduces to a constant symbol of *sort* j_k. A structure $\mathcal{A}$ of $Sort(\Sigma)$ is then of the form:

(1) $$\mathcal{A} = (A_0, A_1, \ldots, A_n, =_{A_0}, \ldots, =_{A_n}, F_0, \ldots, F_m) ,$$

where each relation $=_{A_i}$ is an equality relation on A_i and each F_k is a partial functional on $\mathcal{A}$ of the same arity as $\mathbf{F}_k$. To indicate the dependence on $\mathcal{A}$ we may write $F_k^{\mathcal{A}}$ for F_k.

General Assumptions

(i) For each $\mathcal{A}$, $A_0 = \mathbb{B} = \{tt, ff\}$ and $=_{A_0}$ is the identity relation.
(ii) The constant symbols tt and ff are in the signature.

The formal language of computational procedures over structures of signature σ has variables $a, b, c, \ldots, x, y, z$ of each *sort*, partial function variables φ, ψ, θ of each proper arity σ and constant functional symbols $\mathbf{F}, \mathbf{G}, \mathbf{H}, \ldots$ of each arity $\bar{\sigma} \times \bar{\imath} \rightharpoonup j$ (where $\bar{\sigma}, \bar{\imath}$ may be empty); the latter may degenerate to function symbols ($\bar{\sigma}$ empty) or constant symbols ($\bar{\sigma}$ and $\bar{\imath}$ empty). Each of the following schemata introduces a specified $\mathbf{F}$ (of specified arity) either directly or in terms of previously specified functional symbols. These schemata implicitly tell us what the terms of the language are.

In the LFP schema, $\mathbf{L}$ is an operator symbol of higher type level; it applies to terms of arity $\sigma \times \bar{\imath} \rightharpoonup j$ where $\sigma = (\bar{\imath} \rightharpoonup j)$ and $\bar{\imath}$ is non-empty, and produces in each such case a term of arity σ. This will be interpreted in each structure $\mathcal{A}$ by a specified operator $\mathbf{L}^{\mathcal{A}}$.

General Assumptions (contd.)

(iii) For each $\mathcal{A}$, $\mathbf{L}^{\mathcal{A}}$ is an operator from $\mathcal{A}$-functionals G of arity $\sigma \times \bar{\imath} \rightharpoonup j$, where $\sigma = (\bar{\imath} \rightharpoonup j)$, to A_σ-partial functions of arity σ (for each $\bar{\imath}, j$).

Note that at this stage we don't assume $\mathbf{L}^{\mathcal{A}}$ to be an LFP operator on $\mathcal{A}$.

The Schemata

Conditions on the variables or the arities in the following schemata are explained directly following them.

I.	(Initial functionals)	$\mathbf{F}(\varphi, x) \simeq \mathbf{F}_k(\varphi, x) \quad (k = 0, \ldots, m)$
II.	(Identity functions)	$\mathbf{F}(x) = x$
III.	(Application functionals)	$\mathbf{F}(\theta, x) \simeq \theta(x)$
IV.	(Conditional definition)	$\mathbf{F}(\varphi, x, v) \simeq [\text{if } v = tt \text{ then } \mathbf{G}(\varphi, x) \text{ else } \mathbf{H}(\varphi, x)]$
V.	(Structural)	$\mathbf{F}(\varphi, x) \simeq \mathbf{G}(\varphi_f, x_g)$
VI.	(Individual substitution)	$\mathbf{F}(\varphi, x) \simeq \mathbf{G}(\varphi, x, \mathbf{H}(\varphi, x))$
VII.	(Function substitution)	$\mathbf{F}(\varphi, x) \simeq \mathbf{G}(\varphi, \lambda u.\mathbf{H}(\varphi, x, u), x)$
VIII.	(Least fixed point)	$\mathbf{F}(\varphi, x, u) \simeq [\mathbf{L}(\lambda\theta, w.\mathbf{G}(\varphi, \theta, x, w))](u)$.

In the schemata I, IV–VIII $\varphi = (\varphi_1, \ldots, \varphi_\mu)$ is a sequence of variables of arity $\bar{\sigma} = (\sigma_1, \ldots, \sigma_\mu), \mu \geq 0$, and $x = (x_1, \ldots, x_\nu)$ is a sequence of variables of arity $\bar{\imath} = (i_1, \ldots, i_\nu), \nu \geq 0$. In IV, v is a Boolean (sort 0) variable. In the structural schema V, $f\colon \{1, \ldots, \mu'\} \to \{1, \ldots, \mu\}$ and $g\colon \{1, \ldots, \nu'\} \to \{1, \ldots, \nu\}$, i.e., $\mathbf{G}(\varphi_f, x_g)$ is $\mathbf{G}(\varphi_{f(1)}, \ldots, \varphi_{f(\mu')}, x_{g(1)}, \ldots, x_{g(\nu')})$; this schema thus accounts for expansion, identification, and permutation of individual and function variables. In the schema VII, $\mathbf{G}$ is of arity $(\bar{\sigma}, \tau) \times \bar{\imath} \xrightarrow{\cdot} j$. In the LFP schema VIII, the variable θ is of arity $\bar{\imath}' \xrightarrow{\cdot} j$ and w, u are of arity $\bar{\imath}'$ for some non-empty $\bar{\imath}'$.

Definition. By an *abstract computational procedure* for signature Σ we mean a schematically generated partial functional $\mathbf{F}$.

Now for each $\mathcal{A}$ of signature Σ and each $\mathbf{F}$ generated by the above schemata there is an associated partial functional $\mathbf{F}^{\mathcal{A}}$, where in each case the intended meaning is clear. In particular for VIII, $\mathbf{F}^{\mathcal{A}}(\varphi, x, u) \simeq \psi(u)$ where $\psi = \mathbf{L}^{\mathcal{A}}(\lambda\theta, w.\mathbf{G}^{\mathcal{A}}(\varphi, \theta, x, w))$; thus if $\mathbf{G}^{\mathcal{A}}$ is $\mathcal{A}$-monotonic and $\mathbf{L}^{\mathcal{A}}$ is an LFP operator for $\mathcal{A}$, then ψ is a least solution of

$$\psi(u) \simeq \mathbf{G}^{\mathcal{A}}(\varphi, \psi, x, u)\,. \tag{2}$$

In this recursion, φ, x act as parameters. Writing $G_{\varphi,x}$ for $\lambda\theta, w.\mathbf{G}^{\mathcal{A}}(\varphi, \theta, x, w)$, we may rewrite (2) as:

$$\psi(u) \simeq G_{\varphi,x}(\psi, u), \text{ or } \psi = \widehat{G}_{\varphi,x}(\psi) \tag{3}$$

8 The Main Theorems

The general assumptions of §7 continue in force. However, we now add:

General assumptions (contd.)

(iv) It is now assumed for each $\mathcal{A}$ that $\mathbf{L}^{\mathcal{A}}$ is an LFP-operator on $\mathcal{A}$.

THEOREM 1 Preservation of strong monotonicity. *Suppose each* $\mathbf{F}_k^{\mathcal{A}}(k = 0,\ldots,m)$ *is strongly* $\mathcal{A}$*-monotonic. Then the same holds for each* $\mathbf{F}^{\mathcal{A}}$.

Proof. Recall that a functional F is called strongly $\mathcal{A}$-monotonic if it is $\mathcal{A}$-monotonic and preserves $\mathcal{A}$-equalities. We prove this holds for each $\mathbf{F}^{\mathcal{A}}$ by induction on the generation of $\mathbf{F}$ by the schemata. This is quite straightforward for I–VII. Only VIII requires a little attention. Let $G = \mathbf{G}^{\mathcal{A}}$ there and let $F = \mathbf{F}^{\mathcal{A}}$. Given any $\varphi^{(1)}, \varphi^{(2)}, x^{(1)}, x^{(2)}, u^{(1)}, u^{(2)}$ it must be shown that

$$\begin{aligned}(1)\qquad &\varphi^{(1)} \subseteq_{\mathcal{A}} \varphi^{(2)} \wedge x^{(1)} =_{\mathcal{A}} x^{(2)} \wedge u^{(1)} =_{\mathcal{A}} u^{(2)} \wedge F(\varphi^{(1)}, x^{(1)}, u^{(1)}) \downarrow \Rightarrow \\ &F(\varphi^{(2)}, x^{(2)}, u^{(2)}) \downarrow \wedge F(\varphi^{(2)}, x^{(2)}, u^{(2)}) =_{\mathcal{A}} F(\varphi^{(1)}, x^{(1)}, u^{(1)}).\end{aligned}$$

Let

$$(2)\qquad G^{(i)}(\theta, w) = G(\varphi^{(i)}, \theta, x^{(i)}, w) \qquad \text{for } i = 1, 2.$$

By induction hypothesis each $G^{(i)}$ is strongly $\mathcal{A}$-monotonic. Hence $G^{(1)} \subseteq_{\mathcal{A}} G^{(2)}$ and so by the general hypothesis on $L = \mathbf{L}^{\mathcal{A}}$, we have $L(G^{(1)}) \subseteq_{\mathcal{A}} L(G^{(2)})$. In other words, if we take

$$(3)\qquad \psi^{(i)} = L(\lambda\theta, w.G^{(i)}(\theta, w))$$

then

$$(4)\qquad \psi^{(1)} \subseteq_{\mathcal{A}} \psi^{(2)}.$$

But, by definition,

$$(5)\qquad F(\varphi^{(i)}, x^{(i)}, u^{(i)}) \simeq \psi^{(i)}(u^{(i)}),$$

so that (1) is immediate by (4).

Remark. Actually Theorem 1 shows that as long as we restrict ourselves to structures with strongly $\mathcal{A}$-monotonic initial functionals $F_k^{\mathcal{A}}$, we do not have to give meaning to $\mathbf{L}^{\mathcal{A}}(G)$ in §7 except for strongly $\mathcal{A}$-monotonic G.

THEOREM 2. (Invariance under isomorphism.) *Suppose that* (h, h') *establishes an isomorphism between* $\mathcal{A}$ *and* $\mathcal{A}'$. *Then for each abstract computational procedure* $\mathbf{F}$ *we have* $\mathbf{F}^{\mathcal{A}} \leftrightarrow \mathbf{F}^{\mathcal{A}'}$ *under* (h, h').

Proof. Let $\mathcal{A} = (A_0, \ldots, A_n, =_{A_0}, \ldots, =_{A_n}, F_0^{\mathcal{A}}, \ldots, F_m^{\mathcal{A}})$ and $\mathcal{A}' = (A'_0, \ldots, A'_n, =_{A'_0}, \ldots, =_{A'_n}, F_0^{\mathcal{A}'}, \ldots, F_m^{\mathcal{A}'})$. The notion of isomorphism used here is that of Part I of this paper, [1] §4.3, extended directly to functional structures as follows. First of all, $h = (h_0, \ldots, h_n), h' = (h'_0, \ldots, h'_n)$, where:

(1) (i) $h_i: A_i \to A_{i'}, h'_i: A_{i'} \to A_i$ for $i = 0, \ldots, n$, (ii) $x =_{A_i} y \Rightarrow h_i(x) =_{A'_i} h_i(y)$ and $x' =_{A'_i} y' \Rightarrow h'_i(x') =_{A_i} h'_i(y')$, and (iii) $h'_i(h_i(x)) =_{A_i} x$ for each $x \in A_i$ and $h_i(h'_i(x')) =_{A'_i} x'$ for each $x' \in A'_i$.

Moreover, it is assumed that

(2) $\mathbf{F}_k^{\mathcal{A}} \leftrightarrow \mathbf{F}_k^{\mathcal{A}'}$ under (h, h') for $k = 0, \ldots, m$,

where the notion of correspondence under (h, h') is defined for individuals, partial functions and functionals as follows:

$$\text{(3)} \qquad x \leftrightarrow x' \text{ for } x, x' \text{ of } sort\, i, \text{ if } h_i(x) =_{A'_i} x' .$$

By (1) this is equivalent to $h'_i(x') =_{A_i} x$.
This extends to sequences $x = (x_1, \ldots, x_\nu)$, $x' = (x'_1, \ldots, x'_\nu)$ of the same sorts by

$$\text{(4)} \qquad x \leftrightarrow x' \text{ if } x_k \leftrightarrow x'_k \text{ for } k = 1, \ldots, \nu .$$

Next, given φ of arity $\bar{\imath} \dot{\rightarrow} j$ in $\mathcal{A}$ and φ' of the same arity in $\mathcal{A}'$, define

$$\text{(5)} \qquad (\varphi \leftrightarrow \varphi') \Leftrightarrow \forall x \in A_{\bar{\imath}} \forall x' \in A'_{\bar{\imath}}[(x \leftrightarrow x') \wedge (\varphi(x) \downarrow \vee \varphi(x') \downarrow) \Rightarrow \\ \varphi(x) \downarrow \wedge \varphi(x') \downarrow \wedge (\varphi(x) \leftrightarrow \varphi'(x'))]$$

Then for $\varphi = (\varphi_1, \ldots, \varphi_\mu)$, $\varphi' = (\varphi_1, \ldots, \varphi'_\mu)$, take

$$\text{(6)} \qquad \varphi \leftrightarrow \varphi' \text{ if } \varphi_k \leftrightarrow \varphi'_k \text{ for } k = 1, \ldots, \mu .$$

Finally, for F, F' of arity $\bar{\sigma} \times \bar{\imath} \dot{\rightarrow} j$, in $\mathcal{A}$, $\mathcal{A}'$, resp., we define

$$\text{(7)} \qquad (F \leftrightarrow F') \Leftrightarrow \forall \varphi \in A_{\bar{\varphi}} \forall \varphi' \in A'_{\bar{\varphi}} \forall x \in A_{\bar{\imath}} \forall x' \in A'_{\bar{\imath}}[(\varphi \leftrightarrow \varphi') \wedge (x \leftrightarrow x')] \wedge \\ (F(\varphi, x) \downarrow \vee F'(\varphi', x') \downarrow) \Rightarrow F(\varphi, x) \downarrow \wedge F'(\varphi', x') \downarrow \wedge (F(\varphi, x) \leftrightarrow F(\varphi', x'))] .$$

It is now a routine matter to verify by induction that for each $\mathbf{F}$ generated by the schemata I-VII

$$\text{(8)} \qquad \mathbf{F}^{\mathcal{A}} \leftrightarrow \mathbf{F}^{\mathcal{A}'} .$$

Again only VIII requires a little more work. Thus let $G = \mathbf{G}^{\mathcal{A}}$, $G' = \mathbf{G}^{\mathcal{A}'}$ and suppose $\varphi \leftrightarrow \varphi'$, $x \leftrightarrow x'$ and $u \leftrightarrow u'$. Fix the parameters φ, x, φ', x' and let $\psi = \mathbf{L}^{\mathcal{A}}(\lambda\theta, w.G(\theta, w))$, $\psi' = \mathbf{L}^{\mathcal{A}'}(\lambda\theta', w'.G'(\theta', w'))$. Then ψ, ψ' are LFP's of $\hat{G}$, $\hat{G}'$ resp. Thus we have

(9) $\psi(u) \simeq G(\psi, u)$ and ψ is least such under $\subseteq_{\mathcal{A}}$ and similarly for ψ', G' in $\mathcal{A}'$.

Now we can define equality preserving ψ_1 and ψ'_1 in such a way that

$$\text{(10)} \qquad \psi_1 \leftrightarrow \psi' \text{ and } \psi \leftrightarrow \psi'_1$$

simply transferring ψ' to $\mathcal{A}$ using h, and ψ to $\mathcal{A}'$ by h', resp., i.e., $\psi_1 = \lambda w.\psi'(h(w))$, and $\psi'_1 = \lambda w'.\psi(h'(w'))$. Since $G \leftrightarrow G'$ it follows that ψ_1 is a fixed-point of $\hat{G}$ up to $=_{\mathcal{A}}$ and ψ_1' is a fixed point of $\hat{G}'$ up to $=_{\mathcal{A}'}$. Hence $\psi \subseteq_{\mathcal{A}} \psi_1$ and $\psi' \subseteq_{\mathcal{A}'} \psi_1'$. By chasing this through with (10) we may conclude that

$$(11) \qquad \psi \leftrightarrow \psi',$$

as required.

9 Closure Under Simultaneous LFPs

Without stating the general result, we shall illustrate the argument for this in a special case, for simplicity. Assume given $\mathcal{A}$ and an LFP operator L on $\mathcal{A}$. Suppose given G_0, G_1 both strongly $\mathcal{A}$-monotonic, where G_0 is of arity $(\sigma_0, \sigma_1) \times \bar{\imath}_0 \dot{\rightarrow} j_0$, G_1 is of arity $(\sigma_0, \sigma_1) \times \bar{\imath}_1 \dot{\rightarrow} j_1$, $\sigma_0 = (\bar{\imath}_0 \dot{\rightarrow} j_0)$ and $\sigma_1 = (\bar{\imath}_1 \dot{\rightarrow} j_1)$. We want to construct the simultaneous LFP's ψ_0, ψ_1 of

$$(1) \qquad \begin{cases} \psi_0(u) \simeq G_0(\psi_0, \psi_1, u) \\ \psi_1(v) \simeq G_1(\psi_0, \psi_1, v). \end{cases}$$

Our plan is to solve the second equation uniformly as a function $\widehat{H}$ of ψ_0, then substitute that for ψ_1 in the first equation to solve for ψ_0. More formally, let

$$(2) \qquad \widetilde{G}_0(\theta_0, \theta_1) = \lambda u. G_0(\theta_0, \theta_1, u) \text{ and } \widetilde{G}_1(\theta_0, \theta_1) = \lambda v. G_1(\theta_0, \theta_1, v).$$

Then take

$$\begin{aligned} (3) \qquad &\text{(i)} \quad H(\theta_0, v) \simeq [L(\lambda\theta_1, w. G_1(\theta_0, \theta_1, w))](v),\ \widehat{H}(\theta_0) = \lambda v. H(\theta_0, v), \\ &\text{(ii)} \quad \psi_0 = L(\lambda\theta_0, u. G_0(\theta_0, \widehat{H}(\theta_0), u)), \\ \text{and } &\text{(iii)} \quad \psi_1 = \widehat{H}(\psi_0). \end{aligned}$$

As we have seen, H is strongly A-monotonic, so by construction and hypothesis on L

$$\begin{aligned} (4) \quad &\text{(i)} \quad \widetilde{G}_1(\theta_0, \widehat{H}(\theta_0)) \subseteq_{\mathcal{A}} \widehat{H}(\theta_0) \text{ and (ii) } \widetilde{G}_1(\theta_0, \theta_1) \subseteq_{\mathcal{A}} \theta_1 \Rightarrow \widehat{H}(\theta_0) \subseteq_{\mathcal{A}} \theta_1, \\ &\text{(iii)} \quad \widetilde{G}_0(\psi_0, \widehat{H}(\psi_0)) \subseteq_{\mathcal{A}} \psi_0 \text{ and (iv) } \widetilde{G}_0(\theta_0, \widehat{H}(\theta_0)) \subseteq_{\mathcal{A}} \theta_0 \Rightarrow \psi_0 \subseteq_{\mathcal{A}} \theta_0. \end{aligned}$$

It follows from (4)(iii) that $\widetilde{G}_0(\psi_0, \psi_1) \subseteq_{\mathcal{A}} \psi_0$ and from (4)(i) that $\widetilde{G}_1(\psi_0, \psi_1) \subseteq_{\mathcal{A}} \psi_1$. Now suppose that

$$(5) \qquad \widetilde{G}_0(\theta_0, \theta_1) \subseteq_{\mathcal{A}} \theta_0 \text{ and } \widetilde{G}_1(\theta_0, \theta_1) \subseteq_{\mathcal{A}} \theta_1.$$

Then by (4)(ii) we have $\widehat{H}(\theta_0) \subseteq_{\mathcal{A}} \theta_1$, hence by monotonicity of G_0, $\widetilde{G}_0(\theta_0, \widehat{H}(\theta_0)) \subseteq_{\mathcal{A}} \widetilde{G}_0(\theta_0, \theta_1) \subseteq_{\mathcal{A}} \theta_0$; but then $\psi_0 \subseteq_{\mathcal{A}} \theta_0$ by (4)(iv). Now $\widetilde{G}_1(\psi_0, \theta_1) \subseteq_{\mathcal{A}} \widetilde{G}_1(\theta_0, \theta_1) \subseteq_{\mathcal{A}} \theta_1$ by monotonicity of G_1 so by (4)(ii) again $\widehat{H}(\psi_0) \subseteq_{\mathcal{A}} \theta_1$, i.e., $\psi_1 \subseteq_{\mathcal{A}} \theta_1$. Thus ψ_0, ψ_1 are the $\mathcal{A}$-least solutions of (5), as desired.

The general treatment of simultaneous LFP with parameters follows exactly the same lines.

10 LFP Operators in the Set-Theoretic and Recursion-Theoretic Interpretations

10.1 LFP Operators in the Full Set-Theoretic Interpretation

Given $\mathcal{A}$ and $\mathcal{A}$-monotonic $G: A_\sigma \times A_{\bar{\imath}} \rightharpoonup A_j$ with $\sigma = (\bar{\imath} \rightharpoonup j)$, define, as usual, $\varphi_G^{(\alpha)}$ for $\alpha \in ORD$ by $\varphi_G^{(0)} = \varepsilon$ (the empty function), $\varphi_G^{(\alpha+1)} = \widehat{G}(\varphi_G^{(\alpha)})$ and for limit λ, $\varphi_G^{(\lambda)} = \bigcup_{\alpha<\lambda} \varphi_G^{(\alpha)}$. Then it is proved by induction that each $\varphi_G^{(\alpha)} \in A_\sigma$ and $\beta \leq \alpha \Rightarrow \varphi_G^{(\beta)} \subseteq_{\mathcal{A}} \varphi_G^{(\alpha)}$. Then $L^{\mathcal{A}}(G) = \varphi_G^{(\alpha_0)}$ for the least α_0 with $\varphi_G^{(\alpha_0)} =_{\mathcal{A}} \varphi_G^{(\alpha_0+1)}$. This satisfies the conditions for $L^{\mathcal{A}}$, including that if $G_1 \subseteq_{\mathcal{A}} G_2$ then $L^{\mathcal{A}}(G_1) \subseteq_{\mathcal{A}} L^{\mathcal{A}}(G_2)$ by $(\varphi_{G_1}^{(\alpha)} \subseteq_{\mathcal{A}} \varphi_{G_2}^{(\alpha)}$ for each $\alpha)$.

10.2 LFP Operators in the Set-Theoretic Interpretation on $V = \omega$

Here, in general, $L^{\mathcal{A}}$ is dealt with just as in 10.1. But more can be said if we are dealing with partial recursive (p.r.) functionals of arbitrary partial function arguments. Now it makes sense to consider the same functional acting on $\mathcal{A}$ and on a structure $\mathcal{V}$ all of whose domains $V_i = V = \omega$. In particular, consider p.r. $G: V_\sigma \times V_{\bar{\imath}} \rightharpoonup V_j$ with $\sigma = \bar{\imath} \rightharpoonup j$. Since G is p.r., it is monotonic and $L^{\mathcal{V}}(G) = \varphi_G^{(\omega)}$, a p.r. function. It does not follow that G is $\mathcal{A}$-monotonic, but if it is then we obtain the same result on $\mathcal{A}$ as on $\mathcal{V}$, according to the following.

LEMMA 1. *Suppose $G: V_\sigma \times V_{\bar{\imath}} \rightharpoonup V_j$ is partial recursive, where $\sigma = (\bar{\imath} \rightharpoonup j)$, and that $G: A_\sigma \times A_{\bar{\imath}} \rightharpoonup A_j$ is $\mathcal{A}$-monotonic. Then $L^{\mathcal{A}}(G) = L^{\mathcal{V}}(G) = \varphi_G^{(\omega)}$.*

Proof. By definition, $\varphi^{(0)} = \varepsilon$, $\varphi^{(n+1)} = \widehat{G}(\varphi^{(n)})$ for each $n < \omega$ (we omit the 'G' subscript). By hypothesis $\widehat{G}: V_\sigma \to V_\sigma$ and $\widehat{G}: A_\sigma \to A_\sigma$, so it follows by induction that $\varphi^{(n)} \in V_\sigma$ and $\varphi^{(n)} \in A_\sigma$ for each $n < \omega$. Moreover, by monotonicity of p.r. functionals, we have $m \leq n \Rightarrow \varphi^{(m)} \subseteq_{\mathcal{V}} \varphi^{(n)}$, and by assumed $\mathcal{A}$-monotonicity of G, we have $m \leq n \Rightarrow \varphi^{(m)} \subseteq_{\mathcal{A}} \varphi^{(n)}$. Thus $\varphi^{(\omega)} \in V_\sigma$ and $\varphi^{(\omega)} \in A_\sigma$. Now $\widehat{G}(\varphi^{(\omega)}) \subseteq_{\mathcal{V}} \varphi^{(\omega)}$ by continuity of p.r. functionals, and if $\widehat{G}(\psi) \subseteq_{\mathcal{V}} \psi$ then $\varphi^{(\omega)} \subseteq \psi$, since $\varphi^{(n)} \subseteq_{\mathcal{V}} \psi$ for all n by induction on n. To show $L^{\mathcal{A}}(G) = \varphi^{(\omega)}$, we must prove (i) $\widehat{G}(\varphi^{(\omega)}) \subseteq_{\mathcal{A}} \varphi^{(\omega)}$ and (ii) $\widehat{G}(\psi) \subseteq_{\mathcal{A}} \psi \Rightarrow \varphi^{(\omega)} \subseteq_{\mathcal{A}} \psi$, for all $\psi \in A_\sigma$. For (i), it is to be shown that if $x \in A_{\bar{\imath}}$ and $G(\varphi^{(\omega)}, x) \simeq y$ then $\varphi^{(\omega)}(x) \simeq y$; but by continuity again, $G(\varphi^{(n)}, x) \simeq y$ for some n, i.e., $\varphi^{(n+1)}(x) \simeq y$, so we have the desired conclusion. For (ii) we prove by induction on n that $\varphi^{(n)} \subseteq_{\mathcal{A}} \psi$; if this holds for n then it holds for $n+1$ by $\widehat{G}(\varphi^{(n)}) \subseteq_{\mathcal{A}} \widehat{G}(\psi) \subseteq_{\mathcal{A}} \psi$. Hence $\varphi^{(\omega)} = \bigcup_n \varphi^{(n)} \subseteq_{\mathcal{A}} \psi$.

10.3 LFP Operators in the Extensional Recursion-Theoretic Interpretation

Here again $V = \omega$, and by the explanation of 4.3, PFn consists in this case of all p.r. functions, and $PFnl$ consists of all p.r. functionals restricted to p.r. arguments. The preceding Lemma carries over directly for G p.r.

10.4 LFP Operators in the Intensional Recursion-Theoretic Interpretation

Here PFn consists of indices of p.r. functions; we $z, u, w, \ldots$ to range over these, instead of $\varphi, \psi, \theta, \ldots$. According to the explanation of 4.4, $PFnl$ coincides with PFn in this case, but we use letters $f, g, h, \ldots$ instead of $F, G, H, \ldots$ when thinking of indices as functionals. Now these are not automatically monotonic on $\mathcal{V}$ as are p.r. functionals in the extensional sense. For those that are, we have the following modification of Lemma 1 of 10.2.

LEMMA 2. *Suppose* $g: V_\sigma \times V_{\bar{\imath}} \rightharpoonup V_j$ *is* $\mathcal{V}$*-monotonic, where* $\sigma = \bar{\imath} \dot{\rightarrow} j$*, and also that* $g: A_\sigma \times A_{\bar{\imath}} \rightharpoonup A_j$ *is* $\mathcal{A}$*-monotonic. Then* $L^{\mathcal{V}}(g)$ *exists and* $L^{\mathcal{V}}(g) = L^{\mathcal{A}}(g)$.

Proof. The hypothesis tells us that $\forall z, w \in V_\sigma \forall x \in A_{\bar{\imath}}[z \subseteq_{\mathcal{V}} w \wedge g(z,x) \downarrow \Rightarrow g(w,x) \downarrow \wedge g(z,x) = g(w,x)]$. Recall that $z \subseteq w \Leftrightarrow \forall x \in V_\sigma(z(x) \downarrow \Rightarrow w(x) \downarrow \wedge z(x) = w(x))$. It follows that g is extensional (or "effective") in the recursion-theoretic sense. Hence by the Myhill-Shepherdson theorem [1], g is determined by a partial-recursive functional $G(\varphi, x)$ with $g(z,x) \simeq G(\{z\}, x)$ for all z. Then $L^{\mathcal{V}}(g) = L^{\mathcal{V}}(G)$ and the argument for Lemma 1 carries over directly to the present situation.

Remark. Additional work is required to show that the operator L can itself be given by an index for a p.r. functional; this is done in the Appendix below.

11 Computation on ADTs as Ordinary Computation

We now restrict ourselves to $V = \omega$ and the recursion-theoretic interpretation, which can be regarded extensionally (4.3, 10.3) or intensionally (4.4, 10.4). For simplicity, we work with the extensional interpretation in this section. Thus every structure $\mathcal{A}$ considered is given as

$$(1) \qquad \mathcal{A} = (A_0, A_1, \ldots, A_n =_{A_0}, \ldots, =_{A_n}, F_0, \ldots, F_m)$$

where the F_k are partial recursive functionals on $\mathcal{V}$, hence $\mathcal{V}$-monotonic. Let $\mathcal{A}$ be of signature Σ.

THEOREM 3. *Suppose that each* F_k *in (1) is strongly* $\mathcal{A}$*-monotonic. Then each abstract computational procedure* $\mathbf{F}$ *has* $\mathbf{F}^{\mathcal{V}}$ *a p.r. functional on* $\mathcal{V}$ *which is strongly* $\mathcal{A}$*-monotonic and which meets the conditions for* $\mathbf{F}^{\mathcal{A}}$ *on* $\mathcal{A}$.

Proof. By induction on the generation of $\mathbf{F}$ by the schemata I-VIII. What the assumption amounts to is that we have $\mathbf{F}_k^{\mathcal{A}} = F_k = \mathbf{F}_k^{\mathcal{V}}$, since we are considering F_k as a p.r. functional of given arity on $\mathcal{V}$, which happens to act as an $\mathcal{A}$-functional of the same arity when applied to arguments from $\mathcal{A}$. The induction is straightforward when one takes note of the following points for the schemata VII (function level substitution) and VIII (LFP). In the first case, we have

$$(\text{VII}) \qquad \mathbf{F}(\varphi, x) \simeq \mathbf{G}(\varphi, \lambda u.\mathbf{H}(\varphi, x, u), x)\,.$$

Interpreted in $\mathcal{V}$, $\mathbf{H}^{\mathcal{V}}$ is a p.r. functional, hence for each choice of (p.r.) arguments φ, and each x we have $\lambda u.\mathbf{H}^{\mathcal{V}}(\varphi, x, u)$ also partial recursive. Thus $\mathbf{F}^{\mathcal{V}}$ is a p.r. functional of p.r. arguments. Now $\mathbf{H}^{\mathcal{A}}$ is supposed to be strongly $\mathcal{A}$-monotonic, so $\lambda u.\mathbf{H}^{\mathcal{A}}(\varphi, x, u)$ is also an $\mathcal{A}$-partial function. Thus the interpretation of VII over $\mathcal{V}$ also works as an interpretation over $\mathcal{A}$. In the case of the LFP schema we have

$$\text{(VIII)} \qquad \mathbf{F}(\varphi, x, u) \simeq [\mathbf{L}(\lambda\theta, w.\mathbf{G}(\varphi, \theta, x, w))](u)\,.$$

Now for each choice of (p.r.)arguments φ and each x, let $G_{\varphi,x} = \mathbf{G}^{\mathcal{V}}_{\varphi,x} = \lambda\theta, w.\mathbf{G}^{\mathcal{V}}(\varphi, \theta, x, w)$. Then $\mathbf{G}_{\varphi,x}$ is a p.r. functional (uniformly in φ, x) which is strongly $\mathcal{A}$-monotonic. Hence by Lemma 1 of 10.2 (as carried over to 10.3), $L^{\mathcal{A}}(G_{\varphi,x}) = L^{\mathcal{V}}(G_{\varphi,x})$. This too is a p.r. functional uniformly in φ, x. Hence we can treat $\mathbf{F}^{\mathcal{A}}$ as the p.r. functional $\mathbf{F}^{\mathcal{V}}$ acting on $\mathcal{A}$-arguments. Finally, $\mathbf{F}^{\mathcal{V}}$ is strongly $\mathcal{A}$-monotonic by Theorem 1 of §8. QED

In Part I of this paper ([1]) we explained (absolute) ADTs as collections K of structures $\mathcal{A}$ of a given signature Σ, closed under ismomorphism; K is called *strict* if it is an isomorphism type, otherwise *loose*. In the case of a strict ADT K, an abstract computational procedure on K is identified with a function(al) $\mathbf{F}$ generated by the schemata I-VIII. This gives an actual computational procedure $\mathbf{F}^{\mathcal{A}}$ in whatever way a representative $\mathcal{A}$ of K is chosen in the recursion-theoretic interpretation, i.e., where the basic functionals of $\mathcal{A}$ are all partial recursive (on V). Any such $\mathcal{A}$ constitutes an *implementation* of K; Theorem 2 of §8 shows that the specific procedure $\mathbf{F}^{\mathcal{A}}$ is independent of the choice of implementation, in the sense that it is invariant under isomorphism. It is in this sense that we can speak of computation on ADTs as ordinary computation.

In the case of loose ADTs K we may think of abstract computational procedures $\mathbf{F}$ as providing an $\cong$-invariant *uniform* means for passing from any implementation $\mathcal{A}$ of K to a computational procedure on $\mathcal{A}$.

The picture in both cases is suitably modified for relative ADTs.

12 Examples

There is no space here to illustrate in detail how the above theory may be applied in practice. It suffices to say that all the standard finitary examples of ADTs, such as lists, trees, sets, etc., as described in Part I of this paper all have implementations whose basic F_k are simply partial recursive functions (cf. especially Part I, §§7.2–7.5). An interesting example of a loose finitary ADT whose basic structure makes use of functionals is given in Moschovakis [5], namely where the structures have a finite domain A (say A_1) with the existential quantifier $\exists_A$, given by

$$\text{(1)} \qquad \exists_A(\varphi) \simeq \begin{cases} 0 \text{ if } (\exists x \in A)\varphi(x) = 0 \wedge \varphi\colon A \to \{0,1\} \\ 1 \text{ if } (\forall x \in A)\varphi(x) = 1\,. \end{cases}$$

For any specific choice of finite A, $\exists_A$ is partial recursive, with

$$\text{(2)} \qquad \exists_A(\varphi) \simeq \prod_{x \in A} \varphi(x)\,.$$

Uniform computational procedures over such ADTs are given in terms of a basic functional operator $\exists$. Many kinds of data type queries in practice may be considered as examples of such abstract procedures.

There remain the cases of infinite streams and infinite precision reals which were objects of special attention in Part I. Here, it seems, only the intensional recursion-theoretic interpretation is appropriate. For example, in the case of infinite streams over a given $(A, =_A)$,

$$\text{(3)} \qquad \text{Stream}(A) = (S, A, =_S, =_A, cons, first, rest, sim)$$

where $cons : A \times S \to S$, $first : S \to A$, $rest : S \to S$ and $sim : (\mathbb{N} \to A) \to S$, and we have the effective implementation described in [1]§7.6, given by

$$\text{(4)} \qquad \begin{array}{ll} \text{(i)} & S = (\mathbb{N} \to A) \\ \text{(ii)} & cons(a, z) = \lambda n[\text{if } n = 0 \text{ then } a, \text{ else } z(n-1)] \\ \text{(iii)} & first(z) = z(0) \\ \text{(iv)} & rest(z) = \lambda n.z(n+1) \\ \text{(v)} & sim(z) = z\,. \end{array}$$

Actually, to spell out this construction, we need to consider a structure combining Stream(A), $\mathbb{B}$, and $\mathcal{N} = (\mathbb{N}, sc_{\mathbb{N}}, pd_{\mathbb{N}}, eq_{\mathbb{N}}, 0)$. Note that in (4)(v) we are considering z in two guises, first as an element of $\mathbb{N} \to A$ and then as an element of S. it is because of the first that sim acts as a functional.

Now sim is extensional and monotonic, because it is only defined on total $z{:}\,\mathbb{N} \to A$. But because of the latter, it is not a p.r. functional as it stands, and the theory of §11 does not apply to it. However, we have an immediate extension of sim to the p.r. operation $sim^* = \lambda z.z$ on V. Thus any abstract computational procedure **F** for structures of signature appropriate to streams (with $\mathcal{N}$ included) can be construed as a computational procedure in the usual sense by taking sim^* for sim in this implementation.

The story is similar for the ADT of infinite precision reals, which requires a functional $lim(x, z)$ for passing from Cauchy sequences $x = \langle x_n \rangle (x{:}\,\mathbb{N} \to \mathbb{Q})$ of rationals with modulus of convergence $z(z{:}\,\mathbb{N} \to \mathbb{N})$ to reals $(\mathbb{R})$. The implementation for Bishop constructive reals is described in [1], §7.7. Again, lim as defined there (7.7(3)) only applies to total objects, but it has an immediate p.r. extension $lim^*(x, z)$ applying to arbitrary x and z. Abstract computational procedures for infinite precision reals may then be concretized using lim^* for lim in such an implementation.

Appendix—A Uniform Index for LFP Operators in the Intensional Recursion-Theoretic Interpretations

Here we carry out the work for the Remark of 10.4. For simplicity consider finding a LFP $L^V(g)$ for $g{:}\, V_\sigma \to V_\sigma$ where $V_\sigma = (V \underset{\smile}{\rightarrow} V)$, in the form lg for a suitable index l. (Note that g plays the role of $\widehat{G}$ in the LFP operators.) Let e_0 be an index of the empty function.

LEMMA. *There exists l with the following properties:*
(i) $lg \downarrow$ for all g, and if we set $g_0 = e_0$ and $g_{n+1} \simeq gg_n$ for all $n < \omega$, then
(ii) $(lg\,x) \downarrow \Leftrightarrow \exists n(g_n x \downarrow)$ and
(iii) $lg\,x \simeq y \Rightarrow \exists n(g_n x \simeq y)$.

Proof. Note that g_n is defined iff there exists a sequence number s of length $n+1$ with $(s)_0 = e_0$ and for each $i < n$, $g(s)_i \simeq (s)_{i+1}$, i.e., $\exists y(T_1(g,(s)_i,y) \wedge U(y) = (s)_{i+1})$, so that for each $i \leq n$, $(s)_i = g_i$. Let

$$(1) \qquad R(g,n,x,y) \Leftrightarrow \exists x\{Seq(s) \wedge lh(s) = n+1 \wedge (s)_0 = e_0 \\ \wedge \forall i < n[g(s)_i \simeq (s)_{i+1}] \wedge (s)_n x \simeq y\}$$

so $R(g,n,x,y) \Leftrightarrow g_n$ is defined and $g_n x \simeq y$. Let

$$(2) \qquad S(g,x,y) \Leftrightarrow \exists n R(g,n,x,y);$$

then both R and S are recursively enumerable. Hence we can find r.e. $S^*(g,x,y)$ which uniformizes S, i.e.:

$$(3) \qquad \begin{aligned} &\text{(i)} \quad \exists y S^*(g,x,y) \Leftrightarrow \exists y S(g,x,y), \\ &\text{(ii)} \quad S^*(g,x,y) \Rightarrow S(g,x,y), \text{ and} \\ &\text{(iii)} \quad S^*(g,x,y_1) \wedge S^*(g,x,y_2) \Rightarrow y_1 = y_2 . \end{aligned}$$

In other words, S^* is the graph of a partial recursive function $l^*(g,x)$ with domain $\{(f,x) \mid \exists y S(g,x,y)\} = \{(f,x) \mid \exists n g_n x \downarrow\}$. Moreover $l^*(g,x) \simeq y \Rightarrow \exists n(g_n(x) \simeq y)$. Since S^* is r.e., it follows that l^* is partial recursive. Now define $l = \lambda g \lambda x. l^*(g,x)$. Then l satisfies the conditions of the theorem.

COROLLARY. *If $g: V_\sigma \to V_\sigma$ where $V_\sigma = (V \rightharpoonup V)$ and g is monotonic on V_σ then*
(i) each g_n is defined, $g_n: V \rightharpoonup V$,
(ii) $g_n \subseteq g_{n+1}$ each n, and
(iii) $lg\,x \simeq y \Leftrightarrow \exists n(g_n\,x \simeq y)$.

Proof. Since g is total on V_σ and monotonic we have (i) and (ii) by induction on n. For (iii), suppose $g_n x \simeq y$. Then $g_n x \downarrow$, so $lg\,x \downarrow$ by (ii) above. Let $lg\,x \simeq z$. Then by (iii) above, there exists m with $g_m x \simeq z$. Take $k = \max(n,m)$. Since $g_n \subseteq g_k$ and $g_m \subseteq g_k$ we have $g_k x \simeq y$ and $g_k x \simeq z$ so $y = z$, i.e., $lg\,x \simeq y$. This proves (iii) here.

Remarks. (i) By the Myhill-Shepherdson Theorem it is sufficient for this Corollary to assume that g is total and extensional on V_σ. (ii) What the above proofs do is convert the construction of the sequence $\varphi_G^{(n)}$ for any G given by an index g (for $\widehat{G}$) into effective generation of a sequence of indices g_n for $\varphi_G^{(n)}$. then lg represents $\varphi_G^{(\omega)}$, i.e., $\bigcup_n \varphi_G^{(n)}$, when G is monotonic. The uniformization argument serves to define lg for arbitrary g.

References

[1] S. Feferman. A new approach to abstract data types, I. Informal development. *Mathematical Structures in Computer Science* (to appear).

[2] S. Feferman. Inductive schemata and recursively continuous functionals. *Logic Colloquium '76,* North-Holland, Amsterdam (1977), 373–392.

[3] S.C. Kleene. Recursive functionals and quantifiers of finite types I. *Trans. Amer. Math. Soc.* **91** (1959), 1–52.

[4] S.C. Kleene. Recursive functionals and quantifiers of finite types revisited I. *Generalized Recursion Theory II,* North-Holland, Amsterdam (1978), 185–222.

[5] Y.N. Moschovakis. Abstract recursion as a foundation of the theory of recursive algorithms. *Computation and Proof Theory,* Lecture Notes in Mathematics 1104 (1984), 289–364.

[6] Y.N. Moschovakis. The formal language of recursion, *J. Symbolic Logic* 54 (1989), 1216–1252.

[7] J. Myhill and J. Shepherdson. Effective operations on partial recursive functions. *Zeitschr. Math. Logik u. Grundlag. Math.* 1 (1955), 310–317.

[8] R.A. Platek. *Foundations of Recursion Theory.* Ph.D. Thesis, Stanford University, 1966.

A primitive recursive set theory and AFA : on the logical complexity of the largest bisimulation

Tim Fernando*

fernando@cwi.nl

Abstract. A subsystem of Kripke-Platek set theory proof-theoretically equivalent to primitive recursive arithmetic is isolated; Aczel's (relative) consistency argument for the Anti-Foundation Axiom is adapted to a (related) weak setting; and the logical complexity of the largest bisimulation is investigated.

1 Introduction

As every programmer understands, computing depends on coding: only what can be coded can be computed. Conversely, constraints on coding determine the sense in, and degree to which what can be coded can be computed. To a logician, it is natural to express these constraints in a formal system, and relate computation to proofs in that system.[2] Strong systems such as Zermelo-Fraenkel set theory with Choice, ZFC, allow all kinds of mathematical concepts to be coded as sets, however natural or unnatural these formulations might appear. Over a weaker set theory, coding can have greater computational significance, and even a somewhat odd and unlikely use of the notion of a set can sometimes prove fruitful. An instructive example is the set-theoretic coding of (countable) linguistic notions in Kripke-Platek set theory, KP (see Barwise [5]).[3]

More recently, the theory of non-well-founded sets presented in Aczel [4] has been applied in Barwise and Etchemendy [6] for a direct (i.e., natural?) set-theoretic coding of linguistic concepts with a possibly circular character. In this case, however, it is not clear that there is any need for infinite sets, and various alternatives[4] to Aczel's conception of a non-well-founded set have been put forward that suffice for finite sets. By contrast, the consistency proof for the *Anti-Foundation Axiom*, AFA, given in Aczel [4] is carried out relative to the system ZFC^- of ZFC minus foundation that supports a far richer notion of set than that of finite ones. So horrendously rich

* The author is gratefully indebted to Prof. S. Feferman for supervision, to CWI for refugé, and to the Netherlands Organization for the Advancement of Research (project NF 102/62-356, 'Structural and Semantic Parallels in Natural Languages and Programming Languages') for funding.

[2] One way of measuring the computational character of a formal theory is through the *provably recursive functions* given by its Π^0_2-theorems, which is used implicitly below.

[3] In what follows, KP is, as in Barwise [5], *not* assumed to contain the axiom of infinity; otherwise, see Jäger [12].

[4] Mislove, Moss and Oles [16], Abramsky [1], and Rutten [20].

a notion, in fact, that the problem becomes what notion of computation can be associated with these sets. The question of the computational character of AFA is of particular interest given its origins in the (computational) theory of transition systems (see chapter 8 of Aczel [4]). In that work on Milner's SCCS as well as on transition systems given in Plotkin's SOS-style (Rutten [19]), infinite sets are involved. Furthermore, if transition systems are to be related to first-order models (and some such steps are taken in Fernando [10, 11]), then the question of identifying a weak set theory supporting both transition systems and first-order models arises. In any case, *the present author's interest in analyzing* AFA *lies largely in its relation to the notion of a bisimulation — a notion fundamental to semantic attempts at explicating the dynamic nature of information*. For such semantic investigations, it is natural to appeal not only to the ordinary notions of computability and decidability familiar to computer scientists, but also to subtle, set-theoretic notions.[5]

Now, a logical analysis of AFA might proceed in various ways. Lindström [15] formalizes L. Hallnäs' conception of non-well-founded sets in Martin-Löf type theory, building on a constructive version of ZF given in Aczel [3], that, as it turns out, is equivalent to ZF over classical logic. This equivalence blocks a direct understanding in terms of proof-theoretic measures (that at present fall far short of ZF). And from a classical model-theoretic point of view, it would be natural to replace ZFC by a theory, say KP, with many interesting models, and investigate the question mark ? in the diagram

$$\begin{array}{ccc} Con(\mathsf{ZFC}^-) & \Longrightarrow & Con(\mathsf{ZFC}^- + \mathsf{AFA}) \\ \Downarrow & & \Downarrow \\ Con(\mathsf{KP}^-) & \overset{?}{\Longrightarrow} & Con(\mathsf{KP}^- + \mathsf{AFA}) \end{array}$$

where Con is a consistency statement formulated in terms of models. It bears repeating that the theory KP^- in the diagram might be enriched, so long as the models of interest are not ruled out. As will become clear below, *the issue here is not the consistency of* AFA, *but its computational requirements*. And these requirements are most clearly exposed in a theory more (directly) sensitive to constructive principles than ZFC^-.

This is not to say that ZFC^- is devoid of any intuitions about construction. The "limitations of size" principle behind the set-class distinction has been so widely accepted and developed that it is perhaps not terribly appropriate to apply the label "set theory" to a theory supporting the existence of a universal set. And there are sound foundational reasons to look at finer questions of size (through a theory of "counting") given that the object $\omega = \{0, 1, \ldots\}$ is *infinitely* more interesting (and complicated) than $\Omega = \{\Omega\}$. A comparison of these two sets suggests that some care must be exercised in pushing the intuition that a non-well-founded set is a limit of well-founded sets, particularly when it leads to a universal set (as is the case in Abramsky [1]).

[5] Having said this, it should be pointed out that ordinary recursion-theoretic questions about processes have been studied; the reader might consult Ponse [18] and the references cited therein.

The approach taken below is to carry out Aczel's relative consistency argument for AFA in a weak setting connected with a view of mathematics that, although called finitist, can nonetheless support infinite objects. The reader is referred to Feferman [9] for background on proof-theoretic and foundational reductions related to consistency arguments. For orientation, it is useful to note that the system PRA of primitive recursive arithmetic is commonly associated with finitism, and (reminiscent of KP's suitability for countable syntactic notions) is adequate for formulating elementary syntactic notions (involved, for example, in Gödel's incompleteness theorems). Briefly then, the next section describes a subsystem KP_1 of KP proof-theoretically equivalent to PRA (building on the correspondence between hereditarily finite sets and natural numbers, the theory of primitive recursive set functions in Jensen and Karp [14], and the reduction in Parsons [17] of Σ^0_1-IA to PRA). (The point here is that quantifier complexity for set theory is related but not identical to that for number theory.) Section 3 carries out Aczel's construction of a model of AFA in a primitive recursive framework provided by explicit mathematics (Feferman [8]) where a model of KP_1 can be defined. Complications arising from the problem of preserving restricted schemes of comprehension and collection motivate the discussion in section 4 of computational "counting" principles for the largest bisimulation.

2 A primitive recursive subsystem of KP

The analysis below rests on the well-known correspondence between the natural numbers ω and the hereditarily finite sets HF given by $c : HF \to \omega$

$$\begin{aligned} c\emptyset &:= 0 \\ ca &:= \sum_{x \in a} 2^{cx} \end{aligned}$$

and $d : \omega \to HF$

$$\begin{aligned} d0 &:= \emptyset \\ dn &:= \{di \mid i\text{th bit of } n \text{ is } 1\} . \end{aligned}$$

Note that $\in$ (on HF) is a primitive recursive predicate

$$dm \in dn \Leftrightarrow odd([n/2]^m)$$

and accordingly is defined by

$$t_\in[m, n] = 0$$

for some primitive recursive term $t_\in(x, y)$ in the language $\mathcal{L}(\mathsf{PRA})$ of PRA. Now, we can describe an interpretation $-^*$ of $\mathcal{L}(\in)$ in $\mathcal{L}(\mathsf{PRA})$ by passing syntactically from $x \in y$ to $t_\in(x, y) = 0$, and semantically from an $\mathcal{L}(\mathsf{PRA})$-structure $\mathcal{M} = \langle M, \ldots \rangle$ to an $\mathcal{L}(\in)$-structure $\mathcal{M}^* = \langle M, E \rangle$ where

$$E := \{(m, n) \in M \times M \mid \mathcal{M} \models t_\in[m, n] = 0\} .$$

Observe that by the elementary closure properties of primitive recursive predicates, every Δ_0-formula $\varphi(\overline{x})$ in $\mathcal{L}(\in)$ $-^*$-translates to the form (provably equivalent in PRA) of an equation

$$t_\varphi(\overline{x}) = 0 .$$

Going the other direction, we have an interpretation $-^\circ$ of $\mathcal{L}(\mathsf{PRA})$ in $\mathcal{L}(\in)$ by the usual identification of natural numbers with finite ordinals. Note that the predicate $\omega(x)$ in $\mathcal{L}(\in)$ is Δ_0. Furthermore, the (numerical) primitive recursive functions can be extracted as restrictions to ω of primitive recursive set functions, to which we now turn.

The *primitive recursive set functions* are given in Jensen and Karp [14] as follows. Close the initial functions

$$\begin{aligned} P_{n,i}(\overline{x}) &= x_i \\ S^2(x_0, x_1) &= x_0 \cup \{x_1\} \\ C(x_0, x_1, x_2, x_3) &= \begin{cases} x_0 \text{ if } x_2 \in x_3 \\ x_1 \text{ otherwise} \end{cases} \end{aligned}$$

under substitution

$$F(\overline{x}) = G(H_1(\overline{x}), \ldots, H_k(\overline{x}))$$

and recursion

$$F(x, \overline{w}) = G(\bigcup_{u \in x} F(u, \overline{w}), x, \overline{w}) .$$

The *primitive recursive formulas* are the defining formulas for the set functions above. For example, the defining formula $\Phi(x, \overline{w}, y; \varphi(z, x, \overline{w}, y))$ for a function derived by recursion from a function G with defining formula $\varphi(z, x, \overline{w}, y)$ is

> there is a function h such that $h(x) = y$ and for all u in the domain of h, $u \subseteq$ domain h and
> $$\varphi(\bigcup_{v \in u} hv, u, \overline{w}, hu) .$$

Let PRS be the (classical) first-order theory in the language of set theory consisting of the axioms of extensionality, pairing, union, Δ_0-separation, induction on primitive recursive formulas φ

$$\forall z(\forall v \in z \varphi(v) \supset \varphi(z)) \supset \forall z \varphi(z)$$

and the Σ_1-*recursion rule*

$$\frac{\forall z, x, \overline{w}\ \exists! y\ \varphi(z, x, \overline{w}, y)}{\forall x, \overline{w}\ \exists! y\ \Phi(x, \overline{w}, y; \varphi(z, x, \overline{w}, y))}$$

where $\Phi(x, \overline{w}, y; \varphi(z, x, \overline{w}, y))$ is the defining formula for the function derived by recursion from a Σ_1-formula $\varphi(z, x, \overline{w}, y)$. Transitive models of PRS are *prim-closed* in the sense of Jensen and Karp [14].

Under suitable arithmetization, the collections of proofs in PRA and PRS are primitive recursive. Furthermore, a primitive recursive function can be constructed mapping (provably in PRA) axioms φ of PRS to PRA-proofs of φ^*. Consequently,

Proposition 1. (PRA) PRS $\vdash \varphi$ *implies* PRA $\vdash \varphi^*$.

The converse of Proposition 1 fails because the sets that PRA $-^*$-induces are "finite." (For a counter-example, take the $\mathcal{L}(\in)$-sentence that asserts that every non-empty a set has an $\in$-maximal element

$$\exists x \in a\ x \in a \supset \exists z \in a\ \forall x \in a\ z \notin x\ ;$$

its $-^*$-translation is a theorem of PRA.) We can, however, approximate a converse. As e° is a primitive recursive formula for every $\mathcal{L}$(PRA)-equation e, another inductive argument on the the length of a proof yields

Proposition 2. (PRS) PRA $\vdash \psi$ *implies* PRS $\vdash \psi^\circ$.

Furthermore, every model $\mathcal{M}$ of PRA can be embedded in a model of PRS, namely $\mathcal{M}^*$ via $\pi : \mathcal{M} \cong \mathcal{M}^{*\circ}$

$$\pi 0 := 0$$
$$\pi(n+1) := \sum_{m \leq n} 2^{\pi m}\ ,$$

whence

Proposition 3. PRS *is a conservative extension of* PRA.

Mention of primitive recursive formulas can be avoided altogether by asserting the principle of induction for all Σ_1-formulas. Set PRS$'$ to PRS with primitive recursive induction promoted to Σ_1-induction (Σ_1IA). Now, the $-^*$-translated content of the Σ_1-recursion rule does not change since Parsons [17] proved (in PRA) that if

$$\Sigma_1^0\text{-IA} \vdash \forall n \exists m R(n,m)$$

where R is primitive recursive, then

$$\text{PRA} \vdash R(n, fn)$$

for some primitive recursive function f. The arguments for PRS and PRA adapt readily to yield

Proposition 4. *1.* (PRA) PRS$'$ $\vdash \varphi$ *implies* Σ_1^0-IA $\vdash \varphi^*$.
2. (PRS) Σ_1^0-IA $\vdash \psi$ *implies* PRS$'$ $\vdash \psi^\circ$.
3. PRS$'$ *is a conservative extension of* Σ_1^0-IA.

As with Proposition 1, the converse to part 1 of Proposition 4 fails, which leads us to formulate

Lemma 5. *Let Φ be a primitive recursive collection of $\mathcal{L}(\in)$-formulas for which there is, provably in Σ_1^0-IA, a primitive recursive function f such that for every $\varphi \in \Phi$, $f\varphi$ is a Σ_1^0-IA-proof of φ^*. Then* PRS′ $+ \Phi$ *is proof-theoretically equivalent to* PRA.

Sieg [21] contains a wealth of information concerning Σ_1^0-IA, including "easy and helpful facts" (his words) such as

(a) Π_1^0-IA is equivalent to Σ_1^0-IA (p. 46), and
(b) Σ_1^0-collection[6] is contained in Σ_1^0-IA (p. 53).

Concerning point (b), it is interesting to note that PRS′ is a subsystem of the predicative set theories in Feferman [7], and hence does not imply Δ_0-collection:

$$\forall x \in a \exists y\ \varphi(x,y) \supset \exists z \forall x \in a \exists y \in z\ \varphi(x,y)$$

for Δ_0-formulas $\varphi(x,y)$. (It has the same transitive models as PRS, including sets that are *not* admissible.) Nevertheless, Φ in Lemma 5 can be taken to be Δ_0-collection, by adapting Sieg's argument for (b).[7] It is well-known that in the presence of Δ_0-collection, the distinction between Σ_1- and Σ-formulas (also called generalized or essentially Σ_1-formulas) evaporates. As for point (a), this allows us to conclude that, defining the subsystem KP_1 of KP as $\mathsf{KP}^- + (\Sigma_1 + \Pi_1)$IA (where KP^- is KP minus foundation)[8],

Theorem 6. KP_1 *is proof-theoretically equivalent to* PRA.

3 Aczel's AFA construction in a weak setting

To shed light on the infinitary demands of AFA, it is natural (as argued in section 1) to carry out Aczel [4]'s (relative) consistency argument for the axiom in a weaker

[6] These are arithmetic principles

$$\forall x < a\ \exists y\ \varphi(x,y) \supset \exists z \forall x < a\ \exists y < z\ \varphi(x,y)\ ,$$

where $\varphi(x,y)$ is Σ_1^0.

[7] Assume (in Σ_1^0-IA) that

$$(\forall x \in a \exists y\ \varphi(x,y))^*$$

where φ is Δ_0. Now, calling the formula

$$b \leq a \supset \exists z \forall x < b \exists y < z\ t_\in(x,b) = 0\ \supset\ (t_\in(y,z) = 0 \wedge \varphi^*(x,y))$$

$\psi(b)$, then as $\psi(0)$ and $\psi(b) \supset \psi(b+1)$, it follows by Σ_1^0-IA that (because $t_\in[m,n] = 0$ implies $m < n$)

$$(\exists z \forall x \in a \exists y \in z\ \varphi(x,y))^*\ .$$

[8] The Σ_1-recursion rule is a consequence (relative to KP^-) of $(\Sigma_1 + \Pi_1)$IA. If the existence of the transitive closure of a set is added to KP^- (as in work by Jäger), then Π_1IA is not necessary to justify the rule, although Π_1IA is useful for purposes other than proving the existence of transitive closures (see Barwise [5]), an example of which is given in section 4 below. The author does not see how to derive Π_1IA from Σ_1IA (in particular, how to adapt the argument in Sieg [21] reducing Π_1^0-IA to Σ_1^0-IA).

setting than ZFC$^-$. Accordingly, over a model $\langle S, \approx, \in\rangle$ of KP$^-$ (i.e., KP minus foundation), define the following.

- A *graph* G is a pair $(N_G, \rightarrow_G)$ with $\rightarrow_G \subseteq N_G \times N_G$.
- A *decoration of a graph* G is a function d on N_G such that $da \approx \{db \mid a \rightarrow_G b\}$.
- The *Anti-Foundation Axiom*, AFA, is the assertion that every graph has a unique decoration.
- A *pointed graph (pg)* is a pair (G, a) consisting of a graph G and a set $a \in N_G$.[9]
- A *bisimulation between graphs G and G'* is a set R such that whenever bRb',

$$\forall x \leftarrow_G b\ \exists y \leftarrow_{G'} b'\ xRy \wedge \forall y \leftarrow_{G'} b'\ \exists x \leftarrow_G b\ xRy\ .$$

- Let

$$Bis(R, G, a, G', a') \Leftrightarrow \text{“}R \text{ is a bisimulation between } G \text{ and } G' \text{ such that } aRa'\text{”}\ ,$$

  and let S_0 be the collection of all pg's, and $\approx_0$ be the subcollection of $S_0 \times S_0$ given by

$$(G, a) \approx_0 (G', a') \Leftrightarrow \exists R\ Bis(R, G, a, G', a')\ .$$

The preceding definitions all refer to sets (i.e., objects in S), except for the collections Bis, S_0 and $\approx_0$. These collections will serve as useful abbreviations, but where do they live? Rather than working in a framework where “limitations of size” lead, for example, to complications with quotients[10], it is possible instead to work in the framework of explicit mathematics (Feferman [8]), where

(1) theories of weak proof-theoretic strength can be formulated naturally, and
(2) the problem of quotients can be sidestepped by adopting Bishop's use of “equality” relations.

Concerning point (1), observe that a model of KP$_1$ can be defined (by numerically coding the hereditarily finite sets) in the theory APP + ECA + Obj-ind$_N$ described in Jäger [13] (where it is stated, furthermore, to be proof-theoretically equivalent to PRA). As for point (2), this was anticipated above in isolating the interpretation $\approx$ of equality on S. To go along with $\approx_0$, define the subclass $\in_0$ of $S_0 \times S_0$ as follows

$$(G, a) \in_0 (G', a') \equiv \exists b \leftarrow_{G'} a'\ \ (G', b) \approx_0 (G, a)\ .$$

Theorem 7. [11] *If S, $\approx$ and $\in$ are* (APP + ECA)*-classes such that*

$$\langle S, \approx, \in\rangle \models \mathsf{KP}^-$$

[9] The notion of an *accessible pointed graph* is avoided at some aesthetic cost to push through the argument in KP$^-$.

[10] Lemma 2.17 of Aczel [4] employs Scott's trick plus some choice principle to form a *strongly extensional quotient.*

[11] The author suspects (but has not had the energy to check all the necessary details) that the proof below can be formalized in APP+ECA, plus, if necessary, Obj-ind$_N$. (Hence, the use of explicit mathematics.) But as it stands, the reduction claimed is model-theoretic, not proof-theoretic.

then (APP + ECA)*-classes* S_0, $\approx_0$ *and* $\in_0$ *can be defined (as above) such that*

$$\langle S_0, \approx_0, \in_0 \rangle \models \text{Extensionality} + \text{Pair} + \text{Union} + \text{AFA} \ .$$

Furthermore, the passage from $\langle S, \approx, \in \rangle$ *to* $\langle S_0, \approx_0, \in_0 \rangle$ *preserves satisfaction of (full)* Separation, *(full)* Collection, Infinity, Power, *and* Choice.

Proof. First, observe that ECA supports the (class) definitions of $S_0, \approx_0$, and $\in_0$ from $S, \approx$, and $\in$ above since the only terms that occur *qua* class in the defining formulas are $S, \approx$, and $\in$. Second, to see that $\approx_0$ is an equivalence relation is routine (assuming KP$^-$): clearly, $\approx_0$ is reflexive (since for every pg (G, a), the restriction of $\approx$ to N_G is a bisimulation on G), symmetric (since if R is a bisimulation between G and G', then R^{-1} is a bisimulation between G' and G), and transitive (since if R is a bisimulation between G and G', and R' is a bisimulation between G' and G'', then $R \circ R'$ is a bisimulation between G and G''). Third, although a quotient need not be formed, it is necessary to prove that $\in_0$ respects $\approx_0$. So suppose $(G, a), (G', a'), (G_1, a_1)$ and $(G'_1, a'_1) \in S_0$ satisfy

$$(G_1, a_1) \approx_0 (G, a) \in_0 (G', a') \approx_0 (G'_1, a'_1) \ ,$$

with the object of showing

$$(G_1, a_1) \in_0 (G'_1, a'_1) \ .$$

Since $(G', a') \approx_0 (G'_1, a'_1)$, there is a bisimulation R between G' and G'_1 relating a' to a'_1, and since $(G, a) \in_0 (G', a')$, there is a $b \leftarrow_{G'} a'$ with $(G', b) \approx_0 (G, a)$. Choose a $b_1 \in N_{G'_1}$ such that bRb_1 and $a'_1 \rightarrow_{G'_1} b_1$. Then $(G', b) \approx_0 (G'_1, b_1)$ (via R), whence $(G'_1, b_1) \approx_0 (G_1, a_1)$ (as required) since $(G', b) \approx_0 (G, a) \approx_0 (G_1, a_1)$.

Next, for the axioms, it is helpful to define the (class) map $[-]^0$ from S_0 to S given by

$$[(G, a)]^0 = \{(G, a') \mid a \rightarrow_G a'\} \ ,$$

and associate with every formula φ in the language $\{=, \dot{\in}\}$ of set theory the predicate $[\varphi]_0$ obtained by interpreting the quantifiers over S_0, the equality symbol $=$ by $\approx_0$, and the membership symbol $\dot{\in}$ by $\in_0$. A crucial property of the interpretation $\varphi \mapsto [\varphi]_0$ is that for every formula φ, and quantifier $Q \in \{\forall, \exists\}$,

$$[Qx \dot{\in} y \ \varphi]_0 \equiv Qx \in [y]^0 \ [\varphi]_0 \ .$$

This is a consequence of three facts: (1) $\approx_0$ is an equality for $\langle S_0, \approx_0, \in_0 \rangle$, (2) for every pg (G, a), and every $x \in_0 (G, a)$, there is a $y \in [(G, a)]^0$ where $x \approx_0 y$, and (3) for every pg (G, a), every $x \in [(G, a)]^0$ is $\in_0 (G, a)$.

Now, to establish the analog to Corollary 3.3 and Proposition 3.7 in Aczel [4] (implying the system V_c constructed there is *full*), define the predicate

$$Copy(G, a, A) \Leftrightarrow \forall (G', a') \in S_0 \quad (G', a') \in_0 (G, a) \ \equiv \\ \exists (G'_1, a'_1) \in A \ \ (G'_1, a'_1) \approx_0 (G', a')$$

and assert

Lemma 8. (KP$^-$) *For every set (i.e., member of the class S) $A \subseteq S_0$, there is a $(G,a) \in S_0$ unique up to $\approx_0$ such that $Copy(G,a,A)$.*
Proof. Given such an A, pick an $a \notin \bigcup\{N_{G'} \mid (G',a') \in A\}$ (justified by an argument by contradiction using Δ_0-separation and the Russell set) and define (over $\langle S, \approx, \in \rangle$)

$$N_G := \{a\} \cup \bigcup\{N_{G'} \mid (G',a') \in A\}$$
$$\to_G := \{(a,a') \mid (G',a') \in A\} \cup \bigcup\{\to_{G'} \mid (G',a') \in A\} .$$

Then $Copy(G,a,A)$ holds. Moreover, to prove uniqueness up to $\approx_0$, suppose $Copy(G_1,a_1,A)$ were true. Constructing a bisimulation between G and G_1 relating a to a_1 appears simple enough — take

$$\{(a,a_1)\} \cup \{(b,b_1) \in N_G \times N_{G_1} \mid \exists (G',a') \in A \;\; \psi(G,b,G',a') \wedge \psi(G_1,b_1,G',a')\}$$

where $\psi(X,y,U,v)$ is $\exists R \; Bis(R,X,y,U,v)$. But the existential quantifier in ψ must be either bounded using Power, or else the definition cannot be justified by KP's limited separation principles. Fortunately, a more delicate argument is possible. Given an R satisfying

$$\forall b \leftarrow_G a \; \exists (G',a') \in A \;\; Bis(R,G,b,G',a') \wedge$$
$$\forall (G',a') \in A \; \exists b \leftarrow_G a \;\; Bis(R,G,b,G',a')$$

and an R_1 satisfying

$$\forall (G',a') \in A \; \exists b_1 \leftarrow_{G_1} a_1 \;\; Bis(R_1,G',a',G_1,b_1) \wedge$$
$$\forall b_1 \leftarrow_{G_1} a_1 \; \exists (G',a') \in A \;\; Bis(R_1,G',a',G_1,b_1) ,$$

compose R with R_1 and throw in (a,a_1) to form a bisimulation between G and G_1 relating a to a_1. It remains to show how to construct the required R and R_1. Observe that R can be obtained by applying Σ-collection (as given in Barwise [5], Theorem 4.4, p. 17) first to

$$\forall b \leftarrow_G a \; \exists R' \; \exists (G',a') \in A \;\; Bis(R',G,b,G',a') ,$$

then to

$$\forall (G',a') \in A \; \exists R' \; \exists b \leftarrow_G a \;\; Bis(R',G,b,G',a')$$

(which hold since $Copy(G,a,A)$), and forming the union. Constructing R_1 is similar. ⊣

Preservation of the axioms is proved à la Rieger's theorem (Aczel [4]) by applying Lemma 8 to a suitable A for the existence (or in the case of Extensionality, the uniqueness up to $\approx_0$) of a required $(G,a) \in S_0$.

Extensionality: given (G_0,a_0) and (G_1,a_1) in S_0 such that

$$\forall (G',a') \in S_0 \;\; (G',a') \in_0 (G_0,a_0) \equiv (G',a') \in_0 (G_1,a_1) ,$$

then $Copy(G_0, a_0, [(G,a)]^0)$ and $Copy(G_1, a_1, [(G,a)]^0)$, whence Lemma 8 yields $(G_0, a_0) \approx_0 (G_1, a_1)$.

Pair: given $(G,a), (G',a') \in S_0$, appeal to Lemma 8 with $A = \{(G,a), (G',a')\}$.

Union: given $(G,a) \in S_0$, appeal to Lemma 8 with $A = \bigcup\{[(G,a')]^0 \mid a \rightarrow_G a'\}$.

Separation: given a formula $\varphi(x)$ and $(G,a) \in S_0$, let A be the set

$$\{(G,a') \in [(G,a)]^0 \mid [\varphi((G,a'))]_0\}$$

(which exists, assuming $\langle S, \approx, \in\rangle$ satisfies Separation).

Collection: let $\varphi(x,y)$ be a formula and $(G,a) \in S_0$ such that

$$\forall x \in_0 (G,a)\ \exists y \in S_0\ [\varphi(x,y)]_0\ .$$

Assuming $\langle S, \approx, \in\rangle$ satisfies Collection, there is a set B such that

$$\forall x \in [(G,a)]^0\ \exists y \in B\ [\varphi(x,y)]_0\ .$$

Appeal to Lemma 8 with $A = B \cap S_0$.

AFA: Let g be a pg such that ["g is a graph"]$_0$. A first attempt would be to apply Lemma 8 to $\{B \in S_0 \mid [\psi]_0\}$ where ψ is

$$\exists x \dot{\in} N_g\ \ B = (x, \{z \mid x \rightarrow_g z\})\ .$$

Unfortunately, this class is *not* a set, the problem being that given a pg (G,a), a proper class of pg's (G',a') may satisfy $(G',a') \in_0 (G,a)$. (Similarly, with $\approx_0$.) This suggests a different interpretation of the language of set theory, obtained by interpreting the quantifiers over S_0, $=$ by $\approx$, and $\dot{\in}$ by $\in [-]^0$. Write $[\varphi]^0$ for the result of applying this interpretation to some formula φ.[12] Now, if d is the result of applying Lemma 8 to $\{B \in S_0 \mid [\psi]^0\}$, then it follows that for every $x \in_0 N_g$,

$$\forall y \in_0 d(x)\ [x \rightarrow_g y]_0 \wedge \forall z\ [x \rightarrow_g z]_0 \supset z \in_0 d(x),$$

whence (by Extensionality),

$$\begin{aligned} x &\approx_0 d(x) \\ &\approx_0 \{z \mid x \rightarrow_g z\} \\ &\approx_0 \{d(z) \mid x \rightarrow_g z\}\ , \end{aligned}$$

---

[12] Ideally,

$$[\varphi(\overline{x})]_0 \equiv \exists \overline{y} \approx_0 \overline{x}\ [\varphi(\overline{y})]^0$$

would always hold; however, counter-examples such as $\neg\ x \dot{\in} x'$ are easy to find. Counter-examples not involving negation are more difficult to produce, and the author suspects that the above equivalence might hold for (Σ-formulas) φ constructed "non-negatively" (allowing for bounded quantification, which might be justified by Σ-collection). If so, then separation and collection over non-negative Δ_0-formulas might be true in $\langle S_0, \approx_0, \in_0\rangle$.

that is, ["d is a decoration of g"]$_0$. Furthermore, if $d' \in S_0$ satisfies ["d' is a decoration of g"]$_0$, then $d \approx_0 d'$, since for every $x \in_0 N_g$, there is a bisimulation between $d(x)$ and $d'(x)$ (whence $d(x) \approx_0 d'(x)$).

Infinity: given an $a \in S$ that makes Infinity true (i.e., "$a = \omega$") in $\langle S, \approx, \in \rangle$, apply Lemma 8 to the set A obtained by applying Σ-collection to $\forall n \in a\ \exists b\ P(n, b)$ where

$$P(n,b) \Leftrightarrow \exists f\ function(f) \wedge dom(f) \approx n \ \wedge\ Q(f,b,n) \ \wedge\ \forall k \in n\ Q(f, f(k), k)$$

and

$$Q(f,x,y) \Leftrightarrow (\forall c \in [x]^0\ \exists z \in y\ f(z) \approx c) \ \wedge\ (\forall z \in y\ f(z) \in [x]^0)\ .$$

Power: given $(G, a) \in S_0$, apply Lemma 8 to a set A such that

$$\forall x \in Pow([(G,a)]^0)\ \exists (G', a') \in A\ \ Copy(G', a', x)$$

(again obtained by Σ-collection, noting that $Copy$ can be put in Σ-form[13]).

Choice: choice functions can be produced in S_0 (assuming such exist in S) as in the proof of Rieger's theorem in Aczel [4].
$\dashv$

The problem with preserving Δ_0-separation and Δ_0-collection is the unbounded existential quantifier in the definition of $\approx_0$ (whence the passage from φ to $[\varphi]_0$ does not preserve Δ-formulas). Since a bisimulation R between G and G' relating a to a' where $a \in N_G$ and $a' \in N_{G'}$ can be taken to be a subset of $N_G \times N_{G'}$, the problem is overcome by assuming Power.[14]

Corollary 9. *If S, $\approx$ and $\in$ are* (APP + ECA)*-classes such that*

$$\langle S, \approx, \in \rangle \models \mathsf{KP}^- + \mathsf{Power}$$

then the classes S_0, $\approx_0$ and $\in_0$ in Theorem 7 form a model of $\mathsf{KP}^- + \mathsf{Power} + \mathsf{AFA}$.

As already mentioned, the theory

$$\mathsf{T} := \mathsf{APP} + \mathsf{ECA} + \mathsf{Obj\text{-}ind}_N$$

of explicit mathematics is proof-theoretically equivalent to PRA, according to Jäger [13]. Within T, classes $S, \approx, \in$ can be defined (as in the previous section) such that

$$\langle S, \approx, \in \rangle \models \mathsf{KP}_1 + \mathsf{Power}$$

---

[13] That is, $Copy(G, a, x)$ can be re-expressed as

$$\forall (G', a') \in [(G,a)]^0\ \exists (G_1', a_1') \in A\ \ (G', a') \approx_0 (G_1', a_1') \quad \wedge$$
$$\forall (G', a') \in A\ (G', a') \in_0 (G, a)\ .$$

[14] As Prof. Barwise has pointed out to the author, the notation KP$^-$ + Power might be interpreted as requiring that the powerset predicate is taken to be Δ_0. This is not necessary, because all possible bisimulations referred to in $[-]_0$-translating a Δ_0-formula can be found in a set constructed from (the interpretations of) its free variables (since all quantifiers are bounded).

(where, by Lemma 5, $\mathsf{KP}_1 + \mathsf{Power} \equiv \mathsf{PRA}$). Combined with Corollary 9, this provides an illustration of the finitary character of AFA. From the point of view of admissible sets (and more generally, computational theories based on enumeration), however, Power is an undesirable axiom that is best avoided (or at least weakened to Power or Infinity), and its use for capturing the largest bisimulation suggests a certain impredicativity about AFA somewhat at odds with the claim that the axiom is finitistic. The question arises as to whether the existence of a bisimulation can be expressed by a Δ-predicate over a finitist (i.e., at most primitive recursive) subsystem of KP. A natural attempt at answering this question affirmatively would involve some principle of induction.[15]

4 Induction and the largest bisimulation

In practice, a good deal of the proof-theoretic strength of a set theory lies in its induction principles. Over a universe of possibly non-well-founded sets, however, such principles must be formulated carefully (given the difficulty in applying these globally). The approach taken below is to introduce a unary relation symbol *Ord* plus suitable axioms, and to relativize the induction principles to *Ord*. More precisely, let KP^{Ord} be KP plus

$$\begin{array}{rcl} & Ord(\emptyset) & \\ Ord(x) & \supset & Ord(x \cup \{x\}) \\ \forall y \in x\ Ord(y) & \supset & Ord(\bigcup x) \\ Ord(x) & \supset & \forall y \in x\ (Ord(y)\ \wedge\ \forall z \in y\ z \in x)\ . \end{array}$$

(Note that this is a conservative extension of KP since, under foundation, $Ord(x)$ can be given a Δ_0-definition, as in Barwise [5].) Now, let KP_1^{Ord} be the result of replacing foundation in KP^{Ord} by $(\Sigma_1 + \Pi_1)\mathsf{IA}$ relativized to *Ord*:

$$\forall^{Ord}\alpha(\forall\beta \in \alpha\varphi(\beta) \supset \varphi(\alpha)) \supset \forall^{Ord}\alpha\ \varphi(\alpha)$$

for Σ_1- and Π_1-formulas φ (where $\forall^{Ord}\alpha\ \chi$ is $\forall\alpha\ Ord(\alpha) \supset \chi$). KP_1^{Ord} supports the $(\Sigma_1\textit{-recursion rule})^{Ord}$

$$\frac{\forall z, x, \overline{w}\ \exists! y\ \psi(z, x, \overline{w}, y)}{\forall^{Ord}\alpha\ \forall\overline{w}\ \exists! y\ \Phi(\alpha, \overline{w}, y; \psi(z, \alpha, \overline{w}, y))}$$

where $\Phi(x, \overline{w}, y; \psi(z, x, \overline{w}, y))$ is the defining formula for the function derived by recursion from a Σ_1-formula $\psi(z, x, \overline{w}, y)$ (see section 2). This rule provides a constructive (i.e., ordinal) approach to the least and greatest fixed points of certain inductive definitions.[16]

---

[15] A negative answer for a particular subsystem T might proceed by choosing an appropriate model of T, with respect to which a Δ-definition of the largest bisimulation would lead to a contradiction. The interest in such a result would depend largely on the interest in the model used. The author has tried and failed to push through such an argument for $L(\omega_1^{CK})$.

[16] The reader is referred to Aczel [2] for far more background than is presently required.

Assume $\varphi(R,x)$ is a Δ predicate in which R occurs positively and for which there are Δ-predicates $\varphi_I(\alpha,x)$ and $\varphi_J(\alpha,x)$ satisfying for all α such that $Ord(\alpha)$,

$$\varphi_I(\alpha,x) \equiv \varphi(\exists\beta\in\alpha\ \varphi_I(\beta,-),x)$$

and for $\alpha > 0$,

$$\varphi_J(\alpha,x) \equiv \varphi(\forall\beta\in\alpha\ \varphi_J(\beta,-),x)\ .$$

Implicit here is the assumption that $\exists\beta\in\alpha\ \varphi_I(\beta,-)$ defines a set (i.e., $\{a \mid \exists\beta\in\alpha\ \varphi_I(\beta,a)\}$), as does, for $\alpha>0$, $\forall\beta\in\alpha\ \varphi_J(\beta,-)$. The idea is to characterize the least fixed point I_φ and the greatest fixed point J_φ of $\varphi(R,x)$ by

$$I_\varphi(x) \equiv \exists^{Ord}\alpha\ \varphi_I(\alpha,x) \tag{1}$$
$$J_\varphi(x) \equiv \forall^{Ord}\alpha\ \varphi_J(\alpha,x)\ . \tag{2}$$

Typically, $\Leftarrow$ of (1) and $\Rightarrow$ of (2) are justified by induction principles, while the converses rest on a cardinality argument — i.e., a well-ordering principle that enables induction to be enforced globally (on all sets). But, how are the left hand sides defined in the first place? If we agree that

$$J_\varphi(x) \Leftrightarrow \exists R\ (\forall y\in R\ \varphi(R,y)\ \wedge\ x\in R)\ ,$$

then (2) makes $J_\varphi(x)$ Δ. Note that stipulating

$$I_\varphi(x) \Leftrightarrow \forall R\ (\forall y\ (\varphi(R,y)\supset y\in R)\ \supset\ x\in R)$$

not only fails to lower the complexity of $I_\varphi(x)$ given by (1), but is, in fact, incorrect, since the variable R must range over classes (including proper ones for choices of $\varphi(R,x)$ such as $x=x$). As for the corresponding definition above of $J_\varphi(x)$, the point is that the argument R in $\varphi(R,y)$ must be a set, which in the terminology of Aczel [4] induces a "set continuous" class operator

$$R \mapsto \{a \mid \exists r(\in V)\ r\subseteq R\ \wedge\ \varphi(r,a)\}\ .$$

We now apply these ideas to a concrete case.

Towards an inductive construction of a bisimulation between pointed graphs $(N,\to),a$ and $(N',\to'),a'$, define

$$\varphi(R,a,a';N,\to,N',\to') \Leftrightarrow a\in N\ \wedge\ a'\in N'\ \wedge$$
$$\forall x\leftarrow a\ \exists y\leftarrow' a'\ xRy\ \wedge\ \forall y\leftarrow' a'\ \exists x\leftarrow a\ xRy\ .$$

Lemma 10. *Over* KP_1^{Ord}, *a* Δ *predicate* $\varphi_J(\alpha,a,a';N,\to,N',\to')$ *can be constructed such that*

$$\varphi_J(0,a,a';N,\to,N',\to') \equiv a\in N\ \wedge\ a'\in N'\ ,$$

and for all $\alpha\neq 0$ *such that* $Ord(\alpha)$,

$$\varphi_J(\alpha,a,a';N,\to,N',\to') \equiv \varphi(\forall\beta\in\alpha\ \varphi_J(\beta,-,-';N,\to,N',\to'),a,a';N,\to,N',\to')\ .$$

Consider next what principles are needed to establish the following assertion: for all pointed graphs $(N, \rightarrow), a$ and $(N', \rightarrow'), a'$,

$$\exists R\ Bis(R, (N, \rightarrow), a, (N', \rightarrow'), a') \equiv \forall^{Ord}\alpha\ \varphi_J(\alpha, a, a'; N, \rightarrow, N', \rightarrow')\ , \quad (3)$$

where recall from section 2 that $Bis(R, G, a, G', a')$ says that R is a bisimulation between G and G' relating a to a'. ($\Rightarrow$) follows from Π_1IA, even if relativized to Ord, applied to the slightly modified assertion: for all graphs $(N, \rightarrow)$ and $(N', \rightarrow')$, for every ordinal α, and for all $x \in N$ and $y \in N'$,

$$\exists R\ Bis(R, (N, \rightarrow), x, (N', \rightarrow'), y) \supset \varphi_J(\alpha, x, y; N, \rightarrow, N', \rightarrow')\ .$$

The base case $\alpha = 0$ is trivial, and the induction step is routine. Conversely, ($\Leftarrow$) of (3) is implied by the following assertion ψ: for all graphs $(N, \rightarrow)$ and $(N', \rightarrow')$, there is an ordinal $\hat{\alpha}$ such that

$$\forall x \in N\ \forall y \in N'\ \ \varphi_J(\hat{\alpha}, x, y; N, \rightarrow, N', \rightarrow')\ \supset\ \forall^{Ord}\beta\ \varphi_J(\beta, x, y; N, \rightarrow, N', \rightarrow')\ .$$

The reason is that from ψ, it follows that the set

$$R := \{(x, y) \in N \times N' \mid \varphi_J(\hat{\alpha}, x, y; N, \rightarrow, N', \rightarrow')\}$$

is a bisimulation between $(N, \rightarrow)$ and $(N', \rightarrow')$, and, assuming the right hand side of (3) holds, aRa'. Note that ψ^* (where $-^*$ is the translation from the language of set theory to the language of arithmetic given in section 2) is provable in Σ^0_1-IA, the point being that, under $-^*$, the "closure ordinal" $\hat{\alpha}$ can be computed primitive recursively from $(N, \rightarrow)$ and $(N', \rightarrow')$. By Lemma 5 and the preceding discussion of (3), it follows that

Theorem 11. $\mathsf{KP}_1^{Ord} + \psi$ *is a primitive recursive set theory, relative to which the largest bisimulation is* Δ. *More precisely,* $\mathsf{KP}_1^{Ord} + \psi$ *is proof-theoretically equivalent to* PRA, *and proves (3) for all pointed graphs* $(N, \rightarrow), a$ *and* $(N', \rightarrow'), a'$.

We leave to the interested reader the question of whether $\mathsf{KP}_1^{Ord} + \psi$ is contained in a theory T proof-theoretically equivalent to PRA, lying between KP^- and KP (plus, if necessary, a global well-ordering principle), for which the passage from $S, \approx, \in$ to $S_0, \approx_0, \in_0$ in the previous section sends models of T to models of $T+$AFA. The author suggests taking T to be KP_1^{Ord}, although he has (alas) been unable to determine whether or not this works.

References

1. Samson Abramsky. Topological aspects of non-well-founded sets. Handwritten notes.
2. Peter Aczel. An introduction to inductive definitions. In J. Barwise, editor, *Handbook of Mathematical Logic*. North-Holland, Amsterdam, 1977.
3. Peter Aczel. The type-theoretic interpretation of constructive set theory. In *Logic Colloquium '77*. North-Holland, Amsterdam, 1978.
4. Peter Aczel. *Non-well-founded sets.* CSLI Lecture Notes Number 14, Stanford, 1988.
5. Jon Barwise. *Admissible sets and structures.* Springer-Verlag, Berlin, 1975.

6. Jon Barwise and John Etchemendy. *The liar: an essay on truth and circularity*. Oxford University Press, Oxford, 1987.
7. Solomon Feferman. Predicatively reducible systems of set theory. In *Proc. Symp. Pure Math.*, vol 13, Part II. Amer. Math. Soc., Providence, R.I., 1974.
8. Solomon Feferman. A language and axioms for explicit mathematics. In J.N. Crossley, editor, *Algebra and Logic*, LNM 450. Springer-Verlag, Berlin, 1975.
9. Solomon Feferman. Hilbert's program relativized: proof-theoretical and foundational reductions. *Journal of Symbolic Logic*, 53(2), 1988.
10. Tim Fernando. Transition systems over first-order models. Manuscript, 1991.
11. Tim Fernando. Parallelism, partial execution and programs as relations on states. Manuscript, 1992.
12. Gerhard Jäger. A version of Kripke-Platek set theory which is conservative over Peano arithmetic. *Zeitschr. f. math. Logik und Grundlagen d. Math*, 30, 1984.
13. Gerhard Jäger. Induction in the elementary theory of types and names. Preprint, 1988.
14. R.B. Jensen and C. Karp. Primitive recursive set functions. In *Proc. Symp. Pure Math.*, vol 13, Part I. Amer. Math. Soc., Providence, R.I., 1971.
15. Ingrid Lindström. A construction of non-well-founded sets within Martin-Löf's type theory. *Journal of Symbolic Logic*, 54, 1989.
16. M. Mislove, L. Moss, and F. Oles. Non-well-founded sets obtained from ideal fixed points. In *Fourth annual symposium on Logic in Computer Science*, 1989.
17. Charles Parsons. On a number-theoretic choice schema and its relation to induction. In J. Myhill, editor, *Intuitionism and Proof Theory*. North-Holland, Amsterdam, 1970.
18. Alban Ponse. Computable processes and bisimulation equivalence. Technical Report CS-R9207, Centre for Mathematics and Computer Science, 1992.
19. J.J.M.M. Rutten. Non-well-founded sets and programming language semantics. Technical Report CS-R9063, Centre for Mathematics and Computer Science, 1990.
20. J.J.M.M. Rutten. Hereditarily finite sets and complete metric spaces. Technical Report CS-R9148, Centre for Mathematics and Computer Science, 1991.
21. Wilfried Sieg. Fragments of arithmetic. *Annals of Pure and Applied Logic*, 28, 1985.

This article was processed using the LaTeX macro package with LLNCS style

On Bounded Theories

Jörg Flum
Mathematisches Institut
Universität Freiburg
D-7800 Freiburg
flum@sun1.ruf.uni-freiburg-de

0. Introduction. In several papers the relationship between the computational complexity of a query and its expressibility in first-order logic with a fixed number of bound variables has been investigated (cf. [5], [7]). In [6] Immerman and Kozen study the ***k*-variable property**. Intuitively speaking they say that "a first-order theory Σ satisfies the k-variable property if every first-order formula is equivalent under Σ to a formula with at most k bound variables". (A similar notion had already been studied in [1] and [4] to characterize models of temporal logic having a finite basis for the temporal connectives.) Contrary to intuition their precise notion is not stable. There are theories with the k-variable property for $k = 1$ but not $k = 2$, or for $k = 2$ but not for $k = 3, \ldots$.

We introduce the notion of k-boundedness as formal counterpart of the idea that k bound variables suffice. Intuitively speaking, a theory Σ is ***k*-bounded** if every first-order formula $\varphi(x_1, \ldots, x_n)$ - say with $n \geq k$ - is equivalent under Σ to a formula $\psi(x_1, \ldots, x_n)$ with bound variables among $x_1, \ldots, x_n$ such that of each subformula at most k variables are in the scope of a quantifier of φ.

It turns out that

1. Every k-bounded theory is $(k+1)$-bounded.
2. Every k-bounded theory has the k-variable property.
3. The examples of theories in [6] having the k-variable property are k-bounded.
4. There is an algebraic characterization of k-boundedness in terms of partial isomorphisms (or, equivalently, of a variant of the Ehrenfeucht-Fraisse game).

Moreover the use of ω-saturated structures considerably simplifies the proofs (note that all finite structures are ω-saturated!).

1. Let L be a first-order language with individual variables $x_1, x_2, \ldots$, connectives $\neg, \vee$, and the quantifier $\exists$. Let L_k be the sublanguage of L consisting of all formulas containing only variables among $x_1, \ldots, x_k$. In [6] a first-order theory Σ is said to satisfy the ***k*-variable property** if for all formulas $\varphi \in L$ with free variables among $x_1, \ldots, x_k$, there exists a $\psi \in L_k$ such that $\Sigma \models \varphi \leftrightarrow \psi$. This notion is not stable (but compare 2.h below), e.g. let Σ be the theory of discrete linear orderings without endpoints. Σ has the 1-variable property (hence $var(\Sigma) = 1$ in the sense of [6]) but not the 2-variable property: $\exists x_3(x_1 < x_3 \land x_3 < x_2)$ is not equivalent to a formula in L_2 (clearly one would obtain a stable notion, if one requires that each formula $\varphi(x_1, \ldots, x_n)$ - say for $n \geq k$ - should be equivalent to a conjunction of formulas $\psi(x_{i_1}, \ldots, x_{i_k})$ for $1 \leq i_1 < \ldots < i_k \leq n$, where the (free and bound) variables of $\psi(x_{i_1}, \ldots, x_{i_k})$ are among $x_{i_1}, \ldots, x_{i_k}$; but this notion does not capture the intuitive idea that k quantifiers suffice, because, in general, simple (quantifier free) first-order statements about $x_1, \ldots, x_n$ could not be expressed by such a formula). Moreover, in the proofs given in [6] the authors need the technical restriction that in every model of Σ every finitely generated substructure is finite, an assumption which is not needed in our proofs. Note that, in general, replacing function symbols by relation symbols (for their graphs) increases the number of bound variables needed (compare exercise 3 in [8]): Let $\Sigma(f)$ and $\Sigma(R)$ be the theory of discrete linear orderings without endpoints, where f is the successor function and R is the graph of the successor function respectively. Then $\Sigma(f)$ admits elimination of quantifiers (hence has the k-variable property for each k), while $\Sigma(R)$ does not have the 2-variable property: $\exists x_3(Rx_1x_3 \land Rx_3x_2)$ is not equivalent to a formula in L_2.

Fix natural numbers n and k and set $p := \max\{n, k\}$. A formula φ is said to be an (n, k)-formula, if the free variables of φ are among $x_1, \ldots, x_n$, the bound variables are among $x_1, \ldots, x_p$, and of each subformula at most k variables are in the scope of a quantifier of φ. In particular, then each subformula of φ is in L_p. We give an inductive definition of the notion of (n, k)-formula, which will be useful later on. *The letters M, N,... always denote subsets of $\{1, \ldots, p\}$ of cardinality $\leq k$.*

Definition 1.1. Write $\vdash \varphi, M$, if $\vdash \varphi, M$ can be obtained by finitely many applications of the following rules (a) - (d) (intuitively speaking, $\vdash \varphi, M$ holds if φ is intended as subformula of an (n, k)-formula at a place, where at most the variables x_i with $i \in M$ are in the scope of a quantifier):

(a) If φ is atomic, $\varphi \in L_p$, and M is arbitrary then $\vdash \varphi, M$.

(b) If $\vdash \varphi, M$ then $\vdash \neg\varphi, M$.

(c) If $\vdash \varphi_1, M_1$, $\vdash \varphi_2, M_2$ and $N \subset M_1 \cap M_2$ then $\vdash (\varphi_1 \lor \varphi_2), N$.

(d) If $\vdash \varphi, M$, $i \in M$ and $N \subset M$ then $\vdash \exists x_i\varphi, N$.

φ is an *(n, k)-formula*, if $fr(\varphi) \subset \{x_1, \ldots, x_n\}$ (where $fr(\varphi)$ denotes the set of variables free in φ) and $\vdash \varphi, M$ for some M.

For example, $x < y \land y < z \land \exists x(y < x \land x < z) \land \exists z(x < z \land z < y)$ is a $(3, 1)$-formula (where $x = x_1, y = x_2, z = x_3$).

Definition 1.2. Let Σ be a set of L-sentences.
a) Σ is *(n, k)-bounded*, if for all φ with $fr(\varphi) \subset \{x_1, \ldots, x_n\}$ there exists an (n, k)-formula ψ with the same free variables such that $\Sigma \models \varphi \leftrightarrow \psi$.
b) Σ is *k-bounded*, if Σ is (n, k)-bounded for all n.

As an immediate consequence of the definition, we have:

Lemma 1.3. 1) If Σ is (n,k)-bounded then Σ is $(n,k+1)$-bounded.
2) If Σ is k-bounded then Σ is $(k+1)$-bounded.
3) Σ has the k-variable property iff Σ is (k,k)-bounded. - In particular, k-bounded theories have the k-variable property.

Without loss of generality we assume that the vocabulary (the set of non-logical symbols) of L is finite. By convention, $\star$ will not belong to the universe A of any structure $\mathfrak{A}$. For $\overline{a} \in (A \cup \{\star\})^r, \overline{a} = a_1 \dots a_r$, let $\sup(\overline{a}) := \{i \mid a_i \in A\}$ be the support of $\overline{a}$. For $a \in A$ let $\overline{a}\frac{a}{i}$ denote $a_1 \dots a_{i-1} a a_{i+1} \dots a_r$. Whenever writing $\mathfrak{A} \models \varphi[\overline{a}]$ we presuppose that $fr(\varphi) \subset \{x_i \mid i \in \sup(\overline{a})\}$. By a standard compactness argument one gets:

Lemma 1.4. For a theory Σ the following are equivalent:
(i) Σ is (n,k)-bounded.
(ii) For any (ω-saturated) models $\mathfrak{A}$ and $\mathfrak{B}$ of Σ, $\overline{a} \in (A \cup \{\star\})^n$, $\overline{b} \in (B \cup \{\star\})^n$ with $\sup(\overline{a}) = \sup(\overline{b})$ if

$$\mathfrak{A} \models \varphi[\overline{a}] \quad \text{iff} \quad \mathfrak{B} \models \varphi[\overline{b}]$$

holds for any (n,k)-formula, then this equivalence holds for all first-order formulas, i.e. $(\mathfrak{A}, \overline{a})$ and $(\mathfrak{B}, \overline{b})$ are elementarily equivalent, $(\mathfrak{A}, \overline{a}) \equiv (\mathfrak{B}, \overline{b})$.

We give an algebraic characterization of (n,k)-boundedness in terms of extensions of partial isomorphism. This characterization is helpful for establishing or refuting the (n,k)-boundedness of concrete theories.

Suppose $\overline{a} \in (A \cup \{\star\})^r$, $\overline{b} \in (B \cup \{\star\})^r$. $(\overline{a}, \overline{b})$ is a *partial isomorphism* of $\mathfrak{A}$ to $\mathfrak{B}$, if $\sup(\overline{a}) = \sup(\overline{b})$ and if the map $a_i \mapsto b_i$ for $i \in \sup(\overline{a})$ is well-defined and extends to an isomorphism of the substructures generated by $\overline{a}$ and $\overline{b}$ respectively.

For the rest of the paper fix n, k and let $p := \max\{n,k\}$.

Suppose given structures $\mathfrak{A}$ and $\mathfrak{B}$. Let I be a set of triples $(\overline{a}, \overline{b}, M)$ where $\overline{a} \in (A \cup \{\star\})^p$, $\overline{b} \in (B \cup \{\star\})^p$ and $(\overline{a}, \overline{b})$ is a partial isomorphism of $\mathfrak{A}$ to $\mathfrak{B}$ (and $M \subset \{1, \dots, p\}, \mid M \mid \leq k$). We say that I has the (n,k)-*back and forth* property if the following holds:

(n,k)-*forth*:

If $(\overline{a}, \overline{b}) \in I$, $i \in \{1, \dots, p\}$ and $|M \cup \{i\}| \leq k$ then for all $a \in A$ there is $b \in B$ such that $(\overline{a}\frac{a}{i}, \overline{b}\frac{b}{i}, M \cup \{i\}) \in I$.

(n,k)-*back*:

If $(\overline{a}, \overline{b}) \in I$, $i \in \{1, \dots, p\}$ and $|M \cup \{i\}| \leq k$ then for all $b \in B$ there is $a \in A$ such that $(\overline{a}\frac{a}{i}, \overline{b}\frac{b}{i}, M \cup \{i\}) \in I$.

In case $\sup(\overline{a}) = \sup(\overline{b}) \subset \{1, \dots, n\}$ we write

$$\overline{a} \simeq_{n,k} \overline{b}$$

if there is a set I with the (n,k)-back and forth property and $(\overline{a},\overline{b},\emptyset) \in I$.

Partial isomorphisms in such a set I preserve the validity of (n,k)-formulas. More precisely

Lemma 1.5. Suppose I has the (n,k)-back and forth property. Then for $(\overline{a},\overline{b},M) \in I$ and any formula φ with $\vdash \varphi, M$ we have

$$(+) \qquad \mathfrak{A} \models \varphi[\overline{a}] \text{ iff } \mathfrak{B} \models \varphi[\overline{b}].$$

In particular, if $\overline{a} \simeq_{n,k} \overline{b}$ then $\overline{a}$ and $\overline{b}$ satisfy the same (n,k)-formulas in $\mathfrak{A}$ and $\mathfrak{B}$ respectively.

Proof. We show, by induction on φ, that (+) holds for any φ with $\vdash \varphi, N$ for some N, $M \subset N$. The cases φ atomic and $\varphi = \neg\psi$ are trivial. Let φ be $\varphi_1 \vee \varphi_2$. Since $\vdash \varphi, N$ there are N_i with $\vdash \varphi_i, N_i$ and $N \subset N_i$, in particular $M \subset N_i$, hence (+) holds for $(\varphi_1 \vee \varphi_2)$ by induction hypothesis. Suppose $\varphi = \exists x_i\psi$. Then there is N' with $N \subset N'$, $i \in N'$ and $\vdash \psi, N'$; in particular $M \cup \{i\} \subset N'$. Suppose $\mathfrak{A} \models \exists x_i\psi[\overline{a}]$, say $\mathfrak{A} \models \psi[\overline{a}\frac{a}{i}]$. Then, by the (n,k)-forth property, there is $b \in B$ such that $(\overline{a}\frac{a}{i}, \overline{b}\frac{b}{i}, M \cup \{i\}) \in I$ (note that $|M \cup \{i\}| \leq |N'| \leq k$). Thus, by induction hypothesis, $\mathfrak{B} \models \psi[\overline{b}\frac{b}{i}]$, hence $\mathfrak{B} \models \exists x_i\psi[\overline{b}]$.

For ω-saturated $\mathfrak{A}$ and $\mathfrak{B}$ there is a canonical set I_0 with the (n,k)-back and forth property. In fact, set

$$\begin{aligned} I_0 := \{(\overline{a},\overline{b},M) | \ & M \subset \{1,\ldots,p\}, |M| \leq k, \overline{a} \in (A \cup \{\star\})^p, \overline{b} \in (B \cup \{\star\})^p, \\ & \sup(\overline{a}) = \sup(\overline{b}) \ \text{ and } \ \mathfrak{A} \models \varphi[\overline{a}] \ \text{ iff } \ \mathfrak{B} \models \varphi[\overline{b}] \\ & \text{for all } \ (n,k)\text{-formulas } \varphi \ \text{ with } \ \vdash \varphi, M\}. \end{aligned}$$

Note that if $(\overline{a},\overline{b},M) \in I_0$ for some M then $(\overline{a},\overline{b})$ is a partial isomorphism of $\mathfrak{A}$ to $\mathfrak{B}$. Moreover we have:

Lemma 1.6. I_0 has the (n,k)-back and forth property.

Proof. For example we show the (n,k)-forth property. Suppose $(\overline{a},\overline{b},M) \in I_0$, $1 \leq i \leq p$, $|M \cup \{i\}| \leq k$ and $a \in A$. By ω-saturatedness it suffices to show that for any finitely many formulas $\varphi_1,\ldots,\varphi_r$ with $\vdash \varphi_j, M \cup \{i\}$ and $\mathfrak{A} \models \varphi_j[\overline{a}\frac{a}{i}]$ there is a $b \in B$ such that $\mathfrak{B} \models (\varphi_1 \wedge \ldots \wedge \varphi_r)[\overline{b}\frac{b}{i}]$. Now $\vdash (\varphi_1 \wedge \ldots \wedge \varphi_r), M \cup \{i\}$ and hence $\vdash \exists x_i(\varphi_1 \wedge \ldots \wedge \varphi_r), M$. Since $\mathfrak{A} \models \exists x_i(\varphi_1 \wedge \ldots \wedge \varphi_r)[\overline{a}]$ and $(\overline{a},\overline{b},M) \in I_0$ we have $\mathfrak{B} \models \exists x_i(\varphi_1 \wedge \ldots \wedge \varphi_r)[\overline{b}]$.

Summing up we have

Theorem 1.7. For ω-saturated $\mathfrak{A}$ and $\mathfrak{B}$, $\overline{a} \in (A \cup \{\star\})^p$ and $\overline{b} \in (B \cup \{\star\})^p$ with $\sup(\overline{a}) = \sup(\overline{b}) \subset \{1,\ldots,n\}$ properties (1) - (3) are equivalent:
(1) $\overline{a}$ and $\overline{b}$ satisfy the same (n,k)-formulas.
(2) $(\overline{a},\overline{b},\emptyset) \in I_0$.
(3) $\overline{a} \simeq_{n,k} \overline{b}$.

(1)⇒(2) and (2)⇒(3) are immediate from the definitions of I_0 and $\simeq_{n,k}$ (and Lemma 1.6) respectively. (3)⇒(1) holds by Lemma 1.5.

Theorem 1.7 is the key to the algebraic characterization of (n,k)-boundedness; in fact it implies together with Lemma 1.4:

Theorem 1.8. For a theory Σ the following are equivalent:
(i) Σ is (n,k)-bounded.
(ii) For all ω-saturated models $\mathfrak{A}$ and $\mathfrak{B}$ of Σ, $\overline{a} \in (A \cup \{\star\})^p$, $\overline{b} \in (B \cup \{\star\})^p$ with $\text{sup}(\overline{a}) = \text{sup}(\overline{b}) \subset \{1,\ldots,n\}$, if $\overline{a} \simeq_{n,k} \overline{b}$ then $(\mathfrak{A},\overline{a}) \equiv (\mathfrak{B},\overline{b})$.

(Note that $(\mathfrak{A},\overline{a}) \equiv (\mathfrak{B},\overline{b})$, if there is a set of partial isomorphisms containing $(\overline{a},\overline{b})$ with the (usual) back and forth property, thus (ii) can be formulated in purely algebraic terms.)

The algebraic characterization of (n,k)-boundedness is well-suited for many purposes. However, it lacks the intuitive appeal of the following game-theoretic formulation in terms of a variant of the Ehrenfeucht-Fraisse game:

Given structures $\mathfrak{A}$ and $\mathfrak{B}$ and $\overline{a} \in (A \cup \{\star\})^p$, $\overline{b} \in (B \cup \{\star\})^p$ with $\text{sup}(\overline{a}) = \text{sup}(\overline{b}) \subset \{1,\ldots,n\}$ we introduce the game $G(\mathfrak{A},\mathfrak{B},\overline{a},\overline{b})$. It is played by two players, I and II. There are colourless pebbles $\alpha_1,\ldots,\alpha_p$ for $\mathfrak{A}$ and $\beta_1,\ldots,\beta_p$ for $\mathfrak{B}$; at the beginning α_i is placed on a_i and β_i on b_i. There are k colours. In the course of the game each colour can be used to colour exactly one pair of pebbles (α_i,β_i). Each round consists of a move of Player I followed by a move of Player II. First Player I selects either $\mathfrak{A}$ or $\mathfrak{B}$, say $\mathfrak{A}$. Now either he can take an already coloured pebble and place it on some element of $\mathfrak{A}$ or, in case there is still a colour available, he can colour with it a colourless pair (α_i,β_i) and place α_i on some element of $\mathfrak{A}$. In both cases, Player II answers placing the corresponding β_i on some element of $\mathfrak{B}$. Player I wins the game at round r if the map induced by the pebbles for $\mathfrak{A}$ and $\mathfrak{B}$ after round r is not a partial isomorphism. Thus Player II has a winning strategy if and only if he can always find matching elements to preserve the partial isomorphism. Clearly

Theorem 1.9. For $\mathfrak{A},\mathfrak{B},\overline{a}$ and $\overline{b}$ as given we have:

$$\overline{a} \simeq_{n,k} \overline{b} \quad \text{iff} \quad \text{Player II has a winning strategy for the game } G(\mathfrak{A},\mathfrak{B},\overline{a},\overline{b}).$$

For $n = 0$ we obtain the game introduced in [2], [5] and [8] to check wether $\mathfrak{A}$ and $\mathfrak{B}$ are equivalent in the language L_k; for $n = k$ we obtain the game-theoretic characterization of the k-variable property given in [6].

2. Remarks and Examples. a) (cf. [6], [8]) The theory Σ of linear orderings is 3-bounded: Let $\mathfrak{A}$ and $\mathfrak{B}$ be ω-saturated models of Σ. For $n = k = 3$ define I_0 as above. Set

$$\begin{aligned} J := \{q \mid \ & q \text{ is a partial isomorphism of } \mathfrak{A} \text{ to } \mathfrak{B}, \text{ there are} \\ & r \in \mathcal{N},\ \ a_1 < \ldots < a_r \in A \ \text{ with domain } (q) = \{a_1,\ldots,a_r\} \\ & \text{and } \ (a_i, a_{i+1}, \star) \in I_0 \ \text{ for } \ i = 1,\ldots,r-1\} \end{aligned}$$

(we use two different notations for partial isomorphisms!). J has the back and forth property (use the corresponding property of I_0). Suppose $\overline{a} \cong_{l,3} \overline{b}$, where l is a natural number. Then $(\overline{a}, \overline{b}) \in J$ and therefore $(\mathfrak{A}, \overline{a}) \equiv (\mathfrak{B}, \overline{b})$. Hence, by Theorem 1.8, Σ is $(l,3)$-bounded. Since l was arbitrary, Σ is 3-bounded.

b) Let Σ_k be the theory in a language with a binary relation symbol $\leq$ given by the axioms

$$\begin{array}{l} \forall x \quad x \leq x \\ \forall x \forall y \forall z ((x \leq y \wedge y \leq z) \rightarrow x \leq z) \\ \forall x \exists y_1 \ldots \exists y_k \forall z ((x \leq z \wedge z \leq x) \rightarrow (z = y_1 \vee \ldots \vee z = y_k)) \end{array}$$

Arguing similarly as in a) one easily shows for $k \geq 3$ that Σ_k is k-bounded but not $(k-1)$-bounded.

c) In [6] theories of bounded-degree trees are introduced and it is shown that they have the k-variable property for some k. One can generalize the arguments to obtain the k-boundedness.

d) Let $\Sigma(R)$ be the theory of discrete linear orderings introduced earlier (before definition 1.1). $\Sigma(R)$ is 3-bounded (argue similarly as in a)) but not 2-bounded: Let $\mathfrak{A}$ be an ω-saturated model, $a, b, c, d \in A$, Rab, Rbc and Rcd. Then it is easily seen that $ac \simeq_{2,2} ad$, but $\mathfrak{A} \models \exists x_3 (Rx_1x_3 \wedge Rx_3x_2)[ac]$, while $\mathfrak{A} \not\models \exists x_3 (Rx_1x_3 \wedge Rx_3x_2)[ad]$.

e) Clearly - as usual - instead of sets I with the (n,k)-back and -forth property one can also consider sequences $(I_t)_{t \leq l}$ with the corresponding property (or games $G_l(\mathfrak{A}, \mathfrak{B}, \overline{a}, \overline{b})$ with at most l moves) and (n,k)-formulas of quantifier rank $\leq l$. (One has to pay attention with the function symbols.) For the proof of the corresponding equivalence it ist helpful to introduce formulas that tell us how one can extend $\overline{a}$ in $\mathfrak{A}$, more precisly (cf. [3]): For a structure $\mathfrak{A}$ and $\overline{a} \in (A \cup \{\star\})^p$ and arbitrary M the l-Hintikka formula $\varphi^l_{\overline{a},M}$ is defined by induction (for simplicity assume that L contains no function symbols):

$$\varphi^0_{\overline{a},M} = \bigwedge \{\psi \mid \psi \quad \text{atomic or negated atomic and} \quad \mathfrak{A} \models \psi[\overline{a}]\},$$

$$\varphi^{l+1}_{\overline{a},M} = \varphi^l_{\overline{a},M} \wedge \bigwedge_{\substack{1 \leq i \leq p \\ |M \cup \{i\}| \leq k}} \Big(\bigwedge_{a \in A} \exists v_i \varphi^l_{\overline{a}\frac{a}{i}, M \cup \{i\}} \wedge \forall v_i \bigvee_{a \in A} \varphi^l_{\overline{a}\frac{a}{i}, M \cup \{i\}} \Big).$$

Note that $\vdash \varphi^l_{\overline{a},M}, M$.

f) One can generalize (in a canonical way) the notion of (n,k)-formula to the infinitary language $L_{\infty\omega}$ and prove:

For a theory Σ the following are equivalent:
(i) Each $L_{\infty\omega}$-formula $\varphi(x_1, \ldots, x_n)$ is equivalent to a (n,k)-formula $\psi(x_1, \ldots, x_n) \in L_{\infty\omega}$.
(ii) For any models $\mathfrak{A}$ and $\mathfrak{B}$ of Σ, any $\overline{a}$ and $\overline{b}$ with $\sup(\overline{a}) = \sup(\overline{b})$ and

$\sup(\overline{a}) \subset \{1,\ldots,n\}$, if $\overline{a} \cong_{n,k} \overline{b}$ then the structures $(\mathfrak{A},\overline{a})$ and $(\mathfrak{B},\overline{b})$ are partially isomorphic.

For a proof introduce - as above - for any ordinal ξ the formula $\varphi^{\xi}_{\overline{a},M}$ and use the fact that the class of well-orderings is not $L_{\infty\omega}$-definable. In particular the argument in a) shows that the theory of linear orderings is 3-bounded in $L_{\infty\omega}$.

g) We say that Σ is *k-good*, if for all n each formula $\varphi(x_1,\ldots,x_n)$ can be equivalently rewritten using at most k bound variables (and the same free variables) (cf.[4]). Clearly if Σ is k-bounded then Σ is k-good. On the other hand there is a theory which is 1-good but not k-bounded (not even (k,k)-bounded) for any k. In fact, let L have two unary relation symbols U and S and a binary relation symbol E: think of urelements (U), sets (S) and the ϵ-relation (E), although we do not require extensionality. Consider the class $\mathcal{K}$ of finite L-structures $\mathfrak{A}$ with $A = U^{\mathfrak{A}} \dot{\cup} S^{\mathfrak{A}}$ and $S^{\mathfrak{A}} \neq \emptyset$. For $\mathfrak{A},\mathfrak{B} \in \mathcal{K}$ define $\mathfrak{A} \sqsubseteq \mathfrak{B}$ ("$\mathfrak{A}$ is a strong substructure of $\mathfrak{B}$"), if $\mathfrak{A}$ is a substructure of $\mathfrak{B}$ and

$$\forall b \in S^{\mathfrak{B}} \exists a \in S^{\mathfrak{A}} \forall u \in U^{\mathfrak{A}} (E^{\mathfrak{B}}ua \leftrightarrow E^{\mathfrak{B}}ub) \quad .$$

Then, with respect to $\sqsubseteq$, $\mathcal{K}$ has the amalgamation and the joint embedding property. Let $\mathfrak{A}_0$ be the uniquely determined universal and homogeneous structure given by $\mathcal{K}$ and set $\Sigma := Th(\mathfrak{A}_0)$. Then, for all k, Σ is not k-bounded but Σ is 1-good: e.g. each formula $\varphi(x_1,\ldots,x_n) \wedge Ux_1 \wedge \ldots \wedge Ux_n$ is equivalent to a boolean combination of quantifier-free formulas and formulas of the form

$$\exists x_{n+1}(\bigwedge_{i\in I} Ex_ix_{n+1} \wedge \bigwedge_{i\notin I, 1\leq i\leq n} \neg Ex_ix_{n+1})$$

when $I \subset \{1,\ldots,n\}$. This definition of Σ was suggested by M.Ziegler.

h) Suppose Σ is a theory in a relational first-order language L and all relation symbols of L have arity $\leq k$. Assume that Σ is (k,k)-bounded in any language containing additional unary relational symbols. Then Σ is k-bounded. To prove this replace a variable x_i with $i > k$ by a unary relation symbol P_i "with value $\{x_i\}$" and apply the (k,k)-boundedness of Σ in the language L augmented by these P_i.

References

[1] Amir, A., Gabbay, D.: *Preservation of expressive completeness in temporal models*, Information and Computation 72(1978), 66-83.

[2] Barwise, J.: *On Moschovakis closure ordinals*, J.Symb.Log. 42(1977), 292-296.

[3] Flum, J.: *First-order logic and its extensions*, ISILC Logic Conf. Kiel, Lect. Notes in Math. 499(1975), 248-310.

[4] Gabbay, D.: *Expressive functional completeness in tense logic*, in "Aspects of Philosophical Logic" (Monnich, Ed.), Reidel, Dordrecht (1981), 91-117.

[5] Immerman, N.: *Upper and lower bounds for first-order expressibility*, J.Comp.Syst.Sciences 25(1982), 76-98.

[6] Immerman, N., Kozen, D.: *Definability with a bounded number of bound variables*, Information and Computation 83(1989), 121-139.

[7] Immerman, N., Lander, E.: *Describing graphs: A first-order approach to Graph Canonization*, in: A.Selman (Ed.), Complexity Theory Retrospective, Springer, New York (1990), 59-81.

[8] Poizat, B.: *Deux ou trois choses que je sais de L_n*, J.Symb.Log. 47(1982), 641-658.

The Cutting Plane Proof System with Bounded Degree of Falsity

Andreas Goerdt

FB 17 Mathematik/ Informatik
Universität -GH- Paderborn
Postfach 1621
Warburger Straße 100
D-W-4790 Paderborn
Germany

Abstract

The cutting plane proof system for proving the unsatisfiability of propositional formulas in conjunctive normalform is based on a natural representation of formulas as systems of integer inequalities. We define a restriction of this system, the cutting plane system with bounded degree of falsity, and show the results: This system p-simulates resolution and has polynomial size proofs for the pigeonhole formulas. The formulas from [9] only have superpolynomially long proofs in the system. Our system is the only known system with provably superpolynomial proof size, but polynomial size proofs for the pigeonhole formulas.

1 Introduction

The systematic study of the complexity and proof length of various proof systems for propositional logic (like resolution [3], tableaux calculus [10], Frege systems (or Hilbert style systems, these are the usual systems with modus ponens)) was initiated by Cook/Reckhow in [6]. One motivation for this is the following: As propositional proof systems are just nondeterministic algorithms for the coNP-complete language of unsatisfiable propositional formulas (or equivalently tautologies) the NP $\neq$ coNP assumption implies the existence of hard examples for any such proof system. Hard examples are infinite families of formulas having only proofs of superpolynomial size in the system considered. It is an interesting problem to show the existence of hard examples for increasingly powerful proof systems, without assuming NP $\neq$ coNP. Even for the relatively weak resolution proof system the existence of hard examples was proved by Haken only in 1985 [7]. Ajtai [1] showed this result for bounded depth Frege systems (which are stronger than resolution). Both authors use a family of formulas encoding the pigeonhole principle (saying that a total mapping from a set with $n+1$ elements to a set with n elements is not injective). The existence of hard examples for Frege systems with unbounded formula depth (or for stronger systems) is a well known open problem [4]. Buss [2] proved that Frege systems allow for short (i.e. polynomial) proofs of the pigeonhole principle.

The following results are known about the cutting plane proof system: The cutting plane proof system p-simulates (see definition 2.1 below) resolution and has short proofs for the formulas encoding the pigeonhole principle [5]. In [8] we show that Frege systems p-simulate the cutting plane system. The existence of hard examples for the cutting plane system is an open problem. In the cutting plane system with bounded degree of falsity, considered here, we restrict the form of the inequalities which can occur in proofs. This system is still complete. The restriction imposed on the inequalities allows us to modify the argument of [9] which is a variation of the argument from [7] to show the superpolynomial lower bound for our system.

In section 1 we introduce the cutting plane proof system with bounded degree of falsity. In section 2 we prove the superpolynomial lower bound for this system.

2 The cutting plane proof system with bounded degree of falsity

We introduce some basic facts, define the cutting plane system with bounded degree of falsity, and prove some of its properties.

By the size of a proof in any propositional proof system we mean the length of the proof written out as string over the binary alphabet. So the size also accounts for the size of the formulas occurring in a proof whereas the number of proof steps does not.

The basic notion when comparing the complexitiy of propositional proof systems is that of p-simulation.

Definition (2.1) *([6] definition 1.5)*
Let $\mathcal{P}$ and $\mathcal{Q}$ be two propositional proof systems. System $\mathcal{P}$ p-simulates system $\mathcal{Q}$ iff there is a polynomial time computable function f such that for any proof Q (represented as string over a binary alphabet) of a formula F in the system $\mathcal{Q}$ $f(Q)$ is a proof of F in $\mathcal{P}$. In particular we have: As f is a polynomial time computable function, the size of $f(Q)$ is polynomial in the size of Q.

A literal is a propositional variable like x or a negated propositional variable like $\bar{x}$. We call x positive literal and $\bar{x}$ negative literal. A clause is a disjunction of literals like $L_1 \vee \ldots \vee L_n$. Clauses are also written as $\{L_1, \ldots, L_n\}$. We assume that the L_i are different and w. l.o.g. that our clauses are non-tautological, i.e. they do not contain both x and $\bar{x}$ for one variable x. A truth assignment π assigns 1 or 0 to the propositional variables (1 for true and 0 for false). A propositional formula in conjunctive normalform is a conjunction of clauses like $C_1 \wedge \ldots \wedge C_n$ which is often written as $C_1, \ldots, C_n$. A truth value assignment π satisfies a formula in conjunctive normalform F iff each clause in F contains a literal which has the value 1 under π.

Then we write $\pi \models F$.

The resolution proof rule is the rule

$$\frac{C \cup \mathrm{x} \quad D \cup \bar{\mathrm{x}}}{C \cup D}.$$

A resolution proof of the formulas F in conjunctive normalform is the derivation of the empty clause from the clause of F with the resolution rule. The formula F is unsatisfiable iff there exists a resolution proof of F (see [3]). That is, the resolution proof system is sound and complete.

The cutting plane proof system is based on the representation of clauses as inequalities with variables ranging over 0 and 1. The representation $R(\mathrm{x})$ of the positive literal x is x, the representation $R(\bar{\mathrm{x}})$ of $\bar{\mathrm{x}}$ is $1-\mathrm{x}$. A clause $L_1 \vee \ldots \vee L_n$ is represented as

$$R(L_1 \vee \ldots \vee L_n) = (R(L_1) + \ldots + R(L_n) \geq 1).$$

For example $R(\mathrm{x} \vee \bar{\mathrm{y}} \vee \bar{z}) = \mathrm{x} + 1 - \mathrm{y} + 1 - z \geq 1$ which is equivalent to $\mathrm{x} - \mathrm{y} - z \geq -1$. In the sequel we always assume that our inequalities are in the following normalform: Each summand on the left hand side is an integer multiple of a variable, each variable occurs only in one summand, and the right hand side just is an integer. The representation of the clause $L_1 \vee \ldots \vee L_n$ in this normalform then can be written as $\sum_{i=1}^{n} \epsilon_i \mathrm{x}_i \geq -N + 1$ where

$$\epsilon_i = \begin{cases} -1 & \text{if} \quad L_i = \bar{\mathrm{x}}_i \\ 1 & \text{if} \quad L_i = \mathrm{x}_i \end{cases}$$

and N is the number of negative literals in $L_1 \vee \ldots \vee L_n$. Now, a formula in conjunctive normalform $F = C_1 \wedge \ldots \wedge C_n$ is represented as a set of inequalities,

$R(F) = R(C_1), \ldots, R(C_n), \mathrm{x} \geq 0, -\mathrm{x} \geq -1$

for each occurring variable x.(The last two inequalities say $0 \leq \mathrm{x} \leq 1$.)

Let $I = \sum_{i=1}^{n} v_i \mathrm{x}_i \geq P$ be an inequality and π an assignment assigning 0 or 1 to each x_i. The value of the left hand side of I under π is called $LHS(I, \pi)$. We say π satisfies I iff $LHS(I, \pi) \geq P$. In this case we write $\pi \models I$. We say a partial assignment π of the variables from I hurts a variable x iff x occurs positively in I and $\pi(\mathrm{x}) = 1$ or x occurs negatively x and $\pi(\mathrm{x}) = 0$. If π hurts x in I then $\mathrm{LHS}(I, \pi) > LHS(I, \pi')$ where $\pi'(\mathrm{x}) = 1 - \pi(\mathrm{x})$ and π' is equal to π on the remaining variables.

Obviously holds: π satisfies the propositional formula F iff π satisfies all inequalities of $R(F)$. Now we can define the cutting plane proof system: The formulas of the cutting plane proof system are just inequalities in the normalform described above. The cutting plane system has three kinds of proof rules:

Addition

$$\frac{\sum v_i x_i \geq P \quad \sum w_i x_i \geq Q}{\sum (v_i + w_i) x_i \geq P + Q}$$

Multiplication

$$\text{For } c \in N : \frac{\sum v_i x_i \geq P}{\sum (c \cdot v_i) x_i \geq c \cdot P}$$

Division
For $c \in N \setminus \{0\}$ such that $v_i = c \cdot w_i$ for all i, and not all $v_i = 0$,and $Q = \lceil \frac{1}{c} \cdot P \rceil$:

$$\frac{\sum v_i \cdot x_i \geq P}{\sum w_i \cdot x_i \geq Q}$$

The condition not all $v_i = 0$ is no real restriction. It ensures that we cannot divide by a number which is not contained in the formulas. This could increase the size of a proof without being visible from the formula size. In [5] the c's are explicitly included in the proof.

These rules are correct in the sense that all integers $x_1, \ldots, x_n$ for which the premisses hold the conclusion holds. Note that the rounding step in the division rule is correct because the left hand side always is an integer ($\lceil \frac{1}{c} \cdot P \rceil = \min\{x | x \in N, x \geq \frac{1}{c} \cdot P\}$). A cutting plane proof of a formula in conjunctive normalform consists in the derivation of an unsatisfiable inequality $0 \geq m$ with $m > 0$ from the inequalities representing the formula. Note that our notion of cutting plane proof differs slightly from that in [5], but both notions p-simulate each other and thus can be considered as equivalent in the present context. An example derivation can be found in 1.3 of [8].

Theorem (2.2) *(see [5])*
Let F be a system of inequalities representing a propositional formula in conjunctive normalform. The system F is unsatisfiable iff an unsatisfiable inequality $0 \geq m$ is derivable with the cutting plane proof system from F. This shows: The cutting plane proof system is sound and complete.

Next we define the cutting plane proof system with bounded degree of falsity.

Definition (2.3)
(a) Let $I = \sum_{i=1}^{n} v_i x_i \geq P$ and let π be an assignment of 0 or 1 to the variables of I. The degree of falsity of I under π is given by

$$DGF(I, \pi) = P - LHS(I, \pi).$$

Then we have: π does not satisfy I iff $DGF(I, \pi) > 0$.
The degree of falsity of I is given by

$$DGF(I) = \max\{DGF(I, \pi) | \pi \text{ is a 0,1-assignment}\}.$$

Then we have:

$$DGF(I) = P - \min\{LHS(I,\pi)|\pi \textit{ assignment}\} = P - LHS(I,\rho),$$

where the assignment ρ *is given by*

$$\rho(x_i) = \begin{cases} 0 & \textit{if} \quad v_i \geq 0 \\ 1 & \textit{if} \quad v_i < 0, \end{cases}$$

i.e. ρ *does not hurt any variable occurring in* I.
The degree of falsity of a cutting plane proof P *is given by*

$$DGF(P) = \max\{DGF(I)|I \textit{ is an inequality in } P\}.$$

(b)Let $b : N^+ \rightarrow N^+$. *The cutting plane system with degree of falsity* $b(n)$ *(short* $DGF \;\; b(n)$*) is the following restriction of the cutting plane system: To prove a system of inequalities* F *with* $n \geq 1$ *variables we only allow cutting plane proofs* P *with* $DGF(P) \leq b(n)$.

It is crucial for the results to come that the degree of falsity may increase with the number of variables. Note that the cutting plane proof system with $DGF \;\; b(n)$ is a reasonable proof system in the sense that if $b(n)$ is polynomial time computable we can check whether a proof is of $DGF \;\; b(n)$ in time polynomial in the proof size.

The following corollary follows essentially by checking the degree of falsity of constructions from [5].

Corollary (2.4)
(a) The cutting plane system with $DGF \;\; b(n) = 1$ *p-simulates resolution and hence is complete.*
(b) The cutting plane system with $DGF \;\; b(n) = 3\sqrt{n}$ *has polynomial size proofs for the formulas encoding the pigeonhole principle.*

Proof:

(a) As

$$R(L_1 \vee \ldots \vee L_n) = (\sum_{i=1}^{n} \epsilon_i x_i \geq -N + 1)$$

where $\epsilon_i = 1$ iff $L_i = x_i$ and $\epsilon_i = -1$ otherwise and N is the number of negative literals in $L_1 \vee \ldots \vee L_n$, we get that $DGF(R(L_1 \vee \ldots \vee L_n)) = 1$. Hence, the claim holds for inequalities which represent clauses. Using resolution we can perform the following derivation step:

$$\mathrm{x} \vee L_1 \vee \ldots \vee L_k \vee G_1 \vee \ldots \vee G_m \qquad \bar{\mathrm{x}} \vee L_1 \vee \ldots \vee L_k \vee H_1 \vee \ldots \vee H_l$$

$$L_1 \vee \ldots \vee L_k \vee G_1 \vee \ldots \vee G_m \vee H_1 \vee \ldots \vee H_l,$$

where the L_i are all the common literals of the two clauses whereas the variables underlying the G_i and H_j are pairwise distinct. (Note that we exclude w.l.o.g. clauses containing L and $\bar{L}$ from consideration.) This derivation step can be simulated by the following cutting plane proof:
Let N_1 be the number of negative literals among the L_i, N_2 that among the G_i and N_3 that among the H_i. Then we can derive as follows with $\epsilon_i, \gamma_i, \delta_i \in \{-1, 1\}$ defined as usual:

$$\mathrm{x} + \sum \epsilon_i \mathrm{x}_i + \sum \delta_i \mathrm{y}_i \geq -N_1 - N_2 + 1 \qquad -\mathrm{x} + \sum \epsilon_i \mathrm{x}_i + \sum \gamma_i z_i \geq -1 - N_1 - N_3 + 1$$

$$\sum 2\epsilon_i \mathrm{x}_i + \sum \delta_i \mathrm{y}_i + \sum \gamma_i z_i \geq -2N_1 - N_2 - N_3 + 1$$

$\mathrm{x} \geq 0, -\mathrm{x} \geq -1$ for all variables x

$$\sum 2\epsilon_i \mathrm{x}_i + \sum 2\delta_i \mathrm{y}_i + \sum 2\gamma_i z_i \geq -2N_1 - 2N_2 - 2N_3 + 1$$

Division and rounding

$$\sum \epsilon_i \mathrm{x}_i + \sum \delta_i \mathrm{y}_i + \sum \gamma_i z_i \geq -N_1 - N_2 - N_3 + 1.$$

This derivation simulates the above resolution step and each occurring inequality has $DGF \leq 1$.

(b) For an introduction of the family of formulas $(PHP_n)_{n \in N}$ encoding the pigeonhole principle see [7]. The formula PHP_n uses the set of $n(n+1)$ variables

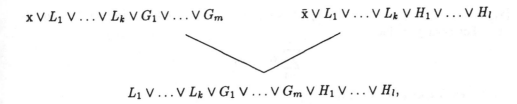

$\{x_{i_j} | i = 1, \ldots, n, j = 1, \ldots, n+1\}$. It is represented by the following set of inequalities. For each j we have

$$\sum_{i=1}^{n} x_{i_j} \geq 1,$$

and for each $i, k_1 \neq k_2$ we have

$$-x_{i_{k_1}} - x_{i_{k_2}} \geq -1$$

and of course $x_{i_j} \geq 0, -x_{i_j} \geq -1$. A polynomial size cutting plane proof of these formulas is constructed as follows: First we derive for each i the inequality

$$\sum_{j=1}^{n+1} -x_{i_j} \geq -1.$$

To this end we show how derive for each k and l with $1 \leq k \leq n, 1 \leq l \leq n$, and $k + l \leq n + 1$ the inequality

$$\sum_{j=k}^{k+l} -x_{i_j} \geq -1.$$

We proceed inductively on l. The inequalities for $l = 1$ belong to PHP_n. To derive

$$\sum_{j=k}^{k+l+1} -x_{i_j} \geq -1$$

we proceed as follows

$$\sum_{j=k}^{k+l} -x_{i_j} \geq -1 \qquad\qquad \sum_{j=k+1}^{(k+1)+l} -x_{i_j} \geq -1.$$

Derivable by ind. hyp.

$$-x_{i_k} + \sum_{j=k+1}^{k+l} -2x_{i_j} - x_{i_{(k+l+1)}} \geq -2$$

$$-x_{i_k} - x_{i_{k+l+1}} \geq -1$$

$$-2x_{i_k} + \sum -2x_{i_j} - 2x_{i_{(k+l+1)}} \geq -3$$

Division by 2 and rounding

$$\sum_{j=k}^{k+l+1} -x_{i_j} \geq -1.$$

The degree of falsity of the inequalities occurring here is bounded by $2(n+1)$. On the other hand we can derive the inequality with degree of falsity $n+1$

$$\sum_{i=1}^{n}\sum_{j=1}^{n+1} x_{i_j} \geq n+1.$$

From this we get with the inequalities derived before

$$\sum_{i=2}^{n}\sum_{j=1}^{n+1} x_{i_j} \geq n, \ldots, \sum_{j=1}^{n+1} x_{n_j} \geq 2, 0 \geq 1,$$

where the degree of falsity always is $\leq n$.

As the number of variables of PHP_n is $n(n+1)$, and $2(n+1) \leq 3 \cdot \sqrt{n(n+1)}$ the claim follows.

■

For the lower bound proof in section 3 we need the following:

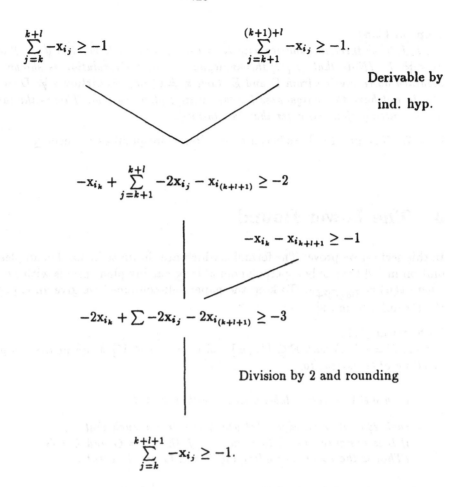

Lemma (2.5)
Let I, J, K be three inequalities with $K = I + J$, then we have: If $\pi \not\models K$ then $\pi \not\models I$ or $\pi \not\models J$. (Note that in [7] the analogous result for resolution is shown: If C is obtained by resolution from D and E, then $\pi \not\models C$ implies either *$\pi \not\models D$* or *$\pi \not\models E$. We do not have this uniqueness for the cutting plane system. This is the fact which makes cutting plane stronger than resolution).*

Proof: The proof follows because addition of inequalities respects $\leq$.

■

3 The Lower Bound

In this section we prove: The formulas which are shown to be hard examples for resolution in [9] have only superpolynomial long cutting plane proofs with DGF $b(n)$ where $b(n) = \frac{n}{(\log n)^2+1}$. To keep the paper self-contained we give an exposition of the formulas from [9].

Definition (3.1)
(a) Let $G = (V, E)$ with $E \subseteq \{\{v, w\} | v \neq w, v, \ v, w \in V\}$ be an undirected graph. A labelling of G assigns to

- *each $v \in V$ a vertex label which is either 0 or 1*
- *each edge of G an edge label which is a literal such that:*
  if L is assigned to e, G to f and $e \neq f$ then $L \neq G$ and $\bar{L} \neq G$.
  (That is the variables underlying L and G are different.)

A labelling is called odd iff the sum modulo 2 of all vertex labels is equal to 1, otherwise, it is called even.
(b) Let G be a graph with a labelling and let v be a vertex of G. Let $L_1, \ldots, L_n$ be the set of literals labelling edges incident with v. The set of clauses belonging to v, Clauses v, is defined by:

$$C \in \textit{ Clauses } v \textit{ iff } C = H_1 \vee \ldots \vee H_n$$

with $H_i \in \{L_i, \bar{L}_i\}$ for all i and

$$|\{H_i | H_i = \bar{L}_i, 1 \leq i \leq n\}| \textit{ is even iff the label of } v \textit{ is } 1$$

and

$$|\{H_i | H_i = \bar{L}_i, 1 \leq i \leq n\}| \textit{ is odd iff the label of } v \textit{ is } 0.$$

Let $\oplus$ be the propositional connective denoting exclusive-or (i.e. sum modulo 2). If v is labelled with 1 then the conjunction of all clauses in Clauses v is equivalent to $L_1 \oplus \ldots \oplus L_n \leftrightarrow 1$. This follows from the observation, that the assignment π does not satisfy $L_1 \oplus \ldots \oplus L_n \leftrightarrow 1$ iff $\{L_i | \pi(L_i) = 1\}$ is even. If v is labelled with 0, the conjunction is equivalent to $L_1 \oplus \ldots \oplus L_n \leftrightarrow 0$.

The propositional formula corresponding to a labelled graph G is the conjunction of all clauses in the set $\{C \in$ Clauses $v | v$ vertex of $G\}$. We denote this formula by $F(G)$.

Example: We consider the labelled graph

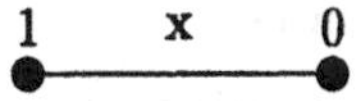

The left vertex gives us the clause x (for $x \leftrightarrow 1$) the right vertex gives us $\bar{x}$ (for $x \leftrightarrow 0$). For the labelled graph

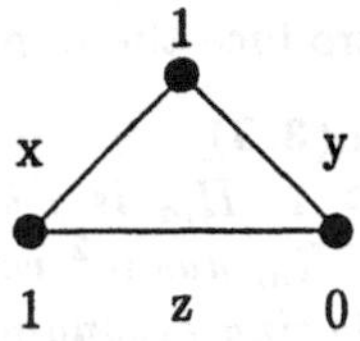

we get $(x \vee z) \wedge (\bar{x} \vee \bar{z})$ (for $x \oplus z \leftrightarrow 1$) and $(x \vee y) \wedge (\bar{x} \vee \bar{y})$ and $(\bar{y} \vee z) \wedge (y \vee \bar{z})$. This formula is satisfied by the assignment π with $\pi(x) = 1, \pi(z) = 0, \pi(y) = 0$.

One can easily tell from the labelling of G, when $F(G)$ is satisfiable.

Corollary (3.2)
Let G be a connected graph with a labelling. The formula $F(G)$ is unsatisfiable iff the labelling of G is odd.

Proof: Let G be a graph with an odd labelling. If $F(G)$ is satisfied by the assignment π, we have that all the propositional formulas of the form $L_1 \oplus \ldots \oplus L_n \leftrightarrow \epsilon$ belonging to the vertices of G are satisfied by π. This implies that the sum modulo 2 of all left hand sides of the above equivalences is equal to that all right-hand sides. This is a contradiction: The sum of all left-hand sides gives 0 because each literal is incident with excactly 2 vertices and hence occurs exactly twice. The sum of all right-hand sides gives 1 because the labelling is odd.

For the other direction the following observation: If G is a graph all of whose vertices are labelled with 0, then $F(G)$ is satisfied by the assignment π with $\pi(L) = 0$ for each literal L. Note that the conjunction of all $C \in$ Clauses v is equivalent to $L_1 \oplus \ldots \oplus L_n \leftrightarrow 0$.

Now let G be a connected graph with an even labelling. Let v, w be two vertices of G which are labelled with 1. Then we have vertices $x_1, \ldots, x_n$ such that we have the situation

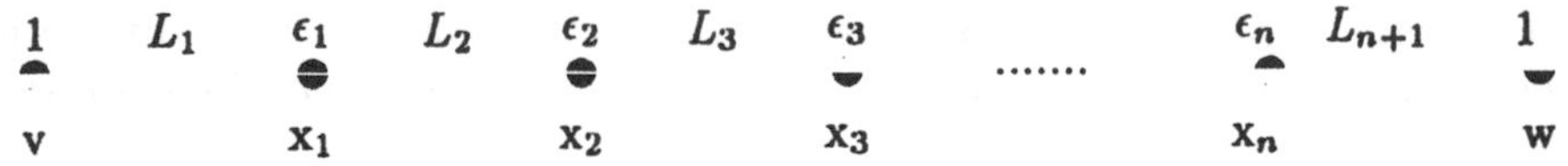

with $n \geq 0$ in G.

If we label v with 0, x_1 with $1 - \epsilon_1$ and L_1 with $\bar{L}_1$ we get a new labelled graph G_1 with $F(G_1) = F(G_2)$. Then we replace $1 - \epsilon_1$ again by ϵ_1 and L_2 by $\bar{L}_2$ and ϵ_2 by $1 - \epsilon_2$, this gives us a labelled graph G_2 with $F(G_2) = F(G_1)$. Analogously we obtain G_{n+1} with $F(G) = F(G_{n+1})$, the labels of v and w are both 0, and the other labels are unchanged. As the labelling of G is even, we can construct in this way a graph H with all labels 0 and $F(G) = F(H)$. The remark above implies that $F(H)$ and hence $F(G)$ is satisfiable. ■

Now we introduce the family of graphs used in our definition of the formulas.

Definition (3.3)
(a) Let $m \geq 1$. H_m is a bipartite graph such that each mode has a degree ≤ 5 and each side of H_m has m^2 vertices. Moreover, there is a constant $d > 0$, such that H_m has the following expanding property: If V_1 is a subset of vertices from one side of H_m with $|V_1| \leqq \frac{m^2}{2}$ and

$$V_2 = \{v | \exists w \in V_1 \cdot v \text{ is connected by an edge with } w\}$$

then $|V_2| \geq (1 + d) \cdot |V_1|$.

By lemma 5.1 from [9] such a graph H_m exists for each m.
(b) The graph G_m is the following modification of H_m: We number the vertices from each side of H_m by $1, 2, \ldots, m^2$. For each side and each i with $1 \leq i \leq m^2 - 1$ we connect vertex i and vertex $i + 1$ by an edge.

We call the original edges from H_m middle edges and the new ones side edges.

(c) We choose an odd labelling of G_m where each edge is labelled with a positive literal. Let Var_m be the set of propositional variables labelling the edges of G_m. We call the literals labelling middle edges middle literals those labelling side edges side literals. We abbreviate $F(G_m)$ by F_m.F_m is unsatisfiable and the number of clauses in F_m is $\leq 128m^2$.

Definition (3.4)
(a) A truth value assignment π of Var_m is critical iff there exists exactly one clause C in F_m with $\pi \not\models C$. If v is a vertex of G_m then π is v-critical iff π is critical and $\pi \not\models C$ for a $C \in$ Clauses v.
(b) A partial truth value assignment S of the variables from Var_m is k-critical for $k \in N$ iff $|$ Def $S| = k$ and Def $S \subseteq$ the set of middle literals of G_m. Let $V(S) =$ the set of vertices incident with an edge labelled with a literal from Def S. $V(S)$ contains $\leq k$ vertices from each side of G_m.

Corollary (3.5)
Let k be less than the number of middle literals of G_m, let S be a k-critical assignment, and v a vertex of G_m. Then S can be extended to a v-critical assignment.

Proof: We show how to extend the k-critical assignment S to a v-critical assignment, where v is an arbitrary vertex of G_m. First we choose a middle literal x $\notin$ Def S. Then we assign arbitrary truth values to all middle literals which are not in Def $S \cup \{x\}$. The side edges and the edge labelled with x can be viewed as a tree with root v. Starting from the leaves of this tree going to the root we assign truth values to the literals labelling the edges of the tree such that the clauses labelling the vertices are satisfied under the assignment. (The truth values assigned are unique.) As this process ends at the root, V, of the tree and F_m is unsatisfiable, there must be $C \in$ Clauses v which does not satisfy the assignment obtained in this way. Hence, this assignment is v-critical. ∎

Definition (3.6)
Let I be an inequality over Var_m and let S be a k-critical assignment of Var_m. We define the set of vertices of G_m, Cover (I, S), by Cover $(I, S) = \{v$ vertex of $G_m | S$ can be extended to a v-critical assignment π with $\pi \not\models I$ and $v \notin V(S)$.$\}$

Now we are ready for the proof of:

Theorem (3.7)
Let $b(n) = \frac{n}{(\log n)^2+1}$ then holds: The cutting plane proof system with DGF $b(n)$ allows for polynomial size proofs of the pigeonhole principle. The formulas F_m only have superpolynomial proofs in the cutting plane system with DGF $b(n)$.

Proof: The first claim follows from corollary 1.4 (b) because $3 \cdot \sqrt{n(n+1)} \leq \frac{n\cdot(n+1)}{(\log(n(n+1)))^2+1}$ for n large enough. The second claim is presented as the rest of this section. ∎

Convention (3.8)
Let $b(n) = \frac{n}{(\log n)^2+1}$. Let D be the constant from definition 3.3 (a). Let $M \in N$ be fixed such that $\frac{D}{16} \cdot M^2$ is an integer. We assume that M is large enough. Let

$$F = \frac{D}{16}, \quad N = M^2$$

and

$$L = \text{ the number of middle literals of } G_M.$$

Then $N \leq L \leq 5 \cdot N$ because the degree of G_M is ≤ 5. The formula G_M has $\leq 7 \cdot N$ variables. Let C be a cutting plane proof of F_M with DGF $b(7N)$.

Definition (3.9)
For each $F \cdot N$-critical assignment S of Var_M let I_S be the first inequality I in C with

$$\text{Cover } (I, S) \geq \frac{N}{4}.$$

The inequalities I_S exist and contain a certain number of variables cf. [9]:

Corollary (3.10)
Let S be $(F \cdot N)$-critical.
(a) The inequality I_S exists.
(b) I_S contains at least $F \cdot N$ middle literals.

Proof: (a) Let J be the final inequality of C, then Cover $(J, S) \geq 2N - 2 \cdot F \cdot N \geq N$ as $D < 8$. Hence, the inequality I_S exists.
(b) If I is an initial inequality of C we have Cover $(I, S) \leq 1$, if I is derived from the inequality J by multiplication or division, we have Cover $(I, S) =$ Cover (J, S) because $\pi = I \Leftrightarrow \pi = J$ for all assignments π. Therefore the inequality I_S must be derived from two inequalities, say J and K, by addition. From lemma 2.5 we get

$$\text{Cover } (J_S, S) \subseteq \text{ Cover } (J, S) \cup \text{ Cover } (K, S),$$

hence $\frac{1}{4} \cdot N \leq$ Cover $(I_S, S) < \frac{1}{2} \cdot N$. Let

$$\text{Cover } (I_S, S) = W_1 \cup W_2$$

where W_1 is contained in one side of G_M and W_2 is contained in the other side. Let $|W_1| \geq |W_2|$, then $|W_1| \geq \frac{1}{8} \cdot N$ because $|W_1| + |W_2| \geq \frac{1}{4} \cdot N$. Let

$$W_3 = \{v | \exists w \in W_1 .\ w \text{ is connected to } v \text{ by a middle edge}\}.$$

Then

$$|W_3| \geq (1 + D) \cdot |W_1| \geq |W_2| + D \cdot |W_1| \geq |W_2| + D \cdot \frac{1}{8} \cdot N.$$

As W_3 and W_2 are contained in the same side of G_M, at least $\frac{D}{8} \cdot N$ vertices of W_3 are not in W_2. Let $Y = W_3 \setminus (W_2 \cup V(S))$. Then $|Y| \geq \frac{1}{8} \cdot D \cdot N - F \cdot N = F \cdot N$. Now let $\mathrm{y} \in Y$ and let x be a propositional variable labelling an edge from y to $w \in W_1$. This x exists because $\mathrm{y} \in W_3$. Then x occurs in I_s. For, let π be a w-critical extension of S with $\pi \not\models I_S$. Then the assignment π' given by

$$\pi'(z) = \begin{cases} \pi(z) & \text{if } z \neq \mathrm{x} \\ 1 - \pi(\mathrm{x}) & \text{if } z = \mathrm{x} \end{cases}$$

is y-critical. Then $\pi' \models I_S$, because otherwise $\mathrm{y} \in W_2$. Hence x must occur in I_S.

■

The following theorem finishes the proof of theorem 3.7.

Theorem (3.11)
There exists $N^{f(N)}$ many different inequalities I_S in C, where $f(N) \to \infty$.

Proof: The number of $F \cdot N$-critical assignments is $\binom{L}{F \cdot N} \cdot 2^{F \cdot n}$. For each $F \cdot N$-critical assignment S we show

$$|\{T | I_T = I_S\}| \cdot N^{f(N)} \leq \binom{L}{FN} \cdot 2^{F \cdot N},$$

where $f(N) \to \infty$. This implies that we have at least $N^{f(N)}$ different inequalities I_S in C.

Let S be an arbitrary $F \cdot N$-critical assignment and let C be a set of $F \cdot N$ middle literals occurring in I_S (G exists by Corollary 2.11(b)). Let

$$p = \frac{F \cdot N}{L}, \quad t = \frac{1}{2} \cdot p$$

and

$$A = \{T \| \text{ Def } T \cap G| \leq t \cdot F \cdot N\},$$
$$B = \{T \| \text{ Def } T \cap G| \leq t \cdot F \cdot N \text{ and } L_T = I_S\},$$

then $\{T | I_T = I_S\} \subseteq A \cup B$. We show

$$\frac{|A|}{\binom{L}{F \cdot N} \cdot 2^{F \cdot N}} \leq N^{-f(N)} \text{ and } \frac{|B|}{\binom{L}{F \cdot N} \cdot 2^{F \cdot N}} \leq N^{-f(N)}$$

where $f(N) \to \infty$, which implies

$$|\{T | L_T = I_S\}| \cdot N^{f(N) - \log 2} \leq \binom{L}{F \cdot N} \cdot 2^{F \cdot N}.$$

The first inequality: As for each $T \in A$ there is a $k \leq t \cdot F \cdot N$ with $|$ Def $T \cap G| = k$, we have

$$|A| \leq \sum_{k \leq t \cdot F \cdot N} \binom{F \cdot N}{k} \cdot \binom{L - FN}{FN - k} \cdot 2^{F \cdot N}$$

and by [11] inequality (7), (3), and $L \geq N$ we get

$$\frac{|A|}{\binom{L}{F \cdot N} \cdot 2^{F \cdot N}} \leq ((\frac{p}{t})^t \cdot (\frac{1-p}{1-t})^{1-t})^{F \cdot N} \leq (\frac{2}{e})^{\frac{1}{2} \cdot p \cdot F \cdot N} \leq (\frac{2}{e})^{\frac{1}{2} \cdot F^2 \cdot N}$$

which is of the required form.

The second inequality relies on the following observation: If T is an $F \cdot N$-critical assignment with $|$ Def $T \cap G| = k$ and T can be extended to an assignment π with $\pi \not\models I_S$, then T hurts $\leq b(7N)$ literals from G w.r.t. I_S, because otherwise we could construct from π an assignment ρ (by changing the values of the hurting literals in Def T) with $\text{DGF}(\rho, I_S) = b(7N) + 1$. Therefore

$$|B| \leq \sum_{k = t \cdot F \cdot N}^{F \cdot N} \binom{F \cdot N}{k} \cdot \binom{L - F \cdot N}{F \cdot N - k} \cdot \sum_{j=0}^{b(7 \cdot N)} \binom{k}{j} \cdot 2^{F \cdot N - k}$$

$$\leq \binom{L}{F \cdot N} \cdot (F \cdot N)^{b(7 \cdot N)+1} \cdot 2^{F \cdot N - t \cdot F \cdot N}.$$

Finally we get

$$\frac{|B|}{\binom{L}{F \cdot N} \cdot 2^{F \cdot N}} \leq \frac{(F \cdot N)^{b(7 \cdot N)+1}}{2^{t \cdot F \cdot N}} = 2^{(\log F \cdot N) \cdot (b(7 \cdot N)+1) - t \cdot F \cdot N}$$

$$\leq 2^{\frac{7 \cdot N}{\log 7 \cdot N} + \log F \cdot N - \frac{1}{10} F \cdot N}$$

because $F = \frac{D}{16} < 7$ and $t = \frac{1}{2} p = \frac{1}{2} \frac{FN}{L} \geq \frac{1}{10} F$.

The final expression is the required form.

References

[1] M. Ajtai, *The complexity of the pigeonhole principle*, Proceedings of the 29th Symposium on Foundations of Computer Science (1988).

[2] S. Buss, *Polynomial size proofs of the propositional pigeonhole principle*, Journ. Symb. Logic 52 (1987) pp.916–927.

[3] C.-L. Chang, R.C.-T. Lee, *Symbolic Logic and mechanical theorem proving*, Academic Press (1973).

[4] P. Clote, *Bounded arithmetic and computational complexity*, Proceedings Structures in Complexity (1990) pp. 186–199.

[5] W. Cook, C.R. Coullard, G. Turan, *On the complexity of cutting plane proofs*, Discr. Appl. Math. 18 (1987) pp. 25–38.

[6] S. A. Cook, R.A. Reckhow, *The relative efficiency of propositional proof systems*, Journ. Sym. Logic 44 (1979) pp. 36–50.

[7] A. Haken, *The intractability of resolution*, Theor. Comp. Sci. 39 (1985) pp. 297–308.

[8] A. Goerdt, *Cutting plane versus Frege proof systems*, CSL (1990) LNCS 533, pp. 174–194.

[9] A. Urquhart, *Hard examples for resolution*, JACM 34 (1) (1987) pp. 209–219.

[10] R.A. Smullyan, *First-order Logic*, Springer Verlag (1968).

[11] V. Chvatal, *Probabilistic methods in graph theory*, Annals of Operations Research 1 (1984) pp. 171–182.

Denotational Versus Declarative Semantics for Functional Programming *

Juan Carlos González Moreno
Dpto. L.S.I.I.S. (U.P.M., Spain)

María Teresa Hortalá González, Mario Rodríguez Artalejo
Dpto. de Informática y Automática. (U.C.M., Spain)

Abstract. ***Denotational semantics*** **is the usual mathematical semantics for functional programming languages. It is *higher order* (H.O.) in the sense that the semantic domain D includes $[D \rightarrow D]$ as a subdomain. On the other hand, the usual *declarative semantics* for logic programs is *first order* (F.O.) and given by the least Herbrand model. In this paper, we take a restricted kind of H.O. conditional rewriting systems as computational paradigm for functional programming. For these systems, we define both H.O. denotational and F.O. declarative semantics as two particular instances of *algebraic semantics* over *continuous applicative algebras*. For the declarative semantics, we prove *soundness* and *completeness* of rewriting, as well as an *initiality* result. We show that both soundness and completeness fail w.r.t. the denotational semantics and we present a natural restriction of rewriting that avoids unsoundness. We conjecture that this restricted rewriting is complete for computing denotationally valid F.O. results.**

1 Introduction

This work stems from our interest in the integration of logic and functional programming. Though this topic has received much attention in the last years [6,2,7], most of the existing research refers to F.O. languages. The semantic foundation of H.O. logic + functional languages does still need further investigation. Existing approaches propose either to translate H.O. syntax into F.O. syntax [3,4] or to use H.O. logics with a F.O. semantics [13,5].

In our previous work [10], we used *conditional narrowing* for a resticted kind of H.O. rewriting systems as a computational paradigm for H.O. *functional logic programming*, and we stablished soundness and completeness results w.r.t. a F.O. declarative semantics induced by a syntactic H.O. into F.O. translation. Subsequently, we realized that the effect of the syntactic translation was equivalent to the direct definition of a F.O. least model semantics, given by an algebra over an infinitary Herbrand universe equiped with a continuous *apply* operation. We also observed that denotational semantics can be obtained in the same way as a least model over a different H.O. domain D which includes $[D \rightarrow D]$ as a subdomain. The aim of this paper is to show that these ideas lead to a new and interesting view of the semantics of functional languages, well suited to their integration with logic languages.

The organization of the paper is as follows. In Section 2 we develop the basic facts about *applicative expressions* and *continuous applicative algebras* (C.A.A.s), which will be the basis of our mathematical semantics. In Section 3 we define a Simple Functional Language (SFL) and explain how to use a kind of infinite H.O. rewriting as its operational semantics. In Section 4 we show that C.A.A.s and a least fixpoint technique can be used to specify both a *F.O. declarative* and a *H.O. denotational* semantics for SFL programs. We also present an *initiality result* for the declarative semantics. In Section 5 we develop several results about *soundness* and *completeness* of rewriting w.r.t. both semantics; they show that declarative semantics characterizes the operational behaviour of SFL programs more adequately than denotational semantics. Section 6 summarizes some conclusions and planned lines of future work.

*Research supported by the PRONTIC project TIC 89/0104

2 Applicative Expressions and Algebras

First we introduce the basic facts about applicative expressions, algebras, expression evaluation and equations.

2.1 Signatures with constructors

A ***signature with constructors*** is any pair $\Sigma = (DC_\Sigma, FS_\Sigma)$ where $DC_\Sigma = \cup DC^n_\Sigma$ and $FS_\Sigma = \cup FS^n_\Sigma$ are *ranked* sets of ***constructors*** and ***function symbols***, respectively. We assume that all the sets DC^n_Σ, FS^n_Σ are mutually disjoint and use the notation *rank*(ϕ) for the rank of any symbol $\phi \in DC_\Sigma \cup FS_\Sigma$.

2.2 Expressions and patterns

Given a countably infinite set of ***variables*** $X, Y, Z \in Var$, we define the set of ***applicative expressions*** $e \in Exp_\Sigma$ of signature Σ by the following syntax:

$$e ::= X \mid c \mid f \mid (e_0 e_1)$$

An expression $(e_0 e_1)$ stands for the ***application*** of e_0 to e_1. As usual, we assume that application associates to the left and omit brackets accordingly. Note that any applicative expression can be written in exactly one of the 6 following forms:

X % *variable* c % *constructor* f % *function symbol*
$(X\ e_1 \ldots e_m)$ % *curried application of a variable*
$(c\ e_1 \ldots e_m)$ % *curried application of a constructor*
$(f\ e_1 \ldots e_m)$ % *curried application of a function symbol*

A curried application of a constructor or function symbol is called *partial*, *exact* or *exceeding* according to the case that the number m of arguments is less than, equal to or greater than the symbol's rank. We define 2 special kinds of expressions as follows:

- ***Patterns*** $s, t \in \mathrm{Ptr}_\Sigma$
  $t ::= X\ \%X \in \mathrm{Var} \mid c\ \%c \in DC_\Sigma \mid f\ \%f \in FS^n_\Sigma, n > 0$
  $\mid (c\ t_1 .. t_m)\ \%c \in DC^n_\Sigma, 1 \le m \le n \mid (f\ t_1 .. t_m)\ \%f \in FS^n_\Sigma, 1 \le m < n$
- ***First order expressions*** $e \in \mathrm{FOExp}_\Sigma$
  $e ::= X\ \%X \in \mathrm{Var} \mid c\ \%c \in DC^0_\Sigma \mid f\ \%f \in FS^0_\Sigma$
  $\mid (c\ e_1 .. e_n)\ \%c \in DC^n_\Sigma, 0 < n \mid (f\ e_1 .. e_n)\ \%f \in FS^n_\Sigma, 0 < n$

We also define the set of ***first order patterns*** as $\mathrm{FOPtr}_\Sigma = \mathrm{Ptr}_\Sigma \cap \mathrm{FOExp}_\Sigma$. Intuitively, the rank of a constructor is the exact number of arguments needed for building a data structure; the rank of a function symbol is the exact number of arguments needed for knowing how to evaluate the function call; patterns are nonevaluable expressions which represent (possibly H.O.) data; F.O. patterns represent data structures without functional components; F.O. expressions avoid H.O. variables and involve only exact curried applications.
Convention: For the rest of the paper, such expressions as "n-ary", "n arguments", etc. must be understood as refering to *curried functions*.

2.3 Continuous applicative algebras

We assume that the reader is familiar with the notion of *Scott domain* [14]. For the partial ordering of a given domain D we use the notation $\sqsubseteq_D$ or simply $\sqsubseteq$ if D is clear by the context.
A ***continuous applicative algebra*** (C.A.A.) of signature Σ is any algebraic structure $\mathcal{A}$ consisting of a *Scott domain* $D_\mathcal{A}$ with ordering $\sqsubseteq_\mathcal{A}$ as carrier, a ***continuous binary operation*** $\circ_\mathcal{A} \in [D_\mathcal{A} \times D_\mathcal{A} \rightarrow D_\mathcal{A}]$ (meant as interpretation of ***application***), which is required to be strict w.r.t. its first argument, and ***interpretations*** $\phi_\mathcal{A} \in D_\mathcal{A}$ for all symbols $\phi \in DC_\Sigma \cup FS_\Sigma$. The notion is similar to the ***applicative structures*** used in combinatory logic [11].
Any $x \in D_\mathcal{A}$ induces n-ary continuous functions $\{x\}^n_\mathcal{A}$ (for all $n \in \mathbb{N}$) defined by

$\{x\}^0_{\mathcal{A}} = x$, $\{x\}^{n+1}_{\mathcal{A}}(y) = \{x \circ_{\mathcal{A}} y\}^n_{\mathcal{A}}$. Here and in the sequel we use infix notation for the application operation. For $\phi \in DC_\Sigma \cup FS_\Sigma$, we abbreviate $\{\phi_{\mathcal{A}}\}^n_{\mathcal{A}}$ as $\phi^n_{\mathcal{A}}$.

2.4 Evaluation of expressions

A *valuation* over a given C.A.A. $\mathcal{A}$ is any mapping η: Var $\rightarrow D_{\mathcal{A}}$. The value $[\![e]\!]^{\mathcal{A}}\eta$ of an expression e in $\mathcal{A}$ under η is defined recursively:

$[\![X]\!]^{\mathcal{A}}\eta = \eta(X)$; $[\![\phi]\!]^{\mathcal{A}}\eta = \phi_{\mathcal{A}}$, for $\phi \in DC_\Sigma \cup FS_\Sigma$; $[\![e_0 e_1]\!]^{\mathcal{A}}\eta = [\![e_0]\!]^{\mathcal{A}}\eta \circ_{\mathcal{A}} [\![e_1]\!]^{\mathcal{A}}\eta$

For *ground* expressions e without occurrences of variables we write simply $[\![e]\!]^{\mathcal{A}}$ for e's value in $\mathcal{A}$.

2.5 Equations and strict equations

We refer to formulas of the form "$e_1 = e_2$" and "$e_1 == e_2$" as *equations* and *strict equations*, respectively, and define their satisfaction under valuations in C.A.A.s as follows:

$\mathcal{A} \models (e_1 = e_2)\eta$ iff $[\![e_1]\!]^{\mathcal{A}}\eta$ and $[\![e_2]\!]^{\mathcal{A}}\eta$ are *equal*

$\mathcal{A} \models (e_1 == e_2)\eta$ iff $[\![e_1]\!]^{\mathcal{A}}\eta$ and $[\![e_2]\!]^{\mathcal{A}}\eta$ are *equal*, *finite* and *total*

where finiteness and totality are understood w.r.t. $\sqsubseteq_{\mathcal{A}}$. In the sequel, we say that an element $x \in D_{\mathcal{A}}$ is *finished* iff x is finite and total.

Now we introduce some special kinds of algebras which will play an important role later.

2.6 Weakly liberal and liberal algebras

Let $\mathcal{A}$ be an arbitray C.A.A. We say that $\mathcal{A}$ is *weakly liberal* iff $c_{\mathcal{A}}$ is a finished for every $c \in DC^0_\Sigma$ and $c^n_{\mathcal{A}}$ preserves finished elements for every $c \in DC^n_\Sigma$, $n > 0$. We say that $\mathcal{A}$ is *liberal* iff it is weakly liberal and, in addition, $\phi_{\mathcal{A}}$ is a finished for every $\phi \in DC^n_\Sigma \cup FS^n_\Sigma$, $n > 0$, and $\phi^m_{\mathcal{A}}$ preserves finished elements for every $\phi \in DC^n_\Sigma \cup FS^n_\Sigma, n > m > 0$. The following is obvious from the definitions:

Proposition 2.1

(a) Let $t \in$ FOPtr$_\Sigma$. Then $[\![t]\!]^{\mathcal{A}}\eta$ is finished, provided that $\mathcal{A}$ is weakly liberal and η assigns finished values to t's variables.

(b) Let $t \in$ Ptr$_\Sigma$. Then $[\![t]\!]^{\mathcal{A}}\eta$ is finished, provided that $\mathcal{A}$ is liberal and η assigns finished values to t's variables. ■

2.7 Minimally free algebras

We say that a C.A.A. $\mathcal{A}$ is *minimally free* iff it satisfies the following conditions:

- For every $c \in DC^n_\Sigma, 1 \leq m \leq n : c^m_{\mathcal{A}}$ is injective.
- For every $c \in DC^n_\Sigma, d \in DC^m_\Sigma, c \not\equiv d$ and for all $0 \leq k \leq n, 0 \leq l \leq m : c^k_{\mathcal{A}}$ and $d^l_{\mathcal{A}}$ have disjoint ranges.
- For all $c \in DC^n_\Sigma$ and $x_i \in D_{\mathcal{A}}$: if $(c^n_{\mathcal{A}}\ x_1..x_i..x_n)$ is finite, then x_i must be finite.
- For every $t \in$FOPtr$_\Sigma$: if $\mathcal{A} \models (X = t)\eta$ and X occurs in t, then $\eta(X)$ must be infinite, unless $X \equiv t$.

2.8 Herbrand algebras

Let $\Sigma_\perp$ be the signature obtained by expanding Σ with a new nullary constructor $\perp$. The ground patterns of signature $\Sigma_\perp$ are called *ground partial patterns* of signature Σ. Over such partial patterns we consider the least partial ordering $\sqsubseteq$ which satisfies:

- $\perp \sqsubseteq t$ for all t.
- $s_i \sqsubseteq t_i (1 \leq i \leq m) \Rightarrow (\phi\ s_1..s_m) \sqsubseteq (\phi\ t_1..t_m)$ for all $m \geq 1$ and $\phi \in DC_\Sigma \cup FS_\Sigma$ whose rank is such that $(\phi\ s_1..s_m)$ and $(\phi\ t_1..t_m)$ are partial patterns.

The set of all partial Σ-patterns, equiped with the partial ordering $\sqsubseteq$, can be completed to a CPO which we call H_Σ, whose elements can be viewed as finitely branching trees in a natural way (internal nodes corresponding to curried applications of constructors and function symbols to an appropriate number of arguments). The idea is the same as in the construction of initial continuous algebras in [9]. It is easy to check that H_Σ is a domain. We say that H_Σ is the *Herbrand domain* of signature Σ. By abuse of notation and terminology, we say that the elements of H_Σ are *patterns*. It turns out that the finished elements of H_Σ can be identified with the ground syntactic patterns in Ptr_Σ. The ordering over H_Σ will be denoted by $\sqsubseteq_H$ in the sequel.
Now we can define: $\mathcal{A}$ is a *Herbrand algebra* iff $D_\mathcal{A}$ is H_Σ, $c_\mathcal{A}$ is the pattern c for all $c \in DC_\Sigma$, $f_\mathcal{A}$ is the pattern f for all $f \in FS_\mathcal{A}$ with rank(f) > 0, and the application operation $\circ_\mathcal{A}$ satisfies the following constraints:

- For any patterns $t_0,\ t_1 \in H_\Sigma$, $t_0 \circ_\mathcal{A} t_1 = (t_0 t_1)$, whenever the syntactic application $(t_0\ t_1)$ is a pattern.
- $t_0 \circ_\mathcal{A} t_1 = \bot$, whenever t_0 is of the form c, rank(c) = 0, or of the form $(c\ s_1..s_n)$, $rank(c) = n > 0$.

Note that this determines the behaviour of $\circ_\mathcal{A}$ up to the following cases:
$-\ f \circ_\mathcal{A} t$, where $rank(f) = 1$ $\quad -\ (f\ t_1..t_m) \circ_\mathcal{A} t$, where $rank(f) = m+1 = n \geq 2$
According to our notational conventions, these values can be written also as $(f^1_\mathcal{A}\ t)$ and $(f^n_\mathcal{A}\ t_1..t_m t)$, respectively. The next Proposition follows easily from this observation and the definitions.

Proposition 2.2
Any Herbrand algebra $\mathcal{A}$ is minimally free, liberal and univocally determined by the family of continuous functions $f^n_\mathcal{A}$, with $n \in \mathbb{N}$ and $f \in FS^n_\Sigma$. ∎

2.9 Scott Algebras

Let $c \in DC^n_\Sigma$. For any Scott domain D, we define $c(D^n)$ as the domain whose elements are formal patterns $(c\ x_1 \ldots x_n)$, partially ordered by: $(c\ x_1..x_n) \sqsubseteq (c\ y_1..y_n)$ iff $x_1 \sqsubseteq_D y_1$ *and*...*and* $x_n \sqsubseteq_D y_n$. For $n = 0$, $c(D^n)$ has a single bottom element. We now define D_Σ as the least solution of the recursive domain equation:

$$D \simeq .. + c(D^n) + .. + [D \to D]$$

(where c ranges over DC^n_Σ and n ranges over $\mathbb{N}$) which is unique up to isomorphism; cfr. [14]. We say that D_Σ is the *Scott domain* of signature Σ and write $\sqsubseteq_D$ for its ordering. Note that we have isomorphic copies of $c(D^n_\Sigma)$ and $[D_\Sigma \to D_\Sigma]$ included in D_Σ as subdomains; let us call them "$c(D^n_\Sigma)$ **in** D_Σ" and "$[D_\Sigma \to D_\Sigma]$ **in** D_Σ", respectively.
By induction on $n \in \mathbb{N}$, we can define the domain $\mathrm{Fun}^n(D_\Sigma)$ of continuous n-ary functions over D_Σ: $\mathrm{Fun}^0(D_\Sigma) = D_\Sigma$; $\mathrm{Fun}^{n+1}(D_\Sigma) = [D_\Sigma \to \mathrm{Fun}^n(D_\Sigma)]$. It is easy to prove that each $\mathrm{Fun}^n(D_\Sigma)$ is isomorphic to a subdomain "$\mathrm{Fun}^n(D_\Sigma)$ **in** D_Σ" of D_Σ. For $n = 1$, we write simply $\mathrm{Fun}(D_\Sigma)$. For any given $x \in D_\Sigma$ and $n \in \mathbb{N}$, we write $\mathrm{fun}^n(x)$ for the n-ary function represented by x, if $x \in \mathrm{Fun}^n(D_\Sigma)$ **in** D_Σ. By convention, $\mathrm{fun}^n(x)$ is the bottom of $\mathrm{Fun}^n(D_\Sigma)$ if $x \notin \mathrm{Fun}^n(D_\Sigma)$ **in** D_Σ. We abbreviate $\mathrm{fun}^1(x)$ as fun(x). Now we can define a binary application operation over D_Σ by $x \circ y = \mathrm{fun(x)\ (y)}$. It can be checked that this $\circ$ is continuous and strict w.r.t. its first argument.
For any $c \in DC^n_\Sigma$ we define $\mathrm{con}^n(c)$ as the element of $\mathrm{Fun}^n(D_\Sigma)$ **in** D_Σ representing the n-ary function which maps each (curried) tuple $x_1..x_n$ into $(c\ x_1..x_n)$ **in** D_Σ. For $n = 0$, $\mathrm{con}^0(c)$ is simply c **in** D_Σ.
Finally, we can define: $\mathcal{A}$ is a *Scott algebra* iff $D_\mathcal{A}$ is D_Σ, $c_\mathcal{A}$ is $\mathrm{con}^n(c)$ for all $c \in DC^n_\Sigma$, $f_\mathcal{A} \in \mathrm{Fun}^n(D_\Sigma)$ **in** D_Σ for all $f \in FS^n_\Sigma$, and $\circ_\mathcal{A}$ is $\circ$.
Let $\mathcal{A}$ be a Scott algebra. For $x,\ y \in \mathrm{Fun}^n(D_\Sigma)$ **in** D_Σ, it is easy to see that

$\{x\}^n_{\mathcal{A}} = \{y\}^n_{\mathcal{A}}$ implies $x = y$. The next proposition follows from this observation and our definitions:

Proposition 2.3
Any Scott algebra $\mathcal{A}$ is minimally free, weakly liberal and univocally determined by the family of continuous functions $f^n_{\mathcal{A}}$, with $n \in \mathbb{N}$ and $f \in FS^n_{\Sigma}$. ■

Note that Scott algebras are *not* liberal. In fact, the only finished elements of D_Σ are values of F.O. ground patterns, which belong to the "$c(D^n_\Sigma)$ in D_Σ" subdomains.

3 A Simple Functional Language and its Operational Semantics

In this Section we use *infinite rewriting*, similarly as in [8,12] for modelling the essential operational features of a *lazy functional language*. Our formalism *will not* include λ-abstractions. However, it will be possible to express H.O. functions through partial application, as done in some existing languages such as Miranda [15].

3.1 Defining rules and SFL programs

A *defining rule* for $f \in FS^n_\Sigma$ is any conditional equation of the form:

$$f\ t_1 \,..\, t_n = r \Leftarrow l_1 == r_1, .., l_m == r_m$$

where $t_1 \,..\, t_n$ is a *linear tuple of F.O. patterns* (i.e. no variable occurs more than one time) and r, l_j, r_j are expressions. We say that $l \equiv f\ t_1 \ldots t_n$ is the *left hand side* (lhs), r is the *right hand side* (rhs) and $l_j == r_j$ are the *condition*. Any variable occurring in the rhs or in the condition is required to occur also in the lhs. Sometimes we shall display a defining rule as $l = r \Leftarrow C$ where C stands for the condition, or simply as $l = r$ if the condition is empty.

Consider a pair of different rules for the same f, renamed apart so that they do not share variables:

$$R_1 : f\ t_1 \,..\, t_n = r_1 \Leftarrow C_1 \qquad R_2 : f\ s_1 \,..\, s_n = r_2 \Leftarrow C_2.$$

We say that this pair of rules is *nonoverlapping* iff for every minimally free algebra $\mathcal{A}$ there is no valuation η: Var $\rightarrow D_{\mathcal{A}}$ such that $\mathcal{A} \models (t_i = s_i)\eta$ for all $1 \le i \le n$, $\mathcal{A} \models (C_1 \cup C_2)\eta$ and $[\![r_1]\!]^{\mathcal{A}}\eta \ne [\![r_2]\!]^{\mathcal{A}}\eta$. Although this condition is not an effective one, there are effective and pragmatically convenient conditions that imply it; see Appendix.

A set $\mathcal{R}$ of defining rules which includes no pair of overlapping rules will be called *nonambiguous* in the sequel. We also say that a nonambiguous set of rules is a *SFL program*. SFL stands for "Simple Functional Language". Let us show some simple examples of SFL programs. In each case, we explain briefly why the rules cannot overlap over any minimally free algebra.

Example 1 *The map function*

map F nil = nil
map F(cons X Xs) = cons X(map F Xs)

The constructors are "nil" and "cons". The two lhs do not overlap.

Example 2 *Parallel and*

and true $Y = Y$ and X true $= X$
and false Y = false   and X false = false

The constructors are "true" and "false". When two lhs overlap, the two rhs become identical.

Example 3 *Less trivial nonambiguity*

$f\ X\ Y = X \Leftarrow g\ X\ == Y\ , g(g\ X) == suc\ Y$
$f\ X\ Y = Y \Leftarrow g\ Y == suc\ X, g(g\ Y) == X$

The only constructor is "suc". The conjunction of the two conditions implies "X is finite and X = suc X", which is impossible in a minimally free algebra.

3.2 Finite rewriting derivations

We use *join conditional rewriting* [7] as the operational semantics of SFL. Hence, the operational interpretation of a defining rule $f\ t_1..t_n = r \Leftarrow l_1 == r_1, .., l_m == r_m$ is a *join conditional* rewriting rule $f\ t_1 .. t_n \rightarrow r \Leftarrow l_1 \downarrow r_1, .., l_m \downarrow r_m$
A condition $l_j \downarrow r_j$ must be satisfied by rewriting l_j and r_j to the same pattern $t \in Ptr_\Sigma$. Formally, finite rewriting with the rules given by a SFL program $\mathcal{R}$ depends on the following notions:

•*One step rewriting*: $e \rightarrow e'$ •*Several steps rewriting*: $e \rightarrow^* e'$ •*Joining*: $l \downarrow r$
We define them by mutual recursion. In what follows, σ denotes any finite substitution of expressions for variables, and $C[e]$ denotes any *context* of the expression e, i.e., any expression with a distinguished occurrence of e as subexpression.

$$\frac{l_1\sigma \downarrow r_1\sigma\ ..\ l_m\sigma \downarrow r_m\sigma}{C[l\sigma] \rightarrow C[r\sigma]} \quad \text{if } l = r \Leftarrow l_1 == r_1, .., l_m == r_m \text{ is a rule in } \mathcal{R}.^1$$

$$\frac{}{e \rightarrow^* e} \qquad \frac{e \rightarrow e' \quad e' \rightarrow^* e''}{e \rightarrow^* e''} \qquad \frac{l \rightarrow^* t \quad r \rightarrow^* t}{l \downarrow r} \quad \text{if } t \in Ptr_\Sigma$$

If $e \rightarrow^* e'$, we say that e *reduces to e' by a finite rewriting derivation.* Note that each top level step of this derivation corresponds to the application of a rule and may depend on finitely many other finite derivations, used to check the conditions by joining. It makes sense to reason inductively over the finite structure of such derivations. For instance, for proving a property of a derivation we can assume as induction hypothesis that the same property holds for the "subderivations" used for checking conditions. We shall use this proof technique later.

3.3 Infinite rewriting derivations

Our notion of *infinite rewriting* differs from [8,12] in some respects. For computing limits we use l.u.b.s in the Herbrand domain instead of Cauchy convergence in a metric space of trees. Moreover, our rules are *H.O.* and *conditional* rather than *F.O.* and *unconditional.* Admitedly, many interesting theoretical questions dealt with in [8,12] become trivial in our setting, due to our restriction to *constructor based* rules. We need an auxiliary notion. The *shell* of an expression e is a pattern $|\, e \,| \in H_\Sigma$ defined by recursion on e's structure:

$|\, c \,| = c$, for all $c \in DC_\Sigma$ $\quad |\, f \,| = f$, for all $f \in FS^n_\Sigma, n > 0$
$|\, (c\ e_1\ ..\ e_m) \,| = (c\ |\, e_1 \,|\ ..\ |\, e_m \,|)$, for all $c \in DC^n_\Sigma, n \geq m \geq 1$
$|\, (f\ e_1\ ..\ e_m) \,| = (f\ |\, e_1 \,|\ ..\ |\, e_m \,|)$, for all $f \in FS^n_\Sigma, n > m \geq 1$
$|\, e \,| = \perp$, in all other cases.

Intuitively, $|\, e \,|$ is the outermost part of e that cannot be further reduced. The following proposition follows easily from the definitions:

Proposition 3.1

(a) For any e, $\mathcal{A}$ and η: $[\![\, |\, e \,|\,]\!]^{\mathcal{A}}\eta \sqsubseteq_{\mathcal{A}} [\![e]\!]^{\mathcal{A}}\eta$

(b) For all e, e' : $e \rightarrow^* e'$ implies $|\, e \,| \sqsubseteq_H |\, e' \,|$ ■

Now, given a ground expression $e \in \mathrm{Exp}_\Sigma$ and a pattern $t \in H_\Sigma$, we can define:

- *e finitely converges to t* (in symbols, $e \rightarrow^*_c t$) iff there is e' such that $e \rightarrow^* e'$ and $|\, e' \,| \sqsupseteq_H t$. Note that every e trivially converges to $\perp$.
- *e converges to t in ω steps* (in symbols, $e \rightarrow^\omega_c t$) iff there is an infinite derivation $e \equiv e_0 \rightarrow e_1 .. \rightarrow e_n \rightarrow\ ..$ such that $lub_n\, |\, e_n \,| \sqsupseteq_H t$
(Note that each top level step $e_n \rightarrow e_{n+1}$ may recursively depend on other finite derivations. The lub is taken w.r.t. $H'_\Sigma s$ ordering).
- *e converges to t in at most ω steps* (in symbols, $e \rightarrow^{\leq\omega}_c t$) iff $e \rightarrow^\omega_c t$ or $e \rightarrow^*_c$t.

[1] Note that $m \geq 0$.

For example, given a SFL program with constructors "pair", "0", "suc", "nil" and "cons" and rules:
$f\ X = pair\ (d\ X)\ (g\ X) \quad g\ X = cons\ X\ (g(suc\ X)) \quad d\ X = i\ (d\ X) \quad i\ Z = Z$
we have (using a friendly notation for lists):

$$f\ 0 \rightarrow_c^\omega pair\ \bot\ [0,\ suc\ 0,\ suc\ (suc\ 0),\ ..\ ]$$

Note that there are many different derivations converging to this limit, as well as derivations that would compute less information in the limit. For instance, the derivation:

$$f\ 0 \rightarrow pair\ (d\ 0)\ (g\ 0) \rightarrow pair\ (i\ (d\ 0))\ (g\ 0) \rightarrow pair\ (d\ 0)\ (g\ 0) \rightarrow\ ..$$

(where $(g\ 0)$ is never reduced) would converge to the limit pattern $(pair\ \bot\ \bot)$. Intuitively, a *fairness* condition is required for infinite derivations to compute as much information as possible. In order to formalize this idea we define some concepts.

An expression e is in *weak head normal form* (WHNF) iff e is either a constructor, or a function symbol of positive rank, or a partial application of a symbol in Σ, or an exact application of a constructor in Σ. An expression e *admits* a WHNF iff it can be reduced to an expression in WHNF by some finite derivation. A *redex* of e is any subexpression of e that has the form of an exact application of a function symbol (or a function symbol of rank 0) and can be rewritten in one top level step. A redex r of e is called *outermost* iff some of the following cases applies:

- e is in WHNF and r is an outermost redex of some of the arguments.
- e is of the form $(f\ e_1\ ..\ e_n..e_m)$, $m \geq n =$ rank(f), and r is $(f\ e_1\ ..\ e_n)$ (i.e. f in case $n = 0$).
- e is of the form $(f\ e_1\ ..\ e_n..e_m)$, $m \geq n =$ rank(f), $(f\ e_1\ ..\ e_n)$ is not a redex and r is an outermost redex of e_i for some $1 \leq i \leq$n.

Note that if r is an outermost redex of e, any derivation starting at e necessarily preserves r as long r itself is not rewritten. Now we can define: An infinite derivation of length ω, say $e_0 \rightarrow e_1 .. \rightarrow e_n \rightarrow ..$ is called *fair* iff for every n and every outermost redex r of e_n that admits a WHNF there is some $m \geq n$ such that r is rewritten in the step $e_m \rightarrow e_{m+1}$. By convention, all finite derivations are regarded as fair.

4 Declarative and Denotational Semantics

In this Section we explain how to relate C.A.A.s to SFL programs by means of the notion of *model* and show that two alternative semantics of SFL programs (*declarative* and *denotational*) correspond to distinguished models, both obtained by the same kind of fixpoint construction.

4.1 Models of SFL programs

Let a C.A.A. $\mathcal{A}$ and a SFL program $\mathcal{R}$ be given. The following notions define the *logical meaning* of rules and programs in C.A.A.s.

- $\mathcal{A}$ *satisfies* (resp., *exactly satisfies*) a rule $l = r \Leftarrow C$ iff every η such that $\mathcal{A} \models C\eta$ verifies $[\![l]\!]^\mathcal{A}\eta \sqsupseteq [\![r]\!]^\mathcal{A}\eta$ (resp. $[\![l]\!]^\mathcal{A}\eta = [\![r]\!]^\mathcal{A}\eta$).
- $\mathcal{A}$ is a *model* (resp. *exact model*) of $\mathcal{R}$ iff $\mathcal{A}$ satisfies (resp. exactly satisfies) all the rules in $\mathcal{R}$.

We use the notations $\mathcal{A} \models \mathcal{R}$, $\mathcal{A} \models_= \mathcal{R}$ for models and exact models, respectively.

4.2 Least Herbrand models and declarative semantics

We define a partial ordering $\sqsubseteq_{HALG}$ over the family HALG$_\Sigma$ of all Herbrand algebras of signature Σ by $\mathcal{A} \sqsubseteq_{HALG} \mathcal{B}$ iff $f_\mathcal{A} \sqsubseteq f_\mathcal{B}$ for every $f \in FS_\Sigma$ and $\circ_\mathcal{A} \sqsubseteq \circ_\mathcal{B}$. With the help of Proposition 2.2 we can show:

Proposition 4.1
HALG$_\Sigma$ is a Scott domain under the ordering $\sqsubseteq_{HALG}$. Moreover, for any $\mathcal{A}$, $\mathcal{B} \in$ HALG$_\Sigma$: $\mathcal{A} \sqsubseteq_{HALG} \mathcal{B}$ iff $f^n_\mathcal{A} \sqsubseteq f^n_\mathcal{B}$ for every $f \in FS_\Sigma$, rank(f) = n. ∎

Let us fix an arbitrary SFL program $\mathcal{R}$. Given an n-ary $f \in FS_\Sigma$, a Herbrand algebra $\mathcal{A}$ and elements $x_1 .. x_n \in D_\mathcal{A}$, we define $R^\mathcal{A}[f\ x_1 .. x_n]$ as the set of all values $[\![r]\!]^\mathcal{A}\eta$ such that η is a valuation of variables over $D_\mathcal{A}$ and there is some rule $f\ t_1 .. t_n = r \Leftarrow C$ in $\mathcal{R}$ satisfiying $\mathcal{A} \models C\eta$ and $[\![t_i]\!]^\mathcal{A}\eta = x_i$ for $1 \leq i \leq$n. Since Herbrand algebras are minimally free (Proposition 2.2) and programs are nonambiguous, this set is either empty or a singleton. Hence, $lub\ R^\mathcal{A}[f\ x_1 .. x_n]$ is well defined (and equals either $\perp$ or the single member of the set). Moreover, it can be checked that the n-ary function $\mathcal{T}_\mathcal{R}(\mathcal{A}, f)$ over $D_\mathcal{A}$ defined by $\mathcal{T}_\mathcal{R}(\mathcal{A}, f)\ x_1 .. x_n = lub R^\mathcal{A}[f\ x_1 .. x_n]$ is continuous. Again by Proposition 2.2, there is a unique Herbrand algebra $\mathcal{T}_\mathcal{R}(\mathcal{A}) = \mathcal{B}$ such that $f^n_\mathcal{B} = \mathcal{T}_\mathcal{R}(\mathcal{A}, f)$ for all $f \in FS_\Sigma$, $n = rank(f)$. We have just obtained an operator $\mathcal{T}_\mathcal{R}$: HALG$_\mathcal{R}$ $\rightarrow$ HALG$_\mathcal{R}$, similar to the "immediate consequences" operator over Herbrand interpretations for F.O. logic programs [1]. We can prove:

Theorem 4.1 $\mathcal{T}_\mathcal{R}$ is continuous and verifies:

(a) For any $\mathcal{A} \in$ HALG$_\mathcal{A}$: $\mathcal{A} \models \mathcal{R}$ iff $\mathcal{T}_\mathcal{R}(\mathcal{A}) \sqsubseteq_{HALG} \mathcal{A}$.

(b) $\mathcal{T}_\mathcal{R}$ has a least fixpoint $\mathcal{H}_\mathcal{R} = lub_k \mathcal{H}^k_\mathcal{R}$, where $\mathcal{H}^0_\mathcal{R}$ is the bottom of HALG$_\Sigma$ and $\mathcal{H}^{k+1}_\mathcal{R} = \mathcal{T}_\mathcal{R}(\mathcal{H}^k_\mathcal{R})$.

(c) $\mathcal{H}_\mathcal{R}$ is the least Herbrand model of $\mathcal{R}$ and an exact model.

Proof Sketch

The continuity of $\mathcal{T}_\mathcal{R}$ is proved by using the fact that the evaluation of a fixed expression in $\mathcal{A}$ under η behaves as a continuous function of $\mathcal{A}$, η. (a) follows easily from the definitions, as in the case of logic programs. (b) holds because $\mathcal{T}_\mathcal{R}$ is continuous. Given any Herbrand model $\mathcal{A} \models \mathcal{R}$ we have $\mathcal{T}_\mathcal{R}(\mathcal{A}) \sqsubseteq_{HALG} \mathcal{A}$ by (a), and this can be used to prove $\mathcal{H}^k_\mathcal{R} \sqsubseteq_{HALG} \mathcal{A}$ (for all k) by induction on k; hence, $\mathcal{H}_\mathcal{R} \sqsubseteq_{HALG} \mathcal{A}$. To complete the proof of (c), we have to show that $\mathcal{H}_\mathcal{R}$ is an exact model of $\mathcal{R}$. We abbreviate $\mathcal{H}_\mathcal{R}$ as $\mathcal{H}$ for simplicity. For any rule $l = r \Leftarrow C$ in $\mathcal{R}$ and any valuation η over $\mathcal{H}$ such that $\mathcal{H} \models C\eta$, we have to show that $[\![l]\!]^\mathcal{H}\eta = [\![r]\!]^\mathcal{H}\eta$. Since the equations in C are strict and $\mathcal{H} = lub_k \mathcal{H}^k$, continuity of expression evaluation guarantees the existence of some k_0 such that $\mathcal{H}^k \models C\eta$ for all $k \geq k_0$. Then $[\![r]\!]^\mathcal{H}\eta = lub_k [\![r]\!]^{\mathcal{H}^k}\eta = lub_{k \geq k_0} [\![r]\!]^{\mathcal{H}^k}\eta =^2 lub_{k \geq k_0} [\![l]\!]^{\mathcal{H}^{k+1}}\eta = lub_k [\![l]\!]^{\mathcal{H}^k}\eta = [\![l]\!]^\mathcal{H}\eta$ which finishes the proof. ∎

We can now define the *declarative meaning* dcl$[\![e]\!]$ of a ground expression e w.r.t. a program $\mathcal{R}$ as $[\![e]\!]^\mathcal{H}$, where $\mathcal{H} = \mathcal{H}_\mathcal{R}$.

4.3 Initiality of the least Herbrand model

Here we show that least Herbrand models are initial objects in a suitable category. Let $\mathcal{A}, \mathcal{B}$ be C.A.A.s of signature Σ. We define a *morphism* $h : \mathcal{A} \rightarrow \mathcal{B}$ as any continuous and strict mapping $h : D_\mathcal{A} \rightarrow D_\mathcal{B}$ which satisfies the following conditions, where x_i range over $D_\mathcal{A}$:

- $h((c^m_\mathcal{A} x_1 .. x_m)) = (c^m_\mathcal{B} h(x_1) .. h(x_m))$ for $c \in DC^n_\Sigma, n \geq m \geq 0$
- $h((f^m_\mathcal{A} x_1 .. x_m)) = (f^m_\mathcal{B} h(x_1) .. h(x_m))$ for $f \in FS^n_\Sigma, n > m \geq 0$
- $h(f_\mathcal{A}) \sqsubseteq_\mathcal{B} f_\mathcal{B}$ for $f \in FS^0_\Sigma$
- $h(x \circ_\mathcal{A} y) \sqsubseteq_\mathcal{B} h(x) \circ_\mathcal{B} h(y)$ for $x,\ y \in D_\mathcal{A}$

It is trivial to check that the functional composition of morphisms is again a morphism. Thus, the liberal models of $\mathcal{R}$ as objects with morphisms as arrows form a category LIBMOD$_\mathcal{R}$.

Theorem 4.2

For any SFL program $\mathcal{R}$, the least Herbrand model $\mathcal{H}_\mathcal{R}$ is initial in LIBMOD$_\mathcal{R}$.

[2](since $\mathcal{H}^{k+1} = \mathcal{H}_\mathcal{R}(\mathcal{H}^k)$)

Proof Sketch
For any (possibly partial and/or infinite) pattern $t \in H_\Sigma$ and any C.A.A. $\mathcal{A}$, we can define the value $[t]^{\mathcal{A}}$ as the lub of all values $[s]^{\mathcal{A}}$ where s ranges over the finite approximations of t in H_Σ. The mapping $ev_{\mathcal{A}} : H_\Sigma \rightarrow D_{\mathcal{A}}$ defined by $ev_{\mathcal{A}}(t) = [t]^{\mathcal{A}}$ turns out to be continuous and strict. It is easily seen that any morphism $h : \mathcal{H}_{\mathcal{R}} \rightarrow \mathcal{A}$ (if existing) must be identical to $ev_{\mathcal{A}}$. Thus, it suffices to check that $ev_{\mathcal{A}}$ is a morphism in the case that $\mathcal{A}$ is a liberal model of $\mathcal{R}$. The proof of this depends on the liberality of $\mathcal{A}$ and the characterization of $\mathcal{H}_{\mathcal{R}}$ as $\mathrm{lub}_k \mathcal{H}^k_{\mathcal{R}}$ given by Theorem 4.1. ∎

4.4 Least Scott models and denotational semantics

The constructions that have led us to least Herbrand models can be performed in an analogous way for Scott models, by using Proposition 2.3 instead of Proposition 2.2. With the natural definition of a partial ordering $\sqsubseteq_{SALG}$, we get the analogon of Proposition 4.1:

Proposition 4.2
SALG_Σ is a Scott domain under the ordering $\sqsubseteq_{SALG}$. Moreover, for any $\mathcal{A}, \mathcal{B} \in \mathrm{SALG}_\Sigma$: $\mathcal{A} \sqsubseteq_{SALG} \mathcal{B}$ iff $f^n_{\mathcal{A}} \sqsubseteq f^n_{\mathcal{B}}$ for every $f \in FS_\Sigma$, rank(f) = n. ∎

Moreover, we can define an operator $\mathcal{S}_{\mathcal{R}}$: $\mathrm{SALG}_{\mathcal{R}} \rightarrow \mathrm{SALG}_{\mathcal{R}}$ and prove the analogon of Theorem 4.1:

Theorem 4.3 $\mathcal{S}_{\mathcal{R}}$ is continuous and verifies:

(a) For any $\mathcal{A} \in \mathrm{SALG}_{\mathcal{A}} : \mathcal{A} \models \mathcal{R}$ iff $\mathcal{S}_{\mathcal{R}}(\mathcal{A}) \sqsubseteq_{SALG} \mathcal{A}$.

(b) $\mathcal{S}_{\mathcal{R}}$ has a least fixpoint $\mathcal{D}_{\mathcal{R}} = lub_k \mathcal{D}^k_{\mathcal{R}}$, where $\mathcal{D}^0_{\mathcal{R}}$ is the bottom of SALG_Σ and $\mathcal{D}^{k+1}_{\mathcal{R}} = \mathcal{S}_{\mathcal{R}}(\mathcal{D}^k_{\mathcal{R}})$.

(c) $\mathcal{D}_{\mathcal{R}}$ is the least Scott model of $\mathcal{R}$ and an exact model. ∎

We can now define the *denotational meaning* dnt[e] of a ground expression e w.r.t. a program $\mathcal{R}$ as $[e]^{\mathcal{D}}$, where $\mathcal{D} = \mathcal{D}_{\mathcal{R}}$. Since $\mathcal{D}$ comes from a least fixpoint construction, it is clear that this corresponds to the usual definition of denotational semantics. Note that $\mathcal{D}$ is not an object of $\mathrm{LIBMOD}_{\mathcal{R}}$, since Scott algebras are not liberal. In fact, an analogon of Theorem 4.2 could neither be obtained for the category of weakly liberal models of $\mathcal{R}$, because $\mathcal{D}'s$ carrier has too many "junk" elements.

5 Soundness and Completeness

In this Section we investigate the relationship between rewriting and the two mathematical semantics defined in Section 4. It will turn out that declarative semantics characterizes the operational behaviour of SFL programs more adequately than denotational semantics.
Let us fix an arbitrary SFL program $\mathcal{R}$. Our subsequent results will refer to rewriting by means of the rules in $\mathcal{R}$. Also, $\mathcal{H}$ and $\mathcal{D}$ will stand for the least Herbrand model and the least Scott model of $\mathcal{R}$, respectively.

5.1 Soundness and completeness w.r.t. declarative semantics

First we establish the *soundness of finite rewriting*:

Lemma 5.1 (*Soundness lemma*)
Finite rewriting is *sound w.r.t. all liberal models* in the following sense: For all ground expressions e, e', r, l and any liberal model $\mathcal{A}$ of $\mathcal{R}$:

(a) $e \rightarrow^* e'$ implies $[e]^{\mathcal{A}} \sqsupseteq_{\mathcal{A}} [e']^{\mathcal{A}}$ and even $[e]^{\mathcal{A}} = [e']^{\mathcal{A}}$ if $\mathcal{A}$ is an exact model.

(b) $r \downarrow l$ implies $\mathcal{A} \models r == l$.

Proof Sketch

We prove (a) and (b) simultaneously by induction on the structure of finite rewriting derivations. The base case for (a) is that e' comes from e in one rewriting step by means of an inconditional rule. In this case (a) is obvious by definition of the notions of model and exact model. The base case for (b) is that $r \equiv l \in \mathrm{Ptr}_\Sigma$. In this case (b) holds because $[r]^{\mathcal{A}}$ is a finished element by the liberality of $\mathcal{A}$ (Proposition 2.1 (b)). The inductive step for (a) follows from the inductive hypothesis of (a) and (b), using again the definitions of model and exact model. For proving the inductive step for (b), we can apply the inductive hypothesis of (a) to get $[l]^{\mathcal{A}} \sqsupseteq_{\mathcal{A}} [t]^{\mathcal{A}} \sqsubseteq_{\mathcal{A}} [r]^{\mathcal{A}}$(where $t \in \mathrm{Ptr}_\Sigma$ is such that $l \rightarrow^* t^* \leftarrow r$); from this we infer $[l]^{\mathcal{A}} == [t]^{\mathcal{A}} == [r]^{\mathcal{A}}$ using again that $[t]^{\mathcal{A}}$ is finished by the liberality of $\mathcal{A}$. ■

With the help of this result we can prove the *soundness of converging derivations*:

Theorem 5.1 (*Soundness w.r.t. liberal models*)

Converging derivations are *sound w.r.t. all liberal models* in the following sense: For every ground expression e, every pattern $t \in H_\Sigma$ and any liberal model $\mathcal{A}$ of $\mathcal{R} : e \rightarrow_c^{\leq\omega} t$ implies $[e]^{\mathcal{A}} \sqsupseteq_{\mathcal{A}} [t]^{\mathcal{A}}$ and even $[e]^{\mathcal{A}} = [t]^{\mathcal{A}}$ if $[t]^{\mathcal{A}}$ is a total element (in particular, if $t \in \mathrm{Ptr}_\Sigma$, by Proposition 2.1 (b)).

Proof

First note that $[t]^{\mathcal{A}}$ makes sense even if t is partial and/or infinite, as discussed in the proof of Theorem 4.2.

We have either $e \rightarrow_c^* t$ or $e \rightarrow_c^\omega$ t. In the first case, there is some e' such that $e \rightarrow^* e'$ and $\mid e' \mid \sqsupseteq_H t$. By Proposition 3.1 (a) we know that $[\ \mid e' \mid\ ]^{\mathcal{A}} \sqsubseteq_{\mathcal{A}} [e']^{\mathcal{A}}$, and the thesis follows from Lemma 5.1. In the second case, there is an infinite derivation $e \equiv e_0 \rightarrow e_1 \,..\, \rightarrow e_n \rightarrow \,..$ such that $lub_n \mid e_n \mid \sqsupseteq_H t$. By Proposition 3.1 (b) we have $\mid e_n \mid \sqsubseteq_H \mid e_{n+1} \mid$ for all n, and we can reason as follows:

$$
\begin{array}{lll}
[t]^{\mathcal{A}} & \sqsubseteq_{\mathcal{A}} lub_n[\mid e_n \mid]^{\mathcal{A}} & \text{(easy to prove from } t \sqsubseteq_H lub_n \mid e_n \mid) \\
 & \sqsubseteq_{\mathcal{A}} lub_n[\![e_n]\!]^{\mathcal{A}} & \text{(by Proposition 3.1 (a))} \\
 & \sqsubseteq_{\mathcal{A}} lub_n[e]^{\mathcal{A}} & ([\![e_n]\!]^{\mathcal{A}} \sqsubseteq_{\mathcal{A}} [e]^{\mathcal{A}} \text{ for all } n\text{, by Lemma 5.1)} \\
 & = [e]^{\mathcal{A}} &
\end{array}
$$

■

As a corollary we obtain:

Theorem 5.2 (*Soundness w.r.t. declarative semantics*)

For any ground expression e and any pattern $t \in H_\Sigma : e \rightarrow_c^{\leq\omega} t$ implies $dcl[\![e]\!] \sqsupseteq_H t$ and even $dcl[\![e]\!] = t$ if t is a total pattern.

Proof

Inmediate from the previous theorem, by observing that $\mathcal{H}$ is a liberal model of $\mathcal{R}$ and that $[t]^{\mathcal{H}} = t$ for all $t \in H_\Sigma$. ■

For any ground expression e and any $k \in \mathbb{N}$, let us abbreviate $[e]^{\mathcal{H}^k}$ as $[e]^k$ and remember that the characterization of $\mathcal{H}$ as $\mathrm{lub}_k\mathcal{H}^k$ implies $dcl[\![e]\!] = lub_k[\![e]\!]^k$. Let us also write o^k for the application operation of $\mathcal{H}^k$. Our completeness results will follow from a key lemma which is the main result in the paper:

Lemma 5.2 (*Completeness Lemma*)

For every ground expression e and every $k \in \mathbb{N}$ there is some finite rewriting derivation $e \rightarrow^* e'$ such that $\mid e' \mid \sqsupseteq_H [e]^k$. This implies $\mid e' \mid = [\![e]\!]^k$ if $[e]^k$ is a total pattern.

Proof

We reason by induction over the well founded ordering:

$(e', k') \prec (e, k)$ iff either $k' < k$ or ($k' = k$ and e' is aproper subexpression of e)

where k, k' are natural numbers and e, e' are ground expressions. Given e and k, e must have one of the forms $(e_0e_1), f$ or c. We go through the three cases.

Case 1: $e \equiv (e_0 e_1)$.
It suffices to consider the nontrivial case $[\![e_0e_1]\!]^k \neq\perp$. For i= 1, 2 we get by induction hypothesis applied to e_i, k : $e_i \rightarrow^* e'_i$ with $|\ e'_i\ | \sqsupseteq_H [\![e_i]\!]^k$. It follows that $[\![e_0e_1]\!]^k \sqsubseteq_H |\ e'_0\ | \circ^k |\ e'_1\ | \neq\perp$ and by strictness of $\circ^k$ w.r.t. its first argument, we also know that $|\ e'_0\ | \neq\perp$. This means that e'_0 must be either a partial application or an exact application or a constructor. This last possibility is excluded because it would imply $|\ e'_0\ | \circ^k |\ e'_1\ | =\perp$. We are left with two subcases:
Subcase 1.1: $e'_0 \equiv (c\ a_1\ ..\ a_m)$ for some $c \in DC_\Sigma$, $rank(c) > m \geq 0$.
The derivation $e_0e_1 \rightarrow e'_0e'_1$ is already sufficient, since $|\ e'_0e'_1\ | = |\ (c\ a_1\ ..\ a_m e'_1)\ | = |\ e'_0\ | \circ^k |\ e'_1\ | \sqsupseteq_H [\![e_0e_1]\!]^k$
Subcase 1.2: $e'_0 \equiv (f\ a_1\ ..\ a_m)$ for some $f \in DC_\Sigma$, $rank(f) > m \geq 0$.
We may have $m+1 <$ rank(f) or $m+1 =$ rank(f). In the first case, we can reason exactly as in Subcase 1.1. So, let us assume that $m+1 = rank(f)$. Then we note that $\perp \neq |\ e'_0\ | \circ^k |\ e'_1\ | = (\{[\![f]\!]^k\}^{m+1}\ |\ a_1\ |\ ..\ |\ a_m\ ||\ e'_1\ |)$. Let us rename e'_1 as a_{m+1} for uniformity of notation. Remembering that $\mathcal{H}^k$ is defined by iteration of the operator $\mathcal{T}_\mathcal{R}$, we can infer that $k > 0$ and that there is some rule $R \equiv f\ t_1\ ..t_m\ t_{m+1} = r \Leftarrow C$ in $\mathcal{R}$ and some valuation η: Var $\rightarrow H_\Sigma$ verifying $|\ a_i\ | = [\![t_i]\!]^{k-1}\eta$ for $1 \leq i \leq m+1$, $\mathcal{H}^{k-1} \models C\eta$ and $[\![e_0e_1]\!]^k \sqsubseteq_H (\{[\![f]\!]^k\}^{m+1}\ |\ a_1\ |\ ..\ |\ a_m\ ||\ e'_1\ |) = [\![r]\!]^{k-1}\eta$. Note that all the variables occurring in r and C must occur in $(f\ t_1\ ..\ t_m\ t_{m+1})$. Because of the linearity of the tuple $t_1\ ..\ t_m t_{m+1}$ it is easy to prove that $|\ a_i\ | = [\![t_i]\!]^{k-1}\eta$ $(1 \leq i \leq m+1)$ implies the existence of a finite substitution σ: Var $\rightarrow$ Exp$_\Sigma$ such that $(f\ t_1\ ..\ t_m\ t_{m+1})\sigma \equiv (f\ a_1\ ..\ a_m\ a_{m+1})$ and $|\ \sigma(X)\ | = \eta(X)$ for every variable X occurring in $(f\ t_1\ ..\ t_m\ t_{m+1})$. This last property of σ can be used to justify the following
Claim: $[\![b\sigma]\!]^{k-1} \sqsupseteq_H [\![b]\!]^{k-1}\eta$ *holds for any expression b which uses only variables occurring in* $(f\ t_1\ ..\ t_m\ t_{m+1})$.
Indeed, if we note as σ^{k-1}, $|\ \sigma\ |$ the valuations defined by $\sigma^{k-1}(X) = [\![\sigma(X)]\!]^{k-1}$ and $|\ \sigma\ |\ (X) = |\ \sigma(X)\ |$, respectively, we can prove the claim as follows:

$$\begin{aligned} [\![b\sigma]\!]^{k-1} &= [\![b]\!]^{k-1}\sigma^{k-1} && \text{(by a well known substitution lemma)} \\ &\sqsupseteq_H [\![b]\!]^{k-1}\ |\ \sigma\ | && \text{(because } \sigma^{k-1}(X) \sqsupseteq_H |\ \sigma\ |\ (X) \text{ for all } X) \\ &= [\![b]\!]^{k-1}\eta && \text{(by coincidence of } |\ \sigma\ | \text{ and } \eta \text{ over } b) \end{aligned}$$

Now, from $\mathcal{H}^{k-1} \models C\eta$ we can infer that any expression b that occurs as member of some strict equation in C must satisfy $[\![b]\!]^{k-1}\eta = t_b$ for some ground $t_b \in$Ptr$_\Sigma$ (since these are the only finished elements in H_Σ). Since t_b is total, the claim implies that also $[\![b\sigma]\!]^{k-1} = t_b$. We can apply the induction hypothesis to $b\sigma$, $k-1$ and conclude that $b\sigma \rightarrow^* t_b$ for each of these expressions b. These derivations establish $l\sigma \downarrow r\sigma$ for all $l == r$ in C. Hence, R can be used to rewrite $(f\ a_1\ ..\ a_m\ a_{m+1})$ to $r\sigma$, and combining this step with the induction hypothesis for $r\sigma$, $k-1$ we get a derivation:

$$(e_0e_1) \rightarrow^* (e'_0e'_1) \equiv (f\ a_1\ ..\ a_m\ a_{m+1}) \rightarrow r\sigma \rightarrow^* e'$$

where $|\ e'\ | \sqsupseteq_H [\![r\sigma]\!]^{k-1}$. Since the claim guarantees that $[\![r\sigma]\!]^{k-1} \sqsupseteq_H [\![r]\!]^{k-1}\eta$ and we noticed above that $[\![r]\!]^{k-1}\eta \sqsupseteq_H [\![e_0e_1]\!]^k$, we are done.
Case 2 : $e \equiv f$, $f \in FS^n_\Sigma$
The trivial derivation $f \rightarrow^* f$ is again sufficient if either $n > 0$ or $[\![f]\!]^k =\perp$. Hence, we assume $n = 0$ and $[\![f]\!]^k \neq\perp$. Similarly as in Subcase 1.2, we can infer that $k > 0$ and that there is some rule $R \equiv f = r \Leftarrow C$ in $\mathcal{R}$ such that $\mathcal{H}^{k-1} \models C$ and $[\![f]\!]^k = [\![r]\!]^{k-1}$. Note that both r and C must be ground. Essentially by the same reasoning as in Subcase 1.2 (simplified by the fact that r and C are ground) we can prove that R can be used to rewrite f to r, and combining this step with the induction hypothesis for r, $k-1$ we get a derivation $f \rightarrow r \rightarrow^* e'$ which obviously verifies $|\ e'\ | \sqsupseteq_H [\![f]\!]^k$.

Case 3 : $e \equiv c$, $c \in DC_\Sigma$. We are done with the trivial derivation $c \rightarrow^* c$. ∎

Using this lemma we can prove:

Theorem 5.3 (*Completeness w.r.t. declarative semantics*)
For any ground expression e and any pattern $t \in H_\Sigma$, $t \sqsubseteq_H dcl[\![e]\!]$ implies that $e \rightarrow_c^{\leq\omega} t$. Moreover, the derivation can be chosen to be *fair*, and finite in case that t is finite.

Proof
If $t = \perp$ the trivial derivation $e \rightarrow^* e$ suffices. So, let us assume $t \neq \perp$, e.g. $t \equiv (\phi\ t_1\ ..\ t_m)$ for some t_i and some $\phi \in DC_\Sigma \cup FS_\Sigma$. The finite pattern $s \equiv (\phi\ \perp\ ..\ \perp)$ - with m arguments - satisfies $dcl[\![e]\!] \sqsupseteq_H s$. Since $dcl[\![e]\!] = lub_k [\![e]\!]^k$, we can find a k such that $[\![e]\!]^k \sqsupseteq_H s$. By lemma 5.2 we obtain $e \rightarrow^* e'$ such that $|\ e'\ | \sqsupseteq_H [\![e]\!]^k \sqsupseteq_H s$. It follows that $e' \equiv (\phi\ e_1\ ..\ e_m)$ for some ground expressions e_i. Moreover, $dcl[\![e]\!] = dcl[\![e']\!] = (\phi\ dcl[\![e_1]\!]\ ..\ dcl[\![e_m]\!])$ by soundness of rewriting (lemma 5.1), since $\mathcal{H}$ is an exact model. By definition of the ordering in H_Σ we can conclude that $dcl[\![e_i]\!] \sqsupseteq_H t_i$ for $1 \leq i \leq$m. Now we are in a position to reiterate the argument for each e_i. If t was finite, this procedure will eventually produce a finite converging derivation $e \rightarrow_c^*$t. If t was infinite, we can continue the procedure indefinitely so that a fair derivation $e \rightarrow_c^\omega t$ is produced. ∎

By combining the soundness and completeness results, we are able to prove:

Theorem 5.4 (*Adequateness of fair rewriting w.r.t. declarative semantics*)
For any ground expression e and any pattern $t \in H_\Sigma$, the following statements are equivalent:

(a) $[\![e]\!]^{\mathcal{A}} \sqsupseteq_{\mathcal{A}} [\![t]\!]^{\mathcal{A}}$ for all *liberal* models $\mathcal{A} \models \mathcal{R}$.

(b) $dcl[\![e]\!] \sqsupseteq_H t$.

(c) $e \rightarrow_c^{\leq\omega} t$ by means of some *fair* rewriting derivation.

Proof
(a) implies (b) because $[\![t]\!]^{\mathcal{H}} \equiv$ t. (b) implies (c) by the completeness theorem 5.3. (c) implies (a) by the soundness theorem 5.1. ∎

5.2 Unsoundness and incompleteness w.r.t. denotational semantics

Unfortunately, both soundness and completeness of rewriting fail w.r.t. denotational semantics. We have:

Theorem 5.5 (*Unsoundness and incompleteness w.r.t. denotational semantics*)
The analoga of Theorem 5.2 and Theorem 5.3 are false w.r.t. denotational semantics.

Proof
Consider the SFL program with constructors "0", "suc", "nil" and "cons" and the following defining rules (written using a friendly syntax for lists):

plus 0 Y = Y   plus (suc X) Y = suc (plus X Y)
times 0 Y = 0   times (suc X) Y = plus (times X Y) Y
even 0 = true   even (suc 0) = false   even (suc (suc X)) = even X
member [X | Xs] Y = true ⇐ X == Y
member [X | Xs] Y = true ⇐ (member Xs Y) == true
funpair N = [(plus N), (times (plus N N))] ⇐ (even N) == true

There is a finite rewriting derivation member (funpair 0) (times 0) $\rightarrow^*$ true. However, $dnt[\![member\ (funpair\ 0)\ (times\ 0)]\!] = \perp$, as a consequence of the fact that the strict equation $(times\ 0) == (times\ 0)$ does not hold in Scott algebras, because the meaning of $(times\ 0)$ is not a finished element. Hence, *soundness fails.*

On the other hand, it is easy to check that dnt[plus 0] = dnt[times (suc 0)]; both denotations are in fact the identity function over the Scott domain of the program's signature. However, the expression (plus 0) is a pattern and cannot be rewritten. Hence, *completeness fails.* ∎

5.3 Restricted rewriting. Its behaviour w.r.t. denotational semantics

Unsoundness of rewriting w.r.t. denotational semantics is due to the failure of the soundness lemma 5.1 for nonliberal algebras. It is possible to recover soundness results w.r.t. weakly liberal algebras (and hence w.r.t. denotational semantics) if the specification of joining is modified as: $\dfrac{l \rightarrow^* t \quad r \rightarrow^* t}{l \downarrow r}$ if $t \in \mathrm{FOPtr}_{\Sigma}$.
Let us call ***restricted rewriting*** to the new rewriting notion obtained in this way. With the help of Proposition 2.1 (a) and following the same ideas as in previous proofs, we can prove:

Theorem 5.6 (***Restricted Soundness w.r.t. weakly liberal algebras***)
The analoga of Lemma 5.1, Theorem 5.1 and Theorem 5.2 hold for restricted rewriting w.r.t. weakly liberal models and denotational semantics. (Note that in the analogon of Theorem 5.2 it is not true that $dnt[\![t]\!] = t$, and a distinction between t and $dnt[\![t]\!]$ must be made) ■

However, rewriting (and a fortiori restricted rewriting) remains incomplete w.r.t. denotational semantics, as shown in Theorem 5.5. We are not aware of any example of incompleteness that does not depend on the fact that syntactically different H.O. patterns may have the same denotational meaning. We have the following:

Conjeture 5.1 (***Restricted completeness w.r.t. denotational semantics***)
For any ground expression e and any *first order* pattern $t \in H_{\Sigma}$, $dnt[\![t]\!] \sqsubseteq_D dnt[\![e]\!]$ implies that $e \rightarrow_{c}^{\leq\omega} t$ by means of *restricted rewriting*. Moreover, the derivation can be chosen to be *fair*, and finite in case that t is finite. ■

Unfortunately, it seems that the proof of the Completeness Lemma 5.2 cannot be straightforwardly adapted for proving the conjecture, since it depends on syntactical manipulations of H.O. patterns which are not possible in the Scott domain D_{Σ}.

6 Conclusions

We have studied a kind of H.O. conditional rewriting systems that correspond to the commonly accepted operational view of H.O. lazy functional programs (without λ-abstractions). As mathematical semantics for these programs, we have considered traditional denotational semantics over a Scott domain as well as an alternative declarative semantics over an infinitary Herbrand domain, inspired by logic programming. We have shown that both semantics are instances of a generic least fixed point construction that can be understood in an algebraic framework.
Our results also show that declarative semantics exhibits a better algebraic and logic behaviour, due to initiality (Th. 4.2), soundness (Th. 5.2), completeness (Th. 5.3) and adequateness (Th. 5.4). Unsoundness w.r.t. the denotational semantics, as shown in Th. 5.5, is due to a rather technical limitation of the denotational domain. More significantly, incompleteness w.r.t. denotational semantics (also shown in Th. 5.5) reflects the ability of the denotational semantics to identify syntactically different patterns that can not be reduced to a common normal form. The fact that this may not happen for F.O. patterns reinforces our belief in Conjecture 5.1. If this were true, denotational semantics could be regarded as an adequate characterization of the F.O. results computable by restricted rewriting.
We conclude that the declarative semantics is a reasonable option for the integration of functional and logic programming, and even an interesting alternative for purely functional programming languages. In a forthcoming paper, we shall extend this research to ***functional logic languages***, along the lines of our previous work [10]. This will require to allow for extra variables in the conditions of rules as well as to establish soundness and completeness for ***conditional narrowing*** instead of rewriting. In other future works we plan to investigate ***type systems*** from the viewpoint of declarative semantics.

References

1. K.R.Apt: Logic Programming. In J.van Leeuwen (ed.), Handbook of Theoretical Computer Science, vol. *B*, Elsevier Science Publishers, 1990.
2. M.Bellia, G.Levi: The Relation between Logic and Functional Languages: a Survey. J. Logic Programming 3, 1986, pp. 217-236.
3. M.Bellia, P.G.Bosco, E.Giovannetti, G.Levi, C.Moiso, C.Palamidessi: A two-level approach to logic and functional programming, in Procs. PARLE '87.
4. M.H.M.Cheng, M.H.van Emden, B.E.Richards: On Warren's Method for Functional Programming in Logic. Procs. 7th Int. Conf. on Logic Programming, MIT Press, 1990, pp. 546-560.
5. W.Chen, M.Kifer, D.S.Warren: HiLog: A First Order Semantics for Higher-Order Logic Programming Constructs. Procs. North American Conf. on Logic Programming '89, MIT Press, 1989, pp. 1090-1114.
6. D. de Groot, G.Lindstrom (eds): Logic Programming: Functions, Relations and Equations. Prentice Hall, 1986.
7. N.Dershowitz, M.Okada: A Rationale for Conditional Equational Programming. Theor. Comp. Sci. 75, 1990, pp. 111-138.
8. N.Dershowitz, S.Kaplan, D.A.Plaisted: Rewrite, Rewrite, Rewrite, Rewrite, Rewrite, ... Theoretical Computer Science 83, 1991, pp. 71-96.
9. J.A.Goguen, J.W.Thatcher, E.G.Wagner,J.B.Wright: On Initial Algebra Semantics and Continuous Algebras. J. ACM 24, 1, 1977, pp. 68-95.
10. J.C.González-Moreno, M.T.Hortalá-González, M.Rodríguez-Artalejo: A Functional Logic Language with Higher Order Logic Variables. Technical Report DIA 90/6, October 1990.
11. J.R.Hindley, J.P.Seldin: Introduction to Combinators and λ-Calculus. Cambridge University Press, 1986.
12. J.R.Kennaway, J.W.Klop, M.R.Sleep, F.J.de Vries: Transfinite Reductions in Orthogonal Term Rewriting Systems. Procs. RTA '91, Springer LNCS 488, 1989, pp. 1-12.
13. D.A.Miller, G.Nadathur: Higher Order Logic Programming. Procs. 3th Int. Conf. on Logic Programming, Springer LNCS 225, 1986, pp. 448-462.
14. D.S.Scott: Domains for denotational semantics. Procs. ICALP '82, Springer LNCS 140, 1982, pp. 577-613.
15. D.A.Turner: Miranda: a non-strict functional language with polymorphic types. Procs. ACM Conf. on Functional Programming Languages and Computer Architectures, Springer LNCS 201, 1985, pp. 1-16.

Appendix A : Effective Test for Nonambiguity

Consider a pair of different rules for the same f, renamed apart so that they do not share variables: $R_1 : f t_1 ... t_n = r_1 \Leftarrow C_1 \quad R_2 : f s_1 ... s_n = r_2 \Leftarrow C_2$.
In this Appendix we present a decidable condition which is sufficient (*though not necessary!*) for a pair like this to be *nonoverlapping*. We say that the pair is *sufficiently nonoverlapping* iff one of the 3 following conditions holds:

(a) The two lhs $(f\ t_1 ... t_n)$ and $(f\ s_1 ... s_n)$ are *not unifiable*. Note that this is a *first order unification problem!*)

(b) The two lhs have a m.g.u. σ s.t. $r_1\sigma \equiv r_2\sigma$. (*syntactical identity!*)

(c) The two lhs have a m.g.u. σ s.t. $C_1\sigma \cup C_2\sigma$ is *sufficiently inconsistent*.

We still have to define the notion of sufficient inconsistency (for finite sets of strict equations) in such a way that it is decidable and implies unsatisfiability over minimally free algebras. For any finite set C of strict equations, let $C_=$ be the corresponding set of equations. Let Exp(C) be the finite set of all expressions which occur as subexpression of some expression in C. Consider the finite set of equations $Eq(C)$ and the finite set of expressions $Fin(C)$ defined inductively as follows:

Construction of Eq(C)

- $e_1 = e_2 \in C_= \Rightarrow\ e_1 = e_2 \in Eq(C)$
- $e_1 = e_2 \in Eq(C) \Rightarrow\ e_2 = e_1 \in Eq(C)$
- $e \in Exp(C) \Rightarrow\ e = e \in Eq(C)$
- $e_1 = e_2, e_2 = e_3 \in Eq(C)\ \Rightarrow e_1 = e_3 \in Eq(C)$
- $e_i = e'_i\ \in Eq(C)$ for $1 \leq i \leq m \Rightarrow (\phi\ e_1..e_m) = (\phi\ e'_1..e'_m) \in Eq(C)$, if $(\phi\ e_1..e_m), (\phi\ e'_1..e'_m) \in Exp(C)$
- $(c\,e_1..e_m) = (c\,e'_1..e'_m) \in Eq(C)$ and $rank(c) = n \geq m \geq 1\ \Rightarrow\ e_i = e'_i \in Eq(C)$ for $1 \leq i \leq m$

Construction of Fin(C)

- e is member of some strict equation in $C \Rightarrow e \in Fin(C)$
- $(c\ e_1...e_n) \in Fin(C)$ and $c \in DC^m_\Sigma \Rightarrow e_i \in Fin(C)$ for $1 \leq i \leq n$

Note that $Eq(C)$ and Fin(C) can be effectively constructed. By definition, we say that the finite set C of strict equations is ***sufficiently inconsistent*** iff $Eq(C)$ includes some equation of one of the two following forms:

(CFL) $(c\ e_1...e_k) = (d\ e'_1...e'_l)$ with $c \in DC^n_\Sigma, d \in DC^m_\Sigma, c \not\equiv d$, $0 \leq k \leq n,\ 0 \leq l \leq m$

(OCK) $X = t$ where t is a F.O. pattern, X occurs in t, $X \not\equiv t$ and either X or t belongs to $Fin(C)$.

We can prove:

Lemma 1
Let C be a sufficiently inconsistent set of strict equations. Then, for every minimally free algebra $\mathcal{A}$, there is no $\eta : Var \to D_\mathcal{A}$ such that $\mathcal{A} \models C\eta$.

Proof
Assume that C is sufficiently inconsistent but verifies $\mathcal{A} \models C\eta$ for some minimally free $\mathcal{A}$ and some η. Then, we can infer that $\mathcal{A} \models C_=\eta$, that $[\![e]\!]^\mathcal{A}\eta$ is finite for every $e \in Fin(C)$, and that $\mathcal{A} \models (e_1 = e_2)\eta$ for every $(e_1 = e_2) \in Eq(C)$. In particular, it follows that $\mathcal{A} \models (e_1 = e_2)\eta$ for some equation $(e_1 = e_2)$ of the form (CFL) or (OCK); this contradicts the assumption that $\mathcal{A}$ is minimally free. ∎

Using this lemma, we can prove:

Proposition 1
Every sufficiently nonoverlapping pair of rules is nonoverlapping.

Proof sketch
The sufficiently nonoverlapping pair will satisfy (a), (b) or (c) above. Let $\mathcal{A}$ minimally free and $\eta : Var \to D_\mathcal{A}$ be given. If we are in case (a), we can prove that $\mathcal{A} \models (t_i = s_i)\eta$ cannot hold for all $1 \leq i \leq n$. If we are in case (b) and $\mathcal{A} \models (t_i = s_i)\eta$ holds for all $1 \leq i \leq n$, we can prove that $[\![r_1]\!]^\mathcal{A}\eta = [\![r_2]\!]^\mathcal{A}\eta$. If we are in case (c) and $\mathcal{A} \models (t_i = s_i)\eta$ holds for all $1 \leq i \leq n$, we can prove that $\mathcal{A} \models (C_1 \cup C_2)\eta$ does not hold, since otherwise $\mathcal{A} \models (C_1\sigma \cup C_2\sigma)\mu$ would hold for some $\mu : Var \to D_\mathcal{A}$, against Lemma 1. The details of this reasoning depend on another lemma which relates the unification of two lhs $(f\ t_1\ ..\ t_n)$ and $(f\ s_1\ ..\ s_n)$ to solving the set of equations $\{t_1 = s_1,\ ..,\ t_n = s_n\}$ in minimally free algebras. ∎

A straightforward generalization of the notion of sufficient nonoverlapping pair of rules can be obtained by replacing the set $Exp(C)$ defined above by some bigger finite set of expressions; e.g., expressions constructed from $Exp(C)$ in some bounded number k of steps by syntactic application. It can be proved that the power of the resulting notion strictly increases with k. However, for any fixed k there are nonoverlapping pairs of rules which are not detected in this way.

On Transitive Closure Logic

Erich Grädel*

Abstract

We present Ehrenfeucht-Fraïssé games for transitive closure logic (FO + TC) and for quantifier classes in (FO + TC). With this method we investigate the fine structure of positive transitive closure logic (FO + pos TC), and identify an infinite quantifier hierarchy inside (FO + pos TC), formed by interleaving universal quantifiers and TC-operators.

It is also shown that transitive closure logic (and its fragments) have the same expressive power as the linear programs in certain extensions of Datalog.

1 Introduction: what is transitive closure logic?

It is well-known that the expressive power of first order logic (FO) is limited by the lack of a recursion mechanism. One of the simplest and most fundamental queries, that are not first-order expressible is the *transitive closure*, denoted TC. It assigns to a given binary relation E on universe U its reflexive transitive closure, i.e. the set of all pairs $(x, y) \in U \times U$ such that there exist $z_0, \dots, z_r \in U$ with $z_0 = x$, $z_r = y$ and $E(z_i, z_{i+1})$ for all $i > r$. It was first shown by Fagin [11] that TC is not expressible in FO; we will explain the proof, an easy application of Ehrenfeucht-Fraïssé games, in section 3.

On the other side, almost every logic or database query language with recursive constructs is strong enough to define transitive closures. In fact, when logics like fixpoint logic or database query languages like datalog are introduced, one of the first examples usually given is a formula (or a program) that defines TC.

It is therefore natural to define a logic that extends FO precisely by the ability to define transitive closures. This logic, called *transitive closure logic* (FO + TC), was introduced by Immerman [17]. An earlier variant is the database query language QBE, introduced by Zloof [27], in which one can ask transitive closure queries (unlike SQL which is based on first-order logic). Transitive closure logic is obtained by augmenting the syntax of first order logic by the following rule for building formulae:

Let $\varphi(\bar{x}, \bar{y})$ be a formula with $2k$ free variables $\bar{x} = x_1, \dots, x_k$ and $\bar{y} = y_1, \dots, y_k$ and let $\bar{u}$ and $\bar{v}$ be two k-tuples of terms. Then

$$[\mathrm{TC}_{\bar{x},\bar{y}}\ \varphi](\bar{u}, \bar{v})$$

is a formula, which says that the pair $(\bar{u}, \bar{v})$ is contained in the reflexive, transitive closure of the binary relation on k-tuples that is defined by φ.

***Address: Mathematisches Institut, Universität Basel, Rheinsprung 21, CH-4051 Basel, Switzerland, email: graedel@urz.unibas.ch**

Of course, it is understood that transitive closure logic is closed under the usual first order operations. We thus can build Boolean combinations of TC-formulae, we can nest TC-operators etc.

Examples. Let E be a binary relation symbol. Then the formula

$$(\forall u)(\forall v)[\mathrm{TC}_{x,y}\ Exy](u,v)$$

expresses that the digraph defined by E is strongly connected. The formula

$$\neg[\mathrm{TC}_{x,y}\ (\exists z)(Exz \land Ezy)](x,y)$$

says that there is no path of even length from x to y.

Immerman was interested in this logic because of its significance for his long term project to provide logical descriptions of complexity classes (see [14, 19] for surveys on this topic). For such descriptions it is necessary to consider ordered structures: Let $\mathcal{O}$ be the class of *finite successor structures*, i.e. structures whose vocabulary contain a binary predicate S, interpreted by a successor relation, and and two constants 0 and e, interpreted by the first and the last element of S.

Immerman proved that on $\mathcal{O}$, transitive closure logic can express precisely the queries that are computable with nondeterministic logarithmic space [17]. More precisely:

Theorem 1 *Let $L \subseteq \mathcal{O}$. The following are equivalent*

(i) $L \in$ NLOGSPACE;

(ii) there is a $\psi \in$ (FO + TC) such that L is the set of finite models of ψ;

(iii) there exists a quantifier-free formula $\varphi(\bar{x},\bar{y}) \in$ FO such that L is the set of finite models of the formula $[\mathrm{TC}_{\bar{x},\bar{y}}\ \varphi](\bar{0},\bar{e})$.

Immerman's original result was weaker; it said that NLOGSPACE is captured by the logic (FO + pos TC), the restriction of (FO + TC) where the operator TC can occur only positively. However, the closure of NLOGSPACE under complementation [18, 25] implies the equivalence of (FO + pos TC) with (FO + TC) on $\mathcal{O}$. See [19] for a direct proof of this fact.

Theorem 1 strongly depends on the presence of the successor relation. Without successor, only the (trivial) implications *(i)* $\Longleftarrow$ *(ii)* $\Longleftarrow$ *(iii)* survive wheras the reverse directions fail. It has been known for a long time that on unordered structures, the usual logics and query languages do not have the ability to count. For instance, the query EVEN (which asks whether the cardinality of a given structure is even) is clearly computable in LOGSPACE but cannot be expressed in first order logic, transitive closure logic or even fixpoint logic. To see that the implication *(ii)* $\Longrightarrow$ *(iii)* fails, observe that for any quantifier-free formula $\varphi(\bar{x},\bar{y})$, the query defined by

$$\psi(\bar{u},\bar{v}) \equiv [\mathrm{TC}_{\bar{x},\bar{y}}\ \varphi](\bar{u},\bar{v})$$

is *monotone*. This means that for all k-tuples $\bar{u}$ and $\bar{v}$ of elements of a structure $\mathfrak{A}$ and for all extensions $\mathfrak{B}$ of $\mathfrak{A}$,

$$\mathfrak{A} \models \psi(\bar{u},\bar{v}) \Longrightarrow \mathfrak{B} \models \psi(\bar{u},\bar{v}).$$

Clearly this does not even hold for all first-order formulae.

The failure of Theorem 1 on unordered structures raises two questions:

1. What is the expressive power of transitive closure logic?

2. What is the fine structure of (FO + TC)?

It seems not possible to answer the first question in terms of classical complexity theory: It is true that every problem expressible in transitive closure logic is included in NLOGSPACE. But while on one side, (FO + TC) can express certain NLOGSPACE-complete queries (like TC itself), there are on the other side very simple properties (like EVEN) that are not described by any formula of transitive closure logic. We will give an answer to this question by relating transitive closure logic to certain variants of Datalog (a now very popular database query language).

Our results on the fine structure of (FO + TC) show that this logic behaves very differently on arbitrary finite structures than it does on ordered structures. We will exhibit an infinite hierarchy of quantifier classes inside the (FO + pos TC). Our techniques include an Ehrenfeucht-Fraïssé game for transitive closure logic and a generalization of the tree structures that were used in [5, 8, 21] to separate fragments of other logics.

2 Quantifier classes in transitive closure logic

In this section we define the notion of a quantifier class in transitive closure logic. Note, that in a formula $\psi \equiv [\mathrm{TC}_{\bar{x},\bar{y}}\ \varphi(\bar{x},\bar{y})](\bar{u},\bar{v})$ where $\bar{x}$ and $\bar{y}$ are k-tuples, the TC-operator can be considered as a generalized quantifier that binds $2k$ variables. So we define that $\mathrm{depth}(\psi) := 2k + \mathrm{depth}(\varphi)$.

Definition 1 For every word p over the alphabet $\{\exists, \forall, \mathrm{TC}, \neg\mathrm{TC}\}$ we define the *quantifier class* $L(p)$ in (FO + TC) inductively as follows:

- $L(\epsilon)$ contains the quantifier-free formulae.
- For a quantifier $Q \in \{\exists, \forall\}$, the class $L(Qp)$ is the closure under conjunctions and disjunctions of the class $L(p) \cup \{(Qx_i)\varphi \mid \varphi \in L(p)\}$.
- $L((\mathrm{TC})p)$ is the closure under conjunctions and disjunctions of the class of formulae $[\mathrm{TC}_{\bar{x},\bar{y}}\ \varphi(\bar{x},\bar{y})](\bar{u},\bar{v})$ where $\varphi \in L(p)$;
- $L((\neg\mathrm{TC})p)$ is the closure under conjunction and disjunction of the class of all formulae $\neg[\mathrm{TC}_{\bar{x},\bar{y}}\ \neg\varphi(\bar{x},\bar{y})](\bar{u},\bar{v})$ where $\varphi \in L(p)$.

Obviously, (FO + TC) $= \bigcup_p L(p)$, and (FO + pos TC) is the union over those classes $L(p)$ where p does not contain the symbols $\neg$TC. Moreover we have:

Proposition 2 *Let σ be a signature that contains at least two constants a, b. Then, on the class of all σ-structures that satisfy $a \neq b$, every formula in* (FO + TC) *is equivalent to a formula in some $L(p)$ where $p \in \{\mathrm{TC}, \neg\mathrm{TC}\}^*$; moreover, every σ-formula in* (FO + pos TC) *is equivalent to a formula in $L(p)$ for some $p \in \{\forall, \mathrm{TC}\}^*$.*

PROOF. Existential quantifiers can always be replaced by TC-operators: A formula $(\exists z)\psi(z)$ is equivalent to

$$[\mathrm{TC}_{x,y}\ (x = a \wedge \psi(y)) \vee (\psi(x) \wedge y = b)](a, b).$$

By dual arguments, universal quantifiers are replaced by $\neg$TC. ■

The words $p \in \{\exists, \forall, \mathrm{TC}, \neg\mathrm{TC}\}^*$ can be partially ordered. We say that $p \leq q$ if p can be embedded into q, possibly by replacing some occurrences of $\exists$ or $\forall$ by TC or $\neg$TC respectively.

3 Ehrenfeucht-Fraïssé games for transitive closure logic

Ehrenfeucht-Fraïssé games [12, 10] provide a powerful tool for proving inexpressibility results for various logics. In their classical form they give a criterion for the indistinguishability of two (classes of) structures by means of first order formulae:

Definition 2 Suppose we have two structures $\mathfrak{A}$ and $\mathfrak{B}$ of the same vocabulary σ. Let $c_1, \ldots, c_s$ and $d_1, \ldots, d_s$ be the interpretations of the constants of σ in $\mathfrak{A}$ and $\mathfrak{B}$, respectively. The *(first order) k-pebble game* on the pair $(\mathfrak{A}, \mathfrak{B})$ is played by Players I and II[1] as follows: There are k pairs $(u_1, v_1), \ldots, (u_k, v_k)$ of pebbles. Each round of the game consists of either an $\exists$-move or an $\forall$-move:

The $\exists$-move. Player I places a yet unused pebble u_i on an element of $\mathfrak{A}$. Player II answers by putting the corresponding pebble v_i on $\mathfrak{B}$.

The $\forall$-move. Similarly but with 'reversed board': Player I places v_i on $\mathfrak{B}$; Player II responds with u_i on $\mathfrak{A}$.

When all pebbles are placed, Player II wins if the pebbles determine a local isomorphism from $\mathfrak{A}$ to $\mathfrak{B}$. More precisely: Let $a_1, \ldots, a_k$ and $b_1, \ldots, b_k$ be the elements carrying the pebbles $u_1, \ldots, u_k$ and $v_1, \ldots, v_k$. If the mapping f with

$$f(a_i) = b_i \quad \text{for } i = 1, \ldots, k \quad \text{and} \quad f(c_i) = d_i \quad \text{for } i = 1, \ldots, s$$

is an isomorphism between the substructures of $\mathfrak{A}$ and $\mathfrak{B}$ that are generated by the pebbled elements and the constants, then Player II wins; otherwise Player I wins.

In the classical Ehrenfeucht-Fraïssé game, Player I is free to choose between $\exists$-moves and $\forall$-moves. The crucial result, relating first order expressibility and pebble games is:

Theorem 3 (Fraïssé, Ehrenfeucht) *Let $\mathfrak{A}$ and $\mathfrak{B}$ be two structures over the same vocabulary and let $k \in \mathbb{N}$. Then the following two statements are equivalent:*

(i) For all first-order sentences φ of quantifier depth at most k

$$\mathfrak{A} \models \varphi \implies \mathfrak{B} \models \varphi.$$

(ii) Player II has a winning strategy for the k-pebble game on $(\mathfrak{A}, \mathfrak{B})$.

Remark. Unfortunately this fundamental theorem is not contained in most textbooks on mathematical logic. One of the few exceptions is [9].

Ehrenfeucht-Fraïssé games give an elegant proof of the well-known fact, that CONNECTIVITY (and thus also TC) is not first order expressible. Indeed, let $\mathfrak{A}^k$ be a cycle of length 2^k and let $\mathfrak{B}^k$ consist of two copies of $\mathfrak{A}^k$. It is an easy exercise to design a winning strategy for Player II for the k-pebble game on $\mathfrak{A}^k$ and $\mathfrak{B}^k$. She just has to maintain the following property of the elements $a_1, \ldots, a_m \in \mathfrak{A}^k$ and $b_1, \ldots, b_m \in \mathfrak{B}^k$ that are pebbled after the m-th move: For all $i, j \leq m$

[1] As usual we denote Player I by male and Player II by female pronouns.

$$d(a_i, a_j) = d(b_i, b_j) \quad \text{or} \quad d(a_i, a_j), d(b_i, b_j) > 2^{k-m}.$$

Immerman [16] considered pebble games in the course of which already placed pebbles are picked up again and put on a different element. On the logical side this corresponds to requantification of variables. The logic characterized in this way is $L^{\omega}_{\infty\omega}$, i.e. infinitary logic with bounded number of variables, introduced by Barwise [3]. Using these games, Kolaitis and Vardi [23] have established e.g. the 0-1 law for infinitary logic. Since many popular logics and database query languages (such as fixpoint logic, Datalog, etc.) can be embedded into $L^{\omega}_{\infty\omega}$, they also provide a tool for proving inexpressibility in these logics [16, 22]. Variants of Ehrenfeucht-Fraïssé games have also been designed for monadic second order logic [11], first-order logic with counting [4], Datalog [24, 22] and for fragments of fixpoint logic.

We will now define an Ehrenfeucht-Fraïssé game for transitive closure logic. It extends the classical Ehrenfeucht-Fraïssé game with k pebbles by the following moves:

The TC-move. Suppose that r pairs of pebbles are already on the board. For some $\ell \leq (k-r)/2$, Player I selects a sequence $\bar{x}_0, \ldots, \bar{x}_m$ of ℓ-tuples in $\mathfrak{A}$ such that $\bar{x}_0$ and $\bar{x}_m$ consist only of constants and already pebbled elements. Player II indicates a similar sequence (not necessarily of the same length) of ℓ-tuples $\bar{y}_0, \ldots, \bar{y}_n$ in $\mathfrak{B}$ where $\bar{y}_0 = f(\bar{x}_0)$ and $\bar{y}_n = f(\bar{x}_m)$ where f is the local isomorphism defined by the constants and the pebbles.

Player I then selects some $i < n$ and places 2ℓ (yet unused) pebbles on $\bar{y}_i$ and $\bar{y}_{i+1}$. Player II selects a $j < m$ and places the corresponding pebbles on $\bar{x}_j$ and $\bar{x}_{j+1}$.

The ¬TC-move is like the TC-move, but with structures $\mathfrak{A}$ and $\mathfrak{B}$ interchanged.

Definition 3 The TC-*game with k pebbles* on a pair of structures $(\mathfrak{A}, \mathfrak{B})$ is played as follows: The Players I and II make an arbitrary sequence of $\exists$-moves, $\forall$-moves, TC-moves and ¬TC-moves till all k pairs of pebbles are on the structures. Note that it is always Player I who decides what kind of move is performed next. The winning conditions for both players are the same as in the classical Ehrenfeucht-Fraïssé game.

For every quantifier class $L(p)$ in transitive closure logic, we also define an $L(p)$-game with k pebbles: This is the TC-game with the restriction that the word q describing the sequence of moves that are performed satisfies $q \leq p$.

Definition 4 Let L be a class of formulae and let $\mathfrak{A}$ and $\mathfrak{B}$ be finite structures of the same vocabulary. We write $\mathfrak{A} \preceq^k_L \mathfrak{B}$ to denote that

$$\mathfrak{A} \models \psi \quad \Longrightarrow \quad \mathfrak{B} \models \psi$$

for every sentence $\psi \in L$ of depth at most k. We write $\mathfrak{A} \equiv^k_L \mathfrak{B}$ to denote that both $\mathfrak{A} \preceq^k_L \mathfrak{B}$ and $\mathfrak{B} \preceq^k_L \mathfrak{A}$.

The next theorem shows that the TC-game does indead characterize the expressive power of transitive closure logic. In fact the correpsondance is much more precise: Every quantifier class $L(p)$ is characterized by the appropriate $L(p)$-game.

Theorem 4 *Let* $L(p) \subseteq (\text{FO} + \text{TC})$ *be a quantifier class and* $k \in \mathbb{N}$. *For every pair* $(\mathfrak{A}, \mathfrak{B})$ *of structures of the same vocabulary, the following are equivalent:*

(i) $\mathfrak{A} \preceq^k_{L(p)} \mathfrak{B}$

(ii) Player II has a winning strategy for the $L(p)$*-game with k pebbles on* $(\mathfrak{A}, \mathfrak{B})$.

PROOF. We prove this theorem by induction on p. It is trivial for $p = \varepsilon$. For the induction steps, it is useful to note the following: If, in a game with k pebbles, ℓ pairs of pebbles have already been placed on elements $\bar{a} \in \mathfrak{A}$ and $\bar{b} \in \mathfrak{B}$, then the remaining part of the game can be considered as a game with $k - \ell$ pebbles on $(\mathfrak{A}, \bar{a})$ and $(\mathfrak{B}, \bar{b})$ (i.e. on the expansions of $\mathfrak{A}$ and $\mathfrak{B}$, by constants $\bar{a}$ and $\bar{b}$).

$\neg(i) \Longrightarrow \neg(ii)$: Suppose that there is a formula $\psi \in L(p)$ of depth k with $\mathfrak{A} \models \psi$, but $\mathfrak{B} \models \neg\psi$. If ψ is a (positive) Boolean combination of some other formulae, then the two structures are distinguished by at least one of these subformulae. We thus may assume that ψ begins with a quantifier, a TC-operator or a negated TC-operator.

If $\psi \equiv (\exists x)\varphi$, then φ has depth at most $k-1$ and belongs to $L(q)$ where $\exists q \leq p$. Player I wins as follows: In his first move he pebbles an element $a \in \mathfrak{A}$ such that $(\mathfrak{A}, a) \models \varphi$. No matter what element $b \in \mathfrak{B}$ Player II selects, it will always be the case that $(\mathfrak{B}, b) \models \neg\varphi$. By induction hypothesis, Player I wins the remaining $L(q)$-game with $k-1$ pebbles on $((\mathfrak{A}, a), (\mathfrak{B}, b))$.

Now let $\psi \equiv [\mathrm{TC}_{\bar{u},\bar{v}}\ \varphi(\bar{u}, \bar{v})](\bar{c}, \bar{d})$. If $\bar{u}$ and $\bar{v}$ are ℓ-tuples, then φ has depth $k - 2\ell < m$. and belongs to $L(q)$ with $(\mathrm{TC})q \leq p$. Note that $\bar{c}$ and $\bar{d}$ are interpreted by constants of $\mathfrak{A}$ resp. $\mathfrak{B}$. Player I wins as follows: Since $\mathfrak{A} \models \psi$, Player I can exhibit a path defined by φ from $\bar{c}$ to $\bar{d}$; whatever sequence in $\mathfrak{B}$ Player II presents, it must contain two subsequent tuples $\bar{u}', \bar{v}'$ such that

$$(\mathfrak{B}, \bar{u}', \bar{v}') \models \neg\varphi.$$

Player I then pebbles $\bar{u}'$ and $\bar{v}'$. Player II now must place her pebbles on two subsequent k-tuples $\bar{u}, \bar{v}$ of Player I's path in $\mathfrak{A}$; whatever choice she makes,

$$(\mathfrak{A}, \bar{u}, \bar{v}) \models \varphi.$$

Again, the induction hypothesis implies that Player I wins the remaining $L(q)$-game with $k - 2\ell$ pebbles on $(\mathfrak{A}, \bar{u}, \bar{v})$ and $(\mathfrak{B}, \bar{u}', \bar{v}')$.

The cases where $p = \forall q$ and $p = (\neg\mathrm{TC})q$ are similar.

$(i) \Longrightarrow (ii)$: We assume (i) and explain the winning strategy for Player II. Suppose that Player I begins with a TC-move. Let q be the maximal suffix of p such that $(\mathrm{TC})q \leq p$. If Player I begins with a TC-move, choosing a path $\bar{u} = \bar{u}_0, \bar{u}_1, \ldots, \bar{u}_r = \bar{v}$ of ℓ-tuples, then Player II selects her path as follows: For $i < r$, let

$$\Phi_i := \{\varphi(\bar{y}, \bar{z}) \mid \varphi \in L(q),\ \mathrm{depth}(\varphi) \leq k - 2\ell \text{ and } \mathfrak{A} \models \varphi(\bar{u}_i, \bar{u}_{i+1})\}.$$

Note that every set Φ_i is finite, up to logical equivalence. (This is well-known and easy to prove for first-order formulae; it extends to (FO + TC) because applying a TC-operator to equivalent formulae preserves equivalence, and because the closure under conjunctions and disjunction of a finite set of formulae remains finite, up to logical equivalence). So we may form the formula

$$\psi(\bar{y}, \bar{z}) :\equiv \bigvee_{i=0}^{r} \bigwedge \Phi_i$$

where $\bigwedge \Phi_i$ denotes the conjunction over all formulae in the set Φ_i. Obviously, this formula has is in $L(q)$, has depth $k - 2\ell$ and

$$\mathfrak{A} \models [\mathrm{TC}_{\bar{y},\bar{z}}\ \psi](\bar{u}, \bar{v}).$$

By assumption, $\mathfrak{B}$ is also a model of $[\mathrm{TC}_{\bar{y},\bar{z}}\ \psi](\bar{u}, \bar{v})$, so there exists a path $\bar{u} = \bar{v}_0, \bar{v}_1, \ldots, \bar{v}_n = \bar{v}$ which is selected by Player II. Now, whatever $j < n$ is chosen by Player I, there exists some $i < r$ such that $\mathfrak{B} \models \bigwedge \Phi_i(\bar{v}_j, \bar{v}_{j+1})$. Player II selects this i and pebbles $\bar{u}_i$ and $\bar{u}_{i+1}$. After this move, corresponding pebbles lie on $\bar{u}_i, \bar{u}_{i+1}$ on $\mathfrak{A}$ and on $\bar{v}_j$ and $\bar{v}_{j+1}$ on $\mathfrak{B}$; for every formula $\varphi(\bar{y}, \bar{z}) \in L(q)$ of depth $k - 2\ell$

$$(\mathfrak{A}, \bar{u}_i, \bar{u}_{i+1}) \models \varphi \implies (\mathfrak{B}, \bar{v}_j, \bar{v}_{j+1}) \models \varphi.$$

By induction hypothesis, Player II wins the remaining $L(q)$-game with $k-2\ell$ pebbles.

If Player I begins with an $\exists$-move, pebbling an element $a \in \mathfrak{A}$, then Player II selects $b \in \mathfrak{B}$ as follows. Let q be maximal with $\exists q \leq p$ and

$$\Phi := \{\varphi(x) \mid \varphi \in L(q),\ \text{depth}(\varphi) \leq k - 1 \text{ and } \mathfrak{A} \models \varphi(a)\}.$$

Again Φ is finite so the conjunction $\bigwedge \Phi$ over all formulae in Φ is again a formula $\psi(x) \in L(q)$ of depth $k - 1$. Player I now chooses b in such a way that $\mathfrak{B} \models \psi(b)$. Now $(\mathfrak{A}, a) \preceq^{k-1}_{L(q)} (\mathfrak{B}, b)$ and by induction hypothesis, Player II wins the remaining $L(q)$ game with $k - 1$ pebbles.

The cases where Player I begins with an $\forall$-move or a $\neg$TC move follow by dual arguments. ∎

Corollary 5 *For all structures $\mathfrak{A}$ and $\mathfrak{B}$ and all $k \in \mathbb{N}$, the following are equivalent:*

(i) Player II has a winning strategy for the TC*-game with k pebbles on* $(\mathfrak{A}, \mathfrak{B})$.

(ii) $\mathfrak{A} \equiv^k_{\mathrm{TC}} \mathfrak{B}$.

Remark. We did not formally define the notion of a winning strategy. In the classical game such a strategy is given by a sequence of classes of partial isomorphisms that satisfy the *back and forth property* (see e.g. [9]). It is not difficult to define generalized back and forth properties that characterize the TC-game.

Remark. There are also other possibilities to define a game for (FO + TC). Here is another form of a TC-move: Suppose that $m \leq k$ pebbles are already on the board.

TC-move: Player I defines a predicate R of some arity $\ell \leq (k - m)/2$ on structure $\mathfrak{B}$. Player II defines a predicate P on $\mathfrak{A}$ of the same arity as R. Player I then pebbles two ℓ-tuples $\bar{u}', \bar{v}'$ of $\mathfrak{B}$ such that $\bar{u}' \in R$ and $\bar{v}' \notin R$. Player II answers with two ℓ-tuples $\bar{u}, \bar{v}$ of $\mathfrak{A}$ such that $\bar{u} \in P$ and $\bar{v} \notin P$. The predicates P and R are then thrown away and the game proceeds.

It is straightforward to show that this game is equivalent to the one defined above. The new move characterizes a TC-operator in the following way: Player I selects R such that for some formula $\varphi(\bar{x}, \bar{y})$ and some ℓ-tuple $\bar{b} \in \mathfrak{B}$, R is the set of ℓ-tuples, that are reachable from $\bar{b}$ via $\varphi(\bar{x}, \bar{y})$.

4 A strict hierarchy inside transitive closure logic

Recall that (FO + pos TC), contains those formulae of (FO + TC) which have only positive occurrences of TC operators. In the sequel we assume that we always have a vocabulary with at least two constants a and b, which are interpreted by distinct elements in every structure under consideration.

Definition 5 We define the following hierarchy inside (FO + pos TC):

- TC(0) is the set of all formulae $[\mathrm{TC}_{\bar{x},\bar{y}}\ \varphi(\bar{x},\bar{y})](\bar{a},\bar{b})$ where φ is a first order formula;
- $\forall$-TC(m) is the universal closure of TC(m), i.e. the set of formulae $(\forall\bar{x})\psi$ where $\psi \in \mathrm{TC}(m)$;
- TC($m+1$) is the set of formulae $[\mathrm{TC}_{\bar{x},\bar{y}}\ \varphi(\bar{x},\bar{y})](\bar{a},\bar{b})$ where $\varphi \in \forall\text{-TC}(m)$.

The following result is implicit in Immerman's paper [17].

Proposition 6 $\bigcup_{m\in\mathbb{N}} \mathrm{TC}(m) = (\mathrm{FO} + \mathrm{pos\ TC})$.

To prove this result it suffices to show that all strata TC(m) are closed under disjunctions, conjunctions, existential quantification and TC-operators. Immerman proved these closure properties on successor structures; the proofs go through for unordered structures. However, Immerman also showed that on successor structures universal quantifiers can be eliminated as well. We will now see that this is not true on unordered structures. For this purpose we use the appropriate Ehrenfeucht-Fraïssé games for TC(m) and $\forall$-TC(m). First, we prove that (FO + pos TC) does not collapse to TC(0):

Theorem 7 CONNECTIVITY *is not expressible in* TC(0). *Thus* $\forall$-TC(0) $\neq$ TC(0).

PROOF. The appropriate game for TC(0) is a TC-game, which begins with a single TC-move, followed by a classical Ehrenfeucht-Fraïssé game, i.e. a game consisting only of $\exists$- and $\forall$-moves. Let $\mathfrak{A}^k$ be a cycle of length 2^{k+2} containing two distinct constants a, b and let $\mathfrak{B}^k$ be the disjoint union of $\mathfrak{A}^k$ with another cycle of length 2^{k+2} (without constants). We argue that the initial TC-move is no help at all for Player I to distinguish $\mathfrak{A}^k$ from $\mathfrak{B}^k$. The winning strategy of Player II is the following: When Player I chooses a path of ℓ-tuples in $\mathfrak{A}^k$, then she takes the isomorphic path in the first copy of $\mathfrak{B}^k$. After Player I has pebbled two adjacent ℓ-tuples in $\mathfrak{B}^k$ she can respond with the two isomorphic ℓ-tuples in $\mathfrak{A}^k$. Thus, after the TC-move Player II can apply her winning strategy for the classical Ehrenfeucht-Fraïssé game.

The second claim follows from the first because CONNECTIVITY clearly is expressible in $\forall$-TC(0). ■

Remark. Theorem 7 was first proved by Martin Otto in a talk in the Basel-Freiburg seminar on finite model theory. His proof is different; it uses Gaifman's Theorem rather than Ehrenfeucht-Fraïssé games.

We now generalize this and show that the TC-hierarchy is strict:

Theorem 8 *For all* $m \in \mathbb{N}$, $\mathrm{TC}(m) \subsetneq \forall\text{–TC}(m) \subsetneq \mathrm{TC}(m+1)$.

PROOF. It is useful to consider first a slight modification of the structures $\mathfrak{A}^k$ and $\mathfrak{B}^k$ of the previous proof. Let P_0 be a new unary predicate; let $\mathfrak{A}^k_0$ and $\mathfrak{B}^k_0$ be obtained from $\mathfrak{A}^k$ and $\mathfrak{B}^k$ by adding a new element c_0, called the root, which is the unique element on which P_0 is true, and which is connected to all points from $\mathfrak{A}^k$ and $\mathfrak{B}^k$ respectively.

For $m > 0$ let $\mathfrak{A}^k_m$ and $\mathfrak{B}^k_m$ be defined as follows: The vocabulary consists of the edge predicate E, unary predicates $P_0, \ldots, P_m$ and the constants a, b. We build $\mathfrak{A}^k_{m+1}$ and $\mathfrak{B}^k_{m+1}$ using copies of $\mathfrak{A}^k_m$ and $\mathfrak{B}^k_m$ (without the constants).

For m even, $\mathfrak{A}^k_{m+1}$ consists of a unique element c_{m+1} (the root) satisfying P_{m+1}, and k copies of $\mathfrak{B}^k_m$ and of one copy of $\mathfrak{A}^k_m$. The root c_{m+1} is connected to the roots of $\mathfrak{A}^k_m$ and the $\mathfrak{B}^k_m$'s. $\mathfrak{B}^k_{m+1}$ is defined similarly except that we don't take a copy of $\mathfrak{A}^k_m$, but $k+1$ copies of $\mathfrak{B}^k_m$. The constants a, b are interpreted in the first copy of $\mathfrak{B}^k_m$.

For odd m, we let $\mathfrak{A}^k_{m+1}$ consist of $k+1$ copies of $\mathfrak{A}^k_m$, connected to the root c_{m+1}, but $\mathfrak{B}^k_{m+1}$ of one copy of $\mathfrak{B}^k_m$ and k copies of $\mathfrak{A}^k_m$.

Note that $\mathfrak{A}^k_m$ is (isomorphic to) a substructure of $\mathfrak{B}^k_m$ for all m, k.

Lemma 9 *For all $m \geq 0$*

(i) Let $\psi \in \mathrm{TC}(m)$. Then for sufficiently large k

$$\mathfrak{A}^k_{2m} \models \psi \Longrightarrow \mathfrak{B}^k_{2m} \models \psi.$$

(ii) Let $\psi \in \forall\text{-}\mathrm{TC}(m)$. Then for sufficiently large k

$$\mathfrak{A}^k_{2m+1} \models \psi \Longrightarrow \mathfrak{B}^k_{2m+1} \models \psi.$$

PROOF. For $\psi \in \mathrm{TC}(0)$ this follows immediately from the proof of Theorem 7. For $\psi \in \forall\text{-TC}(m)$ and k sufficiently large, we prove that Player II wins the $\forall$-TC(m) game with k pebbles on $\mathfrak{A}^k_{2m+1}$ and $\mathfrak{B}^k_{2m+1}$:

Player I begins the game with $\ell \leq k$ $\forall$-moves, i.e. he pebbles ℓ elements of $\mathfrak{B}^k_{2m+1}$; since $\mathfrak{B}^k_{2m+1}$ contains $k+1$ copies of $\mathfrak{B}^k_{2m}$ we may assume that all his pebbles are on the first k copies of $\mathfrak{B}^k_{2m}$ or on the root. Player II can thus place her pebbles on corresponding elements of $\mathfrak{A}^k_{2m+1}$, i.e. she does not put a pebble on the copy of $\mathfrak{A}^k_{2m}$ in $\mathfrak{A}^k_{2m+1}$. She wins the remaining TC(m) game in the following way: Whenever Player I chooses the root or an element in one of the $\mathfrak{B}^k_{2m}$-copies in $\mathfrak{A}^k_{2m+1}$ then she answers with the corresponding element in $\mathfrak{B}^k_{2m+1}$; when Player I plays in the $\mathfrak{A}^k_{2m}$-copy then Payer II answers in the $(k+1)$th copy of $\mathfrak{B}^k_{2m}$ in $\mathfrak{B}^k_{2m+1}$ according to her winning strategy for the TC(m)-game on $(\mathfrak{A}^k_{2m}, \mathfrak{B}^k_{2m})$.

For $\psi \in \mathrm{TC}(m)$, Player II wins the TC(m)-game on $\mathfrak{A}^k_{2m}$ and $\mathfrak{B}^k_{2m}$ by a similar startegy: Player I begins with a TC-move, i.e. he defines, for some $\ell \leq k/2$, a sequence $\bar{u}_0, \ldots, \bar{u}_n$ of ℓ-tuples in $\mathfrak{A}^k_{2m}$. Player II defines a sequence $\bar{v}_0, \ldots, \bar{v}_n$ in $\mathfrak{B}^k_{2m}$ as follows: since $\bar{u}_0$ and $\bar{v}_0$ consist of constants, we can assume that the path is defined up to $\bar{v}_i$ and show how to find $\bar{v}_{i+1}$. To do this only $\bar{u}_i$, $\bar{u}_{i+1}$ and $\bar{v}_i$ must be taken into account, but not the previous elements of the two sequences. Since pair $\bar{u}_i, \bar{u}_{i+1}$ consists of $2\ell \leq k$ elements, there is a $\mathfrak{A}^k_{2m-1}$-copy in $\mathfrak{A}^k_{2m}$ that contains none of these elements. Let $\mathfrak{A}'$ be $\mathfrak{A}^k_{2m}$ without this $\mathfrak{A}^k_{2m-1}$-copy; similarly, let $\mathfrak{B}'$ be $\mathfrak{B}^k_{2m}$ without the single copy of $\mathfrak{B}^k_{2m-1}$. Player II can thus place an ℓ-tuple $\bar{v}_{i+1}$ in $\mathfrak{B}'$ such that $(\mathfrak{A}', \bar{u}_i, \bar{u}_{i+1})$ is isomorphic to $(\mathfrak{B}', \bar{v}_i, \bar{v}_{i+1})$. Since this holds for all i, we will still have this situation at the end of the TC-move. Player II wins the remaining $\forall$-TC($m-1$) game by copying the moves in $\mathfrak{A}'$ resp. $\mathfrak{B}'$ according to the isomorphism from $\mathfrak{A}'$ to $\mathfrak{B}'$, and by answering the moves of Player I in $\mathfrak{A} - \mathfrak{A}'$ resp $\mathfrak{B} - \mathfrak{B}'$ according to her winning strategy for the $\forall$-TC($m-1$)-game on $(\mathfrak{A}^k_{2m_1}, \mathfrak{B}^k_{2m-1})$. ∎

The proof of Theorem 8 is now implied by the following Lemma whose proof is straightforward.

Lemma 10 *For all $m \geq 0$*

(i) There exists a formula $\psi \in \forall\text{–TC}(m)$ *such that for all k*

$$\mathfrak{A}^k_{2m} \models \psi \wedge \mathfrak{B}^k_{2m} \models \neg\psi.$$

(ii) There exists a formula $\psi \in \text{TC}(m+1)$ *such that for all k*

$$\mathfrak{A}^k_{2m+1} \models \psi \wedge \mathfrak{B}^k_{2m+1} \models \neg\psi.$$

■

5 Transitive closure logic and linear Datalog programs

We now relate (FO + TC) and its fragments to linear programs in certain extensions of Datalog. Datalog is a now very popular database query language, which, in its pure form consists of function-free and negation-free Horn clauses. For background on Datalog we refer to [20].

Definition 6 A *Datalog program* Π consists of a finite set of rules of the form

$$H \leftarrow B_1 \wedge \cdots \wedge B_m$$

where H is an atomic formula $S(x_1, \ldots, x_r)$, called the *head* of the rule, and $B_1 \wedge \cdots \wedge B_m$ is a conjunction of atomic formulae $R(x_1, \ldots, x_m)$ and equalities $x_i = x_j$. Every x_i is either a variable or a constant. A predicate that occurs in the head of some rule is called an *intensional database predicate*, abbreviated IDB predicate; a predicate occuring only in the bodies of the rules is called an *extensional database predicate*, or EDB predicate. One of the IDB predicates is the *goal predicate* of the program. The extensional vocabulary of Π is formed by all EDB predicates and by all constants occurring in Π; a finite structure of this vocabulary is called an *extensional database* EDB for Π.

Given any extensional database $\mathfrak{B}$, the program computes intensional relations, by the usual fixpoint semantics (or, equivalently, minimum model semantics). The result of Π on $\mathfrak{B}$ is the value of the goal predicate after the computation is terminated.

Example. The TC query is computed by the following Datalog program:

$$Txx; \qquad Txy \leftarrow Exy; \qquad Txz \leftarrow Txy \wedge Tyz$$

It is easy to see that every query computed by a Datalog program is monotone under homomorphisms and extensions of the EDB. Since one may also want to compute queries that do not have this property, the following extensions of Datalog have been considered:

Definition 7 Datalog($\neq$) extends Datalog by allowing inequalities $x_i \neq x_j$ in the body of the rules. Datalog($\neg$), allows also negated atomic formula $\neg R(x_1, \ldots, x_r)$ provided R is an extensional predicate. Finally, *stratified Datalog* [6, 26], denoted S-Datalog, allows a limitied form of negation over the IDB predicates: The IDB predicates are partitioned into strata, say $I_1, \ldots, I_m$; for simplicity, let I_0 denote the set of EDB predicates. In a rule

$$S(\bar{x}) \leftarrow B_1 \wedge \cdots \wedge B_m$$

where S is an IDB predicate from stratum I_j (with $j > 0$), the body may contain atomic formulae $S'(\bar{y})$ provided the S' does not belong to a higher stratum than S; and it may contain negated atomic formulae $\neg S'(\bar{y})$ provided S' belongs to a stricly lower stratum than S, i.e. $S' \in \bigcup_{k<j} I_k$.

The semantics of Datalog readily extends to Datalog($\neq$) and Datalog($\neg$). The intensional predicates of a stratified Datalog program are computed stratum by stratum; in fact the set of rules whose head is in I_j may be considered as a Datalog($\neg$) program whose EDB predicates are $\bigcup_{k<j} I_k$.

Datalog($\neq$) and Datalog($\neg$) also have monotonicity properties: Queries computed by Datalog($\neg$) are preserved by extensions; queries computable by Datalog($\neq$) are preserved by extensions and one-to-one homomorphisms. This excludes, e.g., that these programs can compute all first order queries. The question has been raised whether Datalog, Datalog($\neq$) and Datalog($\neg$) are strong enough to compute all polynomial time queries that satisfy the respective monotonicity properties. This has recently been answered negatively [2]. Stratified Datalog is a proper extension of first order logic; it has the same expressive power as (stratified) existential fixpoint logic. It had been claimed that S-Datalog has the same expressive power as fixpoint logic, but Dahlhaus [8] and Kolaitis [21] exhibitted fixpoint queries that are not computable by any stratified Datalog program.

Definition 8 A program in (an extension of) Datalog is *linear* if the body of each rule contains at most one IDB predicate.

The datalog program for TC given above is not linear. But there is of course a linear datalog program for TC, namely

$$Txx; \qquad Txz \leftarrow Txy \wedge Eyz.$$

However, we pay a prize for the linearization: the original Datalog program reaches the fixpoint after $O(\log n)$ iterations whereas the linear program requires n iterations (where n is the length of the longest path).

We will now define two more extensions of Datalog, situated between Datalog($\neg$) and S-Datalog. We first want to define Datalog(FO), arguably the simplest extension of Datalog that includes all FO-queries.

Definition 9 Datalog(FO) extends Datalog by allowing rules where the expressions in the body may be atomic formulae or arbitrary first order formulae $\varphi(\bar{x})$ over the EDB predicates. The semantics of Datalog(FO) is defined in the same way as for Datalog.

Intuitively Datalog(FO) programs allow to apply the recursion mechanism of Datalog not just to the EDB predicates but to any predicate computed by a FO-query on the extensional database. Linear Datalog(FO) programs coincide with the bottom level of the TC-hierarchy.

Theorem 11 LinDatalog(FO) = TC(0).

Next, we define a stronger extension, which allows universal projection of already computed relations.

Definition 10 Datalog(FO,$\forall$) extends Datalog(FO) in the following way: Let I_0 be the EDB predicates and let the IDB predicates are partitioned into strata $I_1, \ldots, I_m$. A rule with head $S(\bar{x})$ may contain the following expressions in the body:

- first order formula over the extensional predicates;
- atoms $R(\bar{y})$ provided R does not belong to a higher stratum than S;
- formulae $(\forall \bar{z})R(\bar{y}, \bar{z})$ provided R belongs to a strictly lower stratum than S.

Again, we have fixpoint semantics for Datalog(FO,$\forall$) by computing the the least fixpoints for lower strata before going to a higher one.

Again, we are interested in the linear programs:

Theorem 12 LinDatalog(FO,$\forall$) = (FO + pos TC).

We will prove a more general result, which implies both Theorem 11 and Theorem 12:

Theorem 13 *Let* $m \in \mathbb{N}$. *A query is expressible in* TC(m) *if and only if it is computable by a linear* Datalog(FO,$\forall$)*-program with* $m+1$ *strata.*

PROOF. Let $\varphi(x_1, \ldots, x_k) \in \text{TC}(m)$ with k free variables. We show, by induction on m, that there is a linear Datalog program with $m+1$ strata and goal predicate Z_φ of arity k that is equivalent to φ. First, let $m = 0$, i.e., let

$$\varphi(\bar{x}) \equiv [\text{TC}_{\bar{y},\bar{z}}\ \alpha(\bar{x}, \bar{y}, \bar{z})](\bar{c}, \bar{d})$$

where $\alpha \in \text{FO}$. The desired Datalog(FO) program is

$$\begin{aligned} T(\bar{x}, \bar{y}, \bar{y}) & & \\ T(\bar{x}, \bar{y}, \bar{z}) & \leftarrow & \alpha(\bar{x}, \bar{y}, \bar{w}) \wedge T(\bar{x}, \bar{w}, \bar{z}) \\ Z_\varphi(\bar{x}) & \leftarrow & T(\bar{x}, \bar{c}, \bar{d}). \end{aligned}$$

Next let $\varphi \in \text{TC}(m+1)$; this means that there exists a formula $\alpha(\bar{x}, \bar{u}, \bar{y}, \bar{z}) \in \text{TC}(m)$ such that

$$\varphi(\bar{x}) \equiv [\text{TC}_{\bar{y},\bar{z}}\ (\forall \bar{u})\alpha](\bar{c}, \bar{d}).$$

By induction there exists a program in LinDatalog(FO,$\forall$) with m strata and goal predicate Z_α which is equivalent to α. Define a new stratum, containing the new IDB predicates T and Z_φ, and add the rules

$$\begin{aligned} T(\bar{x}, \bar{y}, \bar{y}) & & \\ T(\bar{x}, \bar{y}, \bar{z}) & \leftarrow & (\forall \bar{u})Z_\alpha(\bar{x}, \bar{u}, \bar{y}, \bar{w}) \wedge T(\bar{x}, \bar{w}, \bar{z}) \\ Z_\varphi(\bar{x}) & \leftarrow & T(\bar{x}, \bar{c}, \bar{d}). \end{aligned}$$

The new program (with goal predicate Z_φ) is equivalent to φ.

Conversely, consider a linear Datalog(FO,$\forall$) program. First we assume, without loss of generality, that every stratum contains precisely one IDB-predicate. Indead, when we have IDB predicates $T_1, \ldots, T_r$, we can take a new predicate T of larger arity and express $T_1, \ldots, T_r$ as specializations of T, by substituting constants at some places or by identifying certain arguments.

Now suppose that stratum $m+1$ has IDB predicate T; we may assume that every rule in this stratum has the same head $T(y_1,\ldots,y_k)$ where $y_1,\ldots,y_k$ are pairwise distinct variables, because a rule

$$T(z_1,\ldots,z_k) \leftarrow \beta$$

with other variables in the head can always be transformed into

$$T(y_1,\ldots,y_k) \leftarrow \beta \wedge (z_1 = y_1) \wedge \cdots \wedge (z_k = y_k).$$

Thus stratum $m+1$ of the program consists of rules of the form

$$T(\bar{y}) \leftarrow \alpha_i(\bar{y},\bar{z}) \quad \text{and}$$
$$T(\bar{y}) \leftarrow T(\bar{x}) \wedge \beta_j(\bar{x},\bar{y},\bar{z})$$

where α_i and β_j are conjunctions of first order formulae on the EDB predicates and of universal projections $(\forall\bar{z})R(\bar{u},\bar{z})$ where R is an IDB predicate from a lower stratum. By induction hypothesis we assume that α_i and β_j are equivalent to formulae in $\forall$-TC$(m-1)$ (or, if $m=0$, to a first order formula).

We now construct a formula $\psi(\bar{v}) \in \mathrm{TC}(m)$ which expresses that $\bar{v} \in T$ after the execution of this program. Let

$$\varphi(\bar{x},\bar{y}) \equiv \bigvee_j (\exists\bar{z})\beta_j(\bar{x},\bar{y},\bar{z}).$$

This formula says that $T(\bar{y})$ is implied by $T(\bar{x})$ via one of the recursive rules of this program. Then the desired formula is

$$\psi(\bar{v}) \equiv (\exists\bar{u})\Big\{\bigvee_i (\exists\bar{z})\alpha_i(\bar{u},\bar{z}) \wedge [\mathrm{TC}_{\bar{x},\bar{y}}\ \varphi(\bar{x},\bar{y})](\bar{u},\bar{v})\Big\}.$$

When we substitute for α_i and β_i the appropriate formulae in TC$(m-1)$, then this formula is in TC(m). ∎

Full transitive closure logic coincides with the linear programs in stratified Datalog, a result that was first proved in [7]:

Theorem 14 S-LinDatalog = (FO + TC).

Finally we would like to mention that not all Datalog programs are linearizable. In fact there even exist Datalog programs that are not expressible in transitive closure logic, i.e . they are not even equivalent to any *stratified* linear Datalog program:

Theorem 15 Datalog − (FO + TC) $\neq \varnothing$.

This result is implicit already in an early paper of Immerman [15]. In more explicit form it appears in [1].

Note. In this paper the question whether (FO + pos TC) is equal to (FO + TC) also on unordered structures has not been adressed. In my talk at CSL'91 it was left as an open problem. After this paper was completed, Greg McColm and I were able to solve this problem: The complement of the transitive closure cannot be expressed in infinitary logic with bounded universal quantification, and therefore neither in (FO + pos TC), nor in Datalog(FO,$\forall$). We also establish a more general hierarchy theorem for quantifier classes in (FO + TC). These results will be presented in [13].

We summarize our results in the following diagram:

$$
\begin{array}{ccccccc}
 & & & & \text{LinDatalog} & \subset & \text{Datalog} \\
 & & & & \cap & & \cap \\
\text{FO} & \subset & \text{TC}(0) & = & \text{LinDatalog(FO)} & \subset & \text{Datalog(FO)} \\
 & & \cap & & \cap & & \cap \\
 & & \vdots & & \vdots & & \vdots \\
 & & \cap & & \cap & & \cap \\
 & & (\text{FO} + \text{pos TC}) & = & \text{LinDatalog}(\text{FO}, \forall) & \subset & \text{Datalog}(\text{FO}, \forall) \\
 & & \cap & & \cap & & \cap \\
 & & (\text{FO} + \text{TC}) & = & \text{S-LinDatalog} & \subset & \text{S-Datalog}) \\
 & & & & & & \cap \\
 & & & & & & (\text{FO} + \text{LFP})
\end{array}
$$

(where $\subset$ means proper inclusion)

Acknowledgement. I would like to thank the participants of the Basel-Freiburg seminar on finite model theory, and especially Martin Otto, for interesting discussions and valuable suggestions.

References

[1] F. Afrati and S. Cosmadakis, *Expressiveness of Restricted Recursive Queries*, Proceedings of 21st ACM Symposium on Theory of Computing (1989), 113–126.

[2] F. Afrati, S. Cosmadakis and M. Yannakakis, *On Datalog vs. Polynomial Time*, Proceedings of 10th ACM Symposium on Principles of Database Systems (1991).

[3] J. Barwise, *On Moschovakis closure ordinals*, J. Symbolic Logic **42** (1977), 292–296.

[4] J. Cai, M. Fürer and N. Immerman, *An Optimal Lower Bound on the Number of Variables for Graph Identification*, Proceedings of 30th IEEE Symposium on Foundations of Computer Science (1989), 612–617.

[5] A. Chandra and D. Harel, *Structure and Complexity of Relational Queries*, J. Comp. Syst. Sciences **25** (1982), 99–128.

[6] A. Chandra and D. Harel, *Horn Clause Queries and Generalizations*, J. Logic Programming **1** (1985), 1–15.

[7] M. Consens and A. Mendelzon, *GraphLog: a Visual Formalism for Real Life Recursion*, Proceedings of 9th ACM Symposium on Principles of Database Systems (1990), 404–416.

[8] E. Dahlhaus, *Skolem Normal Forms Concerning the Least Fixed Point*, in: "Computation Theory and Logic" (E. Börger, Ed.), Lecture Notes in Computer Science Nr. 270, Springer 1987, 101–106.

[9] H. D. Ebbinghaus, J. Flum and W. Thomas, *Mathematical Logic*, Springer-Verlag, 1984.

[10] A. Ehrenfeucht, *An application of games to the completeness problem for formalized theories*, Fund. Math. **49** (1961) 129–141.

[11] R. Fagin, *Monadic generalized spectra*, Zeitschrift für Math. Logik Grundlagen d. Math. **21** (1975), 89–96.

[12] R. Fraïssé, *Sur quelques classifications des systèmes de relations*, Publications Scientifique de l' Université d' Alger, Série A **1** (1954), 35–182.

[13] E. Grädel and G. McColm, *Hierarchies in Transitive Closure Logic, Stratified Datalog and Infinitary Logic*, in preparation.

[14] Y. Gurevich, *Logic and the Challenge of Computer Science*, in: E. Börger (Ed.), Trends in Theoretical Computer Science, Computer Science Press (1988), 1–57.

[15] N. Immerman, *Number of Quantifiers is Better than Number of Tape Cells*, J. Comp. Syst. Sciences **22** (1981), 65–72.

[16] N. Immerman, *Upper and lower bounds for first-order expressibility*, J. Comp. Syst. Sciences **25** (1982), 76–98.

[17] N. Immerman, *Languages that Capture Complexity Classes*, SIAM J. Comput. **16** (1987), 760–778.

[18] N. Immerman, *Nondeterministic space is closed under complementation*, SIAM J. Comput. **17** (1988), 935–939.

[19] N. Immerman, *Descriptive and Computational Complexity*, in: J. Hartmanis (Ed.), Computational Complexity Theory, Proc. of AMS Symposia in Appl. Math. **38** (1989), 75–91.

[20] P. Kannellakis, *Elements of Relational Database Theory*, in: J. van Leeuwen (Ed.), Handbook of Theoretical Computer Science, vol. B, North Holland, Amsterdam 1990, pp. 1073–1156.

[21] Ph. Kolaitis, *The Expressive Power of Stratified Logic Programs*, Information and Computation **90** (1991), 50–66.

[22] Ph. Kolaitis and M. Vardi, *On the Expressive Power of Datalog: Tools and a Case Study*, Proceeding of 9th ACM Symposium on Principles of Database Systems (1990), 61–71.

[23] Ph. Kolaitis and M. Vardi, *0-1 Laws for Infinitary Logics*, Proceedings of 5th IEEE Symposium on Logic in Computer Science (1990), 156–167.

[24] V. Lakshmanan and A. Mendelzon, *Inductive pebble games and the expressive power of datalog*, Proceedings of 8th ACM Symposium on Principles of Database Systems (1989), 301–310.

[25] R. Szelepcsényi, *The Method of Forced Enumeration for Nondeterministic Automata*, Acta Informatica **26**, (1988), 279–284.

[26] A. Van Gelder, *Negation as Failure using tight derivations for general logic programs*, Proceedings of 3rd IEEE Symposium on Logic Programming (1986), 137–146.

[27] M. Zloof, *Query-by-Example: Operations on the Transitive Closure*, IBM Research Report RC5526 (1976).

Some Aspects of the Probabilistic Behavior of Variants of Resolution

P. Heusch E. Speckenmeyer
Universität Düsseldorf

Abstract

The SAT-Problem for boolean formulas in conjunctive normal form often arises in the area of artificial intelligence, its solution is known as mechanical theorem proving. Most mechanical theorem provers perform this by using a variant of the resolution principle. The time and space complexity of resolution strongly depends on the class of formulas. Horn formulas, where each clause may contain at most one positive literal, represent the most important class where resolution proofs with polynomial length exist.

A special sort of Horn formulas, where each clause contains exactly one positive literal, is called logic programs. In logic programs, clauses of length one are also known as facts, clauses of length greater than one are known as rules. We show that there are typically short resolution proofs under unit/Davis-Putnam resolution for some models of formulas, and we study the "density" of rules necessary in another model for deriving the empty clause with probability tending to one.

1 Introduction

The problem of testing satisfiability for boolean formulas in conjunctive normal form (CNF) is known as SAT-Problem, it was the first problem shown to be NP-complete. Its counterproblem, testing whether a given formula is unsatisfiable, can be used for testing whether a goal can be proved given a set of basic facts and rules using them. This is performed by negating the goal i.e. by inserting its complement into the database and showing that the negated goal and the system of rules and facts is unsatisfiable, hence the goal follows. Mostly resolution is used to perform this refutation proof, since it is an easy applicable tool based on syntactical transformations only.

A boolean formula $F = C_1 \bigwedge C_2 \bigwedge \ldots \bigwedge C_r$ in conjunctive normal form over n variables $v_1, \ldots, v_n$ consists of sets $C_1, \ldots, C_r$ of clauses, where each clause C_l is a set of literals $x_{i_1} \ldots x_{i_k}$, each literal x_{i_j} stands either for a variable or its complement. An assignment $t : \{v_1, \ldots, v_n\} \mapsto \{true, false\}$ satisfies a literal x if $x = \overline{v}$ and $t(v) = false$ or $x = v$ and $t(v) = true$, it satisfies a clause if it satisfies at least one literal from the clause, it satisfies a formula if it satisfies all clauses in this formula. We call F a Horn formula, if every clause contains at most one positive literal, we call F a definite Horn formula, if every clause contains exactly one positive literal. In the case of Horn formulas, we call the positive variable the clause head, the negative ones the clause body. Clauses of length one are also called facts. Logic programs consist of a definite Horn formula and a set of negative literals, the goals. For the resolution refutation, they are simply inserted as clauses into the Horn formula.

For a definite Horn formula F over variable set V we may define the corresponding deduction hypergraph $G_F = (V_F, A_F)$ as follows:

- Every node $v_i' \in V_F$ corresponds to a variable $v_i \in V$.

- A_F contains a many–to–one hyperarc $a = (\{v'_{i_1}, v'_{i_2}, \dots, v'_{i_k}\}, v'_j)$, iff F contains the corresponding rule $r = v_j \leftarrow v_{i_1}, \dots, v_{i_k}$. We call $\{v'_{i_1}, v'_{i_2}, \dots, v'_{i_k}\}$ the source, v'_j the sink of a.

In this graph we may now perform the following procedure:

```
procedure MARK
Mark all nodes in V_F that correspond to variables
in V that are facts in F.
Repeat
  If there is a hyperarc s.t.  all the nodes in its
  source are marked but not the node in its sink,
  mark the sink of a
Until no such hyperarc exists anymore
```

It is clear that every node which is marked at the end of this procedure corresponds to a derivable goal in the corresponding definite Horn formula. The same effect would have been achieved by applying to F the unit resolution algorithm which is given below.

Since there are a lot of different resolution schemes that only differ in the way the clauses are selected as candidates for being resolved, we will specify the differences later and give the following scheme:

```
algorithm RESOLUTION
input:  a boolean formula F in CNF
R_0 := F;
j:=0;
WHILE □∉ R_j AND R_j ≠ ∅
  (AND R_j ≠ R_{j-1} in case of unit and unrestricted resolution)
  DO
  ''compute'' R_{j+1} from R_j;
  j:=j+1;
OD;
IF □∈ R_j
  THEN OUTPUT: F contradictory
  ELSE OUTPUT: F satisfiable
FI;
```

Variants of resolution now can be defined by choice of the function ''compute'' only. For some of these variants we give the definition of ''compute'' as follows:

- *unrestricted resolution:* any two resolvable clauses may be resolved.

$$R_{j+1} = R_j \cup \bigcup_{C_i, C_l \in R_j} \{\text{res\_clause}(C_i, C_l)\}$$

  Here we require that whenever possible at least one new clause must be added to R_j.

- *Davis-Putnam resolution:* Choose an arbitrary permutation π of the variables, two clauses may only be resolved in the j-th iteration of the while-loop if they both contain v_{π_j}. For short, we assume $v_{\pi_j} \equiv v_k$.

$$R_{j+1} = R_j \setminus \{c : c \in R_j \wedge \{v_k, \bar{v}_k\} \cap c \neq \emptyset\} \cup \bigcup_{\substack{v_k \in C_i \in R_j \\ \bar{v}_k \in C_l \in R_j}} \{\text{res\_clause}(C_i, C_l)\}$$

- ***unit resolution:*** at least one clause must be of length one.

$$R_{j+1} = R_j \cup \bigcup_{\substack{\sigma_i, \sigma_l \in R_j \\ \sigma_i = \{v_h\} \text{ or } \sigma_i = \{\overline{v}_h\}}} \{\text{res\_clause}(C_i, C_l)\}$$

The resolvent $C'' = \text{res\_clause}(C, C')$ of two clauses C and C' is defined by

$$\text{res\_clause}(C', C) = \begin{cases} D \cup D', & \text{if } C = D \cup \{x\} \text{ and } C' = D' \cup \{\overline{x}\} \text{ and there} \\ & \text{is no } y \neq x \text{ such that } y \in D \text{ and } \overline{y} \in D' \text{ or} \\ & y \in D' \text{ and } \overline{y} \in D \\ \text{undefined}, & \text{else.} \end{cases}$$

For definite Horn formulas however, the most common "resolution algorithm" is the application of backtracking, the so-called *SLD-resolution*, which is used in most Prolog interpreters. It performs by doing a trivial DFS in the deduction hypergraph (for this the direction of the arcs in the hypergraph must be inverted) to determine resolvable clauses and stops when the goal literal could be resolved.

It is not difficult to show that for every unsatisfiable formula a derivation of the empty clause exists, and that every formula the empty clause can be derived from, is contradictory. The resolution proof fails, if the empty clause could not be resolved and no more new clauses are constructible by resolution steps. If the number of iterations in the while-loop of RESOLUTION to derive the empty clause ($\square$) or to derive the empty formula ($\emptyset$) from an input F is at most k, we denote this by $F \overset{\leq k}{\vdash} \square$ or $F \overset{\leq k}{\vdash} \emptyset$ and call k the *depth* of the resolution proof. The number of constructed clauses is called the *length* of the resolution refutation proof. We note that there is always a refutation of linear depth in the number of variables, by choosing Davis-Putnam resolution. However, since the number of clauses may nearly be squared in each iteration, this does not lead to resolution proofs of polynomial length.

The first result on the complexity of resolution was given by Tseitin [Tse68] for regular resolution, a complete variant of resolution, where the existence of an infinite class of unsatisfiable formulas having regular resolution refutation proofs of superpolynomial length has been shown. For unrestricted resolution Haken [Hak85] was first able to prove that there are infinitely many unsatisfiable formulas having resolution proofs of exponential length in the number of variables. But not all resolution variants have equal time complexity; for Davis-Putnam resolution Goerdt [Goe90] proved the existence of a class of formulas with polynomial length unrestricted resolution refutation proofs, where every Davis-Putnam refutation has at least superpolynomial length.

Another result was given by Chvátal and Szemerédi [CS88], where they showed that there are infinitely many boolean formulas over n variables with cn clauses of length $k > 3$, such that a randomly chosen instance is unsatisfiable with probability tending to one for $n \to \infty$, but every resolution proof of such a formula must have length at least $(1+\epsilon)^n$ for some fixed $\epsilon > 0$, if $c * 2^{-k} \geq 0.7$. This also shows that the average time for solving formulas from this class by using resolution must be superpolynomial in the number of literals.

The most important class of formulas for applications, however, the Horn formulas theoretically don't suffer from these problems at all, since there is a linear time algorithm to resolve unsatisfiable formulas to get the empty formula. However, Kleine Büning and Löwen [KL87] were able to prove that the commonly used SLD resolution has average case exponential time complexity even in the case of propositional calculus and acyclic corresponding deduction hypergraphs. This mainly results from the property that facts having been derived in the resolution proof are destroyed immediately after being used. We note, that this is valid only for formulas where the corresponding deduction hypergraph is acyclic, because cycles in this graph lead to non-terminating computations with high probability. Because of this behavior, which is avoided in practice by carefully following design rules for Prolog programs, but which is almost certain to occur when clauses are chosen at random, we use unit resolution. It will always find a refutation in linear time w.r.t. the length of a formula, if such a refutation exists. This is easily achieved by using appropriate data structures [IM87].

In this paper we will study two models of random formulas: the first model, called the constant density model denoted by $cl(n)_p^r$, consists of arbitrary boolean formulas in conjunctive normal form containing variables from a set of n boolean variables where each formula has exactly r clauses (=disjunctions of literals, a literal is a boolean variable v or its complement $\overline{v}$). The clauses are chosen independently according to the following generation procedure: for each variable v and each clause C the following holds: either v is contained positive in C with probability p or $\overline{v}$ is contained in C with probability p or none of both with probability $1-2p$, for some $0 < p \leq 1/2$.

The second model is a model for modified constant density logic, $\tilde{H}(n,k,l,p)$. An instance $F \in \tilde{H}(n,k,l,p)$ is a logic program over n variables, where every rule from the definite Horn part of F has exactly k negative literals, we have l facts; p denotes the density of the rules. Such an instance F is constructed in three steps as follows:

1. Choose l different facts at random from the set V of n different variables.
2. Choose one goal (i.e. a negative clause of length one) at random from V.
3. For every set $\{b_{i_1}, b_{i_2}, \ldots, b_{i_k}\}$ of k variables used as a clause body perform the following two steps n times:
   (a) Choose a variable a from V at random
   (b) Accept the rule $a \leftarrow b_{i_1}, b_{i_2}, \ldots, b_{i_k}$ to be put into F with probability p, independently.

The third model is the unmodified constant density model for Horn formulas, which is denoted by $H(n,k,l,p)$. An instance $F \in H(n,k,l,p)$ is chosen like an instance from $\tilde{H}(n,k,l,p)$, except step 3 is modified as follows:

3′ For every set $\{b_{i_1}, b_{i_2}, \ldots, b_{i_k}\}$ of k variables used as a clause body perform the following steps n times, once for every $v_j \in V$:
   (a) Choose v_j as head of rule $r = v_j \leftarrow b_{i_1}, b_{i_2}, \ldots, b_{i_k}$
   (b) Accept r with probability p to be put into F, independently.

We note that in both models the head variable of a clause may also be included in the set of body variables, which results in a clause becoming a tautology, but that any two body literals from the same clause are different.

In chapter 2, Proposition 1, we show that under unit resolution the number of while-loops of the algorithmic scheme RESOLUTION for solving formulas $F \in H(n,k,k,p)$ is bounded from above by $2k+1$ with probability tending to 1 for growing values of n. The result even holds if $p = p(n)$ is a decreasing function of n, satisfying $p(n) * n^{\frac{1}{1+k}} \rightarrow \infty$ for $n \rightarrow \infty$, see Corollary 1. I.e. in this model unit resolution proofs typically are very fast.

While Corollary 1 gives an answer to the question for which classes $H(n,k,k,p(n))$ of Horn formulas (unit-)resolution solves formulas typically very fast, Proposition 4 gives a partial answer to the question for which classes of $H(n,k,l,p(n))$ of Horn formulas typically there is a resolution refutation proof. More precisely, we show that if we choose p with $p = \frac{c' * \log n}{n^c}$ and we either have $c = 1$, $c' * \binom{l}{k} > 3$ or $c < k$, $l = \epsilon * n$, , $c * e^k > 3$, then every goal can be deduced from any l initial facts with probability tending to one for $n \rightarrow \infty$.

For the model $cl(n)_p^r$ we study the average length of resolution refutation proofs in chapter 3. We sketch a simple approach for showing that Davis-Putnam resolution refutation proofs in this model take polynomial time in the average, if r is polynomially bounded by n, by relating resolution refutation proofs with proofs by the Davis-Putnam procedure for this model, see [GPB82]. A far more complicated approach for showing the same result can be found in [HTL89].

2 Complexity of Horn Formula Derivation

Proposition 1 *Let $0 < p < 1$ and $F \in H(n,k,k,p)$. Then under unit resolution the following holds:*

$$Prob[F \stackrel{\leq 2k+1}{\vdash} \square] \rightarrow 1,$$

for $n \rightarrow \infty$, i.e. nearly all logic program $F \in H(n,k,k,p)$ can be solved with at most $2k+1$ applications of the while-loop in **RESOLUTION** *when using unit resolution.*

Proof Let $F \in H(n,k,k,p)$ contain the facts $s_1 \leftarrow, s_2 \leftarrow, \ldots, s_k \leftarrow$ and the goal $\leftarrow t$. In order to show $F \stackrel{\leq 2k+1}{\vdash} \square$ under unit resolution, it is sufficient to show that there are $z_1, \ldots, z_k \in F$ such that $t \leftarrow z_1, \ldots, z_k \in F$ holds and $z_i \leftarrow s_1, \ldots, s_k \in F$ holds, for all $1 \leq i \leq k$.

From this we obtain:

$$\begin{aligned}
&\mathrm{Prob}[F \stackrel{\leq 2k+1}{\vdash} \square] \\
&\geq \mathrm{Prob}[\exists z_1, \ldots, z_k \in V \setminus \{s_1, \ldots, s_k, t\} : \\
&\qquad (t \leftarrow z_1, \ldots, z_k \in F) \text{ and } (\forall_{1 \leq i \leq k} : z_i \leftarrow s_1, \ldots, s_k \in F)] \\
&= 1 - \mathrm{Prob}[\forall z_1, \ldots, z_k \in V \setminus \{s_1, \ldots, s_k, t\} : \\
&\qquad \neg((t \leftarrow z_1, \ldots, z_k \in F) \text{ and } (\forall_{1 \leq i \leq k} : z_i \leftarrow s_1, \ldots, s_k \in F)] \\
&\geq 1 - (1 - p^{k+1})^{\frac{n-k+1}{k}} \qquad (2) \\
&\rightarrow 1, \text{for } n \rightarrow \infty. \square
\end{aligned}$$

A little bit of calculation shows that formula (2) approaches limit one, if $p = p(n)$ is decreasing in n and is satisfying $p(n) * n^{\frac{1}{1+k}} \rightarrow \infty$, for $n \rightarrow \infty$. We therefore obtain the following stronger result:

Corollary 1 *Let $p(n)$ be a function of n satisfying $p(n) * n^{\frac{1}{k+1}} \to \infty$, for $n \to \infty$. Let $F \in H(n, k, k, p(n))$. Then under unit resolution the following holds :*

$$Prob[F \overset{\leq 2k+1}{\longmapsto} \square] \to 1, \textit{for } n \to \infty. \square$$

We conjecture that the statement of Corollary 1 remains valid for functions of $p(n)$ decreasing even faster than required in the corollary. This is because the lower bound of (2) in the proof of proposition 1 obviously is not optimal.

The last proposition gave us some informations about the existence of short proofs. Another interesting question deals with the existence of any proof in logic programs. In order to determine the probability of the existence of such a proof (for an arbitrary goal g starting with a set of l different facts), we compute the probability of the complementary case, when such a proof doesn't exist. This certainly holds, if a set $U \neq V$ with at least k variables exists, such that there are no clauses with all body literals from U and the head literal from $U \setminus V$. In the deduction hypergraph G_F we then have a corresponding set $U_F \subset V_F$, such that

$$(\mathcal{P}(U_F) \times (V_F \setminus U_F)) \cap A_F = \emptyset,$$

i.e. we have an empty cut in G_F. $\mathcal{P}(U_F)$ denotes the powerset of U_F. So if we can show that for some clause density p the probability that such an empty cuts exists is tending to 0, we have shown that every goal can be deduced with probability tending to 1. This will be formalized in the following proposition:

Proposition 2 *Let F be a logic program from $\tilde{H}(n, k, l, p)$. Then F will be a contradiction with probability tending to one for $n \to \infty$ if $p = \frac{c' * \log n}{n^c}$ satisfies at least one of the following conditions:*

1. $c = 1$, $c' * \binom{l}{k} > 3$
2. $c < k$, $l = \epsilon * n$, $c * \epsilon^k > 3$

Proof We prove the propostion by showing that for the given values of p the number of expected cuts in the deduction hypergraph G_F tends to zero for $n \to \infty$, if at least 1 or 2 holds. By the Markov-Cebyshev inequality this probability is not greater than the number of expected cuts in G_F. The number of expected cuts, however, is computed as follows:

$$\text{Expected number of empty cuts in } G_F =$$
$$\sum_{U \subset V, |U| \geq l} \text{prob}(U \text{ and } V \setminus U \text{ form an empty cut in } G_F)$$

So we have to compute the probability that a certain set U of f nodes induces an empty cut. This probability is computed in the following way: $N = n * \binom{f}{k}$ arcs in G_F with source in U exist. The probability that exactly i of the N arcs have been chosen to be put into F is

$$\binom{N}{i} * (1-p)^{N-i} * p^i.$$

However, not every arc has its head in $V \setminus U$, the probability that the heads of all clauses lie in U and that the cut remains empty is $(f/n)^i$. So the probability that the set U of nodes induces an empty cut in G_F is

$$\sum_i \binom{N}{i} * (1-p)^{N-i} * p^i * \left(\frac{f}{n}\right)^i.$$

We may now calculate this sum with $N = n * \binom{f}{k}$:

$$\begin{aligned}
&\text{prob}(\exists \text{ empty cut in } F \in \tilde{H}(n,k,l,p)) \\
&\leq \sum_{l \leq f = |U| < |V|} \sum_i \binom{N}{i} * (1-p)^{N-i} * (p * \frac{f}{n})^i \\
&= \sum_{f=l}^{n-1} \binom{n}{f} ((1 - \frac{p(n-f)}{n})^n)^{\binom{f}{k}} \\
&\approx \sum_{f=l}^{n-1} \binom{n}{f} (e^{-p})^{(n-f)*\binom{f}{k}} \\
&= \sum_{f=l}^{n-1} \binom{n}{f} \left(\frac{1}{n}\right)^{\frac{c'*(n-f)*\binom{f}{k}}{n^c}}
\end{aligned}$$

The first equality is an application of the binomial theorem. We note that the approximation holds because $p = \frac{c' * \log n}{n^c}$ and $(1 - \frac{x}{n})^n \approx e^{-x}$, hence $(1 - \frac{p(n-f)}{n})^n \approx e^{-p(n-f)}$. The last equality sign holds because of the definition of p, for $e^{-p} = e^{\frac{c' * \log n}{n^c}} = n^{-c'/n^c}$. In order to prove the proposition we simply have to show that this last sum converges to zero, if either 1 or 2 holds. This is done by proving that for $f < \epsilon * n$ the terms $\binom{n}{f}\left(\frac{1}{n}\right)^{\left(\frac{c'*(n-f)*\binom{f}{k}}{n^c}\right)}$ are geometric series, if $\epsilon < 0.5$. So for $f < 0.5 * n$ the sum is convergent iff

$$\binom{n}{l}\left(\frac{1}{n}\right)^{\frac{c'*(n-l)*\binom{l}{k}}{n}} \to 0$$

which holds when $\binom{l}{k} * c' > 3$. For $f > \epsilon * n$ we have:

$$\begin{aligned}
&\sum_{f=\epsilon * n}^{n-1} \binom{n}{f} * \left(\frac{1}{n}\right)^{\frac{c'*(n-f)*\binom{f}{k}}{n^c}} \\
&\leq \sum_{f=\epsilon * n}^{n-1} \binom{n}{f} * \left(\frac{1}{n}\right)^{\frac{c'*(n-f)*(n*2\epsilon)^k}{n^c}} \\
&= \sum_{f=\epsilon * n}^{n-1} \binom{n}{f} * \left(\frac{1}{n}\right)^{c'*(n-f)*(2\epsilon)^k * n^{k-c}} \\
&\leq \sum_{f=0}^{n} \binom{n}{f} * \left(\frac{1}{n}\right)^{c'*(f)*(2\epsilon)^k * n^{k-c}} - 1 \\
&\approx e^{-n^{k-c}} - 1 \\
&\to 0, \text{ for } n \to \infty
\end{aligned}$$

This finishes our proof □

3 Complexity of Resolution for Arbitrary Formulas

Formulas with constant density have been studied by different authors mostly because they have the property that any two occurences of literals in clauses (either positive or negative) are mutually independent, which makes some problems much easier to analyse. In this chapter, we will sketch a proof that in the constant density model nearly every proof of satisfiability can be done in polynomial time for $n \to \infty$, because formulas in this class are satisfiable with probability tending to 1 for big n. For refutation proofs this means a high probability of unsuccessful refutations, and this can be shown by starting with the following:

Lemma 1 *Let $F \in cl(n)^r_p$ and $S = (v_{i_1}, \ldots, v_{i_l})$ be a sequence of l different variables. Then the following holds: if there is a truth assignment t of the variables of S satisfying F, then Davis-Putnam Resolution resolving clauses in the order given by S stops after at most l iterations, i.e. there is $l' \leq l$ such that $R_{l'} = \emptyset$*

Proof Let $t = \{x_{i_1}, \ldots, x_{i_l}\}$ be a truth assignment of the variables in S with $x_{i_j} \in \{v_{i_j}, \overline{v}_{i_j}\}$, where $x_{i_j} = v_{i_j}$ means $t(v_{i_j}) = true$ and $x_{i_j} = \overline{v}_{i_j}$ means $t(v_{i_j}) = false$, such that $t(F) = true$, i.e. $\forall C \in F : t \cap C \neq \emptyset$. By induction on $|t|$ we show that $R_{|t|} = \emptyset$ holds, $\forall F \in cl(n)^*_p = \bigcup_{r>0} cl(n)^r_p$.

Induction begin: Let $|t| = 1$, then $S = (v_{i_1})$ and $t = \{x_{i_1}\}$, because $\forall C \in F$: $t \cap C \neq \emptyset$, each clause C contains x_{i_1} and no clause contains $\overline{x}_{i_1}$, so $R_1 = \emptyset$.

Induction hypothesis: $\forall F \in cl(n)^*_p$ satisfying $\exists t = \{x_{i_1}, \ldots x_{i_j}\}$, $j < l$, such that $t(F) = true$, $R_j = \emptyset$ holds.

Induction step Now let $t = \{x_{i_1}, \ldots x_{i_l}\}$ be a satisfying truth assignment of $F \in cl(n)^*_p$. Form R_1 by resolution according to v_{i_1}. Obviously, $t' = \{x_{i_2}, \ldots x_{i_j}\}$ is a satisfying truth assignment of R_1. By the induction hypothesis, $R_{|t'|}(R_1) = \emptyset$. Because $R_{|t'|}(R_1) = R_{|t|}$, the lemma follows.□

By means of lemma 1 we can show:

Proposition 3 *Let $0 < p \leq 1/2$, $\epsilon > 0$, $l_0 \geq \frac{\log(\epsilon)-\log(r)}{\log(1-p)}$. Applying the Davis-Putnam variant of* RESOLUTION *to $F \in cl(n)^r_p$, the following holds:*

$$Prob[F \vdash^{\leq l_0} \emptyset] \geq \exp(-\epsilon)$$

for sufficiently large values of r.

Proof By Lemma 1 it is sufficient to determine the probability of a partial truth assignment $t' = \{x_{i_1}, \ldots x_{i_l}\}$ of $v_1, \ldots, v_l$ to satisfy all clauses $c \in F \in cl(n)^r_p$.

Let $C \in F \in cl(n)^r_p$ and $t = \{x_{i_1}, \ldots x_{i_j}\}$ be an arbitrary truth assignment of $v_1, \ldots, v_l$.

Then $\text{Prob}[t \cap C = \emptyset] = (1-p)^l$, hence $\text{Prob}[\forall C \in F : t \cap C \neq \emptyset] = [1-(1-p)^l]^r$. For $l = l_0 = \frac{\log(\epsilon)-\log(r)}{\log(1-p)}$ we obtain by simple calculation $(1-p)^l = \frac{\epsilon}{r}$. Therefore

$$[1-(1-p)^l]^r = [1-\frac{\epsilon}{r}]^r \to \exp(-\epsilon) \tag{1}$$

for $r \to \infty$. Because of lemma 1 and (1)

$$\text{Prob}[\exists l \leq l_0 : R_l = \emptyset] \geq$$

$$\text{Prob}[\exists \text{truth assignment } t \text{ of } v_1, \ldots v_l \text{ with } t(F) = true] \geq$$

$$\text{Prob}[t = \{x_1, \ldots, x_l\} \text{satisfies: } \forall C \in F : t \cap C \neq \emptyset] \to \exp(-\epsilon)$$

This proves proposition 1.□

Corollary 2 *Let* $0 < p \leq 1/2$ *and* $r = n^k$. *Then*

$$Prob[\exists l \leq \frac{-(k+1)}{\log(1-p)} * \log(n) : R_l = \emptyset] \rightarrow 1,$$

for growing values of n, *i.e. if we restrict to the class of formulas with the number of clauses polynomially bounded by the number of variables, then the number of resolution iterations performed in the while-loop of* RESOLUTION *is bounded by* $O(\log n)$ *with probability tending to one.*

Proof By Proposition 1:

$$\text{Prob}[\exists l \leq \frac{\log(\epsilon) - \log(n^k)}{\log(1-p)} : R_l = \emptyset] \geq \exp(-\epsilon).$$

Since $(\log(\epsilon) - \log(n^k))/\log(1-p) < -(k+1)/\log(1-p) * \log(n)$ holds for each $\epsilon > 0$ and n sufficiently large, and $\exp(-\epsilon) \rightarrow 1$ for $\epsilon \rightarrow 0$, we get the result.□

Without proof, we mention the following:

Proposition 4 *Let* $0 < p \leq 1/2$, $r = n^k$. *Then Davis-Putnam Resolution solves formulas from* $cl(n)_p^r$ *in the average in time* $O(r^2 * n^2 * \log n)$.

To give an idea, why proposition 2 is true, note that the expected number of while loops, which have to be performed by Davis-Putnam Resolution is bounded by $O(\log n)$. This can easily be obtained from Corollary 1. Now, choose two arbitrary clauses $C, C' \in F \in cl(n)_p^r$. Then the probability that C, C' contain at least two complementary pairs of literals, i.e. $x, y \in C$ and $\overline{x}, \overline{y} \in C'$ is going to 1 very fast, for growing values of n. I.e. the expected number of clauses added to R_{j+1} as resolvents of clauses from R_j is going down to 0 for growing values of n. Because each iteration of the while-loop can be performed in $O(r^2 * n^2)$ steps, the time bound follows.

References

[CS88] Vašek Chvátal and Endre Szemerédi. Many hard examples for resolution. *Journal of the ACM*, 35(4):759–768, October 1988.

[Goe90] Andreas Goerdt. Davis-putnam resolution versus unrestricted resolution. In Egon Börger, Hans Kleine Büning, and Michael M. Richter, editors, *Workshop on Computer Science Logic 89*, pages 143–162, Berlin, Heidelberg, New York, Tokyo, 1990. Springer Verlag.

[GPB82] A. Goldberg, P. Purdom, and C. Brown. Average time analysis of simplified davis-putnam procedures. *Inform. Proc. Letters*, 15:72–75, 1982.

[Hak85] Armin Haken. The intractability of resolution. *Theoretical Computer Science*, 39:297–308, 1985.

[HTL89] T.C. Hu, C.Y. Tang, and R.C.T. Lee. An average case analysis of a resolution principle algorithm in mechanical theorem proving. Technical report, National Tsing Hua University, Hsinchu, 1989.

[IM87] A. Itai and J. Makowsky. On the complexity of derivations in the propositional calculus. *J. of Logic Programming*, 4:105–117, 1987.

[KL87] Hans Kleine Büning and Ulrich Löwen. SLD-resolution needs expo time. Technical report, Universität-Gesamthochschule Duisburg, 1987.

[Tse68] G. S. Tseitin. On the complexity of derivations in the propositional calculus. In *Studies in Constructive Mathematics and Mathematical Logic*, pages 115–125, New York, London, 1968. Consultants Bureau.

Safe Queries in Relational Databases with Functions

Joram Hirshfeld

School of Mathematical Sciences
Raymond and Beverly Sackler
Faculty of Exact Sciences
Tel Aviv University
Ramat-Aviv, 69978 Israel
email: yoram@taurus.bitnet

Abstract. We define the class of syntactically safe queries in first order languages *with function symbols*. We prove using ideas from model theory that every model independent query is equivalent to a safe query. This answers a question raised by Topor in [Topor 1987].

In relational databases the data is kept in tables R_i of n-tuples of names. To turn this into a part of first order logic *the database* is defined as a collection of atomic sentences, i.e. atomic closed formulas $R_i(t_1, \ldots, t_{n_i})$ where R_i becomes a relation symbol in the language and t_j are closed terms. (Real databases are better represented by many sorted logic since the different attributes in the tables may come naturally from different domains. But following the common practice we ignore this fact and stick to plain first order predicate calculus). With this definition of databases it is most natural to use the logic for asking questions about the base and *a query* is simply a first order formula.

In which model should such a query be evaluated? It turns out that given a database and a query there is indeed a minimal model among the models corresponding to the database and the query but it is not evident whether this model is the intended interpretation.

It is widely agreed upon in the field that rather than decide upon a particular interpretation we should stick to the subclass of queries which have the same answer in all the models corresponding to the database and the query, i.e. the *domain independent queries* [Ullman 89], [Topor 87], [Di Paola 69]. There is however, a big obstacle to choosing the class of domain independent formulas as the class of proper database queries: there is no algorithm to check given a formula if it is in the class or not [Di Paola 69]. It is therefore natural to look for the second best thing, a subclass of the class of domain independent formulas which is easily syntactically defined and has the same expressive power as the domain independent queries. Specifically, the class is *safe* – every formula in the class is domain independent - and the class is *adequate* – every domain independent formula is equivalent (in all the models) to a formula in the class.

Are there safe adequate classes of formulas? Before we answer this question we return to the definition of a relational database above and make another distinction; if all the names t_i in the tables are simple names (i.e. the language includes constant symbols but no function symbols) then we say that the database is *pure* (and that the first order corresponding language is pure). If on the other hand the tables contain also function symbols (as in Loves (John, Father of (John))) then the database is a *generalized* database.

For pure databases the answer is easy: there are many adequate syntactically safe query languages [Topor 87]. On the other hand, none of these languages extends naturally to generalized databases (with function). Furthermore, the proof that these languages are adequate depends strongly on the fact that the minimal model is defined in every other model by a very simple formula. This is no more true for generalized databases. Thus papers [Topor 87] and [Topor-Sonenberg 88] end with the question as to how to generalize their results to databases with function symbols.

The main purpose of this paper is to define a natural safe query language for generalized databases and prove that it is adequate. This will be done in three steps: in section 1 we shall describe the theory for pure databases. In section 2 we shall extend the definitions to databases with functions, suggest a query language and state the main theorem. In section 3 we shall outline the proof of the main theorem which is quite complicated, involves some auxiliary notions and includes lemmas which are of independent interest.

1 Pure Relational Databases

The syntax of pure database query languages is given by the following definitions:

Definitions

a) A *database* D is a collection of atomic sentences of the form $R_i(c_1, \ldots, c_{n_i})$ where $c_1, \ldots, c_{n_i}$ are all constants.

b) A *query* (to the database) is a formula $\varphi(x_1, \ldots, x_n)$ in a first order language whose relation symbols are among those of the database relations. (No function symbols are allowed in the query but it is usually assumed that there is no harm in allowing the query to include also constants which are not mentioned in D. Note also that we shall freely uses both names "query" and "formula").

c) *An answer* to the query $\varphi(x_1, \ldots, x_n)$ is a table of n-tuples.

Clearly the answer is intended to be the collection of tuples that satisfy $\varphi(x_1, \ldots, x_n)$ with respect to D. The main question is where to evaluate the formula $\varphi(x_1, \ldots, x_n)$. What interpretation has the user in mind when he asks the query $\varphi(x_1, \ldots, x_n)$? Before we address this question we shall describe all the possible models and to this end we also define the language that corresponds to a pair D (database) and φ (query). Thus here is the semantics of pure database query languages:

Definitions

a) Given a database D and a query φ the corresponding first order language $L(D, \varphi)$ has as relation symbols those relations that explicitly occur in φ and has the constant symbols that occur in φ or in some table R which is mentioned in φ.

There may be many relations and constants in the database which are not in $L(D,\varphi)$. It is natural to think of $L(D,\varphi)$ as the relations and constants of the database which are relevant to the query. Yet it is questionable whether the intentional notion of "the data relevant to the query" can be formally determined.

b) *A domain* is the set of constants in some first order language (later when we shall allow also function symbols a domain will be the set of closed terms in the language). A domain is a *(D,φ)-domain* if the language includes $L(D,\varphi)$. *The minimal (D,φ)-domain* is the set of constants (closed terms) in $L(D,\varphi)$

c) *A model* (an interpretation) is a domain together with some relations defined on the domain. a model is a (D,φ) *-model* if its domain is a (D,φ)-domain and if its relations are exactly the relations of D which are mentioned in φ. *The minimal (D,φ)-model*is the unique (D,φ)-model whose domain is the minimal (D,φ)-domain The minimal (D,φ)-model is denoted by $\mathrm{Dom}(D,\varphi)$.

Remarks

1) It is clear that for every n, $0 \leq n \leq \infty$ there is up to isomorphism a unique (D,φ)-model D_n that has n elements more than $\mathrm{Dom}(D,\varphi)$. These models are linearly ordered:

$$\mathrm{Dom}(D,\varphi) \subseteq D_1 \subseteq \cdots \subseteq D_n \subseteq \cdots \subseteq D_\infty$$

D_∞ can be called *the universal (D,φ)-model.*

2) Note the special features of these models that arise from databases: Each relation is finite (even in D_∞) and if $D \models R(a_1,\ldots,a_n)$ then $a_1 \ldots a_n$ are in $\mathrm{Dom}(D,\varphi)$. In other words the atomic relations other than equality are *domain independent* – they have the same scope in all the models D_n.

d) A query $\varphi(x_1,\ldots,x_n)$ is *domain independent* if for every database D it has the same answer table in all the (D,φ)-models that correspond to D and φ.

Note that in particular the evaluation of a domain independent query φ can be done in the minimal (D,φ)-model $\mathrm{Dom}(D,\varphi)$ so that there is no questioning that the answer is finite and computable.

The class of domain independent queries seemed to most authors to be what we really mean when we think of database query languages. In [Topor 87] Topor says: "It has long been recognized that only certain formulas make 'reasonable' queries. Informally, a query is only regarded as 'reasonable' if it yields the same answer whatever the domain of the interpretation". Ullman puts it in even stronger words but mentioned also the difficulties [Ullman 89 Pg. 151]: "...What we really would like...called domain independence, is a semantic notion. It will be seen that it is impossible to tell given a formula whether the formula satisfies the property....we shall therefore look for approximations to the ideal and that is where we shall find the notion of safety."

The reason why Ullman had to give up this ideal is the following theorem which in turn follows from Trakhtenbrot's theorem on undecidability of finite satisfiability:

Theorem [Di Paola 69]

The class of domain independent queries is undecidable.

Since we cannot supply the database user with a query language that cannot be described (decided) we must find an alternative. Here is a partial list of names of query languages which were suggested as alternatives to the ideal of domain independent queries:

Safe Queries [Ullman 89]

Allowed queries [Topor 87]

Range Separable queries, range restricted queries, evaluable queries [Demolombe 82].

We shall not give the definitions here. The collector of database dialects is referred to [Topor 87] for a starting point of his search. Instead we shall give the following definition which seems to emphasize what is behind those languages.

Definition Let L be a language of queries (i.e., a class of formulas).

a) L is (syntactically) *safe* if L is defined using syntactical properties and it follows from these properties that every formula in L is domain independent.

b) L is also *adequate* if for every domain independent query φ there is some query φ' in L such that φ' is in the language of φ and φ and φ' are equivalent (i.e. for every database D and for every (D, φ)-model, $M \models \varphi \leftrightarrow \varphi'$).

With this definition we can restate the result in [Topor 87].

Theorem

There is a safe adequate language for pure databases (without functions). Moreover, all the languages in the list above have this property.

Since it is almost immediate from their definitions that these languages are safe it remains to show that they are adequate. For this it is assumed that φ is domain independent, that $R_1, \ldots, R_k$ are the relations mentioned in φ and that $c_1, \ldots, c_m$ are the constants mentioned in φ. Let $D_\varphi(x)$ be the disjunction of all the formulas

$$\exists y_i \cdots \exists y_{j-1} \exists y_{j+1} \cdots \exists y_{n_i} \; R_i(y_1, \ldots, y_{j-1}, x, \; y_{j+1}, \ldots, y_{n_i})$$

and the formulas

$$x = c_\ell$$

where R_i ranges over the relations in φ, $1 \leq j \leq n_i$, where n_i is the arity of R_i and c_ℓ ranges over the constants in φ. Then it is easy to see that in a (D, φ)-model M the formula $D_\varphi(x)$ describes inside M the minimal (D, φ)-domain. We denote now by φ_D the restriction (relativization) of φ to the predicate $D_\varphi(x)$. Then it is elementary logic that

$$M \models \varphi_D(\bar{x}) \quad \text{iff} \quad \text{Dom}(D, \varphi) \models \varphi(\bar{x}).$$

It is also easy to see that φ_D is domain independent and that therefore φ is domain independent iff it is equivalent everywhere to φ_D.

Finally it is not difficult to check that each of the languages above includes all the formulas φ_D, or some obvious logical equivalent formula. It follows that for every domain independent query they include some equivalent formula.

By definition all the safe adequate languages are equivalent. The software engineer may now adopt one which is to his liking and adapt it for practical use. Some of these languages are more natural to phrase a query and others consist of formulas that are more straightforward to evaluate, i.e: "In choosing the safe adequate language there may be a tradeoff between efficiency and user friendliness".

2 Generalized Databases

In general databases the relations may include terms constructed through the use of functions. Thus the database D in first order logic is a finite collection of atomic sentences - sentences of the form $R(t_1, \ldots, t_n)$ where $t_1 \cdots t_n$ are variable free terms. The definition of queries and of the language $L(D, \varphi)$ remains as before and now $L(D, \varphi)$ includes also function symbols that occur in φ or in some table R of D which is mentioned in φ.

Following [Kifer 88] we accept the common assumption that no element has two names. This in turn assures us that given a database D and a query φ there is a minimal (D, φ)-model $\mathrm{Dom}(D, \varphi)$ corresponding to this pair, namely the Herberand model (the free term model) of $L(D, \varphi)$ with the relations as prescribed by the database D.

Note however that $\mathrm{Dom}(D, \varphi)$ is usually infinite. This seems like a very high price to pay for the possibility to talk about father of(x) as a function. Note also that the assumption that elements do not have more than one name is not as harmless as it is in pure databases, where a name is only a name. The fact that for every x and y fatherof(x) is different from sonof(y) is not natural and this assumption should probably be relaxed, in future research.

Again we get up to isomorphism an increasing sequence of (D, φ)-models:

$$\mathrm{Dom}(D, \varphi) \subseteq D_1 \subseteq \cdots \subseteq D_\infty$$

where D_n is obtained from $\mathrm{Dom}(D, \varphi)$ by adding a new elementary constants to $L(D, \varphi)$ and forming the term model with the relations as prescribed by D.

The definition of a domain independent query φ remains as before - for every database D the answer to φ in all the (D, φ)-models is the same. Note however that since $\mathrm{Dom}(D, \varphi)$ is infinite it is not apriori clear that every domain independent query has a finite computable answer. That this is indeed the case was conjectured in [Kifer 88] and proved in [Avron-Hirshfeld 91]*

We want to answer a question in [Topor 87] and [Topor-Sonenberg 88] and generalize the results of section 1 about safe and adequate query languages to general databases. We encounter two difficulties:

a) none of the languages suggested there extends naturally to databases with functions and what is more disturbing:

b) The minimal (D, φ)-domain is infinite and is not defined by a formula inside the bigger (D, φ)-models. Hence even when we find an interesting language a simple relativization argument will not work.

In view of these two points it is not surprising that our definition of the suggested language is somewhat intricate. Our idea is not to try and define directly by induction the set of well behaved formulas, but only a small natural subclass of *absolute* formulas. These absolute formulas will then serve as a means to restrict the range of the variables in a general formula. Thus the language that we suggest may be thought of as a typed language - the absolute formulas are the types and every formula is a good query provided its variables are typed.

---

* I was informed by the referee that Kifer already proved this result in a Stony Brook report which is a revision of [Kifer 88]. I thank the referee for this and for his corrections and remarks.

Definition: absolute formula

(a) Every atomic relation $R(t_1(\overline{x}),\ldots,t_k(\overline{x}))$ is *absolute*.
$x = t$ is absolute if t is variable-free. (Otherwise it is not absolute.)

(b) If $A(\overline{x})$ and $B(\overline{x})$ are absolute then $A \wedge B$ is absolute. If in addition they share exactly the same free variables then $A(\overline{x}) \vee B(\overline{x})$ is absolute (but not for example $R(x) \vee S(y)$)

(c) If $A(\overline{x})$ is absolute then so is also $\exists y A(\overline{x})$

(d) (closure under the application of definable functions:) If $\varphi(x_1,\ldots,x_n)$ is absolute (with $x_1,\ldots,x_n$ free in φ) and if $t_1,\ldots,t_k$ are terms whose variables are among $x_1,\ldots,x_n$ then the following formula is absolute:

$$\psi(y_1,\ldots,y_k) \equiv \exists x_1,\ldots,\exists x_n\,[\varphi(x_1,\ldots,x_n) \wedge y_1 = t_1 \wedge \cdots \wedge y_k = t_k]\ .$$

Definition: A Secured Query

A formula is *secured* if all its variable are relativized to absolute formulas. Formally:

a) Every quantifier free formula is *bounded*.

b) If φ is bounded then so are the following:

$$\exists x[A(x) \wedge \varphi] \quad \text{and} \quad \forall x[A(x) \longrightarrow \varphi]$$

where A is absolute and may include additional free variables only if they are also free in φ (and will therefore be taken care of later).

c) If $\varphi(x_1,\ldots,x_n)$ is bounded and if $A(x_1,\ldots,x_n)$ is absolute then $A \wedge \varphi$ is *secured*. (φ and A have exactly the same free variables.)

Note that from the logical point of view a secured query is jut a first order formula that obeys some additional structural properties. From query language point of view a secured query is just any first order query that has the range of its variables specified.

We may now state the main theorem:

Main Theorem

a) *Every secured query is domain independent.*

b) *Every domain independent query is equivalent to some secured query.*

Hence the secured queries form a safe and adequate language.

3 The Ingredients of the Proof

The proof of the main theorem is quite involved and will only be outlined. It introduces the auxiliary notion of equivalence between term tuples. Some of the lemmas are of interest in their own right.

The first half of the theorem is easy:

Lemma 1 (safety)

Every secured formula and in particular every absolute formula is domain independent.

The proof is by structural induction. ■

Next we introduce some tools to analyze the structure of terms.

a) *The term tree*: Given a number n there corresponds to every term t a structure tree $T_n(t)$ of depth n. The nodes are labeled both by their corresponding subterm and by the main operation in this subterm, (for the leaves only the subterm will be of interest).

Examples

i) For $n = 0$ $T_0(t)$ has a single node labeled by t.
ii) Let t be $F(a, f(g(b)))$ then $T_2(b)$ is

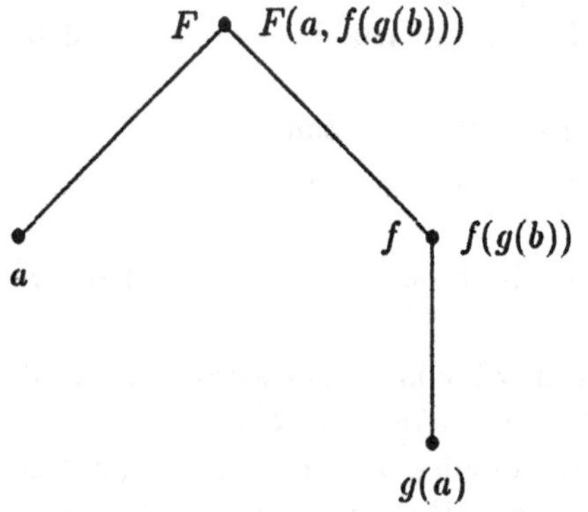

b) *The subterm set*: $S_n(t)$ is the collection of terms on the nodes of $T_n(t)$.

If A is a set of terms then $S_n(A)$ is the union of the sets $S_n(t)$ with t ranging over A.

c) *The superterm set*: Let A be a set of terms in some language L. We denote by A' the set A together with all the terms $F(t_1, \ldots, t_k)$ where $t_1, \ldots, t_k$ are in A. A' includes also all the elementary constants in L (the 0-ary function). For every set of terms A we define by induction:

$$S^0(A) = A \ ., \quad S^{n+1}(A) = (S^n(A))' \ .$$

The operation of subterm set and superterm set do not take us out of the absolute predicates:

Lemma 2 (absoluteness)

Let $\varphi(x)$ be an absolute formula. Then for every n there are absolute formulas $\underline{\varphi_n}(x)$ and $\overline{\varphi}_n(x)$ such that for every database D if A is the set of terms that satisfy $\varphi(x)$ in Dom(D, φ) *then $S_n(A)$ and $S^n(A)$ are the sets of terms that satisfy $\underline{\varphi_n}(x)$ and $\overline{\varphi}_n(x)$ respectively.*

The proof is direct. ■

The main observation in the paper is that in the setting of (D,φ)-models if $\varphi(t_1,\ldots,t_k,x)$ has a solution x then it also has a solution in a small subdomain which is definable from $(t_1,\ldots,t_k)$ using the operations S^n and S_n above. From this it will follow by induction that quantifiers can be bounded (lemma 5). This observation is proved in two steps: first we define the following algebraic equivalence and show that it has a "bounded search property" (lemma 3) and then we show that the algebraic equivalence indeed reflects a logical property (lemma 4).

d) *n-equivalence*: Two k-tuples of terms are n-equivalent

$$\langle t_1,\ldots,t_k\rangle \sim_n \langle s_1,\ldots,s_k\rangle$$

if

i) For every i $T_n(t_i)$ and $T_n(s_i)$ are similar trees: the branching is the same and the nodes (other than leaves of depth n) are labeled with the same main operation.

ii) The correspondence $t_i \longleftrightarrow s_i$ is one to one and well defined (i.e.: $t_i = t_j$ iff $s_i = s_j$). Moreover it extends to a well defined and one to one isomorphism between $S_n(t_1,\ldots,t_k)$ and $S_n(s_1,\ldots,s_k)$ (i.e. if a term on some node of $T_n(t_i)$ equals a term on a node of $T_n(t_j)$ then the corresponding nodes in $T_n(s_i)$ and $T_n(s_j)$ are also equal, and vice versa).

Finally we take also the database into account:

e) *n-Equivalence with respect to a database:* Let D be a database and let $r_1,\ldots,r_m$ be a fixed listing of the terms which are explicitly mentioned in D (e.g. if D is just $\{R(f(g(a)),b)\}$ then the list is $\langle f(g(a)),b\rangle$ and does not include a or $g(a)$). Let $\langle t_1,\ldots,t_k\rangle$ and $\langle s_1,\ldots,s_k\rangle$ be k-tuples of terms. We say that they are n-equivalent with respect to D and write

$$\langle t_1,\ldots,t_k\rangle \sim_{n,D} \langle s_1,\ldots,s_k\rangle$$

if

$$\langle t_1,\ldots,t_k,\ r_1,\ldots,r_m\rangle \sim_n \langle s_1,\ldots,s_k,\ r_1,\ldots,r_m\rangle\ .$$

With these notions we come to the main lemma:

Lemma 3 (weak homogeneity)

a) If $\langle t_1,\ldots,t_k\rangle \sim_{2n,D} \langle s_1,\ldots,s_k\rangle$ then for every t there is some s such that $\langle t,t_1,\ldots,t_k\rangle \sim_{n,D} \langle s,s_1,\ldots,s_k\rangle$.

b) If there are only finitely many such terms s then each such s satisfies $s \in S^n(S_{2n}(s_1,\ldots,s_k))$

c) Assume that the language has at least one function symbol. Then even if there are infinitely many terms s satisfying the equivalence there is an easily computable number $\tilde{n}$ depending only on the language signature and on n such that at least one such s can be found in

$$S^n\left(S_{2n}(s_1,\ldots,s_k)\cup S^{\tilde{n}}(s_1,\ldots,s_k)\right)$$

(actually $\tilde{n}$ can be chosen as $n\cdot w^n$ where w is the highest arity of a function in the language. It is also worth noting that one can construct an algorithm by which this s will have a much simpler form in almost all cases.)

The proof checks which leaves of $T_n(t)$ are in $S_{2n}(t_1,\ldots,t_k)$ and carefully constructs a similar tree for s using the corresponding elements from $S_{2n}(s_1,\ldots,s_k)$. Part (b) corresponds to the case that there is no degree of freedom – all the leaves of $T_n(t)$ occur in $S_{2n}(t_1,\ldots,t_k)$. Part (c) utilizes the fact that at least one of the s_i is not a subterm of any other and that therefore $S^{\tilde{n}}(s_i)$ can supply terms which are general enough to serve as leaves of $T_n(s)$. ■

Next we connect equivalence with satisfiability:

Lemma 4 (truth lemma)

For every formula $\varphi(\overline{x})$ there is an easily computed number n_φ such that for every database D and for every pair

$$\langle t_1,\ldots,t_k\rangle \sim_{n_\varphi,D} \langle s_1,\ldots,s_k\rangle$$

we have

$$\mathrm{Dom}(D,\varphi) \quad \models \quad \varphi(t_1,\ldots,t_k) \longleftrightarrow \varphi(s_1,\ldots,s_k)\,.$$

(Actually $n_\varphi = 2^q \cdot d$ where d is the depth of the deepest term occurring in φ and q is the quantifier rank.)

The proof is by easy induction using Lemma 3(a). ■

Lemma 5 (boundedness lemma)

Let φ be a formula and assume for simplicity that φ is in normal form $Q_1y_1\ldots Q_\ell y_\ell\, A(x_1,\ldots,x_k,y_1,\ldots,y_\ell)$. Let $\psi(x)$ be an absolute formula. Then there are absolute formulas $\psi_1(x),\ldots,\psi_\ell(x)$ such that for every database D, and for every tuple $(t_1,\ldots,t_k)$ satisfying $\psi(t_i)$ in Dom(D,φ) *we have*

$$\mathrm{Dom}(D,\varphi) \models \varphi(t_1,\ldots,t_k) \longleftrightarrow \varphi'(t_1,\ldots,t_k)$$

where φ' is obtained from φ by relativizing Q_i to $\psi_i(x)$.

($\psi,\psi_1,\ldots,\psi_\ell$ describe an increasing sequence of finite sets in Dom(D,φ).)

The proof is by induction on the number of quantifiers. Concentrating on the case where the first quantifier is existential there is by lemmas 3 and 4 a bound on the search of a term witnessing the existence of x. This bound yields ψ_1 by lemma 2 and the induction does the rest. ■

One more piece of information is missing:

Lemma 6 (finiteness lemma)

If $\varphi(\overline{x})$ is domain independent then its answer is finite in every (D,φ) model.

This was conjectured in [Kifer 88] and proved in [Avron-Hirshfeld 91] (see the footnote in section 2).

The Proof of the Main Theorem

Lemma 1 shows that every secured query is domain independent. It remains to show that every domain independent query is equivalent to a secured formula. Let $\varphi(\overline{x})$ be a given domain independent query. By Lemma 6 it has only finitely many solutions and by lemma 3(b) (with a small trick) and lemma 2 there is an absolute formula $\psi(x)$ such that all the solutions satisfy $\psi(x)$. In particular

$$\mathrm{Dom}(D,\varphi) \models \varphi(\overline{x}) \longleftrightarrow (\psi(x_1) \wedge \ldots \wedge \psi(x_n) \wedge \varphi(\overline{x}))$$

But for all $x_1, \ldots, x_n$ in the range of $\psi(x)$ we have

$$\mathrm{Dom}(D,\varphi) \models \varphi(\overline{x}) \longleftrightarrow \varphi'(\overline{x})$$

where φ' is the formula from Lemma 5. Hence

$$\mathrm{Dom}(D,\varphi) \models \varphi(x) \longleftrightarrow (\psi(x_1) \wedge \ldots \wedge \psi(x_n) \wedge \varphi'(\overline{x}))$$

and the right handside formula is secured. Since this formula is domain independent by lemma 1 and $\varphi(\overline{x})$ is domain independent by assumption we have

$$M \models \varphi(\overline{x}) \longleftrightarrow (\psi(x_1) \wedge \ldots \wedge \psi(x_n) \wedge \varphi'(\overline{x}))$$

for every (D,φ)-model M. ∎

There should be a simpler proof which systematically associates types (simple ranges) to quantifiers in a domain independent query to obtain a variant of a secured formula. We did not find this proof yet.

4 Concluding Remarks

a) We return to the initial question: given a database D and a query φ is there an intended interpretation, a model in which the user expects φ to be evaluated? The common answer in the field as we represented in section 1 and followed after that says "no". The query is not meant for a particular model and if it may depend on the model "it is not a reasonable query" [Topor 87].

This approach does not seem right to me. I believe that behind every query there is an intended interpretation. Sometimes it is the minimal (D,φ)-model and often it is a larger model which includes names in the database which are not in $L(D,\varphi)$ but happens to be of relevance to the query. Admittedly there is not a formal way to recognize this model by staring at the query. It depends on the specific database, and the same query to the same database may have a different intended interpretation for different users. Worse yet the intended interpretation may be not entirely clear to the user when he asked the query. Nevertheless, I believe that when the user comes up with a "bad" query he should not be sent home to come back with a domain independent query but should be interactively helped to clarify over which domains the different variables in the query should range. This approach seems very promising but needs further developing.

b) Besides the fact that domain independence is probably not what the user has in mind it is also a bad notion from the logical point of view: it is not robust. Note that $\varphi(x)$ may be non-domain independent while $\varphi(x) \wedge (a = a)$ is. This is the case if $\varphi(x)$ is $\exists y(y \neq b \wedge x = b)$. This example may be sharpened: if the function h does not occur in φ then φ and $\varphi \wedge (h(a) = h(a))$ have an entirely different class of (D, φ)-models (all the countable models for the conjunction are isomorphic in $L(D, \varphi)$). It seems that from the logical point of view starting with $L(D, \varphi)$ on one hand and assuming only term domains on the other leads to curious logical phenomena. The class of domains that correspond to φ is (nontrivially) different when φ is evaluated by itself and when it is evaluated as part of $\varphi \wedge \psi$ or $\varphi \vee \psi$. The reader is invited to modify the example above to obtain queries φ and ψ each of which may be sometimes true and sometimes false and such that $\varphi \wedge \psi$ is domain independent because it is always true (!). Or (replacing φ and ψ by their negation) such that $\varphi \vee \psi$ is domain independent because it is always false.

References

Avron-Hirshfeld 91: A. Avron and J. Hirshfeld, On first order database query languages. Proceeding of the sixth symposium on Logic in Computer Sciences (1991) 226-231.

Demolombe 82: R. Demolombe, Syntactical characterization of a subset of domain independent formulas. Technical report, ONERA-CERT, Toulouse (1982).

Di Paola 69: R.A. Di Paola, The recursive unsolvability of the decision problem for the class of definite formulas. J. ACM 16(2) (1969) 324-327.

Kifer 88: M. Kifer, On safely, domain independence and capturability of database queries. Proc. International Conference on database and knowledge bases. Jerusalem (1988) 405-415.

Topor 87: R.W. Topor, Domain independence formulas and databases. Theoretical Computer Science 52 (1987) 281-306.

Topor-Sonenberg 88: R.W. Topor and E.A. Sonenberg, On domain independent databases, In Foundation of deductive database and logic programming. Editor J. Minker.(1987) 217-240.

Ullman 89: J.D. Ullman, Principles of database and knowledge base systems Volume 1. Computer Science Press (1989).

Logical Inference and Polyhedral Projection *

J. N. HOOKER
GSIA, Carnegie Mellon University, Pittsburgh, PA 15213 USA
Matematisk Institut, Århus Universitet, DK-8000 Århus C, Denmark

September 1991
Revised February 1992

Abstract

We explore connections between polyhedral projection and inference in propositional logic. We formulate the problem of drawing all inferences that contain a restricted set of atoms (i.e., all inferences that pertain to a given question) as a logical projection problem. We show that polyhedral projection partially solves this problem and in particular derives precisely those inferences that can be obtained by a certain form of unit resolution. We prove that this unit resolution algorithm is exponential in the number of atoms in the restricted set but is polynomial in the problem size when this number of fixed. We also survey a number of new satisfiability algorithms that have been suggested by the polyhedral interpretation of propositional logic.

1 Introduction

The inference problem in propositional logic is closely connected with polyhedral theory. In the last few years this connection has suggested a number of new inference algorithms that have substantially advanced the state of the art. The most straightforward connection lies in the integer programming formulation of an inference problem, but there are less obvious links with cutting plane theory, combinatorial optimization and linear complementarity.

We will briefly survey these ideas below, but our main purpose here is to explore the connections between logical inference and the *projection* of polyhedra. We do this in order to address a more general form of the inference problem that can likewise be viewed as a kind of projection problem. The simplest form of the inference problem asks whether a given proposition can be inferred from a set of premises. But the more general problem, and the one more relevant to many applications, is to infer from a set of premises *everything* that is pertinent to a given question.

In an expert system for medical diagnosis, for instance, we add the clinical observations for a particular patient to a set of rules that presumably guide physicians when they diagnose illness. The object is to infer what is wrong with the patient.

*Supported in part by the Air Force Office of Scientific Research, Grant number AFOSR-91-0287.

The simplest kind of inference problem asks, "can we infer that the patient has appendicitis?" But the more general inference problem asks, "What illnesses can we infer that the patient has?"

A natural way to formulate the more general inference problem is as a *discrete* or *logical projection problem*. We can suppose that among the atomic propositions that occur in the medical knowledge base, some are propositions to the effect that the patient has a certain disease, such as "the patient has appendicitis," or "the patient has an intestinal blockage." The diagnostic problem, then, is draw all inferences that involve only these disease propositions. We might infer, for instance, that "the patient has appendicitis," or "the patient has appendicitis or an intestinal blockage," or "the patient has appendicitis only if he has no intestinal blockage."

This if the knowledge base contains atomic propositions $x_1, \ldots, x_n$, we can let $x_1, \ldots, x_k$ be the disease propositions. We wish to find all implied propositions that contain no atoms other than $x_1, \ldots x_k$. We will see that this is a problem of projecting the knowledge base onto the variables $x_1, \ldots, x_k$, and that it is closely related to the problem of projecting a polyhedron.

The general logical projection problem can be solved by a variation of the well-known resolution procedure, but there is no known practical procedure for solving large instances. They can be *partially* solved in reasonable time, however, by projecting a polyhedron. That is, by computing the projection of a polyhedron, we obtain *some* of the inferences involving $x_1, \ldots, x_k$. Our main result will be that polyhedral projection obtains precisely the inferences that can be derived by a modification of unit resolution, which is a form of resolution that solves the inference problem for Horn clauses. The modification we use is "unit K-resolution," which becomes ordinary unit resolution if all atoms in $\{x_1, \ldots, x_k\}$ are erased. Unsurprisingly, unit K-resolution completely solves the logical projection problem for Horn clauses.

Although unit K-resolution is considerably simpler than general resolution, we will prove that it can generate exponentially many inferences involving $x_1, \ldots, x_k$ even when n is proportional to k. However, if k is fixed, the running time of unit K-resolution is polynomial (quadratic) in the problem size. Thus unit resolution can be a practical method of computing inferences that involve $x_1, \ldots, x_k$ when k is small.

H. P. Williams first noted the connection between logical inference and polyhedral projection. He pointed out that the resolution method of theorem proving is closely related to the Fourier-Motzkin method for polyhedral projection [52]. In fact, one can obtain a resolvent by strengthening the result of a Fourier-Motzkin step in a certain way. But an iteration of the Fourier-Motzkin method projects only onto $\{x_1, \ldots, x_{n-1}\}$. To project a polyhedron onto $\{x_1, \ldots, x_k\}$ in one iteration, one must use a more general procedure. Our strategy here is to strengthen the result of the more general procedure in the way that Williams strengthened Fourier-Motzkin. We then appeal to some cutting plane theory we develop elsewhere [28] to show that this yields precisely the inferences that can be obtained by unit K-resolution. This in turn leads to the above results.

In Section 2 below we introduce some basic concepts involving logical inference, resolution, and cutting planes. Section 3 surveys some recent inference methods related to the polyhedral interpretation of logic. The rest of the paper focuses on projection. Section 4 defines logical projection, shows that general "K-resolution"

(resolution on variables in $\{x_{k+1}, \ldots, x_n\}$) solves the logical projection problem, and shows that unit K-resolution solves the problem for Horn clauses. Section 5 explains the basic ideas of polyhedral projection, and Section 6 proves the remaining results cited above.

2 Some Fundamentals

2.1 Propositional Logic

A *clause* in propositional logic, such as $x_1 \wedge \neg x_2 \wedge \neg x_3$, is a disjunction of *literals*, which are variables or their negations. A *unit clause* contains exactly one literal. Clause C *absorbs* clause D when all the literals of C occur in D; C logically implies D if and only if C absorbs D. A formula is in *conjunctive normal form* or *clausal form* when it is a conjunction of clauses. The problem of checking the joint satisfiability of a set of clauses is the original NP-complete problem [11].

Any formula of propositional logic can be put in clausal form [18]. This may require exponential time if no new variables are added but only linear time otherwise [6].

Formula A *implies* formula B if and only if $A \wedge \neg B$ is unsatisfiable. Thus the satisfiability problem and the inference problem are essentially the same in propositional logic. A and B are *equivalent* if they imply each other.

2.2 The Resolution Procedure

Quine [43, 44] proved that the satisfiability problem can be solved by the *resolution* procedure, which Robinson generalized to first-order logic [45]. When two clauses have the property that exactly one atomic proposition x_j occurs posited in one and negated in the other, the *resolvent* of the clauses is a disjunction of all the literals occurring in the clauses except x_j and $\neg x_j$. For instance, (3) below is the resolvent of (1) and (2), which are the *parents* of (3).

$$x_1 \vee x_2 \vee x_4 \tag{1}$$

$$\neg x_1 \vee \neg x_3 \vee x_4 \tag{2}$$

$$x_2 \vee \neg x_3 \vee x_4 \tag{3}$$

The resolvent of x_j and $\neg x_j$ is the *empty clause*, which is unsatisfiable by definition. *Unit resolution* is resolution in which at least one of the parents is a unit clause.

The *resolution procedure* begins with a set S of clauses and removes all clauses that are dominated by others. It then finds all resolvable pairs of clauses whose resolvent is not absorbed by any clause in S. It generates these resolvents, removes from S all clauses absorbed by the resolvents, and adds the resolvents to S. The process is repeated until a) the empty clause is generated, in which case the original clauses are unsatisfiable, or b) no further such resolutions are possible, in which case the clauses are satisfiable. The procedure takes exponential time in the worst case [20] and can be very slow even for random problems [24].

A clause C is a *prime implication* of a set S of clauses if S implies C and implies no other clause that absorbs C. Thus every implication of S is absorbed by some prime implication of S. Quine proved,

Theorem 1 (Quine [43, 44]) *If a set S of clauses is satisfiable, the resolution procedure generates precisely the set of prime implications of S. It generates the empty clause if and only if S is unsatisfiable.*

Horn clauses have received much attention because unit resolution alone is enough to check whether a set of Horn clauses are satisfiable, and unit resolution runs in linear time on Horn clauses [14] (since there is no need to resolve on negative unit clauses). A Horn clause is any clause that contains at most one positive literal.

2.3 The Integer Programming Formulation of the Satisfiability Problem

A clause such as $x_1 \wedge \neg x_2 \wedge \neg x_3$ can be written as a linear inequality in binary variables, in this case $x_1 + (1 - x_2) + (1 - x_3) \geq 1$ or $x_1 - x_2 - x_3 \geq -1$. Here a variable x_j is regarded as true when it takes the value 1 and false when it takes the value 0. Corresponding to a set S of clauses, then, is a system of 0-1 *clausal* inequalities, $Ax \geq a$, $x_j \in \{0,1\}$, all j. Clearly S is satisfiable if and only if the system is soluble. Its *linear relaxation* is obtained by replacing $x_j \in \{0,1\}$ with $0 \leq x_j \leq 1$.

2.4 Resolution and Cutting Planes

A *cutting plane* (or *cut*) for an integer programming problem is an inequality satisfied by all (0-1) solutions of the problem. One way to generate cutting planes is to take nonnegative linear combinations of the inequalities in the linear relaxation, including the bounds $0 \leq x_j \leq 1$, and to round up any fractions that result in the coefficients or right-hand side. Such a cut is called a *rank one* cut. For instance, (4e) below is a positive linear combination of inequalities (a)-(d), where each receives a weight 1/2.

$$
\begin{array}{rcrl}
x_1 + x_2 + x_4 & \geq & 1 & (a) \\
-x_1 - x_3 + x_4 & \geq & -1 & (b) \\
x_2 & \geq & 0 & (c) \\
- x_3 & \geq & -1 & (d) \\
x_2 - x_3 + x_4 & \geq & -1/2 & (e)
\end{array}
\qquad (4)
$$

We can round up the fraction in (4e) to obtain a cutting plane. Chvátal [10] proved that if we generate all rank one cuts in this manner, add them to the constraint set, generate all rank one cuts for the resulting set of inequalities, and so on, then we can generate all cuts in finite time.

A remarkable fact about resolvents is that they are rank one cutting planes. Note for instance that inequalities (4a) and (4b) represent the clauses (1) and (2), and that (4c) and (4d) are bounds of the form $0 \leq x_j \leq 1$. The resulting rank 1 cut represents the resolvent of (1) and (2).

3 A Survey of Inference Methods

Algorithms inspired by the polyhedral interpretation of logic have, in just the last three or four years, brought orders-of-magnitude speedups in the solution of inference problems. The most straightforward approach is to solve the integer programming formulation of an inference problem, which has been well known for some years. It was noted by G. Dantzig in his classic 1963 book on linear programming [12], used by R. M. Karp [39] in 1972 to prove the NP-completeness of integer programming, and elaborated by H. P. Williams [52, 53, 54] among others.

But C. E. Blair, R. Jeroslow and J. K. Lowe were apparently the first, in 1985, to solve nontrivial inference problems by this approach [6] (see also Williams [54]). They initially used a standard branch-and-bound code to solve the integer programming problem, with encouraging results. Hooker [26] showed in the same year that when one uses resolvents as cutting planes in an integer programming framework, one can solve a class of random satisfiability problems orders of magnitude more rapidly than by using resolution alone. This is primarily because resolvents that are separating cuts are relatively easy to find.

The branch-and-bound approach of Blair, Jeroslow and Lowe is very similar to the Davis-Putnam-Loveland (DPL) procedure, which also generates an enumeration tree [13, 40]. They differ only in that branch-and-bound solves the linear relaxation of the satisfiability problem at each node of the enumeration tree, whereas DPL applies unit resolution. Using unit resolution is in a sense equivalent to solving the linear relaxation because it detects satisfiability in the same instances. It may therefore seem that there is no point in solving the linear relaxation. But it has the advantage that it can sometimes, by luck, find integer solutions. When this happens the search can be terminated early with a satisfying solution.

A difficulty with the branch-and-bound approach, however, is that it takes longer to solve a linear programming problem than to apply unit resolution. Jeroslow and Wang [35] found that they could increase computation speed an order of magnitude by replacing the linear programming algorithm with unit resolution and a heuristic method that, like linear programming, sometimes finds a satisfying solution early in the search. The Jeroslow-Wang algorithm also uses a branching heuristic that branches on variables in an order that is likely to lead one to a satisfying solution early in the search, if one exists.

But it turns out that, aside from its branching heuristic, the Jeroslow-Wang algorithm is essentially the DPL procedure with a different order for traversing the tree (DPL is depth-first). In fact, more recent testing [23] shows that the DPL procedure usually runs as fast as the Jeroslow-Wang procedure, provided the former is equipped with the same branching heuristic and efficient data structures.

Hooker and Fedjki [31] restored the competitiveness of linear-programming-based methods with a branch-and-cut algorithm that uses cutting planes at several nodes of the branch-and-bound tree. The generation of cutting planes is based on the theoretical result that all clauses that are rank one cutting planes can be generated by unit resolution [28]. In [31] this forms the basis of a practical algorithm for finding separating cuts.

The branch-and-cut approach in effect strengthens the relaxation solved at each node by adding cuts. Billionnet and Sutter [3] strengthened the relaxation in a different way, namely by replacing unit resolution with a resolution procedure that generates all resolvents that are no longer than their parents.

One can also investigate the results of using a relaxation that is *weaker* than the linear relaxation. This was done by Gallo and Urbani [17], who replaced the linear relaxation with a Horn relaxation. They split each clause, for instance $x_1 \vee x_2 \vee \neg x_3 \vee \neg x_4$, into a the conjunction of a non-Horn part $x_1 \vee x_2 \vee \neg y$ and a Horn part $y \vee \neg x_3 \vee \neg x_4$. Only the Horn parts are checked for joint satisfiability. The Horn relaxation is very weak, but since unit resolution can check a Horn problem for satisfiability in linear time (as opposed to quadratic time in general), the large search trees are often offset by rapid processing of each node. In fact, the Horn relaxation method appears at this writing to be an order or magnitude faster than any other method on relätively easy problems, although it can be much slower on harder problems [23].

There are several mathematical programming algorithms for inference other than branch-and-cut. One is that of Harche and Thompson, which reinterprets the search process as a column subtraction procedure in the simplex tableau for the linear relaxation of the problem [23]. Of several new methods recently tested, this method was the only one to solve all the problems in a benchmark set collected by F. J. Radermacher [41, 49], including several of the famous pigeon hole problems [50]. But the column subtraction method is generally slower than others on easier problems.

Other mathematical programming approaches include the interior point method of Kamath, Karmarkar, Ramakrishnan and Resende [36, 37], and the heuristic developed by G. Patrizi and G. Spera, who solve a linear complementarity formulation of the satisfiability problem as a parametric linear programming problem [42, 46].

A new approach to propositional logic that draws heavily on ideas of combinatorial optimization is the decomposition approach of K. Truemper, which relies on his characterization of classes of polynomially soluble satisfiability problems [48]. Truemper's method is distinguished by the fact that after analyzing one problem, it generates a solution program that can solve similar problems very rapidly.

A satisfiability problem can always be expressed as a problem of checking whether a logic circuit has the same output for every input. In this form it can be solved with the approach of Hooker and Yan, which is a nonnumeric algorithm that results from applying Benders decomposition to the integer programming formulation of the problem [32]. This technique has been applied to important problem of verifying logic circuits, with better results than other methods on certain types of circuits and worse results on others.

Results from graph theory have been used to solve certain classes of inference problems in propositional logic [19, 21, 22]. Optimization techniques can be used to direct an expert system's inquiries for more information [51]. Hooker used cutting plane theory to extend resolution to a larger class of logical formulas [27] and even to general 0-1 linear inequalities [29].

Another approach to propositional logic is to identify classes of problems for which inference is easy; Horn problems are a simple example. Following some initial work of Arvind and Biswas [2] as well as Yamasaki and Doshita [55], Gallo and Scutellà [16] defined a hierarchy of inference problems soluble in polynomial time. Chandru and Hooker [8] used the theory of integer programming to extend Horn problems to a much larger set of problems that can be solved with unit resolution.

Mathematical programming techniques are also useful in other types of logics. The only known practical methods for inference in probabilistic logic are column-generation techniques for a linear programming formulation of the problem [1, 33, 38]. R Jeroslow, G. Gallo, R. Rago and J. Hooker have developed new methods for inference in first-order predicate logic [15, 34, 30]. Chandru and Hooker have proposed a set covering model for inference in Dempster-Shafer theory [9]. Inductive logic has been addressed with pseudo-boolean optimization [4, 5], integer programming [47] and interior point methods [37].

For more complete surveys of the research described above, see [7, 9, 23, 25].

4 Logical Projection

4.1 The Projection Problem

Given a vector $t = (t_1, \ldots, t_n)$ and an index set $K = \{1, \ldots, k\}$ with $k \leq n$, we say that $t_K = (t_1, \ldots, t_k)$ is the *projection* of t onto K and that t is an *extension* of t_K. The projection onto K of a set T of such vectors is $T_K = \{t_K | t \in T\}$. That is, T_K is the set of vectors in $\{0,1\}^k$ that have extensions in T.

If S is a set of logical clauses containing atomic propositions $x_1, \ldots, x_n$, let the *satisfaction set* $T(S)$ be the (possibly empty) set of vectors of satisfying truth assignments to $x_1, \ldots, x_n$. A *logical projection* of S onto K is a set $\overline{S}$ of clauses, containing only atoms $x_1, \ldots, x_k$, for which $T(\overline{S}) = T(S)_K$.

A logical projection in sense says everything that can be inferred from S about the variables $x_1, \ldots, x_k$, because of the following fact.

Lemma 1 *Any logical projection $\overline{S}$ of S onto K is equivalent to the set S_K of all clauses, containing only $x_1, \ldots, x_k$, that can be inferred from S.*

Proof. We wish to show $T(S_K) = T(\overline{S})$. It suffices to show $T(S)_K = T(S_K)$, since by definition $T(S)_K = T(\overline{S})$. Take any point $\bar{t} = (t_1, \ldots, t_k) \in T(S)_K$. Then some extension t of $\bar{t}$ satisfies S and so every implication of S. Thus in particular $\bar{t} = t_K$ satisfies every implication of S containing only $x_1, \ldots x_k$. Conversely, suppose that $\bar{t} = (t_1, \ldots, t_k)$ satisfies S_K. Then some extension t of $\bar{t}$ satisfies S, since otherwise one could infer from S a clause involving $x_1, \ldots, x_k$ that excludes $\bar{t}$. But this means that $\bar{t} = t_K \in T(S)_K$. □.

4.2 Resolution and Logical Projection

One way to solve the projection problem is by the resolution procedure. In particular, it can be solved with a series of *K-resolutions*, which are resolutions on variables in $\{x_{k+1}, \ldots, x_n\}$. To show this we must appeal to two elementary lemmas. We prove the first lemma elsewhere [28].

Lemma 2 *If clause C is implied by a set S of clauses, then it is implied by a subset S^+ of S in which every variable in C always occurs with the sign it has in C.*

Let the *K-resolution procedure* be the same as the resolution procedure defined earlier, except that all the resolutions are K-resolutions. The second lemma is easily proved by induction.

Lemma 3 *Let S' be a result of applying the K-resolution procedure to a satisfiable set S. Then if clause C can be obtained from S through a finite series of K-resolutions, then some clause in S' absorbs C.*

Now we can show that K-resolution generates a projection set.

Theorem 2 *Let S' be the result of applying the K-resolution procedure to a satisfiable set S of clauses. Then the set $\overline{S}$ obtained by deleting from S' all clauses with variables in $\{x_{k+1},\ldots,x_n\}$ is a logical projection of S onto K.*

Proof. It suffices by Lemma 1 to show that $\overline{S}$ is equivalent to S_K. Clearly any clause $C \in \overline{S}$ is implied by S_K, since it is implied by S and contains only variables in $\{x_1,\ldots,x_k\}$.

It remains to show that any clause $C \in S_K$ is implied by $\overline{S}$. Let $x_{j_1},\ldots,x_{j_t}$ be the variables in $\{x_1,\ldots,x_k\}$ that do not occur in C, and let v be any boolean function defined on $J = \{j_1,\ldots,j_t\}$. Then for any such v we can let

$$C(v) = C \vee \bigvee_{j\in J} x_j^{v(j)},$$

where $x_j^1 = x_j$ and $x_j^0 = \neg x_j$. It suffices to show that $\overline{S}$ implies $C(v)$ for every function v, since in this case $\overline{S}$ implies C.

To show this, note that since S implies C it implies $C(v)$ for any v. Thus by Lemma 2 some subset S^+ of S implies $C(v)$, where every variable in $C(v)$ always has in S^+ with the same sign it has in $C(v)$. Theorem 1 says that some series of resolutions generates $C(v)$ from S^+. But since the series can contain no resolutions on variables in $\{x_1,\ldots,x_k\}$, it is a series of K-resolutions. Thus by Lemma 3 $C(v)$ is absorbed by a clause in S', which is also a clause in $\overline{S}$. This completes the proof. □

As an example, consider the following set S of clauses.

$$\begin{array}{lllll} x_1 \vee x_2 \vee & x_3 & & \\ \neg x_1 \vee x_2 & & & \vee \neg x_4 \\ \neg x_1 & \vee\ x_3 & & \\ \neg x_1 & \vee \neg x_3 \vee & x_4 & \\ x_1 & \vee \neg x_3 & & \end{array} \tag{5}$$

The first iteration of $\{1,2\}$-resolution yields resolvents $x_1 \vee x_2$, $\neg x_1 \vee x_4$ and $\neg x_1 \vee x_2 \vee \neg x_3$, while the second yields $\neg x_1 \vee x_2$. So $\overline{S} = \{x_1 \vee x_2, \neg x_1 \vee x_2\}$ is a logical projection onto $\{1,2\}$.

4.3 Horn Clauses

Unit resolution not only checks a set of Horn clauses for satisfiability but solves the logical projection problem as well. Let *unit K-resolutions* be K-resolutions in which at least one of the parents contains only one variable not in $\{x_1,\ldots,x_k\}$. Let the *unit K-resolution procedure* be the same as the resolution procedure except that all resolutions are unit K-resolutions. We will need the following lemma, whose proof is similar to that of Lemma 3.

Lemma 4 *Let S' be a result of applying the unit K-resolution procedure to a satisfiable set S. Then if clause C can be obtained from S through a finite series of unit K-resolutions, then some clause in S' absorbs C.*

Theorem 3 *Let S' be the result of applying the unit K-resolution procedure to a satisfiable set S of Horn clauses. Then the set $\overline{S}$ obtained by deleting from S' all clauses with variables in $\{x_{k+1}, \ldots, x_n\}$ is a logical projection of S onto K.*

Proof. Again we show that $T(\overline{S}) = T(S_K)$. As in the proof of Theorem 2, it is clear that any clause in $\overline{S}$ is implied by S_K. To show that any clause $C \in S_K$ is implied by $\overline{S}$, it again suffices to show that $\overline{S}$ implies $C(v)$ for any v.

Since S implies $C(v)$, by Lemma 2 some subset S^+ of S implies $C(v)$, where every variable in $C(v)$ always occurs in S^+ with the same sign it has in $C(v)$. For any clause $D \in S^+$ let D_0 be the portion of D with variables in $\{x_{k+1}, \ldots, x_n\}$. Then $S_0^+ = \{D_0 | D \in S^+\}$ must be unsatisfiable. For if some $\{t_{k+1}, \ldots, t_n\}$ satisfied S_0^+, then $(t_1, \ldots, t_k, t_{k+1}, \ldots, t_n)$ would satisfy S^+ and falsify $C(v)$ for appropriate choice of $t_1, \ldots, t_k$, which is contrary to the fact that S^+ implies $C(v)$.

But if S_0^+ is unsatisfiable, then by Theorem 1 there is a series of unit resolutions that begins with the clauses in S_0^+ and generates the empty clause. Since each variable in $\{x_1, \ldots, x_k\}$ has the same sign in every occurrence, there is a parallel series of unit K-resolutions that begins with S^+ and generates a clause that absorbs $C(v)$. Thus by Lemma 4 $\overline{S}$ is absorbed by a clause in S', which is a clause in $\overline{S}$. □

5 Polyhedral Projection

5.1 The Projection Problem

Given a polyhedron $P = \{x | Ax \geq a\}$, the ***polyhedral projection problem*** is to find a system $By \geq b$, with $y = (x_1, \ldots, x_k)$, whose satisfaction set is the projected polyhedron P_K. A ***projection cut*** is any inequality that is satisfied by every point in P_K.

It is well known that one way to obtain a projection system is to take all non-negative linear combinations of the inequalities $Ax \geq a$ that cause the coefficients of $x_{k+1}, \ldots, x_n$ to vanish. Let us write the system $Ax \geq a$ as

$$A_1 y + A_2 z \geq a, \tag{6}$$

where $z = (x_{k+1}, \ldots, x_n)$. Then the projection must satisfy all inequalities $u^T A_1 y \geq u^T a$, where u is any vector in the polyhedral cone $U = \{u | u^T A_2 = 0, u \geq 0\}$. This is an infinite system, but an equivalent finite system can be obtained by using only the vectors u that are extreme rays of U.

Suppose for instance we want to project the system below onto $\{1, 2\}$. Note that this system is the linear relaxation of the clauses (5).

$$\begin{array}{rcl} x_1+x_2+x_3 & & \geq 1 \\ -x_1+x_2 & -x_4 & \geq -1 \\ -x_1 \quad +x_3 & & \geq 0 \\ -x_1 \quad -x_3 & +x_4 & \geq -1 \\ x_1 \quad -x_3 & & \geq 0 \\ 0 \leq x_j \leq 1, & j=1,\ldots,4. & \end{array} \tag{7}$$

The polyhedral cone is,

$$U = \{u | u_1 + u_3 - u_4 - u_5 + u_8 - u_{12} = -u_2 + u_4 + u9 - u_{13} = 0\}, \tag{8}$$

where $u_6, \ldots, u_9$ correspond to the constraints $x_j \geq 0$ and $u_{10}, \ldots, u_{13}$ to $-x_j \geq -1$. U has 14 extreme rays, which give rise to only two inequalities that are not already implied by the bounds $0 \leq x_j \leq 1$, namely $2x_1 + x_2 \geq 1$ and $-3x_1 + 2x_2 \geq -2$. These inequalities and the bounds describe the projection of (7) onto $\{1,2\}$. It is in general very difficult to find the extreme rays of a polyhedral cone, however, since it essentially requires enumeration of the basic solutions of $u^T A_2 = 0$.

5.2 Fourier-Motzkin Elimination

Another approach to computing the polyhedral projection is *Fourier-Motzkin elimination*, which is first applied to $Ax \geq a$ to "eliminate" x_n. The result of this elimination is a projection onto $\{1, \ldots, n-1\}$. The procedure is applied to this new system to obtain a projection onto $\{1, \ldots, n-2\}$, and so on.

We can for instance eliminate x_4 from (7) by writing the two inequalities containing x_4 in the form

$$1 - x_1 + x_2 \geq x_4 \tag{9}$$

$$x_4 \geq -1 + x_1 + x_3 \tag{10}$$

By pairing the expression on the left with that on the right we obtain an inequality without x_4:

$$1 - x_1 + x_2 \geq -1 + x_1 + x_3, \tag{11}$$

or

$$-2x_1 + x_2 - x_3 \geq -2. \tag{12}$$

(12) and the inequalities of (7) containing no x_4 comprise a projection system for $\{1,2,3\}$.

Now, to eliminate x_3, we write (12) and the inequalities in (7) containing x_3 as below:

$$\begin{array}{rcl} & x_3 & \geq 1 - x_1 - x_2 \\ & x_3 & \geq x_1 \\ x_1 \geq & x_3 & \\ 2 - 2x_1 + x_2 \geq & x_3 & \end{array}$$

By pairing we obtain four inequalities (after simplifying):

$$\begin{aligned} 2x_1 + x_2 &\geq 1 \\ 0 &\geq 0 \\ -x_1 + 2x_2 &\geq -1 \\ -3x_1 + x_2 &\geq -2, \end{aligned}$$

the second and third of which are redundant of the bounds $0 \leq x_j \leq 1$. We therefore obtain the same projection set for $\{1, 2\}$ as before.

In fact it is easy to see that Fourier-Motzkin elimination is a special case of the projection method described in the previous section. For instance, in combining (9) and (10) to obtain (11), we in effect take a linear combination of (9) and (10) that causes x_4 to vanish (in this case with unit weights). The computational problem is again difficult, however, since the number of inequalities tends to explode after a few iterations.

6 Inference by Polyhedral Projection

6.1 Polyhedral Projection as an Inference Method

Polyhedral projection can be used as an inference method, because a logical projection defines a set of points contained in the polyhedral projection. To make this more precise, let P be the polyhedron described by the linear relaxation of a set S of clauses. A logical projection of S is satisfied by precisely the points in $T(S)_K$. Then if $\text{int}(P_K)$ is the set of integer points in P_K, $T(S)_K \subset \text{int}(P_K)$. This is true simply because $T(S) \subset P$.

In the example of the previous section, the projection cuts $2x_1 + x_2 \geq 1$ and $-3x_1 + x_2 \geq -2$ are valid logical inferences as well. We can check this by noting that all points satisfying the logical projection $\{x_1 \vee x_2, \neg x_1 \vee x_2\}$, namely $(0,1), (1,1)$, also satisfy these inequalities. In other words, $T(S)_K = \{(0,1),(1,1)\}$ is a subset of $\text{int}(P_K)$.

In this particular case $T(S)_K = \text{int}(P_K)$, so that polyhedral projection solves the logical projection problem. But this is not true in general. Consider the clauses,

$$\begin{aligned} &x_1 \vee x_2 \vee x_3 \\ &x_1 \vee x_2 \vee \neg x_3 \\ &x_1 \vee \neg x_2 \vee x_3 \\ &x_1 \vee \neg x_2 \vee \neg x_3 \end{aligned} \tag{13}$$

The logical projection onto $\{1\}$ is the clause x_1, so that $T(S)_K$ contains the single point $x_1 = 1$. But the polyhedral projection P_K is the entire interval $[0, 1]$, which contains $x_1 = 0$ as well. Thus $T(S)_K = \{1\} \neq \text{int}(P_K) = \{0,1\}$. The polyhedral projection therefore "loses information." It produces a valid inference but does not infer everything that can be inferred. In this case it infers the trivial fact that x_1 is either true or false.

6.2 Resolution and Fourier-Motzkin Elimination

We have seen that polyhedral projection can "lose information" but that resolution does not. H. P. Williams [52] observed that the Fourier-Motzkin method of polyhedral projection can be strengthened with a certain reduction and rounding operation so that it does not lose information. The reason is simply that, when so strengthened, a Fourier-Motzkin step becomes a resolution step.

Consider for instance the inequality (12) that a Fourier-Motzkin step obtained from (9) and (10). We can add to it the bounds $x_2 \geq 0$ and $-x_3 \leq -1$ (this is an instance of *reduction*) and divide the sum by 2 to obtain $-x_1 + x_2 - x_3 \geq -3/2$. We can *round* the $-3/2$ up to -1, since any 0-1 point satisfying this inequality satisfies $-x_1 + x_2 - x_3 \geq -1$. But this is equivalent to the clause $\neg x_1 \vee x_2 \vee \neg x_3$, which is the resolvent of the clauses represented by (9) and (10). Since a Fourier-Motzkin step simply adds two inequalities so as to cancel a term, it is clear that by appending a reduction and rounding step we obtain resolution.

6.3 Unit Resolution and Polyhedral Projection

We saw in the previous section that by applying reduction and rounding to an inequality that results from a Fourier-Motzkin step, we obtain a resolvent. But the Fourier-Motzkin method is rather limited, since in one iteration it projects only onto $\{1, ..., n-1\}$. This raises the question as to what inference procedure one might obtain by applying reduction and rounding to an inequality resulting from the more general method that projects onto $K = \{1, ..., k\}$ in one iteration. The answer, we will show, is that one obtains the unit K-resolution procedure.

We rely on a result we prove elsewhere with the help of cutting plane theory [28].

Theorem 4 *Any clause C that is a rank 1 cut for the linear relaxation of a set S of clauses is absorbed by some clause that can be obtained from S by a series of unit K-resolutions, where K contains the indices of the variables in C.*

We can now prove our claim.

Theorem 5 *Let a set S of clauses be written in the form (6), and let $u \in \{u | u^T A_2 = 0, u \geq 0\}$. Then any clause obtained by applying reduction and rounding to $u^T A_1 \geq u^T a$ is implied by clauses that can be obtained from S by a series of unit K-resolutions.*

Proof. Let clause C be the result of applying reduction and rounding to $u^T A_1 \geq u^T a$, and let $C(v)$ for a function v be defined as in the proof of Theorem 2. It suffices to show that, for each v, $C(v)$ is absorbed by a clause obtained by unit K-resolution.

It is clear from the definition of a rank 1 cut that if C is obtained by applying reduction and rounding to an inequality I that is a nonnegative linear combination of the inequalities in (6), then C is a rank 1 cut for (6). The same is true of $C(v)$, since one can add multiples of bounds $0 \leq x_j \leq 1$ to I to obtain an inequality that, after reduction and rounding, yields $C(v)$ rather than C. Then by Theorem 4 a series of K-resolutions yields a clause that absorbs $C(v)$. □

We are now in a position to show that unit K-resolution has the same deductive power as polyhedral projection. It will be convenient to use the following lemma, which says that unit K-resolution does not cut off any integer points in the polyhedral projection.

Lemma 5 *Let S be a set of clauses, P the polyhedron described by its linear relaxation, and $\overline{S}$ the result of applying the unit K-resolution procedure to S and deleting all clauses with variables in $\{x_{k+1}, \ldots, x_n\}$. Then int(P_K) $\subset T(\overline{S})$.*

Proof. It suffices to show that any point $(t_1, \ldots, t_n)$ that satisfies the inequalities representing parents of a unit K-resolvent, where $t_1, \ldots, t_k \in \{0,1\}$, also satisfies the inequality representing the resolvent. Without loss of generality we may suppose that the two parents have the form $ay + c'z + x_n \geq \alpha$ (with $c'_n = 0$) and $by - x_n \geq \beta$, and that the resolvent is $cy + c'z \geq \gamma$. We can suppose without loss of generality that $a, b \geq 0$, since a_j and b_j do not have opposite signs, and if $a_j < 0$ or $b_j < 0$ we can replace x_j with $1 - x_j$. Now there are two cases. a) t has the property that $t_j = 0$ when $c_j = 1$ and $t_j = 1$ when $c_j = -1$, for $j \in \{1, \ldots, k\}$. Then $(t_{k+1}, \ldots, t_n)$ must satisfy $c'z + x_n \geq \alpha$ and $-x_n \geq \beta$ and hence their sum $c'z \geq \gamma$. t therefore satisfies the resolvent. b) t does not have this property, in which case it obviously satisfies the resolvent. □

Theorem 6 *Let S be a satisfiable set of clauses, P the polyhedron described its linear relaxation, and $\overline{S}$ the result of applying the unit K-resolution procedure to S and deleting all clauses with variables in $\{x_{k+1}, \ldots, x_n\}$. Then int(P_K) $= T(\overline{S})$.*

Proof. Lemma (5) implies that int(P_K) $\subset T(\overline{S})$. It remains to show that $T(\overline{S}) \subset$ int(P_K).

We take a point $t \notin$ int(P_K) and will show that $t \notin T(\overline{S})$. Thus t is cut off by some projection cut $u^T A_2 y \geq u^T a$. This cut must therefore logically imply a clause C that t falsifies. We will exhibit another projection cut $v^T A_2 y \geq v^T a$ that yields C when one applies the reduction and rounding procedure. It then follows from Theorem 5 that C is implied by clauses obtainable by unit K-resolution, which means by Lemma 4 that $t \notin T(\overline{S})$.

For convenience write $u^T A_2 y \geq u^T a$ as $by \geq \beta$. Again we can suppose without loss of generality that $b \geq 0$. Let $C = \bigvee_{j \in J} x_j$, where $J \subset \{1, \ldots, k\}$. Then since $bx \geq \beta$ implies C, we must have $\sum_{j \notin J} b_j < \beta$. Thus if we subtract, from $bx \geq \beta$, b_j times the bound $x_j \geq 0$ for each $j \notin J$, we obtain an inequality $cx \geq \gamma$ with $\gamma > 0$. From this it is clear how to alter u to obtain a vector v so that $v^T A_2 \geq v^T a$ is $(c/c_{max})y \geq \gamma/c_{max}$, where $c_{max} = \max_j\{c_j\}$. But rounding and reduction applied to the latter inequality yields C. □

6.4 Complexity of Inference by Polyhedral Projection

Although unit K-resolution is much simpler than general resolution, its complexity can grow exponentially with k.

Theorem 7 *The unit K-resolution procedure applied to a clause set S with n variables generates a set of clauses whose size, in the worst case, grows exponentially with k, even if $n = 2k - 1$ and $|S| = 2k$.*

Proof. Let S consist of clauses $x_1^v \vee x_{k+1}$, clauses $x_j^v \vee \neg x_{j+k-1} \vee x_{j+k}$ for $j = 2, \ldots, k-1$, and clauses $x_j^v \vee \neg x_{2k-1}$, where $v = 0, 1$. (Recall that $x_j^1 = x_j$, $x_j^0 = \neg x_j$.) Then for any 0-1 sequence $v_1, \ldots, v_k$, we can perform unit K-resolutions to obtain $x_1^{v_1} \vee \ldots \vee x_k^{v_k}$. That is, we can resolve $x_1^{v_1} \vee x_{k+1}$ with $x_2^{v_2} \vee \neg x_{k+1} \vee x_{k+1}$, their resolvent $x_1^{v_1} \vee x_2^{v_2} \vee x_{k+2}$ with $x_3^{v_3} \vee \neg x_{k+2} \vee x_{k+3}$, and so on. Thus we obtain 2^k distinct clauses in the projection set, none of which absorbs another. □

It may therefore be practical in many situations to use unit K-resolution only for fairly small k. Fortunately, the complexity of unit K-resolution is polynomial when k is fixed.

Theorem 8 *The complexity of unit K-resolution is at worst quadratic in the number of literals (i.e., literal occurrences) when k is fixed.*

Proof. Let S be the set of clauses to which unit K-resolution is applied. Also let the K-*part* of a clause be the portion of the clause with variables in $\{x_1, \ldots, x_k\}$. Label each literal occurring in the non-K-part of a clause in S. When a clause with only one varable in $\{x_{k+1}, \ldots, x_n\}$ is resolved with another clause C, let the literals in the non-K-part of the resolvent inherit their labels in C.

Now take any two clauses D and E (not necessarily distinct) whose variables belong to $\{x_1, \ldots, x_k\}$. Consider the group G_D of all clauses with K-part D that are generated by the unit k-resolution procedure, and the group G_E of all clauses whose K-part is E. Let L_1 and L_2 be arbitrary labels of literals in S. Since unit K-resolution only checks a pair of clauses for sign alteration when one of them belongs to a clause with one variable in $\{x_{k+1}, \ldots, x_n\}$, on at most one occasion does it compare a literal with label L_1 in G_D with a literal with label L_2 in G_E. Thus the number of comparisons between literals in G_D with those in G_E is at most quadratic in the number of literals in S. Since the number of pairs D, E is bounded by 2^{2k}, the theorem follows. □

In particular ordinary unit resolution ($k = 0$) has quadratic complexity.

References

[1] Andersen, K. A., and J. N. Hooker, Bayesian logic, to appear in *Decision Support Systems.*

[2] Arvind, V. and S. Biswas, An $O(n^2)$ algorithm for the satisfiability problem of a subset of propositional sentences in CNF that includes all Horn sentences, *Information Processing Letters* **24** (1987) 67-69.

[3] Billionnet, A., and A. Sutter, An efficient algorithm for the 3-satisfiability problem, Research Report 89-13, Centre d'études et de recherche en informatique, 292 rue Saint-Martin, 75141 Paris Cedex 03, 1989.

[4] E. Boros, P. Hammer and J. N. Hooker, Boolean regression, working paper 1991-30, Graduate School of Industrial Administration, Carnegie Mellon University, Pittsburgh, PA 15213 USA.

[5] E. Boros, P. L. Hammer, and J. N. Hooker, Predicting cause-effect relationships from incomplete discrete observations, working paper 1991-22, Graduate School of Industrial Administration, Carnegie Mellon University, Pittsburgh, PA 15213 USA.

[6] Blair, C., R. G. Jeroslow, and J. K. Lowe, Some results and experiments in programming techniques for propositional logic, *Computers and Operations Research* **13** (1988) 633-645.

[7] Chandru, V., and J. N. Hooker, Logical inference: A mathematical programming perspective, in S. T. Kumara, R. L. Kashyap, and A. L. Soyster, eds., *Artificial Intelligence: Manufacturing Theory and Practice*, Institute of Industrial Engineers (1988) 97-120.

[8] Chandru, V., and J. N. Hooker, Extended Horn sets in propositional logic, *Journal of the ACM* **38** (1991) 203-221.

[9] Chandru, V., and J. N. Hooker, *Optimization Methods for Logical Inference*, Wiley, to appear.

[10] Chvátal, V., Edmonds polytopes and a hierarchy of combinatorial problems, *Discrete Mathematics* **4** (1973) 305-337.

[11] Cook, S. A., The complexity of theorem-proving procedures, *Proceedings of the Third Annual ACM Symposium on the Theory of Computing* (1971) 151-158.

[12] Dantzig, G. B., *Linear Programming and Extensions*, Princeton University Press (1963).

[13] Davis. M., and H. Putnam, A computing procedure for quantification theory, *Journal of the ACM* **7** (1960) 201-215.

[14] Dowling, W. F., and J. H. Gallier, Linear-time algorithms for testing the satisfiability of propositional Horn formulae, *Journal of Logic Programming* **1** (1984) 267-284.

[15] G. Gallo and G. Rago, A hypergraph approach to logical inference for datalog formulae, working paper, Dip. di Informatica, University of Pisa, Italy (September 1990).

[16] Gallo, G., and M. G. Scutella, Polynomially soluble satisfiability problems, *Information Processing Letters* **29** (1988) 221-227.

[17] Gallo, G., and G. Urbani, Algorithms for testing the satisfiability of propositional formulae, *Journal of Logic Programming* **7** (1989) 45-61.

[18] Genesereth, M. R., and N. J. Nilsson, *Logical Foundations of Artificial Intelligence*, Morgan Kaufmann (Los Altos, CA, 1987).

[19] Glover, F., and H. J. Greenberg, Logical testing for rule-based management, *Annals of Operations Research* **12** (1988) 199-215.

[20] Haken, A., The intractability of resolution, *Theoretical Computer Science* **39** (1985) 297-308.

[21] Hansen, P., A cascade algorithm for the logical closure of a set of binary relations, *Information Processing Letters* **5** (1976) 50-55.

[22] Hansen, P., B. Jaumard and M. Minoux, A linear expected-time algorithm for deriving all logical conclusions implied by a set of boolean inequalities, *Mathematical Programming* **34** (1986) 223-231.

[23] Harche, F., J. N. Hooker and G. L. Thompson, A computational study of satisfiability algorithms for propositional logic, working paper 1991-27, Graduate School of Industrial Administration, Carnegie Mellon University, Pittsburgh, PA 15213 USA, 1991.

[24] Hooker, J. N., Resolution vs. cutting plane solution of inference problems: Some computational experience, *Operations Research Letters* 7 (1988) 1-7.

[25] Hooker, J. N., A quantitative approach to logical inference, *Decision Support Systems* 4 (1988) 45-69.

[26] Hooker, J. N., Resolution vs. cutting plane solution of inference problems: some computational experience, *Operations Research Letters* 7 (1988) 1-7.

[27] Hooker, J. N., Generalized resolution and cutting planes, *Annals of Operations Research* 12 (1988) 217-239.

[28] Hooker, J. N., Input proofs and rank one cutting planes, *ORSA Journal on Computing* 1 (1989) 137-145.

[29] Hooker, J. N., Generalized resolution for 0-1 linear inequalities, to appear in *Annals of Mathematics and AI.*

[30] Hooker, J. N., New methods for inference in first-order predicate logic, working paper 1991-11, Graduate School of Industrial Administration, Carnegie Mellon University, Pittsburgh, PA 15213 USA, 1991.

[31] Hooker, J. N. and C. Fedjki, Branch-and-cut solution of inference problems in propositional logic, to appear in *Annals of Mathematics and AI.*

[32] Hooker, J. N., and H. Yan, Verifying logic circuits by Benders decomposition, working paper 1991-29, Graduate School of Industrial Administration, Carnegie Mellon University, Pittsburgh, USA, August 1988.

[33] Jaumard, B., P. Hansen and M. P. Aragaö, Column generation methods for probabilistic logic, to appear in *ORSA Journal on Computing.*

[34] Jeroslow, R. E., Computation-oriented reductions of predicate to propositional logic, *Decision Support Systems* 4 (1988) 183-197.

[35] Jeroslow, R. E., and J. Wang, Solving propositional satisfiability problems, *Annals of Mathematics and AI* 1 (1990) 167-187.

[36] Kamath, A. P., N. K. Karmarkar, K. G. Ramakrishnan, and M. G. C. Resende, Computational experience with an interior point algorithm on the satisfiability problem, in R. Kannan and W. R. Pulleyblank, eds., *Integer Programming and Combinatorial Optimization*, University of Waterloo Press (Waterloo, Ont., 1990) 333-349.

[37] Kamath, A. P., N. K. Karmarkar, K. G. Ramakrishnan, and M. G. C. Resende, A continuous aproach to inductive inference, manuscript, AT&T Bell Labs, Murray Hill, NJ 07974 USA, 1991.

[38] Kavvadias, D., and C. H. Papadimitriou, A linear programming approach to reasoning about probabilities, to appear in *Annals of Mathematics and Artificial Intelligence.*

[39] Karp, R. M., Reducibility among combinatorial problems, in R. E. Miller and J. W. Thatcher, eds., *Complexity of Computer Computations*, Plenum Press (1972) 85-103.

[40] Loveland, D. W., *Automated Theorem Proving: A Logical Basis*, North-Holland (1978).

[41] Mitterreiter, I., and F. J. Radermacher, Experiments on the running time behavior of some algorithms solving propositional logic problems, working paper, Forschungsinstitut für anwendungsorientierte Wissensverarbeitung, Ulm, Germany (1991).

[42] Patrizi, G., The equivalence of an LCP to a parametric linear program with a scalar parameter, to appear in *European Journal of Operational Research*.

[43] Quine, W. V., The problem of simplifying truth functions, *American Mathematical Monthly* 59 (1952) 521-531.

[44] Quine, W. V., A way to simplify truth functions, *American Mathematical Monthly* 62 (1955) 627-631.

[45] Robinson, J. A., A machine-oriented logic based on the resolution principle, *Journal of the ACM* 12 (1965) 23-41.

[46] Spera, C., Computational results for solving large general satisfiability problems, technical report, Centro di Calcolo Elettronico, Università degli Studi di Siena, Italy, 1990.

[47] Triantaphylou, E., A. L. Soyster, and S. R. T. Kumara, Generating logical expressions from positive and negative examples via a branch-and-bound approach, manuscript, Industrial and Management Systems Engineering, Pennsylvania State University, University Park, PA 16802 USA, 1991.

[48] Truemper, K., Polynomial theorem proving: I. Central matrices, technical report UTDCS-34-90, Computer Science Dept., University of Texas at Dallas, Richardson, TX 75083-0688 USA (1990).

[49] Truemper, K., and F. J. Radermacher, Analyse der Leistungsfähigkeit eines neuen Systems zur Auswertung aussagenlogisher Probleme, technical report FAW-TR-90003, Forschungsinstitut für anwendungsorientierte Wissensverarbeitung, Ulm, Germany (1990).

[50] Tseitin, G. S., On the complexity of derivations in the propositional calculus, in A. O. Slisenko, ed., *Structures in Constructive Mathematics and Mathematical Logic, Part II* (translated from Russian, 1968) 115-125.

[51] Wang, J., and J. Vande Vate, Question-asking strategies for Horn clause systems, working paper, Georgia Institute of Technology, Atlanta, GA, 1989.

[52] Williams, H. P., Fourier-Motzkin elimination extension to integer programming problems, *Journal of Combinatorial Theory* 21 (1976) 118-123.

[53] Williams, H. P., *Model Building in Mathematical Programming*, Wiley (1985).

[54] Williams, H. P., Linear and integer programming applied to the propositional calculus, *International Journal of Systems Research and Information Science* 2 (1987) 81-100.

[55] Yamasaki, S. and S. Doshita, The satisfiability problem for a class consisting of Horn sentences and some non-Horn sentences in propositional logic, *Information and Control* 59 (1983) 1-12.

Stable Logic

Brigitte Hösli

Institut für theoretische Informatik
ETH Zürich, CH-8092 Zürich
e-mail: hoesli@inf.ethz.ch

Abstract: This work investigates a 3-valued semantics, where the third value has the intention "unimportant" or "insignificant". If "true" and "unimportant" are the distinguished values, then we can define a tautology in this logics also as follows: a formula is a tautology iff every subformula, which arises by elimination of propositional variables, is a classical tautology. So our system is stable against loosing or missing information and therefore it is called stable.

There are many connections to other non-classical logics. So we can see these truth-tables as counterpart to Bočvar's one. Furthermore they are related to those of RM_3, the strongest logic in the family of relevance logics. And finally we can derive them from the interpretation of the multiplicative connectives in the phase semantics.

The classical sequent calculus missing the weakening rule on the right side is sound and complete w.r.t. this semantics, where "true" and "unimportant" are distinguished; the calculus missing the weakening rule on the left side is sound and complete w.r.t. this semantics, where "true" is the only distinguished value.

As in classical propositional logic, where boolean algebra is a counterpart to the two-valued semantics, we define an algebra as a counterpart to this three-valued one. A subclass of the algebra can be interpreted as an algebra of pairs of sets, which gives a very graphic representation of our three-valued connectives.

1 A 3-valued semantics

The following three-valued truth-tables are an extension of the classical ones and the third value "i" is a neutral element of both operation "$\wedge$" and "$\vee$".

$\neg$	t	f	i
	f	t	i

$\vee$	t	f	i
t	t	t	t
f	t	f	f
i	t	f	i

$\wedge$	t	f	i
t	t	f	t
f	f	f	f
i	t	f	i

$\rightarrow$	t	f	i
t	t	f	f
f	t	t	t
i	t	f	i

In three-valued semantics it is always helpful to know the intention of the third value. So the intention by Kleene is "still undefined" and it is "paradox" by Bočvar [Kl] and here the intention is *"insignificant"* or *"unimportant"*.

Examples:
Let V be a three-valued valuation and $V(A) = f$, $V(B) = i$, $V(C) = t$, then $V(A \vee B) = f$, $V(C \wedge (B \vee A)) = f$ and $V((C \wedge B) \vee (C \wedge A)) = t$.

Definition of tautologies:
A formula X is a *R-tautology* if $V(X) \neq f$ for all valuation V. (That means "t" and "i" are the distinguished values.)

Examples:
$A \vee \neg A$, $A \wedge B \rightarrow A \vee B$, $A \wedge (B \vee C) \rightarrow (A \wedge B) \vee (A \wedge C)$ are R-tautologies.
$A \vee \neg A \vee B$, $A \rightarrow A \vee B$, $(A \wedge B) \vee (A \wedge C) \rightarrow A \wedge (B \vee C)$ are no R-tautologies.

An equivalent definition:
A formula X is a tautology iff every subformula of X - arisen by discarding some variables - is a classical tautology. (Note that discarding variables $A_1, ..., A_n$ corresponds to assigning the value i to $A_1, ..., A_n$.)

Therefore the tautologies are well-balanced formulas, the truth-value of which is not depending of a certain variable (in contrast to the classical semantics) and remains the same when variables get lost. Hence, our tautologies are stable against loosing or missing variables respectively informations.

Relations to other non-classical logics:

- We can interpret our truth-tables as a counterpart to *Bočvar's* ones, because the third truth-value always succeeds there, but it never succeeds here in presence of true or false [Kl].
- The strongest logic in the family of relevance logics RM_3 has a three-valued semantics too and its truth-table for the implication exactly corresponds to the table here, but its tables for the conjunction and disjunction correspond to Kleene's tables [AB].
- The *nonsense logic* of Hałkowska, which is three-valued, has the same truth-table for the disjunction as here, but the table for the conjunction corresponds to Bocvar's one.
- That the interpretation of the connectives is a special case of Girard's *phase semantics* will be explained later [Gi].

Notation:
We define the value of the conjunction of a finite set of formulas as follows:

$$V(\Gamma_\wedge) = \begin{cases} V(X_1 \wedge ... \wedge X_n) & \text{if } \Gamma \equiv X_1, ..., X_n \\ t & \text{if } \Gamma \equiv \emptyset \end{cases}$$

and the value of the disjunction as follows:

$$V(\Delta_\vee) = \begin{cases} V(Y_1 \vee ... \vee Y_m) & \text{if } \Delta \equiv Y_1, ..., Y_m \\ f & \text{if } \Delta \equiv \emptyset \end{cases}$$

Definition of consequences:

1. Δ is a *R-consequence* of Γ ($\Gamma \models_R \Delta$), if $V(\Delta_\vee) \neq f$ for all valuations V where $V(\Gamma_\wedge) \neq f$. That means "t" and "i" are the distinguished values.
2. Δ is a *L-consequence* of Γ ($\Gamma \models_L \Delta$), if $V(\Delta_\vee) = t$ for all valuations V where $V(\Gamma_\wedge) = t$. That means "t" is the only distinguished value.
3. Δ is a *RL-consequence* of Γ ($\Gamma \models_{RL} \Delta$), if Δ is a R-consequence and a L-consequence of Γ.

Examples:
$A \wedge B \models_R A$, but $\not\models_R A \wedge B \rightarrow A$ (Note: deduction theorem is not valid.)

2 The calculus

At first we consider a sequent calculus of classical logic. As usual the rules divide into logical rules (which introduce the connectives), the cut rule and structural rules (which have a more practical importance, but nevertheless the structural rules have essential consequences: e.g. we obtain classical predicate logic only by adding the weakening and the contraction rules to linear logic, which is decidable.)

Now we investigate the consequence of the weakening rule. What happens when this rule is missing on the right (or left) side? Before looking for such a calculus we simplify the problem by concentrating on propositional logic, using literals (i.e. prop. variables A,B,... and their negations $\neg A$,$\neg B$,...) and defining the negation and implication (put $\neg\neg X :\equiv X$ for literals X and define $\neg(X \wedge Y) :\equiv \neg X \vee \neg Y$, $\neg(X \vee Y) :\equiv \neg X \wedge \neg Y$, $X \rightarrow Y :\equiv \neg X \vee Y$ for arbitrary X,Y). And we regard the left and the right side of a sequent as sets, therefore we can omit the structural rules for contraction and exchange. The essence of weakening nevertheless is the same.

The calculus C

Axiom: $\Gamma \supset \Gamma$ ($\Gamma \neq \emptyset$, Γ a finite set of literals)

Rules: structural rules

$$\frac{\Gamma \supset \Delta}{\Gamma, X \supset \Delta} \text{ W-l} \qquad \frac{\Gamma \supset \Delta}{\Gamma \supset \Delta, X} \text{ W-r}$$

logical rules

$$\frac{\Gamma, X \supset \Delta \quad \Theta, Y \supset \Delta}{\Gamma, \Theta, X \vee Y \supset \Delta} \ \vee\text{-l} \qquad \frac{\Gamma \supset \Delta, X, Y}{\Gamma \supset \Delta, X \vee Y} \ \vee\text{-r}$$

$$\frac{\Gamma, X, Y \supset \Delta}{\Gamma, X \wedge Y \supset \Delta} \ \wedge\text{-l} \qquad \frac{\Gamma \supset \Delta, X \quad \Gamma \supset \Pi, Y}{\Gamma \supset \Delta, \Pi, X \wedge Y} \ \wedge\text{-r}$$

cuts

$$\frac{\Gamma, X \supset \Delta, \neg X}{\Gamma \supset \Delta, \neg X}\ \text{D-l} \qquad \frac{\Gamma, \neg X \supset \Delta, X}{\Gamma, \neg X \supset \Delta}\ \text{D-r}$$

$$\frac{\Gamma, X \supset \Delta \quad \Gamma \supset \Delta, X}{\Gamma \supset \Delta}\ \text{cut}$$

The structural and logical rules in the calculus are similar to Gentzen's rules [Ta] (of course we need structural rules only for weakening and logical rules only for conjunction and disjunction). So the ∨-r rule has a modified form so that it does not imply weakening. The cut has a very restricted form and the "discarding rules" are new. Without these rules we have no chance to reduce the right or the left side of a sequent. They are similar to the negation rules in Gentzen's sequent calculus, but here the negation is not introduced, it already exists. (If weakening is allowed, then these rules coincide to the negation rules.)

One consequence of the weakening rule is immediate: The weakening rule on the left side is responsible that the corresponding semantics is *monotone*; the weakening rule on the right side is responsible that the corresponding logic is *not paraconsistent.*

Definition of derivations:
Let C_R, C_L be the calculus C without W-r, W-l respectively, and C_{RL} the calculus C with neither W-r nor W-l.

A sequent $\Gamma \supset \Delta$ is *derivable in a calculus* $*$, if there exists a finite list where the last sequent is $\Gamma \supset \Delta$ and every sequent in the list is an axiom of $*$ or it is the lower sequent of a rule of $*$ and the upper sequents of the same rule are predecessors of it.

If $*$ is C_R, then Δ must not be empty, if $*$ is C_L, then Γ must not be empty, if $*$ is C_{RL}, then Δ and Γ must not be empty. (These modifications are necessary because the definitions $V(\emptyset_\vee) := f$ and $V(\emptyset_\wedge) := t$. It is possible to define $V(\emptyset_\vee) = V(\emptyset_\vee) = i$ and to discard the above restrictions.)

Remark:
It is easy to prove that the calculus C is sound and complete w.r.t. classical propositional logic. ($\Gamma \supset \Delta$ is derivable in C $\iff$ $\Gamma \models \Delta$)

3 Soundness and completeness

Theorem:

1. The calculus C_R is sound and complete w.r.t. the three-valued semantics where t and i are the distinguished values $\Gamma \supset \Delta$ is derivable in C_R $\iff$ $\Gamma \models_R \Delta$

2. The calculus C_L is sound and complete w.r.t. the three-valued semantics where t is the single distinguished value $\Gamma \supset \Delta$ is derivable in C_L $\iff$ $\Gamma \models_L \Delta$

3. $\Gamma \supset \Delta$ is derivable in C_{RL} $\iff$ $\Gamma \models_{RL} \Delta$

Proof of 1:

"$\Longleftarrow$": Verification is evident; we have to examine the soundness of the axiom and the rules.
"$\Longrightarrow$": Let $\Gamma \supset \Delta$ not be derivable in the calculus C_R.
We will show that $\Gamma \not\models_R \Delta$. So we have to find a valuation V so that $V(X) \neq f$ for all $X \in \Gamma$ and $V(Y) \neq t$ for all $Y \in \Delta$ and $V(Y) = f$ for some $Y \in \Delta$.
The proof is carried out in four steps:

1. We extend Γ to Π and Δ to Σ, such that $\Pi \supset \Sigma$ is also not derivable.
2. We define V on the literals.
3. We prove the following properties:
   If $X \in \Pi$ and $V(X) = i$ then $\neg X \in \Pi$
   If $X \in \Pi$ then $V(X) \neq f$
   If $X \in \Sigma$ then $V(X) \neq w$
4. We show that $V(X) = f$ for some $X \in \Delta$.

We first need to prove some lemmata:

Lemma 1:
If $\Gamma \supset \Delta$ and $\Gamma' \supset \Delta'$ are derivable in C_R, Then $\Gamma, \Gamma' \supset \Delta, \Delta'$ is derivable too.

Proof: Induction on the sum of the lengths of the derivations of $\Gamma \supset \Delta$, $\Gamma' \supset \Delta'$.

Lemma 2:
$\supset X, \neg X$ is derivable in C_R for any formula X.

Proof: Induction on construction of the formula:

- $X \equiv A$:

$$\dfrac{\dfrac{A, \neg A \;\supset\; A, \neg A}{\neg A \;\supset\; A, \neg A}\; D\text{-}l}{\supset\; A, \neg A}\; D\text{-}l$$

- $X \equiv U \vee V$: We know that $\supset U, \neg U$ and $\supset V, \neg V$ are derivable from the induction hypothesis. Hence,

$$\dfrac{\dfrac{\begin{matrix}\vdots \\ \supset\; U, \neg U\end{matrix} \qquad \begin{matrix}\vdots \\ \supset\; V, \neg V\end{matrix}}{\supset\; U, V, \neg U \wedge \neg V}\; \wedge\text{-}r}{\supset\; U \vee V, \neg U \wedge \neg V}\; \vee\text{-}r$$

- $X \equiv U \wedge V$: similar

q.e.d.

Lemma 3:
$\Gamma \supset \Gamma$ for arbitrary non-empty Γ is derivable in C_R.

Proof: see lemma 2 and the rules D-r, W-l, and use $\neg\neg X \equiv X$ (by definition).

Lemma 4:
The rules $\wedge$-l and $\vee$-r are invertible, i.e.

a) if $\Gamma, X \wedge Y \supset \Delta$ is derivable, then $\Gamma, X, Y \supset \Delta$ is derivable too.
b) if $\Gamma \supset \Delta, X \vee Y$ is derivable, then $\Gamma \supset \Delta, X, Y$ is derivable too.

Proof: We extend the claim as follows:

c) If $\Gamma, \neg X \wedge \neg Y \supset \Delta, X \vee Y$ is derivable, then also $\Gamma, \neg X, \neg Y \supset \Delta, X, Y$.

and prove all three propositions by simultaneous induction on the length of the derivation ($=:n$).

- $n = 1$: trivial
- $n > 1$
  a) If $X \wedge Y$ is not the main formula of the last inference, then the lemma follows by induction hypothesis. In the other case the last rule is $\wedge$-l or D-r or W-r. $\wedge$-l: When the formula $X \wedge Y$ is already in the upper sequent of the rule, the lemma follows by induction hypothesis, otherwise the lemma is evident. D-r: Hence $\Gamma, X \wedge Y \supset \neg X \vee \neg Y, \Delta$ is derivable and therefore $\Gamma, X, Y \supset \neg X, \neg Y, \Delta$ is derivable too (see the induction hypothesis of c) and the lemma follows with the D-r rule. W-r: evident.
  b) similarly
  c) We only have to regard the rules $\wedge$-l and D-r and W-r, where $\neg X \wedge \neg Y$ is the main formula and the rules $\vee$-r and D-l, where $X \vee Y$ is the main formula. In all five cases the lemma follows from induction hypothesis. By $\wedge$-l from b) or c), by D-r from c), by W-r from b), by $\vee$-r from a) or c), by D-l from c)

q.e.d.

Lemma 5:
If $\Gamma, X, \neg X, Y, \neg Y, X \vee Y \supset \Delta$ is derivable in C_R, then also $\Gamma, X, \neg X, Y, \neg Y \supset \Delta$.

Proof: Similarly to the last proof we insert two propositions and prove all three sentences simultaneously by induction on the length of the derivation.

If $\Gamma, X, \neg X, Y, \neg Y \supset \Delta, \neg X \wedge \neg Y$ is der. in C_R, then also $\Gamma, X, \neg X, Y, \neg Y \supset \Delta$.
If $\Gamma, X, \neg X, Y, \neg Y, X \vee Y \supset \Delta, \neg X \wedge \neg Y$ is der. in C_R, then $\Gamma, X, \neg X, Y, \neg Y \supset \Delta$.
Observing that we only have to consider derivations where every sequent has the formulas $X, \neg X, Y, \neg Y$ in the left part, it is enough to investigate these cases with $X \vee Y$ as main formula and the proof is evident.
q.e.d.

Lemma 6:
If $\Gamma, \neg X, X \vee Y, Y \supset \Delta$ is derivable in C_R then also $\Gamma, \neg X, X \vee Y \supset \Delta$.

Proof: See lemma 3 and W-r, the postulate and lemma 1 and 2. The following derivation prove the claim

$$\dfrac{\dfrac{\dfrac{\begin{matrix}\vdots\\ \Gamma,\neg X,X\vee Y\supset\neg X\end{matrix}\qquad \dfrac{\begin{matrix}\vdots\\ \Gamma,\neg X,X\vee Y,Y\supset\Delta,Y,\neg Y\end{matrix}}{\Gamma,\neg X,X\vee Y\supset\Delta,Y,\neg Y}\;D\text{-}l}{\Gamma,\neg X,X\vee Y\supset\Delta,Y,\neg X\wedge\neg Y}\;\wedge\text{-}r}{\Gamma,\neg X,X\vee Y\supset\Delta,Y}\;D\text{-}r\qquad \Gamma,\neg X,X\vee Y,Y\supset\Delta}{\Gamma,\neg X,X\vee Y\supset\Delta}\;cut$$

q.e.d.

Now we complete the proof of the first part of the theorem by executing the four steps:

1. We extend Γ to Π and Δ to Σ, such that $\Pi\supset\Sigma$ is also not derivable:

1.	$X\in\Sigma$	$\Longrightarrow$	$\neg X\in\Pi$	D-l
2.	$X\vee Y\in\Sigma$	$\Longrightarrow$	$X,Y\in\Sigma$	$\vee$-r
3.	$X\wedge Y\in\Sigma$	$\Longrightarrow$	$X\in\Sigma$ or $Y\in\Sigma$	$\wedge$-r
4.	$X\vee Y\in\Pi$	$\Longrightarrow$	$X\in\Pi$ or $Y\in\Pi$	$\vee$-l
5.	$X\wedge Y\in\Pi$	$\Longrightarrow$	$X,Y\in\Pi$	$\wedge$-l
6.	$X\vee Y,X,\neg X\in\Pi$	$\Longrightarrow$	$Y\in\Pi$	lemma 6
7.	$X,\neg X,Y,\neg Y,X\vee Y\in\Pi$	$\Longrightarrow$	$\neg(X\vee Y)\in\Pi$	lemma 4
8.	$X,\neg X,Y,\neg Y,X\wedge Y\in\Pi$	$\Longrightarrow$	$\neg(X\wedge Y)\in\Pi$	lemma 5

$\Pi\supset\Sigma$ is not derivable because $\Gamma\supset\Delta$ is not derivable and every extending principle conserves the non-derivability, see the right column.

2. We define the valuation V as follows:

$$V(A):=\begin{cases} i & \text{if } A,\neg A\in\Pi\\ f & \text{if } \neg A\in\Pi \text{ and } A\notin\Pi\\ t & \text{otherwise}\end{cases}$$

3. We show the properties: (a) $X\in\Pi$ and $V(X)=i\Longrightarrow\neg X\in\Pi$
(b) $X\in\Pi\Longrightarrow V(X)\neq f$
(c) $X\in\Sigma\Longrightarrow V(X)\neq t$

Proof: We prove (a) and (b) by simultaneous induction on the structure of the formula X, whereas (c) directly follows from (b) and principle 1.

- X is a literal: The proposition follows from the definition of V.
- $X\equiv Y\wedge Z$: If $X\in\Pi$ and $V(X)=i$, then $Y,Z\in\Pi$ (principle 5) and $V(Y)=V(Z)=i$ (see the truth-table for "$\wedge$"). Therefore $\neg Y,\neg Z\in\Pi$ follows from the induction hypothesis and $\neg X\in\Pi$ from principle 8. - If $X\in\Pi$, then $Y,Z\in\Pi$ (principle 5) and $V(Y)\neq f$, $V(Z)\neq f$ (induction hypothesis). Hence $V(X)\neq f$ (see the truth-table).
- $X\equiv Y\vee Z$: If $X\in\Pi$ and $V(X)=i$, then $X\in\Pi$ or $Y\in\Pi$, say $Y\in\Pi$ (principle 4) and $V(Y)=i$ (see the truth-table for "$\vee$"). Therefore

$\neg Y \in \Pi$ follows from the induction hypothesis and $Z \in \Pi$ from principle 6. Therefore $V(Z) = i$ and $\neg Z \in \Pi$ (induction hypothesis) and $\neg X \in \Pi$ (principle 7.). - If $X \in \Pi$, then $X \in \Pi$ or $Y \in \Pi$, say $Y \in \Pi$ (principle 4) and $V(Y) \neq f$ (induction hypothesis). But if $V(Y) = w$, then $V(X) \neq f$, else $V(Y) = i$ and $\neg Y \in \Pi$ (induction hypothesis) and $Z \in \Pi$ (principle 6). By induction hypothesis, it follows $V(Z) \neq f$ and therefore $V(X) \neq f$.

q.e.d.

4. We show that $V(X) = f$ for some $X \in \Delta$:
   We first prove that there is a formula $X \in \Sigma$ such that $V(X) = f$.
   In order to do this we modify the principles 2 and 3, replacing them by the following properties:
   If $\Pi \supset \Theta, Y \wedge Z$ is not derivable in C_R, then $\Pi \supset \Theta, Y$ or $\Pi \supset \Theta, Z$ is not derivable in C_R either.
   If $\Pi \supset \Theta, Y \vee Z$ is not derivable in C_R, then $\Pi \supset \Theta, Y, Z$ is not derivable in C_R either.
   Here we replace formulas whereas we introduce formulas in the original version of these principles. If we use these ones on Σ we obtain a subset Σ' of Σ which exactly consists in the literals of Σ and $\Pi \supset \Sigma'$ is not derivable. Therefore there is a literal L in Σ', which is not in Π, and hence $V(L) = f$ and $L \in \Sigma$.
   Now we prove that there is already a formula $X \in \Delta$ with $V(X) = f$.
   In order to extend Δ to Σ, we use the principles 2 and 3. We define the sequence Σ_0 ; Σ_1 ; ... ; Σ_k where Σ_0 is defined as Δ and Σ_{i+1} arises from Σ_i by using principle 2 or 3 and $\Sigma_k \equiv \Sigma$. We prove: if there exists $X \in \Sigma_{i+1}$ with $V(X) = f$ then there exists $Y \in \Sigma_i$ with $V(Y) = f$.
   Principle 2: Let $X \vee Y \in \Sigma_k$ and $X \in \Sigma_{k+1} \setminus \Sigma_k$ and $V(X) = f$. $V(Y) = i$ or $V(Y) = f$, because $Y \in \Sigma_{k+1}$. Hence $V(X \vee Y) = f$.
   Principle 3: Let $X \wedge Y \in \Sigma_k$ and $X \in \Sigma_{k+1} \setminus \Sigma_k$ and $V(X) = f$. $V(X \wedge Y) = f$ not depending on the truth-value of Y.

   q.e.d.

q.e.d. (proof of 1)

Proof of 2:
Of course it is possible to prove this claim similar to the last one, but we can use the following duality: $X_1, ..., X_n \models_R Y_1, ..., Y_m \iff \neg Y_1, ..., \neg Y_m \models_L \neg X_1, ..., \neg X_n$ and (with induction on the length of the derivation)
$X_1, ..., X_n \supset Y_1, ..., Y_m$ is der. in $C_R \iff \neg Y_1, ..., \neg Y_m \supset \neg X_1, ..., \neg X_n$ is der. in C_L

q.e.d.

Proof of 3: The structure of the proof is similar to the proof of 1.

4 The cut-rule

Proposition:
The cut can not be eliminated in the calculus C_R, but it is possible to replace it by the rub-rule which has the subformula property. (I.e. every formula of the upper sequent in the rule is a subformula in the lower sequent.)

$$\frac{\Delta, \neg X, X \vee Y, Y \supset \Gamma}{\Delta, \neg X, X \vee Y \supset \Gamma} \text{ rub}$$

Proof:
The sequent $A, \neg A, \neg B, A \vee (B \wedge C) \supset B, A, \neg A$ is derivable in the calculus C_R, because $B \vee A \vee \neg A$ is a R-consequence of $A, \neg A, \neg B, A \vee (B \wedge C)$. Now we show indirect by induction on the length of the derivation that this sequent is not derivable in the calculus C_R without cut.

But considering the proof of the completeness we see that a restriction of the cut-rule is enough. Namely we can replace the cut by the rub rule.

5 The connection to the linear logic

A first evident parallel between our logic and linear logic [Gi] is that *the weakening rule is missing* in both systems. But by adding this rule to our calculus, we obtain classical logic, in contrast to linear logic. There we need the weakening rule as well as the contraction rule in order to obtain classical logic.

An other relationship is that *our three-valued semantics is a special case of the phase semantics.* So we can derive our truthtables from the interpretation of the multiplicative connectives in the phase semantics. Moreover we can derive Kleene's truthtables too, namely from the additive connectives. In order to see this connection, we shortly repeat the most important definitions of the phase semantics:

A *phase space* $\underline{P}$ consists in a commutative monoid $\langle P, \cdot, 1\rangle$ and a subset $\perp$ of P.
$\underline{P} = \langle P, \cdot, 1, \perp\rangle$

If G is a subset of P, then its *dual* $G^\perp$ is $\{p \in P \mid \forall q(q \in G \rightarrow p \cdot q \in \perp)\}$.

A *fact* of $\underline{P}$ is a subset G of P such that $G^{\perp\perp} = G$.

The following sets are facts: $\perp$, $\mathbf{1} := \perp^\perp$, $\top := \emptyset^\perp$, $\mathbf{0} := \top^\perp$.

We define: $G \cdot H := \{p \cdot q \mid p \in G \text{ and } q \in H\}$
$G + H := (G^\perp \cdot H^\perp)^\perp$
$G \star H := (G \cdot H)^{\perp\perp}$
$G \sqcap H := G \cap H$
$G \sqcup H := (G \cup H)^{\perp\perp}$

A *phase structure* $\underline{S}$ for the propositional language consists in a phase space $\underline{P}$ and, for each propositional variable A, a fact $S(A)$ of $\underline{P}$. With each proposition X we associate its interpretation $S(X)$ in a completely straightforward way. $\underline{S} = \langle \underline{P}, S\rangle$

X is *valid* in $\underline{S}$ when $1 \in S(X)$.

X is a *linear tautology* when X is valid in any phase structure $\underline{S}$.

If we restrict the definitions of a phase space and a phase structure we obtain the requested results. So we only investigate phase spaces of the form $\langle P, \cdot, 1, \{1\}\rangle$ and phase structures $\underline{S}$ with $S(A) \in \{\top, \bot, \mathbf{1}, \mathbf{0}\}$ and see that $\bot = \mathbf{1} = \{1\}$, $\top = P$ and $\mathbf{0} = \emptyset$. Furthermore $\mathbf{1}$ corresponds to the truth-value "undefined", $\top$ to "true" and $\mathbf{0}$ to "false" and the connective $+$ corresponds to our "or", $\star$ to our "and" and $\sqcap$ corresponds to Kleene's "and", $\sqcup$ to Kleene's "or" and the values i and t must be distinguished, because $1 \in \mathbf{1}$ and $1 \in \top$.

Summary:
If X is a linear tautology and doesn't contain additive connectives, then X is a tautology in $\mathbf{C}_{\boldsymbol{R}}$.

6 An alternative interpretation of the 3-valued logic

In classical propositional logic, boolean algebra gives us the chance to an alternative interpretation of formulas. If we take - for example - the power set algebra $\mathcal{B} = \langle\{A \mid A \subseteq U\}, \cup, \cap, \mathcal{C}, U, \emptyset\rangle$ we can regard a formula as a set and in particular the disjunction as the union, the conjunction as the cut and the negation as the complement.

If we look for an analogous interpretation of our three-valued logic, we have to assign a pair of sets (A_1, A_2) to each variable. A_2 represents the domain where the variable is true and A_1 represents the domain where it is not false. Therefore A_2 must be a subset of A_1 and the interpretation of the negation is immediately clear.

Now the question is how to define the interpretation of the disjunction and conjunction. If we transfer the boolean operation component-wise, we obtain an interpretation of Kleene's connectives. But we have seen, in comparing our semantics to the phase-semantics, that Kleene's connectives correspond to the additive ones (in the phase-semantics) and our connectives correspond to the multiplicative ones. So we have to find a weaker interpretation for the conjunction and a stronger one for the disjunction. We therefore define:

Definition:
An algebra of the form $\mathcal{P} = \langle \mathrm{M}, \triangledown, \vartriangle, \sqsubset, (U,U), (\emptyset,\emptyset)\rangle$ is called a *pair algebra* if
$\mathrm{M} \subseteq \{(A_1, A_2) \mid A_2 \subseteq A_1 \subseteq U\}$
$(A_1, A_2) \triangledown (B_1, B_2) = (((A_1 \cap B_1) \cup (A_2 \cup B_2)), (A_2 \cup B_2))$
$(A_1, A_2) \vartriangle (B_1, B_2) = ((A_1 \cap B_1), ((A_2 \cup B_2) \cap (A_1 \cap B_1))$
$\sqsubset (A_1, A_2) = (\mathcal{C}(A_2), \mathcal{C}(A_1))$
$(U,U), (\emptyset,\emptyset) \in \mathrm{M}$
M is closed under $\triangledown, \vartriangle, \sqsubset$

By the way if we want to interpret the exponential connectives of linear logic, we need the following definition:

$!(A_1, A_2) = (A_1, \emptyset)$ and $?(A_1, A_2) = (U, A_2)$

We can immediately see that the operations ∇ and Δ have a common root; namely the cut in the first component and the union in the second one. In order that the result is an element of the algebra, we can reduce the domain where the expression is true or increase the domain where it is not false. So we obtain the interpretation of the conjunction or disjunction.

Now we understand also the very strange property that both operations have the same neutral element; namely $(U, \emptyset)$:

$$(U, \emptyset) \,\Delta\, (A_1, A_2) = (A_1, A_2) = (U, \emptyset) \,\nabla\, (A_1, A_2)$$

So $(U, \emptyset)$ corresponds to the third truth-value and it's evident that (U, U) corresponds to "true" and $(\emptyset, \emptyset)$ to "false".

The relation between pair algebra and three-valued logic can be expressed as follows:

Proposition:
A formula X is a R-tautology if and only if $V(X) = (U, D)$ with $D \subseteq U$ for all pair algebras and all assignments V.

Of course the power set algebra is a special case of the pair algebra because $(A, A) \,\Delta\, (B, B) = (A \cap B, A \cap B)$. Considering the mapping: $(A_1, A_2) \longmapsto (A_1, A_1) = (A_1, A_2) \,\Delta\, (U, U)$ we see that every pair algebra has boolean subalgebras.

But generally a subalgebra is not determined by its cardinality and the cardinalities of its boolean subalgebras.

It's possible to generalize the pair algebra, because there is a bigger class of algebras (with the same signature) which describe the class of R-tautologies.

Definition:
An algebra of the form $\mathcal{T} = \langle T, *, +, ', 1, 0\rangle$ is called a *T-algebra* if followings equations are valid for all $x, y, u, v \in T$:

T_1	$(x')' = x$
T_2	$1' = 0$
T_3	$(x + y)' = x' * y'$
T_4	$x + x = x$
T_5	$x + y = y + x$
T_6	$x + (y + z) = (x + y) + z$
T_7	$x + 1 = 1$
T_8	$(x + x') * 1 = 1$
T_9	$((u + v) + (x * y)) * (u + x) * (v + y) = (u + v) + (x * y)$
T_{10}	$(u + v) * (u + x) * (v + x') = (u + x) * (v + x')$

It is easy to see that every pair algebra and in particular every boolean algebra is a T-algebra. But there are (even finite) T-algebras which are not isomorphic to a pair algebra.

Proposition:
A formula X is a R-tautology if and only if $V(X) * 1 = 1$ for all T-algebras and all assignments V.

Proof: $\langle\{f,i,t\},\wedge,\vee,\neg,t,f\rangle$ is a T-algebra. If $V(X) \wedge t = t$ then $V(X) \neq f$ and also X is a R-tautology.

The other direction is more difficult to prove. We have to show that the following Tait-calculus is sound and complete w.r.t. the 3-valued semantics. (Negation is constructive and involutive; for all valuations $V(1) = t$; Γ and Δ are finite sequences of formulas; $V(X_1, ..., X_n) := V(X_1 \vee ... \vee X_n)$)

Axioms:
$$X, \neg X$$
$$1, \Gamma$$

Rules:
$$\frac{\Gamma, X, Y, \Delta}{\Gamma, Y, X, \Delta} \qquad \frac{\Gamma, X, X}{\Gamma, X}$$

$$\frac{\Gamma, X, Y}{\Gamma, X \vee Y} \qquad \frac{\Gamma, X \quad \Delta, Y}{\Gamma, \Delta, X \wedge Y}$$

$$\frac{\Gamma \quad \Delta}{\Gamma, \Delta} \qquad \frac{\Gamma, X \quad \Delta, \neg X}{\Gamma, \Delta}$$

The soundness is evident and the completeness is shown as above.
Then we can show by induction on the length of proofs that, that for every derivable Γ, $V(\Gamma) * 1 = 1$.
The first axiom corresponds to T_8, the second to T_7. The $\wedge$-rule corresponds to T_9 and the cut to T_{10}. Verification of the other rules is evident.

References:

[AB] A.R. Anderson and N.D. Belnap, ***Entailment*** Vol.1, Princeton University Press, Princeton, New Jersey, 1975.

[Gi] J-Y. Girard, ***Linear Logic, Theoretical Computer Science***, vol.50(1987), pp. 1-101.

[Kl] S.C. Kleene,***Introduction of Metamathematics***, D. Van Nostrand Company, Princeton, 1952.

[Ta] G. Takeuti, ***Proof Theory***, North-Holland, Amsterdam, 1975.

A Transformational Methodology for Proving Termination of Logic Programs

M. R. K. Krishna Rao[1] D. Kapur[2] R. K. Shyamasundar[1]

Abstract

An approach for proving termination of well-moded logic programs is given by transforming a given logic program into a term rewriting system. It is proved that the termination of the derived rewriting system implies the termination of the corresponding logic program for well-moded queries under any selection rule implied by the given modings. The approach is mechanizable using termination orderings proposed in the term rewriting literature. Unlike Ullman and van Gelder's approach and Plümer's method, no preprocessing is needed, and the approach works well even in the presence of mutual recursion. This approach has been used recently to show termination of the Prolog implementation of compiler for ProCoS level 0 language PL_0 developed at Oxford University.

1 Introduction

The problem of showing termination of logic programs recently attracted considerable attention and many approaches have been proposed for analyzing the termination property of logic programs (see [2, 14] for a survey). Some of these approaches use simple notions like *level mappings* and *models* to get a simple characterizations (though not mechanizable) and some use *linear-inequalities* to arrive at techniques for the automatic checking of termination of logic programs. In this paper, we take a different approach and propose a transformational approach by reducing the termination problem of logic programs to that of term rewriting systems. The main advantage of this approach is that it is mechanizable using various termination techniques and implementations developed in the term rewriting literature [5, 9]. To motivate the approach, we briefly discuss relevant issues in the termination of logic programs and other approaches.

A general technique for proving termination of any program is to prove that values of some arguments in a recursive call decrease under some well-founded ordering. The presence of local variables used for sideways information passing in logic programs poses a few problems for such an approach. For example, to prove the termination of quick-sort program containing the following clause,

$$qs([H|L],\ S) \leftarrow split(L, H, A, B),\ qs(A, A1),\ qs(B, B1), append(A1, [H|B1], S)$$

[1]Computer Group, Tata Institute of Fundamental Research, Colaba, Bombay 400 005, INDIA.
[2]Department of Computer Science, State University of New York at Albany, NY 12222.
Kapur was partially supported by NSF Grant nos. CCR-8906678 and INT-9014074.
e-mail: krishna@tifrvax.bitnet, kapur@cs.albany.edu, shyam@tifrvax.bitnet

it is sufficient to prove that the first argument of *qs* decreases after each recursive call, i.e., $A < [H|L]$ and $B < [H|L]$ under some well-founded ordering. Proofs of these two conditions are based on the semantics of the split procedure.

For handling *local* variables such as $A, B, A1, B1$ in the above example (see [13] for a discussion of the problem of local variables), Ullman and van Gelder [15] proposed a methodology for proving the termination of a logic program for a query with a given moding information, using inter-argument inequalities of the predicates. They gave an algorithm to generate a set of inequalities of the form $p_i + c \geq p_j$, whose satisfaction is a sufficient condition for the termination of the program under the assumption that logic programs do not contain any function symbol other than '|', the cons operator. Ullman and van Gelder's algorithm needs the following two properties: 1) *uniqueness* property of clauses, which intuitively implies that no variable can appear in input positions of more than one literal in the body of a clause. 2) an atom $p(\cdots)$ in the body of a clause should unify with the heads of all the clauses defining predicate p. Their method must ensure that the clauses satisfy these properties.

Plümer [13] extended Ullman and van Gelder's method to study termination of well-moded programs, by generalizing the form of inequalities to $\Sigma p_i + c \geq \Sigma p_j$ and allowing other function symbols. Plümer assumed that the programs are *normalized* (i.e., no variable occurs more than once in any literal in the program) and do not contain mutual recursion, and he proposed some preprocessing to normalize programs and eliminate mutual recursion. Plümer's method also needs a program to satisfy a property called *the existence of an admissible solution graph for each clause*, which is normally violated by the clauses, in which a variable occurs in input positions of more than one atom in the body. Consider the following example given in [13] with the modings: $add^{(in,in,out)}$, $mult^{(in,in,out)}$

$add(0, Y, Y) \leftarrow$
$add(s(X), Y, s(Z)) \leftarrow add(X, Y, Z)$
$mult(0, Y, 0) \leftarrow$
$mult(s(X), Y, Z) \leftarrow mult(X, Y, Z1), add(Z1, Y, Z)$

It is easy to observe that the last clause violates this property (the variable Y occurs in the input positions of *add* and *mult*) and Plümer's method can not be used to prove its termination.

In this paper, we describe a method that overcomes the drawbacks of the above methods. Our approach is to reduce the termination problem of logic programs to that of term rewriting systems. We develop an algorithm to transform a given well-moded logic program into a term rewriting system and prove that termination of the derived rewrite system implies termination of the logic program for all well-moded queries. Our method consists of two steps: (a) transform the given logic program into a rewrite system and (b) prove the termination of the resulting rewrite system using various techniques available in the literature on term rewriting systems. The transformation removes the *local variables* present in the logic program and makes the termination analysis simple.

The transformation is purely syntactical and easily mechanizable. Our approach thus facilitates the applicability of numerous termination methods proposed in the literature on term rewriting, for studying termination of logic programs. The approach does not require any preprocessing and works well for the programs violating the properties of 'uniqueness' and 'admissible solution graph' or containing mutual recursion. It is thus applicable to a wider class of logic programs.

In [11], we have successfully used this approach for showing termination of the Prolog implementation of compiler for **ProCoS** level 0 language PL_0 developed at Oxford University [7, 8]. Plümer's approach is not applicable on the compiler as the Prolog program has many clauses which do not have admissible solution graphs [13]. This shows the practicality and advantages of the proposed approach.

2 Well-moded logic programs and termination of rewrite Systems

In this section, we give definitions of well-moded programs and a brief overview of the termination results of rewrite systems. We assume that moding information is already available; the reader may refer to Debray and Warren [4] for a discussion on automatic inference of moding information. We follow the notations of [12] for logic programming and of [6] for term rewriting.

Definition 1: A *mode* m of an n-ary predicate p is a function from $\{1, \cdots, n\}$ to the set $\{in,\ out\}$. And $\{i \mid m(i) = in\}$ is the set of input positions of p and $\{o \mid m(o) = out\}$ is the set of output positions of p.

Let $Var(L)$ denote the set of variables occurring in a literal L, $in(L)$ denote the set of variables occurring in the input positions of L and $out(L) = Var(L) - in(L)$.

Definition 2: Let C be a clause $A \leftarrow B_1, \cdots, B_k$ and X be a variable in C. The head A is a *consumer* (*producer*) of X if $X \in out(A)$ ($X \in in(A)$, respectively). A literal B_i is a *consumer* (*producer*) of X if $X \in in(B_i)$ ($X \in out(B_i)$, respectively).[1]

Definition 3: The producer-consumer relation of a clause C: $A \leftarrow B_1, \cdots, B_k$ is defined as $\{ \langle B_i,\ B_j \rangle \mid B_i$ and B_j are producer and consumer of a variable X in C respectively $\}$.

Definition 4: A clause C is *well-moded* if (a) its producer-consumer relation is *acyclic* and (b) every variable in C has a producer. A program P is *well-moded* if every clause in it is well-moded. A *well-moded query* is a well-moded clause without the head.

Example 1: Let us consider the following quick-sort program with modings $q^{(in,out)}$, $s^{(in,in,out,out)}$, $a^{(in,in,out)}$. Here 'c' stands for 'cons'.

1. $q(nil, nil) \leftarrow$
2. $q(c(H, L), S) \leftarrow s(L, H, A, B), q(A, A1), q(B, B1), a(A1, c(H, B1), S)$
3. $s(nil, Y, nil, nil) \leftarrow$

[1] This notion of *producer* is similar to the notion of *generator* used in [3] for studying AND/OR parallelism in logic programming.

4. $s(c(X,Xs),Y,c(X,Ls),Bs) \leftarrow X \leq Y, s(Xs,Y,Ls,Bs)$
5. $s(c(X,Xs),Y,Ls,c(X,Bs)) \leftarrow X > Y, s(Xs,Y,Ls,Bs)$
6. $a(nil,X,X) \leftarrow$
7. $a(c(H,X),Y,c(H,Z)) \leftarrow a(X,Y,Z)$

The producer-consumer relation of all clauses (except second) is empty. And for the second clause it is $\{ \langle s(L,H,A,B), q(A,A1) \rangle, \langle s(L,H,A,B), q(B,B1) \rangle, \langle q(A,A1), a(A1,c(H,B1),S) \rangle, \langle q(B,B1), a(A1,c(H,B1),S) \rangle \}$. It is easy to see that all the variables in each clause are having a producer and the producer-consumer relation of each clause is acyclic. So the quick-sort program is well-moded. □

Definition 5: A *computation rule* (or *selection rule*) [12] is a function from a set of goals to a set of atoms such that the value of the function for a goal is an atom, called *selected* atom, in that goal.

If the same goal occurs at different places in an SLD-derivation, the selection rule selects the same atom every time [12]. This notion can be extended to clauses as follows: given a clause, the selection rule gives an evaluation order among the literals in the body of the clause. It can be captured by a partial order (if $l_i < l_j$ in the partial order, it means that l_i should be selected before selecting l_j).

Definition 6: A selection rule is *implied* by the moding information of a well-moded program if and only if its partial order for each clause in the program is an *extension* of the producer-consumer relation of that clause.

Example 2: Prolog's left-to-right selection rule is implied by the moding information of the quick-sort, since its evaluation order is a linear extension of the producer-consumer relation of all the clauses.

Lemma 1: Let P be a well-moded logic program, S be a selection rule implied by the moding information of P and $\leftarrow q_1(\cdots),\ldots,q_n(\cdots)$ be a well-moded query Q. *If $q_i(\cdots)$ is a minimal element under the partial order of S* (denoted by $<_S$) *then all the input terms of $q_i(\cdots)$ are ground.*

Proof:(by contradiction). Let us assume that a variable (say X) occurs in an input position of $q_i(\cdots)$ (So $q_i(\cdots)$ is consumer of X). Since Q is a well-moded query, every variable in Q has a producer. So X has a producer (say $q_j(\cdots)$) and hence $\langle q_j(\cdots), q_i(\cdots) \rangle$ is in the producer-consumer relation of Q. Since S is implied by the moding information, $<_S$ is an extension of the producer-consumer relation of Q. So it should be the case that $q_j(\cdots) <_S q_i(\cdots)$, contradicting the assumption that $q_i(\cdots)$ is minimal. □

Lemma 2: Let P be a well-moded logic program, S be a selection rule implied by the moding information of P and $G_0, G_1, \ldots, G_n$ be a SLD-derivation (with a selection rule S) of P starting with a well-moded query G_0. *Then every selected atom of each goal G_i has ground input terms.*[2]

Proof: Induction on i. (See [10]). □

It may be noted that some predicates may be used in different modes in a single

[2]Such an SLD-derivation is said to be *data-driven*.

program. We use different subscripts to such a predicate to differentiate between different usages with different modings.

We now briefly review basic concepts related to termination of term rewriting systems (see [5, 6] for more details).

Definition 7: Let $\mathcal{F}$ be a set of function symbols and $\mathcal{X}$ be a set of variables. A term rewriting system $\mathcal{R}$ over the set of terms $\mathcal{T}(\mathcal{F}, \mathcal{X})$ is a finite set of *rewrite rules* of the form $l \rightarrow r$, where l and r are terms in $\mathcal{T}(\mathcal{F}, \mathcal{X})$ such that $Var(r) \subseteq Var(r)$ and l is not a variable.

A rule $l \rightarrow r$ applies to term t in $\mathcal{T}$ if a subterm s of t matches with l through some substitution σ, i.e. $s = l\sigma$ and the rule is applied by replacing the subterm s in t by $r\sigma$. If the resulting term is u, we say t is rewritten to u in one step and denote $t \Rightarrow_{\mathcal{R}} u$ (or simply, $t \Rightarrow u$).

Definition 8: A term rewriting system $\mathcal{R}$ is *terminating* if and only if it does not admit any infinite chain of the form $t_0 \Rightarrow t_1 \Rightarrow \cdots \Rightarrow t_i \Rightarrow t_{i+1} \Rightarrow \cdots$. If $\mathcal{R}$ admits such an infinite chain, then it is *nonterminating*.

The termination of term rewriting systems in general is undecidable. Many methods have been developed for proving termination of classes of term rewriting systems; Dershowitz [5] is a good survey of these methods. A general strategy is to show that every term resulting from rewriting is smaller than the term being rewritten in some well-founded ordering on terms. Many well-founded orderings discussed in [5] exploit the syntactic structure of terms. Simplification orderings [5] such as *lexicographic recursive path orderings, interpretation orderings* are good examples of this and are implemented in theorem provers based on rewriting paradigm, such as RRL [9].

3 Transforming a logic program into a rewrite system

In this section, we give the transformation algorithm and prove that the termination of the derived rewrite system implies the termination of the corresponding logic program for all well-moded queries under any selection rule implied by the moding.

The transformation is based on the idea of eliminating local variables by introducing Skolem functions. For each n-ary predicate p having a moding with k output positions, we introduce k new function symbols $p^1, \ldots, p^k$ of arity $n-k$. These k-function symbols correspond to the k output positions of the predicate p. (If $k = 0$, we introduce a n-ary function symbol p^0.) And then, we construct a set of rewrite rules to compute these new functions. In the following, we informally explain the transformation through a series of examples.

Example 3: Consider the multiplication program discussed in the introduction, with modings: $a^{(\text{in},\text{in},\text{out})}$ and $m^{(\text{in},\text{in},\text{out})}$.

$a(0, Y, Y) \leftarrow$
$a(s(X), Y, s(Z)) \leftarrow a(X, Y, Z)$
$m(0, Y, 0) \leftarrow$
$m(s(X), Y, Z) \leftarrow m(X, Y, Z1),\ a(Z1, Y, Z)$

Using the moding information, the following rewrite rules are derived from the above clauses.

1. Since the output of predicate a for inputs 0 and Y is Y, we get $\mathbf{a^1(0,Y) \rightarrow Y}$.
2. The output of m for inputs 0 and Y is 0. We get $\mathbf{m^1(0,Y) \rightarrow 0}$.
3. From the second clause, the output of a for inputs $s(X)$ and Y is $s(Z)$, where Z is the output of a for the inputs X and Y. We get $\mathbf{a^1(s(X),Y) \rightarrow s(a^1(X,Y))}$.
4. From the last clause, the output of m for inputs $s(X)$ and Y is Z, where Z is the output of a for the inputs $Z1$ and Y, where $Z1$ is the output of m for inputs X and Y. So we get $\mathbf{m^1(s(X),Y) \rightarrow a^1(Z1,Y)}$, where $\mathbf{Z1 = m^1(X,Y)}$. The resulting rule is $\mathbf{m^1(s(X),Y) \rightarrow a^1(m^1(X,Y),Y)}$.

When a nonvariable term appears in an output position of a body literal, we may need to introduce an *inverse function* as illustrated in the following example.

Example 4: Consider the following clause with the modings $a^{(in,out)}$, $b^{(in,out)}$, $c^{(in,out)}$.
$a(X,Y) \leftarrow b(X,f(Z)), c(Z,Y)$
The output of predicate a for input X is Y, where Y is the output of predicate c for input Z. Now, one would have to answer the question: *What is the value of the input argument of predicate c* (i.e., Z) ?. Z is equal to $f^{-1}(f(Z))$, and $f(Z)$ is the output of b for input X. We get $\mathbf{a^1(X) \rightarrow c^1(f^{-1}(b^1(X)))}$. And to reduce $f^{-1}(f(Z))$ to Z, a rewrite rule $\mathbf{f^{-1}(f(X)) \rightarrow X}$ is added. □

In the above two examples, all the variables occurring in output positions in the body are also occurring either in output positions of the head or in input positions of some other literal in the body. And the rewrite systems derived in both the examples capture the termination of the corresponding logic programs correctly. Basically, the above transformation is capturing the data flow in the program execution. When there are some variables occurring only in output positions of a literal in the body, we need to *add additional rewrite rules* as illustrated in the following example.

Example 5: Consider the following (nonterminating) logic program with the moding: $r^{(in,in,out,out,out)}$, $p^{(in,out)}$, where $p \in \{ a, b, c, d \}$ and the rewrite system one derives following the above procedure.

Program	**Rewrite System**
$a(X,f(X)) \leftarrow$	$a^1(X) \rightarrow f(X)$
$b(X,X) \leftarrow$	$b^1(X) \rightarrow X$
$c(X,Y) \leftarrow a(X,Z), r(X,Z,Y,Z_1,Z_2)$	$c^1(X) \rightarrow r^1(X,a^1(X))$
$d(X,Y,g(X,Y)) \leftarrow$	$d^1(X,Y) \rightarrow g(X,Y)$
$r(X,Y,Z_1,Z_2,Z_3) \leftarrow b(X,Z_1), c(Y,Z_2), d(X,Y,Z_3)$	$r^1(X,Y) \rightarrow b^1(X)$
	$r^2(X,Y) \rightarrow c^1(Y)$
	$r^3(X,Y) \rightarrow d^1(X,Y)$

This system is terminating (using recursive path ordering [5] with precedence $r^3 > d^1 > g$, $r^2 > c^1 > a^1 > f$, $c^1 > r^1 > b^1$) while the above program is nonterminating. For the query $\leftarrow c(t,Y)$, where t is a ground term, the program has an infinite SLD-derivation. In the rewrite system, we rewrite $c^1(t)$ for some ground term t through a derivation to $c^1(t) \Rightarrow r^1(t,a^1(t)) \Rightarrow r^1(t,f(t)) \Rightarrow b^1(f(t)) \Rightarrow f(t)$, whereas in evaluating $c(t,Y)$, in a

SLD-derivation, one gets $r(t, f(t), Y, Z1, Z2)$, $b(t, Y)$, $c(f(t), Z1)$, $d(t, f(t), Z2)$ leading to an infinite computation. Computation of the query in the rewriting system involves only a partial computation of r (i.e. computation of r^1 only).

To capture all computation paths of an SLD-derivation, it is necessary to include additional rewrite rules reflecting the (possible) nontermination due to the computation of ***unnecessary*** values (the computation of these values does not provide any information to the head). In the above example, we include the following rewrite rules.

$c^1(X) \rightarrow \#(r^2(X, a^1(X)))$, $\qquad c^1(X) \rightarrow \#(r^3(X, a^1(X)))$

When these rewrite rules are included, the resulting rewrite system is nonterminating. □

3.1 Formal description of the transformation

Though input and output positions of a predicate can mix together in all possible ways, for notational convenience, we write all input positions first followed by all output positions. We write $p(t_{i_1}, \ldots, t_{i_j},\ t_{o_1}, \ldots, t_{o_k})$ to denote an atom $p(\cdots)$ containing the terms $t_{i_1}, \ldots, t_{i_j}$ in input positions and $t_{o_1}, \ldots, t_{o_k}$ in output positions. For each clause c, the transformation procedure needs the following:

1. $Prod(X) = \{ \langle p^l(t_{i_1}, \ldots, t_{i_j}),\ t_{o_l} \rangle \mid X$ occurs in l^{th} output position of atom $p(t_{i_1}, \ldots, t_{i_j},\ t_{o_1}, \ldots, t_{o_k})$ in the body $\}$ is the set of producers of variable X.

2. $Consvar = \{X \in Var(c) - in(head) \mid X \in out(head)$ or X occurs in an input position of an atom in the body $\}$ is the set of variables consumed at least once.

3. $Unsry = \{ p^l(t_{i_1}, \ldots, t_{i_j}) \mid Var(t_{o_l}) \cap Consvar = \phi \} \cup \{ q^0(s_{i_1}, \ldots, s_{i_k}) \mid$ predicate q does not have output positions $\}$, where $p(\cdots)$ and $q(\cdots)$ are atoms in the body. This set corresponds to the set of computations (we call them ***unnecessary computations***) which do not contribute to the outputs of the head directly or indirectly.

```
algorithm TRANSFORM (P : in; R_P : out);
begin
  R_P := φ;                    {* R_P contains rewrite rules *}
  for each clause c: a(t_i1,···,t_ik,t_o1,···,t_ok') ← B1,...Bn  ∈ P do
    begin
      Compute Consvar and Unsry;
      Compute Prod(X) for every variable in Var(c) − in(a(···));
      for j := 1 to k' do
          begin T := {t_oj};        { * T is a set of terms *}
                ELIMINATE-LOCAL-VARIABLES(T);
                R_P := R_P ∪ { a^j(t_i1,···,t_ik) → t | t ∈ T }
          end;
      T := Unsry;
      ELIMINATE-LOCAL-VARIABLES(T);
      R_P := R_P ∪ { a^k'(t_i1,···,t_ik) → #(t) | t ∈ T }
    end
end TRANSFORM.
```

```
procedure ELIMINATE-LOCAL-VARIABLES(T)
begin V := Var(T) - in(A);
  while V ≠ φ do
    begin
      for each X ∈ V do
        begin T' := φ;
          for each ⟨ p^l(···), t ⟩ ∈ Prod(X) do
            if t = X then T' := T' ∪ T.{X/p^l(···)}        (* Replace local var X by its producer-term. *)
            else if t = f(X) then
                begin
                  T' := T' ∪ T.{ X/f^-1(p^l(···)) };     (* Introduce inverse functions *)
                  R_P := R_P ∪ { f^-1(f(X)) → X }
                end;
          T := T'
        end;
      V := Var(T) - in(A)
    end;
end ELIMINATE-LOCAL-VARIABLES;
```

Example 6: Let us illustrate the transformation[3] with the quick-sort program, from which it derives the following rewrite system. We explain how rule 2 is derived from the second clause and other rules can be derived in similar fashion. The head $q(c(H,L),S)$ contains $c(H,L)$ in input position and variable S in output position. Left-hand-side of the rewrite rule will be $q^1(c(H,L)$ and to construct right-hand-side term, algorithm TRANSFORM calls ELIMINATE-LOCAL-VARIABLES with argument $T = \{S\}$. Values of T in various iterations of the while loop in ELIMINATE-LOCAL-VARIABLES are given below.
Ite. 1 $T = \{ a^1(A1, c(H, B1)) \}$ *variable S is replaced by its producer.*
Ite. 2 $T = \{ a^1(q^1(A), c(H, q^1(B))) \}$ *local var A1 and B1 are replaced by their producers .*
Ite. 3 $T = \{ a^1(q^1(s^1(L,H)), c(H, (q^1(s^2(L,H)))) \}$ *local var A and B are replaced.*
Since there are no local variables in T after 3rd iteration, ELIMINATE-LOCAL-VARIABLES returns this T to TRANSFORM which produces the rewrite rule 2 given below.

1. $q^1(nil) \rightarrow match^1(nil)$
2. $q^1(c(H,L)) \rightarrow a^1(\ q^1(s^1(L,H)),\ c(H, q^1(s^2(L,H)))\ )$
3. $s^1(nil,Y) \rightarrow match^1(nil)$

3'. $s^2(nil,Y) \rightarrow match^1(nil)$

4. $s^1(c(X,Xs),Y) \rightarrow match^1(c(X, s^1(Xs,Y)))$

4'. $s^2(c(X,Xs),Y) \rightarrow s^2(Xs,Y)$

5. $s^1(c(X,Xs),Y) \rightarrow s^1(Xs,Y)$

5'. $s^2(c(X,Xs),Y) \rightarrow match^1(c(X, s^2(Xs,Y)))$

6. $a^1(nil,X) \rightarrow match^1(X)$
7. $a^1(c(H,X),Y) \rightarrow match^1(c(H, a^1(X,Y)))$ □

Example 7: Let us consider the following program.
$p \leftarrow q, p$

[3] Our transformation is more general compared to the one proposed in [1], which can handle well-moded logic programs satisfying the conditions that (a) all the predicates have moding (in, ..., in, out), i.e., only one output, and (b) only *variables* occur in output positions of body atoms. The transformation in [1] is a special case of our transformation (both the transformations derive the same (equivalent) rewrite systems from programs satisfying the above two conditions).

Here we have propositions (so no in/out arguments). So, the associated function symbols $\{ p^0, q^0 \}$ are of arity zero (i.e., *constants*). The set $Unsry = \{ p^0, q^0 \}$ and the last statement of TRANSFORM constructs the following two rewrite rules capturing nontermination.

$p^0 \rightarrow \#(q^0)$, $\qquad p^0 \rightarrow \#(p^0)$ □

The correctness of the transformation follows from the following properties:

Property 1: For every atom $p(t_{i_1}, \cdots, t_{i_k}, t_{o_1}, \cdots, t_{o_{k'}})$ in the body of a clause C, the terms $p^j(t_{i_1}, \cdots, t_{i_k})\sigma$, $1 \leq j \leq k' > 0$ (for $k' = 0$, $p^0(t_{i_1}, \cdots, t_{i_k})\sigma$) occur as subterms of the right side of some rule derived from C. (Substitution σ replaces local variables by terms corresponding to their producers).

Property 2: $Var(r) \subseteq Var(l)$ for each rewrite rule $l \rightarrow r \in R_P$.

Lemma 4: The program TRANSFORM terminates.

Proof: The procedure ELIMINATE-LOCAL-VARIABLES is called $k' + |Unsry|$ times by TRANSFORM for each clause, where k' is number of output positions in the head and $|Unsry|$ is the cardinality of set $Unsry$ which is finite.

The main step in the while loop of ELIMINATE-LOCAL-VARIABLES is the following: *application of substitution $\{X/p^j(\cdots)\}$ to the terms in T, where $p^j(\cdots)$ is a producer of X.* That is, an occurrence of a variable in a term corresponding to its consumer is replaced by its producer. Since the producer-consumer relation of a well-moded clause is acyclic, the while loops terminate. From this, termination of ELIMINATE-LOCAL-VARIABLES and TRANSFORM follows. □

3.2 Main result

We associate a *rewrite-tree* with the computation resulting from a well-moded query and establish a relationship between the SLD-tree and rewrite-tree of a query.

Definition 9: Let Q be a well-moded query $\leftarrow q_1(\cdots), \ldots, q_n(\cdots)$ to a well-moded program P and R_Q be the set of rewrite rules derived from the clause $Q \leftarrow q_1(\cdots), \ldots, q_n(\cdots)$. The *rewrite-tree* RT_Q of Q is defined as follows:

1. Root(RT_Q) = Q^0
2. Children of a node $t \in RT_Q$ are $\{ s \mid t \Rightarrow_{R_Q \cup R_P} s \}$.

A rewrite-tree RT_Q contains all the (rewriting) derivations of the rewrite system $R_Q \cup R_P$ starting from an initial term Q^0. The following theorems establish the relationship between SLD-derivations and RT_Q.

Theorem 1: *Let Q be a well-moded query to a well-moded program P. Corresponding to every atom $p(t_{i_1}, \cdots, t_{i_k}, t_{o_1}, \cdots, t_{o_l})$ in the SLD-tree starting with Q, there are terms $\{ p^j(t_{i_1}, \cdots, t_{i_k})\sigma \mid 1 \leq j \leq l \neq 0\}$ occurring as subterms of nodes in the rewrite-tree RT_Q of Q. If $l = 0$, the term $p^0(t_{i_1}, \cdots, t_{i_k})\sigma$ occurs in RT_Q.* (Substitution σ replaces local variables by terms corresponding to their producers).

Proof: Induction over the depth d of the goals in the SLD-tree.

Basis: $d = 0$. The goal at depth 0 (the root of the SLD-tree) is Q itself. The truth of the theorem in this case follows from property 1 of the rewrite rules R_Q.

Induction hypothesis: Let the above theorem be true for $d \leq m - 1$.

Induction step: Now, we prove that the theorem holds for $d = m$. Consider a goal $\leftarrow q_1(\cdots), \ldots, q_n(\cdots)$ at depth $m - 1$. The goals below this (i.e., at depth m) are of the form

(a) $\leftarrow q_1(\cdots)\sigma, \ldots, q_{j-1}(\cdots)\sigma,\ \ q_{j+1}(\cdots)\sigma, \ldots, q_n(\cdots)\sigma$

(the resolution step here used a unit clause $H_1 \leftarrow$, with mgu σ) or

(b) $\leftarrow q_1(\cdots)\sigma, \ldots, q_{j-1}(\cdots)\sigma, q_{j_1}(\cdots)\sigma, \ldots, q_{j_J}(\cdots)\sigma, q_{j+1}(\cdots)\sigma, \ldots, q_n(\cdots)\sigma$,

(input clause in the resolution step is $H_2 \leftarrow q_{j_1}(\cdots), \ldots, q_{j_J}(\cdots)$, and mgu is σ).

In case (a) the terms corresponding to $q_i(\cdots)\sigma$, $i \neq j$, can be constructed from the terms corresponding to $q_i(\cdots)$ (whose existence is guaranteed by the hypothesis) by reducing them using the rewrite rules derived from the unit clause $H_1 \leftarrow$.

In case (b) the terms corresponding to $q_i(\cdots)\sigma$, $i \geq 1$, can be constructed from the terms corresponding to $q_i(\cdots)$ (whose existence is guaranteed by the hypothesis) by reducing them using the rewrite rules derived from the above input clause. The terms corresponding to $q_{j_i}(\cdots)\sigma$, $1 \leq i \leq J$, can be constructed by reducing the terms corresponding to $q_j(\cdots)$ using rewrite rules derived from the input clause (we get the terms corresponding to all $q_{j_i}(\cdots)\sigma$, $1 \leq i \leq j$ due to the property 1). □

Theorem 2: *Let Q be a well-moded query to a well-moded program P. Corresponding to every resolution step in SLD-derivation (under a selection rule implied by the modings) starting with Q, there are reduction steps in the rewrite-tree of Q.*

Proof: Let us consider a resolution step in which $p(t_{i_1}, \cdots, t_{i_k},\ t_{o_1}, \cdots, t_{o_l})$, $l \neq 0$ is resolved using an input clause C (case $l = 0$ can be handled similarly). By Theorem 1, corresponding to this atom there are terms $p^j(t_{i_1}, \cdots, t_{i_k})\sigma$, $1 \leq j \leq l$ occurring as subterms of nodes in the rewrite tree of Q. By Lemma 2, input terms $t_{i_1}, \ldots, t_{i_k}$ of the selected atom $p(\cdots)$ are ground. So, $p^j(t_{i_1}, \cdots, t_{i_k})\sigma = p^j(t_{i_1}, \cdots, t_{i_k})$, $1 \leq j \leq l$.

Let θ be the mgu in the resolution step and $p(s_{i_1}, \cdots, s_{i_k},\ s_{o_1}, \cdots, s_{o_l})$ be the head of the input clause used. Corresponding to this clause, we have rewrite rules $p^j(s_{i_1}, \cdots, s_{i_k}) \rightarrow r_j$. By the definition of SLD-resolution, $p(s_{i_1}, \cdots, s_{i_k},\ s_{o_1}, \cdots, s_{o_l})\theta$ $= p(t_{i_1}, \cdots, t_{i_k},\ t_{o_1}, \cdots, t_{o_l})\theta$. Therefore, $p^j(s_{i_1}, \cdots, s_{i_k})\theta = p^j(t_{i_1}, \cdots, t_{i_k})\theta =$ $p^j(t_{i_1}, \cdots, t_{i_k})$, $1 \leq j \leq l$ (since $t_{i_1}, \cdots t_{i_k}$ are ground). The terms $p^j(t_{i_1}, \cdots, t_{i_k})$, $1 \leq j \leq l$, in RT_Q match with left-hand-sides of rewrite rules derived from the clause C and can be rewritten. □

The following theorem establishes the relationship between the termination of a given well-moded logic program and that of the derived rewrite system.

Theorem 3: *A well-moded logic program terminates for all well-moded queries under any selection rule implied by the modings of the predicates, if the derived term rewriting system terminates.*

Proof: Follows from Theorem 2 and König's lemma. □

Example 8: The termination of the multiplication program in Example 3 can be proved by showing the termination of the rewriting system derived from the program using the recursive path ordering [5] with precedence $s < a^1 < m^1$.

$a^1(0,Y) \to Y,$ $\quad a^1(s(X),Y) \to s(a^1(X,Y))$
$m^1(0,Y) \to 0,$ $\quad m^1(s(X),Y) \to a^1(m^1(X,Y),Y)$

Proving that $r <_{rpo} l$ for rules 1 and 3 is easy. Consider rule 2. Since $s < a^1$, to prove that $s(a^1(X,Y)) <_{rpo} a^1(s(X),Y)$, it is enough to prove $a^1(X,Y) <_{rpo} a^1(s(X),Y)$.
And to prove $a^1(X,Y) <_{rpo} a^1(s(X),Y)$, we need to prove that $\{ X,Y \} \ll_{rpo} \{ s(X),Y \}$, which is indeed the case because $X <_{rpo} s(X)$.
Consider rule 4. Since $a^1 < m^1$, to prove that $a^1(m^1(X,Y),Y) <_{rpo} m^1(s(X),Y)$, it is enough to prove $m^1(X,Y) <_{rpo} m^1(s(X),Y)$ and $Y <_{rpo} m^1(s(X),Y)$. The proof of $m^1(X,Y) <_{rpo} m^1(s(X),Y)$ follows from $\{ X,Y \} \ll_{rpo} \{ s(X),Y \}$. And $Y <_{rpo} m^1(s(X),Y)$ by the subterm property of recursive path ordering. So the program terminates for all well-moded queries. □

The termination of the rewriting system corresponding to the quick-sort program in example 6 can be proved using an interpretation ordering. See [10] for details.

The converse of Theorem 3 is however not true, i.e. termination of the derived term rewriting system is not a necessary condition for termination of the logic program.

Example 9: Let us consider the following well-moded logic program with the modings $p^{(in,out)}$, $tc^{(in,out)}$ and its corresponding rewrite system

$p(a,b) \leftarrow$	$p^1(a) \to b$
$p(b,c) \leftarrow$	$p^1(b) \to c$
$tc(X,Y) \leftarrow p(X,Y)$	$tc^1(X) \to p^1(X)$
$tc(X,Y) \leftarrow p(X,Z), tc(Z,Y)$	$tc^1(X) \to tc^1(p^1(X))$

The above program terminates for the query $\leftarrow tc(a,X)$, while the rewriting system (derived from it) has an infinite derivation

$$tc^1(a) \Rightarrow, tc^1(p^1(a)) \Rightarrow tc^1(p^1(p^1(a))) \Rightarrow \cdots \Rightarrow tc^1(p^{1^i}(a)) \Rightarrow \cdots$$ □

4 Termination under a specific selection rule

The previous section dealt with the termination of logic programs under all *implied selection* rules. In other words, the termination of the program is proved for all data-driven evaluations. Some programs (example 10), however, terminate under some of the implied selection rules while they do not terminate under other implied selection rules; it is important to study termination under a given selection rule.

Example 10: *Reachability Problem:* given a graph $G(V, Ed)$ with set of vertices V and set of edges Ed represented as list of pairs $f(X,Y)$.($f(X,Y)$ stands for the existence of an edge from X to Y in the graph). The problem is to find all the vertices reachable from a given vertex. The modings are: $r^{(in,out,in,in)}$, $e^{(in,in,out)}$, $nm^{(in,in)}$

$r(X,Y,Ed,V) \leftarrow e(X,Ed,Y)$
$r(X,Y,Ed,V) \leftarrow e(X,Ed,Z),\ nm(Z,V),\ r(Z,Y,Ed,[Z|V])$
$e(X,[f(X,Y)|T],Y) \leftarrow$

$e(X, [f(X1, Y)|T], Z) \leftarrow X \neq X1,\ e(X, T, Z)$
$nm(X, [\,]) \leftarrow$
$nm(X, [H|T]) \leftarrow X \neq H,\ nm(X, T)$

This program terminates for the query $\leftarrow r(a, Y, [f(a, b),\ f(b, c),\ f(c, a),\ f(c, d)], [\,])$ under Prolog's left-to-right selection rule, even though it has an infinite *data-driven* evaluation (if we select the atom $r(Z, Y, Ed, [Z|V])$ before $nm(Z, V)$ it goes into a loop).

To deal with termination under a particular selection rule, we modify the transformation as follows. Each predicate symbol p of arity k with moding (in,in,...,in), will be replaced by a new predicate symbol p_{new} of arity k+1. And associate the moding (in,in,...,in,out) with p_{new}. The output of p_{new} is the truth of $p(\cdots)$. The predicate (say q of arity n) to be selected after the predicate replaced in the above step, will be replaced by a new predicate q_{new} of arity n+1. The extra argument is an input argument passed by the previous atom (the truth of the previous atom) p_{new}. Correspondingly, the changes are made consistently throughout the program[4]. We illustrate the method on the above example. For details, see [10].

Example 10 contd.: Since nm has the moding (in,in), we replace this by nm_{new} and the predicate r in the next atom by r_{new}. Replace $\neq$ by a new predicate $noteq$ and consequently e by e_{new}. And make changes in the program as follows.

$r_{new}(True, X, Y, Ed, V) \leftarrow e_{new}(True, X, Ed, Y)$
$r_{new}(True, X, Y, Ed, V) \leftarrow e_{new}(True, X, Ed, Z),\ nm_{new}(True, Z, V, T),$
$\qquad r_{new}(T, Z, Y, Ed, [Z|V])$
$e_{new}(True, X, [f(X, Y)|T], Y) \leftarrow$
$e_{new}(True, X, [f(X1, Y)|T], Z) \leftarrow noteq(X, X1, S),\ e_{new}(S, X, T, Z)$
$nm_{new}(True, X, [\,], True) \leftarrow$
$nm_{new}(True, X, [H|T], S1) \leftarrow noteq(X, H, S),\ nm_{new}(S, X, T, S1)$

We derive the following rewrite system.

$r^1_{new}(True, X, Ed, V) \rightarrow e^1_{new}(True, X, Ed)$
$r^1_{new}(True, X, Ed, V) \rightarrow r^1_{new}(\ nm^1_{new}(True, e^1_{new}(True, X, Ed), V),$
$\qquad e^1_{new}(True, X, Ed), Ed, [e^1_{new}(True, X, Ed)|V])$
$e^1_{new}(True, X, [f(X, Y)|T]) \rightarrow Y$
$e^1_{new}(True, X, [f(X1, Y)|T]) \rightarrow\ e^1_{new}(noteq(X, X1), X, T)$
$nm^1_{new}(True, X, [\,]) \rightarrow True$
$nm^1_{new}(True, X, [H|T]) \rightarrow\ nm^1_{new}(noteq(X, H), X, T)$

The rewrite system terminates over ground terms because the first two rules can be applied only when the first argument of r^1_{new} is $True$, which is passed by nm^1_{new} and the value of nm^1_{new} is $True$ only when X is not a member of the visited vertices. Once all the vertices in a cycle are visited, nm^1_{new} will not pass $True$ to the r^1_{new} avoiding a cycle.

[4] It may be necessary to increase the arity of the predicates at most by two; one for input (the truth of previous atom) and one for output (the truth of the atom itself, which is to be passed to its next atom).

5 Conclusion

The transformation method has been used to prove termination of many programs including *append, split, fair-split, quick-sort and merge-sort.* It has many advantages over the methods of [13] and [15]. No preprocessing is needed and the method is applicable to a larger class of logic programs, i.e. programs violating the properties of *uniqueness* and *existence of admissible solution graph* can be handled. For example, the Prolog implementation of compiler (for **ProCos** level 0 language PL_0) in [8] has many clauses which do not have admissible solution graphs [13]. Hence, the methods of [13] and [15] cannot prove its termination, where as our has been successfully used in proving it termination [11]. This shows the practicality and advantages of the transformation method. The following clause [8] with moding: $c^{(in,in,out,out,in,in)}$, $ce^{(in,in,out,out,in,in)}$, $mtrans^{(in,in,out,out)}$, $psi^{(in,in,out)}$, and $flatten^{(in,out)}$ does not have admissible solution graph (note the occurrence of variables `Psi` and `Psioutputbuf` in input positions of more than one atom in the body).

```
c(Output!E, S, F, M, Psi, Omega) :-
                   ce(E, S, L1, M1, Psi, Omega),
                   psi(Psi, outputbuf, Psioutputbuf),
                   psi(Psi, Output, PsiOutput),
                   mtrans(stl(Psioutputbuf), L1, L2, M2),
                   mtrans(ldlp(Psioutputbuf), L2, L3, M3),
                   mtrans(ldc(PsiOutput), L3, L4, M4),
                   mtrans(ldc(4), L4, L5, M5),
                   mtrans(out, L5, F, M6),
                   flatten([M1, M2, M3, M4, M5, M6], M), !.
```

Plümer remarked in his thesis that the non-existence of an admissible solution graph for the second clause in the multiplication program is possibly due to the *non-linear* relationship between the arguments of m^1. Since the predicate-inequalities of [13] as well as [15] are linear, those methods may not work for the problems which have *inherent* non-linearity. We do not have such restrictions as we are transforming the logic programs into general term rewriting systems (without any restrictions) for studying the termination of logic programs.

Our method is more powerful than Plümer's method in another sense: we study termination under all the *implied* selection rules whereas Plümer's method works only for Prolog's selection rule. Further, our definition of *well-moded* programs is more general than that of Plümer (he considers a program well-moded, only if the Prolog's selection rule is implied by the moding information). The following *permutation* program (with moding $p^{(out,in)}$, $a1^{(in,in,out)}$, $a2^{(out,out,in)}$) terminates for all well-moded queries under right-to-left selection rule. This program is well-moded by our definition and its termination can be proved for all the selection rules implied by the above moding information.

$p([\,],[\,]) \leftarrow$
$p(Xs,[X|Ys]) \leftarrow a1(X1s,[X|X2s],Xs),\ a2(X1s,X2s,Zs),\ p(Zs,Ys)$

This program is not well-moded according to Plümer's definition; termination of this program (under right-to-left selection rule) for the queries of the form $\leftarrow p(X,t)$, where t is a list, cannot be shown by Plümer's method or van Gelder's method since these methods use only the left-to-right selection rule of Prolog.

Acknowledgements: We thank Profs. K.R. Apt and J.W. Klop of CWI for their helpful comments. The first author thanks CWI for partially supporting his visit.

References

[1] F. Alexandre, K. Bsaics and A. Quere (1991) *On using mode input-output for transforming logic programs*, Workshop on Logic Program Synthesis and Transformations LOPSTR'91. Proceedings to appear in Springer Workshops in Computer Science series.

[2] K.R. Apt and D. Pedreschi (1991), *Reasoning about Termination of Prolog Programs*, Technical Report, University of Pisa.

[3] J.S. Conery and D.F. Kibler (1985), *AND parallelism and Nondeterminism in Logic Programs*, New Generation Computing, 3, pp. 43-70.

[4] S. K. Debray and D. S. Warren (1988), *Automatic mode inference for logic programs*, Journal of Logic Programming 5, pp. 207-229.

[5] N. Dershowitz (1987), *Termination of rewriting*, Journal of Symbolic Computation, 3, pp. 69-116.

[6] N. Dershowitz and J.-P. Jouannaud (1990), *Rewrite Systems*, In J. van Leeuwen, editor, *Handbook of Theoretical Computer Science B: Formal Methods and Semantics*, North-Holland, pp. 243-320.

[7] H. Jifeng and C. A. R. Hoare (1989), *Operational Semantics for ProCoS level 0 language*, ProCoS Project Document, OU HJF 1/3, Oxford University.

[8] H. Jifeng, P. Pandya and J. Bowen (1990), *Compiling specification for* ProCoS *Programming language Level 0*, ProCoS Workshop, Malente, April 1990.

[9] D. Kapur and H. Zhang (1989), *An Overview of Rewrite Rule Laboratory (RRL)* , Proc. of Rewrite Techniques and Applications, Springer-Verlag LNCS 355, pp 559-563.

[10] M.R.K. Krishna Rao, D. Kapur and R.K. Shyamasundar (1991), *A Transformational Methodology for Proving termination of Logic Programs* (extended version), Technical report, Tata Institute of Fundamental Research, Bombay 400 005, India.

[11] M. R. K. Krishna Rao, R. K. Shyamasundar and P. Pandya (1992), *Termination proof for ProCoS level 0 language* PL_0 *compiler*, technical report, Tata Institute of Fundamental Research, Bombay, India (in preparation).

[12] J. W. Lloyd (1987), *Foundations of Logic Programming*, Springer-Verlag.

[13] L. Plümer (1990), *Termination proofs for Logic Programs*, Ph. D. thesis, University of Dortmund, Also appears as Springer Verlag LNCS vol. 446.

[14] R. K. Shyamasundar, M. R. K. Krishna Rao and D. Kapur (1992), *Rewriting Concepts in the Study of Termination of Logic Programs*, Proc. ALPUK'92 Conf., London, April 1992.

[15] J.D. Ullman and A. van Gelder (1988), *Efficient Tests for Top-Down Termination of Logical Rules*, JACM, 35(2), pp. 345-373.

Plausibility Logic *

Daniel Lehmann

Institute of Computer Science, Hebrew University
91904 Jerusalem Israel

Abstract. This is an effort towards an abstract presentation of the formal properties of the way we tend to jump to conclusions from less than fully convincing information. In [6], such properties were presented as families of binary relations between propositional formulas, i.e., built out of pre-existing propositional logic. Though the family of cumulative relations is easily amenable to an abstract presentation that does not use the propositional connectives, as was noticed in [8] and [9], no such presentation is known for the more attractive family of preferential relations. Plausibility Logic is a step towards such an abstract presentation. It enables the definition of connectives: each connective is defined by introduction rules only. It provides a nonmonotonic presentation of the Gentzen's consequence relation of classical logic. But, no representation theorem is known for Plausibility Logic and it does not enjoy Cut Elimination.

1 Introduction

Logicians have long wondered "what is a logic?". A. Tarski proposed to start from any set (of formulas) and consider a logic to be a set of pairs containing a finite set of formulas on the left and a single formula on the right: if a formula a may be logically deduced in the given logic from a set X of assumptions then the pair $\langle X, a\rangle$ is in the logic. Soon after, G. Gentzen preferred to stretch this setting a bit and allow not only a single formula on the right but any finite set of formulas on the right. This is the setting considered here. The pair $\langle X, Y\rangle$, i.e. the sequent $X \vdash Y$, is in the logic iff, from the assumptions X one may, in the logic, jump to the conclusion that at least one of the formulas of Y is true. Notice that the formulas on the left, i.e. the assumptions, are understood conjunctively and those on the right, i.e. the conclusions, are understood disjunctively. Since then, some have stretched this setting further, either allowing X and Y to be infinite sets (see for example [10]), or allowing them to be multisets (see for example [1] or [4]) or even sequences. The extension of the present work to infinite sets should be straightforward. The consideration of multisets or sequences, though promising for analyzing theory revision, will not be touched upon.

* Some results contained here have been obtained since the Conference oral presentation. This work was partially supported by grant 351/89 from the Basic Research Foundation, Israel Academy of Sciences and Humanities and by the Jean and Helene Alfassa fund for research in Artificial Intelligence. Part of this work was performed while the author was visiting the Laboratoire d'Informatique Théorique et de Programmation, Université Paris 6.

2 Plausibility Logics

Let L be any set. The elements of L will be called formulas and denoted by small letters from the beginning of the alphabet: a, b, and so on. Finite sets of formulas will be denoted by capital letters from the end of the alphabet: X, Y, and so on. As is customary in logic, the comma will be used to represent the union of sets and formulas will be identified with the corresponding singletons. For example, X, Y represents the union $X \cup Y$, and X, a represents the set $X \cup \{a\}$. a, X, a is equal to and may be replaced by X, a.

A number of interesting properties of logics will be described now. Let $\vdash$ be a logic, i.e. a binary relation between finite sets of formulas. The first of those properties deals with the status of assumptions.

Definition 1. A logic $\vdash$ satisfies Inclusion iff, for any finite set X of formulas and any formula a:

$$\text{(Inclusion)} \quad X, a \vdash a$$

A logic satisfies Inclusion iff one is always ready to jump to the conclusion that any one of the assumptions is true. This seems very reasonable.

The properties we shall describe now are *monotonicity* or *weakening* properties. They essentially say that one may add formulas on the left or on the right of the symbol $\vdash$.

Definition 2. A logic $\vdash$ satisfies Left Monotonicity iff, for any finite sets X and Y of formulas and any formula a:

$$\text{(Left Monotonicity)} \quad \text{If } X \vdash Y, \text{then} X, a \vdash Y.$$

This is the property of Monotonicity of classical logic that we think is not enjoyed by "jumping to conclusions". Plausibility logics will not be required to be left monotonic. The dual property of right monotonicity seems, on the contrary, to be enjoyed by the logics of interest to us.

Definition 3. A logic $\vdash$ satisfies Right Monotonicity iff, for any finite sets X and Y of formulas and any formula a:

$$\text{(Right Monotonicity)} \quad \text{If } X \vdash Y, \text{then} X \vdash a, Y.$$

But there is a weak form of left monotonicity that seems to be enjoyed by the logics of interest. It was first proposed in [3].

Definition 4. A logic $\vdash$ satisfies Cautious Left Monotonicity iff, for any finite sets X and Y of formulas and any formula a:

$$\text{(Cautious Left Monotonicity)} \quad \text{If } X \vdash a \text{ and } X \vdash Y \text{ then } X, a \vdash Y.$$

A discussion of why "jumping to conclusions" is expected to satisfy Cautious Left Monotonicity may be found in [6].

Last, but not least, we must consider Cut rules. The Cut rule satisfied by classical logic and also linear logic is not expected to be satisfied.

Definition 5. A logic $\vdash$ satisfies Cut iff, for any finite sets X, X', Y and Y' of formulas and any formula a:

$$\text{(Cut)} \quad \text{If } X, a \vdash Y \text{ and } X' \vdash a, Y' \text{ then } X, X' \vdash Y, Y'.$$

The following example will show why nonmonotonic logics are not expected to satisfy Cut. Suppose Paul is a very reliable and conscientious worker. Without any assumption, we may jump to the conclusion that he will be at work tomorrow, and even on the assumption that it will be raining tomorrow we may well be willing to jump to this conclusion. Take therefore b to be "it will rain tomorrow" and a to be "Paul will be at work tomorrow". We accept $b \vdash a$. Let, now, c be "it will be very hot tomorrow". We obviously accept $c, a \vdash a$, by Inclusion. Cut would force us to accept also $b, c \vdash a$, which means that on the assumptions that tomorrow it will both rain and be very hot, we shall jump to the conclusion that Paul will be at work. But it may well be that, in our area, in any normal circumstances, it *never* rains on a very hot day, but that the conjunction of heat and rain is the sign of some out of the ordinary, ominous event (e.g. a nuclear explosion) that will convince even Paul to stay at home or in some other sheltered place. We, therefore, propose a weaker form of Cut, obtained by requiring X and X' to coincide. In fact, in the presence of the rules of Plausibility Logic, the Rule of Cut implies Left Monotonicity. We shall now present the form of Cut that we want to accept in Plausibility Logic. Remark that it enables us to "cut" a set of assumptions at once. As far as I know, though (or since) this form is obviously equivalent to Cut in the presence of Left Monotonicity, it has never been considered before.

Definition 6. A logic $\vdash$ satisfies Cautious Cut iff, for any natural number n, any formulas $h_1, \ldots, h_n$ and any finite sets X and Y_i, $i = 0, \ldots, n$ of formulas:

$$\begin{array}{lll} \text{(Cautious Cut)} & \text{If} & X, h_1, \ldots, h_n \vdash Y_0 \\ & \text{and} & X \vdash h_i, Y_i \text{ for } i = 1, \ldots, n \\ & \text{then} & X \vdash Y_0, Y_1, \ldots, Y_n. \end{array}$$

An equivalent (in the presence of Right Monotonicity) form of Cautious Cut is:

$$\text{If } X, h_1, \ldots, h_n \vdash Y \text{ and } X \vdash h_i, Y \text{ for } i = 1, \ldots, n \text{ then } X \vdash Y. \tag{1}$$

The most useful case of Cautious Cut is the case $n = 1$, that will be named Unit Cautious Cut.

$$\text{(Unit Cautious Cut)} \quad \text{If } X, h \vdash Y \text{and } X \vdash h, Y' \text{ then } X \vdash Y, Y'.$$

In the presence of Right Monotonicity, this is equivalent to the following form:

$$\text{If } X, h \vdash Y \text{and } X \vdash h, Y \text{ then } X \vdash Y. \tag{2}$$

One of my surprises in this work is that Unit Cautious Cut is not as strong as the general case of Cautious Cut, even in the presence of all the other rules of Plausibility Logic. This was proved by K. Schlechta (private communication, February 1992). It seems therefore necessary for us to consider the elimination of a set of hypothesis at one go and cannot eliminate them in turn. This must say something interesting about the natural deduction (à la Prawitz) system that must correspond to Plausibility Logic. One may easily check that the following multi-hypothesis Cut is trivially equivalent to Cut, without any need for structural rules.

$$\text{(Multi Cut)} \quad \text{If} \quad X_0, h_1, \ldots, h_n \vdash Y_0 \\ \text{and} \quad X_i \vdash h_i, Y_i \text{ for } i = 1, \ldots, n \\ \text{then} \quad X_0, X_1, \ldots X_n \vdash Y_0, Y_1, \ldots, Y_n.$$

An even weaker version of Cut corresponds to the Cut property of cumulative logic.

Definition 7. A logic $\vdash$ satisfies Cumulative Cut iff, for any finite sets X and Y of formulas and any formula a:

$$\text{(Unit Cumulative Cut)} \quad \text{If } X, a \vdash Y \text{ and } X \vdash a \text{ then } X \vdash Y.$$

One may see that, in the presence of Cautious Left Monotonicity and Right Monotonicity, Unit Cumulative Cut implies the following

$$\text{(Cumulative Cut)} \quad \text{If} \quad X, h_1, \ldots, h_n \vdash Y \\ \text{and} \quad X \vdash h_i \text{ for } i = 1, \ldots, n \\ \text{then} \quad X \vdash Y.$$

We may now define plausibility logics.

Definition 8. A logic $\vdash$ is said to be plausibility iff it satisfies Inclusion, Cautious Left Monotonicity, Right Monotonicity and Cautious Cut.

If we restrict ourselves to sequents the right part of which contains at most one element, we obtain a system that may rightly be called intuitionistic plausibility logic. In this framework, there is no distinction between Cautious and Cumulative Cut. Intuitionistic plausibility logic is essentially the framework called cumulative logic in [8] (without truth-functional connectives). Its relation to the cumulative relations of [6] is not clear, since I do not know how to describe cumulative relations as the cumulative logic of [8] (without connectives) in which have been defined connectives that may be reasonably described by introduction rules, because disjunction does not seem to be characterizable by such rules. The study of intuitionistic plausibility logic with a proper disjunction has not been undertook yet.

In [6] preferential consequence relations have been introduced and studied. They provide examples of plausibility logics in the following way.

Theorem 9. *Let L be an arbitrary set and F the set of all propositional formulas on the set L of variables. Let $\mid\!\sim$ be a preferential relation on F, in the sense of [6]. The logic $\vdash$ defined by:*

$$a_0, a_1 \ldots a_n \vdash b_0, b_1 \ldots b_m \text{ iff } a_0 \wedge a_1 \wedge \ldots \wedge a_n \mid\!\sim b_0 \vee b_1 \vee \ldots \vee b_m \tag{3}$$

is a plausibility logic.

Using theorem 9 and results of [6], one easily sees that Cautious Cut is strictly stronger than Cumulative Cut and that there are plausibility logics that do not satisfy Left Monotonicity, i.e., Left Monotonicity is not a derived rule of plausibility logic. The question of whether all plausibility logics on L may be obtained in the way described in equation (3) from some preferential consequence relation on F will be answered in the negative by Corollary 36.

3 Models

Plausibility logics may be generated by models similar to the preferential models of [6].

Definition 10. A *plausibility* model W is a triple $\langle S, l, \prec \rangle$ where S is a set, the elements of which will be called states, $l : S \mapsto 2^L$ assigns a set of formulas to each state and $\prec$ is a binary relation on S satisfying the following *smoothness condition*: $\forall X \subseteq_f L$ (i.e. X is a *finite* set of formulas), the set of states

$$\widehat{X} \stackrel{\text{def}}{=} \{s \mid s \in S,\ l(s) \models a, \text{for every formula } a \in X\}$$

is smooth, i.e. for any s in $\widehat{X}$ that is not minimal in $\widehat{X}$, there exists a state t minimal in $\widehat{X}$, such that $t \prec s$. If $\prec$ is a strict partial order on S, the model W will be said to be ordered.

Theorem 11. *Any plausibility model defines a plausibility logic $\vdash$ by: $X \vdash Y$ iff for every minimal state s of $\widehat{X}$, $l(s) \cap Y \neq \emptyset$.*

Proof. Obvious. The smoothness property is used only for establishing Left Cautious Monotonicity. ∎

The question of whether any plausibility logic may be generated by a plausibility model is open.

4 The meta-notion of equivalence

The first notion we want to study concerning plausibility logics is the notion of logical equivalence. This is a meta-notion. Given a logic, formulas a and b should be considered to be logically equivalent if one may freely replace one by the other. Since a formula may appear on the left or on the right of the symbol $\vdash$, there seems to be two natural notions of logical equivalence. Fortunately, for plausibility logics, those two notions coincide. For this to hold, Cautious Cut is essential and Cumulative Cut would not do.

Theorem 12. *Let $\vdash$ be a plausibility logic. Let a and b be formulas. The following three conditions are equivalent. They characterize logical equivalence of a and b, that will be denoted $a \equiv b$.*

1. *For any set X, $X, a \vdash b$ and $X, b \vdash a$.*
2. (Left Equivalence) *For any sets X and Y, $X, a \vdash Y$ iff $X, b \vdash Y$.*
3. (Right Equivalence) *For any sets X and Y, $X \vdash a, Y$ iff $X \vdash b, Y$.*

Proof. It is very easy to see, using Inclusion, that both properties 2 and 3 imply 1. Property 1 implies 2 by Cautious Monotonicity and Cumulative Cut. Property 1 implies 3 by Cautious Cut. ∎

One may see, using the cumulative models of [6] and Theorem 9 that, if we have only Cumulative Cut, and not the stronger Cautious Cut, the notion of right equivalence is indeed strictly stronger than that of left equivalence. My conclusion is therefore that there is only one reasonable notion of equivalence, and this seems to fit our everyday experience of this notion. I shall now try to describe the meaning of the usual propositional connectives in our setting, beginnning with negation.

5 Negation

Let us now suppose that there is a unary connective $\neg$ and that the language L is closed under this connective. So, for every formula a, there is a formula $\neg a$. I propose the following two properties to express the idea that $\neg a$ is the negation of a. The first property says that it is inconsistent to assume both a and $\neg a$, i.e., from the assumptions a and $\neg a$ one may jump to any conclusion.

Definition 13. We shall say that $\neg$ satisfies Exclusion (with respect to $\vdash$) iff:

$$\text{(Exclusion) For any formula } a, \text{we have } a, \neg a \vdash \ .$$

Exclusion implies, by Right Monotonicity, that $a, \neg a \vdash Y$, for any Y, and, as will be shown in Theorem 16, that $X, a, \neg a \vdash Y$ for any X and any Y. This seems to me quite a natural property of negation, corresponding to the idea that one cannot (without contradiction), at the same time assume both a formula and its negation.

The second property says that one may always jump to the conclusion that either a or $\neg a$ is true.

Definition 14. We shall say that $\vdash$ and $\neg$ satisfy Certainty iff:

$$\text{(Certainty) For any set } X \text{ and any formula } a, \text{we have} X \vdash a, \neg a.$$

It is perhaps worth noticing that Certainty is implied by the following stronger condition that expresses the idea of negation as failure.

$$X \not\vdash a \Rightarrow X \vdash \neg a$$

Definition 15. Let $\vdash$ be a plausibility relation on a language L closed under $\neg$. The connective $\neg$, will be called a negation for $\vdash$ iff the two properties of Exclusion and Certainty are satisfied. Since it is easy to see that, in the presence of Left Cautious Monotonicity, Right Monotonicity and Cautious Cut, the two properties of Exclusion and Certainty imply Inclusion, one may omit this property.

One may easily see that, for any plausibility logic that is defined by (3), propositional negation is a negation. Here are some properties of negations.

Theorem 16. *Let $\vdash$ be a plausibility logic and $\neg$ a negation. For any sets X, Y of formulas and any formula a:*

1. $X, a, \neg a \vdash$,
2. *if* $X, a \vdash Y$, *then* $X \vdash \neg a, Y$,
3. *if* $X, \neg a \vdash Y$, *then* $X \vdash a, Y$,
4. *if* $X \vdash a, Y$ *and* $X \vdash \neg a, Y'$, *then* $X \vdash Y, Y'$,
5. *if* $a \equiv b$, *then* $\neg a \equiv \neg b$,
6. $a \equiv \neg\neg a$.

Proof. 1. By induction on the size of X. If X is empty, by Exclusion. Otherwise $X = X' \cup \{b\}$. By the induction hypothesis $X', a, \neg a \vdash$ and therefore, by Right Monotonicity $X', a, \neg a \vdash b$. By Cautious Monotonicity, we conclude $X, a, \neg a \vdash$.
2. Suppose $X, a \vdash Y$. By Certainty, $X \vdash a, \neg a$. We conclude by Unit Cautious Cut.

3. Suppose $X, \neg a \vdash Y$. By Certainty, $X \vdash a, \neg a$. We conclude by Unit Cautious Cut.
4. Suppose $X \vdash a, Y$ and $X \vdash \neg a, Y'$. Since $X, a, \neg a \vdash$ by Exclusion, we may apply Cautious Cut (this is in fact the only time, in this section, that we need more than Unit Cautious Cut) to obtain $X \vdash Y, Y'$.
5. By Theorem 12, part 1, it is enough to prove that, if $a \equiv b$, $X, \neg a \vdash \neg b$. But $X, a, \neg a \vdash$ and, since $a \equiv b$, $X, b, \neg a \vdash$ and we conclude by part 2.
6. By Theorem 12, part 1, it is enough to prove $X, a \vdash \neg\neg a$ and $X, \neg\neg a \vdash a$. The first claim follows from $X, a, \neg a \vdash$ by 2 and the second one follows from $X, \neg a, \neg\neg a \vdash$ by part 3.

∎

Corollary 17. *$\neg$ is a negation iff it satisfies Exclusion and property 2 of Theorem 16.*

It is always preferable to describe the meaning of connectives solely by introduction rules, one on the left and one on the right (see for example [5]). Our study, so far suggests the following rules for negation. They are enough to guarantee that $\neg$ is a negation.

$$\frac{}{X, a, \neg a \vdash}\mathcal{L}\neg \qquad\qquad \frac{X, a \vdash Y}{X \vdash \neg a, Y}\mathcal{R}\neg$$

It is also clear that there is at most one negation for plausibility logic (up to logical equivalence).

Lemma 18. *Let $\vdash$ be a plausibility relation on a language L closed under $\neg_0$ and $\neg_1$. If both $\neg_0$ and $\neg_1$ are negations, then for any a $\neg_0 a \equiv \neg_1 a$.*

Proof. We shall use property 1 of Theorem 12. By part 1 of Theorem 16 we have $X, \neg_1 a, a \vdash$. By part 2 of the same theorem, $X, \neg_1 a \vdash \neg_0 a$. And symmetrically. ∎

The next theorem shows that any plausibility logic may be conservatively extended to a richer language with a negation. This means that the connective negation may always be defined in a proper way.

Theorem 19. *Let L be any language, and $L^{\neg}$ be the closure of L under the unary connective $\neg$. If $\vdash$ is a plausibility logic on L, there is a plausibility logic on $L^{\neg}$ for which $\neg$ is a negation and that is a conservative extension of $\vdash$.*

Proof. By property 6 of Theorem 16, one may eliminate double negations and consider $L^{\neg}$ to be the union of L and the set of negated formulas L^{-}. Given a set X of formulas of $L^{\neg}$, we shall define X^{+} to be $X \cap L$ and X^{-} to be $X \cap L^{-}$. Suppose $\vdash$ is a relation on L. Let $\vdash'$ be defined on $L^{\neg}$ by: $X \vdash' Y$ iff at least one of the following holds:

1. $X^{+} \cap X^{-} \neq \emptyset$,
2. $X \cap Y \neq \emptyset$,
3. $Y^{+} \cap Y^{-} \neq \emptyset$,
4. $X^{+} \vdash$ or
5. $X = X^{+}$ and $X^{+} \vdash Y^{+}$.

One shows that, if $\vdash$ is a plausibility logic, then $\vdash'$ is a plausibility logic for which $\neg$ is a negation, and is a conservative extension of $\vdash$. ∎

My conclusion is that there seems to be only one reasonable way of defining negation and that all properties of classical negation seem to be enjoyed by any negation.

6 Conjunction

Let us now suppose that there is a binary connective $\wedge$ and that the language L is closed under this connective. So, for every formulas a and b, there is a formula $a \wedge b$. There are two possible definitions of what it is to be a proper conjunction. The first one describes the way we expect conjunction to act on the left of the $\vdash$ symbol: we expect $\wedge$ to act as the comma does. This first definition defines what has been called an internal (multiplicative in the terminology of [4]) conjunction in [2]. The second definition describes the way we expect conjunction to behave on the right of the $\vdash$ symbol: we jump to the conclusion $a \wedge b$ iff we are ready to jump to both the conclusions a and b. Such a conjunction has been called an external (or additive) conjunction.

Definition 20. The connective $\wedge$ is said to be an *internal* conjunction iff, for any sets X, Y and any formulas a, b, $X, a, b \vdash Y$ iff $X, a \wedge b \vdash Y$. The connective $\wedge$ is said to be an *external* conjunction iff, for any sets X, Y and any formulas a, b, $X \vdash a \wedge b, Y$ iff $X \vdash a, Y$ and $X \vdash b, Y$.

In classical logic both notions are equivalent. In linear logic they are not, and one has to deal with two different conjunctions. In plausibility logic, as in classical logic, both notions are equivalent. Notice that, if $\wedge$ is an internal conjunction and if $a \equiv a'$ and $b \equiv b'$, then $a \wedge b \equiv a' \wedge b'$ (use property 2 of Theorem 12). One may also easily see that, for any plausibility logic that is defined by (3), propositional conjunction is an external and internal conjunction.

Theorem 21. *Let L be closed under $\wedge$ and let $\vdash$ be a plausibility logic. The three following propositions are equivalent.*

1. *$\wedge$ is an external conjunction.*
2. *$\wedge$ is an internal conjunction.*
3. *For any finite set X and any formulas a and b,*

$$X, a, b \vdash a \wedge b \tag{4}$$

$$X, a \wedge b \vdash a \tag{5}$$

$$X, a \wedge b \vdash b \tag{6}$$

Proof. Suppose $\wedge$ is an external conjunction. We shall show that 3 holds. Equation (4) holds since $X, a, b \vdash a$ and $X, a, b \vdash b$. Equations (5) and (6) hold because $X, a \wedge b \vdash a \wedge b$.

Suppose 3 holds. We shall show that $\wedge$ is an internal conjunction. Suppose $X, a, b \vdash Y$. By equation (4) and Cautious Left Monotonicity, we conclude that

$X, a, b, a \wedge b \vdash Y$. Using (5), (6) and Cautious Cut we conclude that $X, a \wedge b \vdash Y$. Notice that we could have used Cumulative Cut here, instead of Cautious Cut. Suppose now that $X, a \wedge b \vdash Y$. Using (5) and Cautious Left Monotonicity, we conclude $X, a \wedge b, a \vdash Y$. From (6), we obtain $X, a, a \wedge b \vdash b$. By Cautious Left Monotonicity, we have $X, a, b, a \wedge b \vdash Y$. We conclude using (4) and Cautious Cut (or Cumulative Cut).

Suppose $\wedge$ is an internal conjunction. Let us prove it is an external conjunction. Notice that, since $X, a, b \vdash a$, we have $X, a \wedge b \vdash a$. If $X \vdash a \wedge b, Y$, then, by Unit Cautious Cut, we have $X \vdash a, Y$. Similarly, if $X \vdash a \wedge b, Y$, then, $X \vdash b, Y$. Suppose now that $X \vdash a, Y$ and $X \vdash b, Y$. Since, $X, a, b \vdash a \wedge b$ we conclude by Cautious Cut that $X \vdash a \wedge b, Y$. Notice that, here, Unit Cautious Cut is not enough. ∎

From now on we shall refer to $\wedge$ as a conjunction, if it satisfies the conditions of Theorem 21. The following two introduction rules characterize conjuctions.

$$\frac{X, a, b \vdash Y}{X, a \wedge b \vdash Y}\mathcal{L}\wedge \qquad\qquad \frac{X \vdash a, Y \quad X \vdash b, Y'}{X \vdash a \wedge b, Y, Y'}\mathcal{R}\wedge$$

Similarly to what was done in Theorem 19, one may show that any plausibility logic may be conservatively extended with a conjunction. Unfortunately Section 9 will show that one cannot always extend conservatively a plausibility logic with both negation and conjunction.

Theorem 22. *Let L be any language, and $L^{\wedge}$ be the closure of L under the binary connective $\wedge$. If $\vdash$ is a plausibility logic on L, there is a plausibility logic on $L^{\wedge}$ for which $\wedge$ is a conjunction and that is a conservative extension of $\vdash$.*

Proof. The construction is easy: treat conjunctions on the left side of $\vdash$ as a comma, and conjunctions on the right as denoting the set implied by the fact it is an external conjunction. ∎

7 Disjunction

Let us now consider disjunction. In plausibility logic, there is only one interesting definition for a disjunction, the internal form. The existence of an external disjunction implies the monotonicity of the logics. Therefore, we define disjunction in the following way.

Definition 23. The connective $\vee$ is said to be a disjunction iff, for any sets X, Y and any formulas a, b, one has $X \vdash a, b, Y$ iff $X \vdash a \vee b, Y$.

Theorem 24. *Let L be closed under $\vee$ and let $\vdash$ be a plausibility logic. The connective $\vee$ is a disjunction iff, for any X, Y, a and b, the three following properties are satisfied:*

$$X, a \vdash a \vee b \tag{7}$$

$$X, b \vdash a \vee b \tag{8}$$

$$\text{if } X, a \vdash Y \text{ and } X, b \vdash Y, \text{ then } X, a \vee b \vdash Y. \tag{9}$$

Proof. Suppose $\vee$ is a disjunction. We have $X, a \vdash a, b$ by Inclusion and Right Monotonicity. Since $\vee$ is a disjunction, we have (7). Similarly for (8). Suppose now that $X, a \vdash Y$ and $X, b \vdash Y$. From (7) and the first hypothesis, we conclude, by Cautious Monotonicity, that

$$X, a, a \vee b \vdash Y. \tag{10}$$

But

$$X, a \vee b \vdash a, b, \tag{11}$$

by Inclusion and the fact that $\vee$ is a disjunction. We conclude from (10) and (11), by Cautious Cut, that

$$X, a \vee b \vdash b, Y. \tag{12}$$

From $X, b \vdash Y$ and $X, b \vdash a \vee b$, we conclude, by Cautious Monotonicity that:

$$X, b, a \vee b \vdash Y. \tag{13}$$

We conclude $X, a \vee b \vdash Y$ by Unit Cautious Cut from (13) and (12).

Suppose now that $\vee$ satisfies (7), (8) and (9). Let us show it is a disjunction. Suppose, first, that $X \vdash a, b, Y$. From this and (7), by Unit Cautious Cut, we have $X \vdash a \vee b, b, Y$. We conclude $X \vdash a \vee b, Y$, by using Unit Cautious Cut and (8). Suppose now that $X \vdash a \vee b, Y$. Since $X, a \vdash a, b$ and $X, b \vdash a, b$, one may use (9) to show that $X, a \vee b \vdash a, b$. By Unit Cautious Cut we conclude that $X \vdash a, b, Y$. ∎

We may now characterize disjunction by introduction rules.

$$\frac{X, a \vdash Y \qquad X, b \vdash Y'}{X, a \vee b \vdash Y, Y'} \mathcal{L}\vee \qquad\qquad \frac{X \vdash a, b, Y}{X \vdash a \vee b, Y} \mathcal{R}\vee$$

Notice that, if $\vee$ is a disjunction and if $a \equiv a'$ and $b \equiv b'$, then $a \vee b \equiv a' \vee b'$ (use property 3 of Theorem 12). We may also notice that if $\vee_0$ and $\vee_1$ are both disjunctions, then for any a, b, $a \vee_0 b \equiv a \vee_1 b$.

The notion of an *external* disjunction, i.e. satisfying $X, a \vee b \vdash Y$ iff $X, a \vdash Y$ and $X, b \vdash Y$ seems unnatural in this setting, since it appears to embody some measure of monotonicity. Indeed, a plausibility logic that possesses such an external disjunction is monotonic, i,e., satisfies Left Monotonicity, at least under some mild assumption. Before we prove this result, we need a definition.

Definition 25. A formula t is said to be a logical truth (for $\vdash$) iff $X, t \vdash Y$ iff $X \vdash Y$, for any X, Y.

Theorem 26. *Let $\vdash$ be a plausibility logic with a logical truth. Then $\vdash$ possesses an external disjunction iff it satisfies Left Monotonicity and possesses a disjunction.*

Proof. Suppose $\vdash$ is a monotonic plausibility logic, and $\vee$ is a disjunction. We shall show, without using the existence of a logical truth, that $\vee$ is an external disjunction. One half of our claim is proved by Theorem 24, equation 9. We only need to show that, if $X, a \vee b \vdash Y$, then $X, a \vdash Y$ and $X, b \vdash Y$. But, by Left Monotonicity, we have $X, a \vee b, a \vdash Y$ and we conclude $X, a \vdash Y$ by Unit Cautious Cut, using equation 7. Similarly for $X, b \vdash Y$.

Suppose now that $\vdash$ is a plausibility logic with a logical truth and $\vee$ an external disjunction. By Theorem 24, it is a disjunction. Let us show that $\vdash$ is monotonic. Suppose $X \vdash Y$. We must show that $X, a \vdash Y$. But we have $X, t \vdash Y$ and also $X, t \vdash t \vee a$. By Right Cautious Monotonicity we conclude that $X, t, t \vee a \vdash Y$, and $X, t \vee a \vdash Y$. Since $\vee$ is an external disjunction, we have $X, a \vdash Y$. ∎

8 The logic behind Plausibility Logic

We shall now see that the logic behind Plausibility Logic is indeed classical logic, as the reader may have guessed. We shall consider the set of sequents that may be deduced from an empty set of sequents by the rules of Plausibility Logic, i.e. the intersection of all plausibility logics, and show that this is exactly classical logic.

Theorem 27. *Let L be a propositional language. The intersection of all plausibility logics ($\vdash$) on L for which $\neg$ is a negation, $\wedge$ is a conjunction and $\vee$ is a disjunction is classical logic, i.e. the set of sequents valid in classical logic.*

Proof. It is clear that any sequent valid for Plausibility Logic is valid in classical logic. One only has to show that any sequent valid classically is also valid in Plausibility Logic. First, one easily sees, by induction on the length of proofs, that Left Monotonicity is valid in this minimal plausibility logic: if $X \vdash Y$, then $X, a \vdash Y$. We only have to check that the classical introduction rules are valid in this minimal plausibility logic. The only rule to differ from the classical case is the left introduction rule for negation. Let us show it is valid. Suppose $X \vdash a, Y$. By Left Monotonicity (see just above), we have $X, \neg a \vdash a, Y$. But $X, \neg a \vdash \neg a$ by Inclusion. By Theorem 16, part 4, we conclude $X, \neg a \vdash Y$. ∎

The following is an easy corollary.

Corollary 28. *Any plausibility logic that possesses a negation, a conjunction and a disjunction defines, by $a \mathrel{|\!\sim} b$ iff $a \vdash b$, a preferential consequence relation.*

Proof. The only rule that causes problem is Left Logical Equivalence, and it is taken care of by Theorem 27. ∎

The following results show that, as in the case of classical logic, both negation and conjunction, and negation and disjunction are expressive enough to express the third connective. It is not a corollary of the previous results, since it concerns any plausibility logic.

Theorem 29. *If $\vdash$ is a plausibility logic, $\neg$ a negation and $\wedge$ a conjunction, then a disjunction may be defined by $a \vee b = \neg(\neg a \wedge \neg b)$.*

Proof. Suppose $X \vdash a, b, Y$. We want to show that $X \vdash \neg(\neg a \wedge \neg b), Y$. Notice that $X, \neg a, \neg b \vdash \neg a$ imply $X \vdash \neg a, \neg(\neg a \wedge \neg b)$. We therefore conclude, by Theorem 16, part 4 that $X \vdash b, \neg(\neg a \wedge \neg b), Y$. It will be enough to show that $X \vdash \neg b, \neg(\neg a \wedge \neg b)$ and conclude similarly. But this follows from the fact that we have $X, \neg a, \neg b \vdash \neg b$.

Suppose now that $X \vdash \neg(\neg a \wedge \neg b), Y$. We want to show that $X \vdash a, b, Y$. It is enough to show that $X \vdash a, b, \neg a \wedge \neg b$. But this follows from $X \vdash a, b, \neg a$ and $X \vdash a, b, \neg b$ and Theorem 21. ∎

Lemma 30. *If* $\vdash$ *is a plausibility logic,* $\neg$ *a negation and* $\vee$ *a disjunction, then* $X, a, b \vdash \neg(\neg a \vee \neg b)$.

Proof. We have $X, a, b, \neg a \vdash$ and $X, a, b, \neg b \vdash$. From Theorem 24 we see that we have $X, a, b, \neg a \vee \neg b \vdash$. We conclude by Theorem 16, property 2. ∎

Lemma 31. *If* $\vdash$ *is a plausibility relation,* $\neg$ *a negation and* $\vee$ *a disjunction, then* $X, \neg(\neg a \vee \neg b) \vdash a$ *and* $X, \neg(\neg a \vee \neg b) \vdash b$.

Proof. From $X, \neg(\neg a \vee \neg b) \vdash a, \neg a, \neg b$, we prove $X, \neg(\neg a \vee \neg b) \vdash a, \neg a \vee \neg b$. But we have $X, \neg(\neg a \vee \neg b), \neg a \vee \neg b \vdash$ and we conclude by Cautious Cut. ∎

Theorem 32. *If* $\vdash$ *is a plausibility relation,* $\neg$ *a negation and* $\vee$ *a disjunction, then a conjunction may be defined by* $a \wedge b = \neg(\neg a \vee \neg b)$.

Proof. Obvious, using property 3 of Theorem 21 and Lemmas 30 and 31. ∎

9 Cut Elimination

It will now been shown that no Cut Elimination theorem holds for plausibility logic. The existence of a disjunction validates a rule that does not involves disjunction. The rule of interest is a special case of the rule Loop of [6].

$$\text{If } c \vdash a \quad a \vdash b \text{ and } b \vdash c \text{ then } a \vdash c. \tag{14}$$

First a lemma will be useful.

Lemma 33. *If* $\vdash$ *is a plausibility logic with disjunction, if* $a \vdash a'$ *and* $b \vdash b'$ *then* $a \vee b \vdash a' \vee b'$.

Proof. By Right Monotonicity $a \vdash a', b'$ and $b \vdash a', b'$. By $\mathcal{L}\vee$, $a \vee b \vdash a', b'$. One concludes by $\mathcal{R}\vee$. ∎

Theorem 34. *In any plausibility logic that has a disjunction, the rule (14) is valid.*

Proof. 1. $a \vdash a \vee c$ by Inclusion and $\mathcal{R}\vee$
2. $c \vdash a$ Hypothesis
3. $a \vdash b$ Hypothesis
4. $b \vdash c$ Hypothesis
5. $c \vdash c$ Inclusion
6. $a \vdash a$ Inclusion
7. $a \vee c \vdash a$ by $\mathcal{L}\vee$ from (6) and (2)
8. $a \vee c \vdash b \vee c$ by Lemma 33, from (3) and (5)
9. $b \vee c \vdash c$ by $\mathcal{L}\vee$ from (4) and (5)
10. $b \vee c \vdash a \vee c$ from (9) by Right Monotonicity and $\mathcal{R}\vee$
11. $b \vee c \vdash c$ by $\mathcal{L}\vee$ from (4) and (5)
12. $b \vee c, a \vee c \vdash c$ by Cautious Left Monotonicity from (11) and (10)
13. $a \vee c \vdash c$ by Unit Cumulative Cut from (12) and (8)

14. $a \vee c, a \vdash c$ by Cautious Left Monotonicity from (13) and (7)
15. $a \vdash c$ by Unit Cumulative Cut from (14) and (1).
∎

Theorem 35. *There is a plausibility logic that does not contain the rule (14).*

Proof. Let W be the plausibility model (see Definition 10) containing four states: s_i, for $i = -1, 0, 1, 2$. The label of s_{-1} is empty. The label of s_0 contains a and b. The label of s_1 contains b and c. The label of s_2 contains c and a. The relation $\prec$ is defined in the following way: $s_{-1} \prec s_j$ for any $j = 0, 1, 2$, $s_0 \prec s_1$, $s_1 \prec s_2$ and $s_2 \prec s_0$. Notice that $\prec$ is not transitive. One may check, by exhausting all possibilities, that W satisfies the smoothness property. Let $\vdash$ be the plausibility logic defined by W (Theorem 11). One sees easily that $\vdash$ provides a counter-example to the rule Loop.
∎

We conclude that, for a plausibility logic with disjunction (or with negation and conjunction), there is no Cut Elimination theorem and no conservative extension theorem for disjunction (or negation and conjunction together) similar to Theorems 19 and 22. We also answer negatively the question asked at the end of Section 2.

Corollary 36. *There is a plausibility logic that cannot be generated as described in equation (3).*

Proof. Any logic defined in this way satisfies rule (14). ∎

A most intriguing open question is whether there is a natural version of the rule Loop, of which (14) is a special case, such that all plausibility logics satisfying it enjoy Cut Elimination and are represented by ordered plausibility models.

10 Implication

It seems natural to define implication by the following introduction rules.

$$\frac{}{X, a, a \to b \vdash b}\mathcal{L}\to \qquad\qquad \frac{X, a \vdash b, Y}{X \vdash a \to b, Y}\mathcal{R}\to$$

We do not expect rule $\mathcal{R}\to$ to be a double arrow rule, i.e. that if $X \vdash a \to b, Y$ then $X, a \vdash b, Y$. Indeed, one may easily show that any plausibility logic that satisfies this last property and has a logical contradiction (i.e., a formula f such that $X \vdash f, Y$ iff $X \vdash Y$) satisfies Left Monotonicity.

Theorem 37. *If $\neg$, $\vee$ and $\to$ are, respectively, a negation, a disjunction and an implication then $a \to b \equiv \neg a \vee b$.*

Proof. We shall first prove that $X, a \to b \vdash \neg a \vee b$. It is enough to show that we have $X, a \to b \vdash \neg a, b$. But $X, a \to b, a \vdash b$ and we may apply Theorem 16, part 2.

We shall now prove that $X, \neg a \vee b \vdash a \to b$. By $\mathcal{R}\to$ it is enough to show $X, \neg a \vee b, a \vdash b$. By $\mathcal{L}\vee$ it is enough to show $X, \neg a, a \vdash b$ and $X, b, a \vdash b$, which are obvious. ∎

11 Quantifiers

The following introduction rules seem to characterize the meaning of quantifiers in plausibility logic.

$$\frac{X, a \vdash Y \quad x \text{ not free in X or Y}}{X, \exists x a \vdash Y}\mathcal{L}\exists \qquad \frac{X \vdash a[t/x], Y}{X \vdash \exists x a, Y}\mathcal{R}\exists$$

$$\frac{}{X, \neg a[t/x], \forall x a \vdash}\mathcal{L}\forall \qquad \frac{X, \neg a \vdash Y \quad x \text{ not free in X or Y}}{X \vdash \forall x a, Y}\mathcal{R}\forall$$

The rules $\mathcal{L}\neg$, $\mathcal{R}\neg$, $\mathcal{L}\forall$, $\mathcal{R}\forall$, $\mathcal{L}\exists$ and $\mathcal{R}\exists$ ensure that the usual duality holds: $\forall x a \equiv \neg\exists x \neg a$ and $\exists x a \equiv \neg\forall x \neg a$.

12 Conclusions

Plausibility logics seem to strike a very intriguing middle ground between classical (monotonic) logics and very general logics such as linear logics. One, in the light of all the above, would also like to look for an abstract presentation of rational logics ([7]).

13 Acknowledgements

Arnon Avron read a preliminary version of this work and his comments are gratefully acknowledged.

References

1. Alan Ross Anderson and Nuel D. Belnap, Jr. *Entailment, The Logic of Relevance and Necessity*, volume 1. Princeton University Press, 1975.
2. Arnon Avron. Simple consequence relations. *Information and Computation*, 92:105–139, 1991.
3. Dov M. Gabbay. Theoretical foundations for non-monotonic reasoning in expert systems. In Krzysztof R. Apt, editor, *Proc. of the NATO Advanced Study Institute on Logics and Models of Concurrent Systems*, pages 439–457, La Colle-sur-Loup, France, October 1985. Springer-Verlag.
4. Jean-Yves Girard. Linear logic. *Theoretical Computer Science*, 50:1–102, 1987.
5. Jean-Yves Girard. *Proofs and Types*, volume 7 of *Cambridge Tracts in Theoretical Computer Science*. Cambridge University Press, 1989.
6. Sarit Kraus, Daniel Lehmann, and Menachem Magidor. Nonmonotonic reasoning, preferential models and cumulative logics. *Artificial Intelligence*, 44(1–2):167–207, July 1990.
7. Daniel Lehmann. What does a conditional knowledge base entail? In Ron Brachman and Hector Levesque, editors, *Proceedings of the First International Conference on Principles of Knowledge Representation and Reasoning*, Toronto, Canada, May 1989. Morgan Kaufmann.

8. David Makinson. General theory of cumulative inference. In M. Reinfrank, J. de Kleer, M. L. Ginsberg, and E. Sandewall, editors, *Proceedings of the Second International Workshop on Non-Monotonic Reasoning*, pages 1–18, Grassau, Germany, June 1988. Springer Verlag. Volume 346, Lecture Notes in Artificial Intelligence.
9. David Makinson. General patterns in nonmonotonic reasoning. In D. M. Gabbay, C. J. Hogger, and J. A. Robinson, editors, *Handbook of Logic in Artificial Intelligence and Logic Programming Vol. 2, Nonmonotonic and Uncertain Reasoning*. Oxford University Press, due 1991. in preparation.
10. Krister Segerberg. *Classical Propositional Operators*. Oxford Logic Guides. Clarendon Press, Oxford, 1982.

Towards Kleene Algebra with Recursion

Hans Leiß
Universität München, CIS
Leopoldstraße 139
D-8000 München 40
leiss@sparc1.cis.uni-muenchen.de

Abstract

We extend Kozen's theory *KA* of Kleene Algebra to axiomatize parts of the equational theory of context-free languages, using a least fixed-point operator μ instead of Kleene's iteration operator *.

Although the equational theory of context-free languages is not recursively axiomatizable, there are natural axioms for subtheories $KAF \subseteq KAR \subseteq KAG$: respectively, these make μ a least fixed point operator, connect it with recursion, and express S. Greibach's method to replace left- by right-recursion and vice versa. Over *KAF*, there are different candidates to define * in terms of μ, such as tail-recursion and reflexive transitive closure. In *KAR*, these candidates collapse, whence *KAR* uniquely defines * and extends Kozen's theory *KA*.

We show that a model $\mathcal{M} = (M, +, 0, \cdot, 1, \mu)$ of *KAF* is a model of *KAG*, whenever the partial order $\leq$ on M induced by $+$ is complete, and $+$ and $\cdot$ are Scott-continuous with respect to $\leq$. The family of all context-free languages over an alphabet of size n is the free structure for the class of submodels of continuous models of *KAF* in n generators.

1 Introduction

Regular algebra is the equational theory of the algebra of regular languages, initiated by Kleene[6]. Redko[13] showed that this theory is not axiomatizable by finitely many equations between regular expressions. Recently, two finite axiomatizations by other means have been given. Pratt[12]'s theory of *Action Logic*, *ACT*, enriches Kleene's set $\{+, 0, \cdot, 1, ^*\}$ of regular operations by left and right *residuals* $\leftarrow$ and $\rightarrow$, and is axiomatized by finitely many equations. Iteration * is characterized in *ACT* by its monotonicity properties and the equation $(x \rightarrow x)^* = x \rightarrow x$ of 'pure induction'. Kozen[8]'s theory of *Kleene Algebra*, *KA*, sticks to Kleene's regular operations, but characterizes * by universal Horn-axioms. Both *ACT* and *KA* are complete for regular algebra.

One of the motivations behind these axiomatizations was to provide an alternative to the largely combinatorial constructions of current treatments of the theory of regular languages. We believe that reasoning about context-free languages could also profit from algebraic and logical means based on axiomatic theories. In particular, it seems that in studying context-free languages, the algebraic tool of formal power series could sometimes be replaced by simpler logical methods exploiting properties of a least-fixed point operator. The aim of this paper is to stimulate research in this direction by giving some basic considerations and problems.

Recall that *regular expressions* are inductively defined by

$$r \;:=\; 0 \;\mid\; 1 \;\mid\; x \;\mid\; a \;\mid\; (r+r) \;\mid\; (r\cdot r) \;\mid\; r^*,$$

where a ranges over a finite list or *alphabet* Σ of constants, and x over an infinite list of variables. With a least-fixed-point operator μ, we define *μ-regular expressions* by

$$r \;:=\; 0 \;\mid\; 1 \;\mid\; x \;\mid\; a \;\mid\; (r+r) \;\mid\; (r\cdot r) \;\mid\; r^* \;\mid\; \mu x\, r.$$

There are two *standard interpretations* of regular expressions:

In the *language interpretation*, $\mathcal{L}_\Sigma$, variables range over the universe of all subsets (or *formal languages*) of the set Σ^* of finite sequences (or *words*) of elements of Σ, 0 denotes the empty set, 1 the singleton set containing the empty word ϵ only, the constant a denotes $\{a\}$, $+$ is set union, $\cdot$ is element-wise concatenation, and * is the union of all finite concatenations of a language with itself. By $\mathcal{REG}_\Sigma$ we understand the subclass of all regular languages over Σ, which are those elements of $\mathcal{L}_\Sigma$ that are the value of a closed regular expression, i.e. one without free variables.

In the *relation interpretation*, $\mathcal{R}_K$, variables range over the universe of all binary relations on a set K, with the empty relation as 0, the identity on K as 1, union and composition as $+$ and $\cdot$, and reflexive transitive closure as *. As an interpretation for the constants a we can take any relations $R_a \subseteq K \times K$; however, as there is no canonical choice for these, it seems natural here to consider *pure* expressions only, i.e. those containing no constants except 0 and 1.

Since all operations are monotone with respect to set and relation inclusion, respectively, we can extend these standard interpretations to μ-regular expressions, letting μ pick the least fixed point of monotone functions. The fundamental theorem of recursion theory on the natural numbers, saying that *(i)* a partial function f is definable by a system of Gödel-Herbrand-Kleene-equations iff *(ii)* f is μ-recursively definable iff *(iii)* f is computable by a Turing-machine, has the following analogue concerning the definability of formal languages[1]:

Theorem 1.1 *For every language $A \subseteq \Sigma^*$, the following conditions are equivalent:*

(i) A is definable by a system of regular equations,

(ii) A is definable by a μ-regular expression,

(iii) A is accepted by a pushdown-store automaton.

[1]In characterizations of context-free languages in the literature, condition (ii) often is missing. This is true at least for Hopcroft/Ullman[5], Lewis/Papadimitriou[9], and Harrison[4]. Salomaa[15] has (ii), but in a somewhat less persipcious notation. Essentially, the pushdown store is just the procedure return stack by which the finite automata for the r_i organize the 'calling' of each other, i.e. their transitions labelled with variables. This offers a nice and simple way to introduce pushdown-store automata and to explain why an implementation of recursion μ needs a stack.

Here, *definable by a system of regular equations* means to be a component of the least solution of a system

$$\begin{array}{rcl} x_1 & = & r_1(x_1,\ldots,x_m) \\ \vdots & \vdots & \vdots \\ x_m & = & r_m(x_1,\ldots,x_m), \end{array}$$

where each r_i is a regular expression whose free variables are among the pairwise distinct recursion variables $x_1,\ldots,x_m$. Context-free grammars can be seen as systems $\{x_i = r_i\}_{1\leq i\leq m}$ of regular equations where the r_i are *-free and in disjunctive normal form. One can use more restrictive normal forms, such as Greibach's[3], or more general expressions, such as μ-regular ones, without changing the class of languages that are definable by systems of equations.

In the present paper, we will use μ-regular expressions as a naming system for context-free languages[2] to axiomatize pieces of their equational theory. This will be done in three steps:

First, we add to the algebraic properties of $+, 0, \cdot, 1$ the assumption that μ picks the least pre-fixed point of definable functions $\lambda x.r$, for each μ-regular expression $r(x, y_1,\ldots,y_n)$:

$$\forall y_1 \ldots \forall y_n\ (r[\mu x.r/x] \leq \mu x.r\ \wedge\ \forall x(r \leq x \rightarrow \mu x.r \leq x)). \tag{1}$$

To express minimality, we need universal Horn-axioms, with conditional (in)equations in the quantifier free part. The resulting theory will be called *Kleene algebra with least fixed points*, or *KAF* for short. (Actually, this is a misnomer, if we drop * from the language).

In a second step, we extend *KAF* by equational axioms to obtain a theory of *Kleene algebra with recursion*, *KAR*, that can model iteration * by recursion μ, a suitably restricted form of the least fixed point operator. In *KAR*, we relate μ with the algebraic operations $+$ and $\cdot$ by

$$\forall a, b.\ (\mu x(b + ax) = \mu x(1 + xa) \cdot b \quad \wedge \quad \mu x(b + xa) = b \cdot \mu x(1 + ax)). \tag{2}$$

(In the following, we often write a, b for free variables, while a_i will be a constant of Σ that may be added to the pure language, without adding any relations between these.) Using these equations and taking $\mu x(1 + ax)$ as a^*, the least fixed point properties for $\mu x(b + ax)$ and $\mu x(b + xa)$ in *KAF* are just the properties of * taken as axioms in Kozen's[8] theory *KA* of Kleene algebra. Hence, *KAR* is indeed a theory of *Kleene algebras*, with iteration generalized to recursion. While *KAR* proves all equations between regular expressions that are valid in the language interpretation, it cannot prove all valid equations between μ-regular expressions, nor can it be completed to do so.

In spite of this limitation, in a third step we look for larger fragments of the equational theory of context-free languages that have natural axioms. We note that

[2]Niwinski[10] investigates the hierarchy of ω-languages obtained by nested least and largest fixed point operators.

S.Greibach's way to eliminate left recursion in context-free grammars relies on an equivalence between grammars that can compactly be expressed as an equation schema between μ-regular expressions r, s, possibly containing x:

$$\mu x(s + rx) = \mu x(\mu y(1 + yr) \cdot s) \quad \wedge \quad \mu x(s + xr) = \mu x(s \cdot \mu y(1 + ry)). \tag{3}$$

Adding this schema to *KAF* gives a theory *KAG*, which extends *KAR*. It remains to be seen to what extent common reasoning about the equivalence between context-free grammars can be carried out within *KAG*.

Equations (2) can also be read as claiming continuity properties about $+$ and $\cdot$. We consider continuous models of *KAF* and relate these to Conway's[2] notion of ***Standard Kleene Algebra***. It will be shown that all continuous models of *KAF*, in particular the standard interpretations, satisfy the identities (3). Moreover, an equation between μ-regular expressions that is valid in the interpretation by context-free languages holds universally in the continuous models of *KAF*.

2 Kleene algebras with least fixed points

Kleene[6] introduced regular expressions and showed that their equivalence (i.e. equality in the language interpretation) is decidable via finite automata. Several attempts have been made to give a complete axiomatization of this equivalence, notably by Salomaa[14], Conway[2], Pratt[12], and Kozen[8]. The main obstacle was Redko's[13] result that there can be no finite axiomatization by means of equations between regular expressions. Recently, Kozen[8] presented the following axiomatization using Horn-formulas.

Definition 2.1 (D. Kozen) The theory of *Kleene algebras*, *KA*, stated in the language $\{+, 0, \cdot, 1, ^*\}$, consists of (1) the theory of idempotent semirings, which says that

- $+$ is associative, commutative, idempotent, and has 0 as neutral element,
- $\cdot$ is associative and has 1 as neutral element from both sides,
- 0 is an annihilator for $\cdot$ from both sides, i.e. $\forall x(0 \cdot x = 0 = x \cdot 0)$,
- $\cdot$ distributes over $+$ from both sides,

and (2) the following assumptions about *, universally quantified over a, b:

$$1 + aa^* \leq a^*, \quad \text{and} \quad \forall x(b + ax \leq x \rightarrow a^*b \leq x), \tag{4}$$

$$1 + a^*a \leq a^*, \quad \text{and} \quad \forall x(b + xa \leq x \rightarrow ba^* \leq x). \tag{5}$$

We use $a \leq b$ as shorthand for $b = (a + b)$. Since $+$ is idempotent, associative, and commutative, the relation $\leq$ is reflexive, transitive, and anti-symmetric, i.e. a partial ordering.

Several other notions of Kleene algebra have been studied in the literature; c.f. Conway[2], Büchi[1], and Pratt[11, 12]. Kozen[8], Theorem 5.5, shows that *KA* is equationally complete with respect to the language interpretation: for all closed regular expressions r and s over Σ,

$$KA \vdash r = s \qquad \text{if and only if} \qquad \mathcal{REG}_\Sigma \models r = s,$$

where $\vdash$ is provability in first-order logic.

A model of the equational theory of $\mathcal{REG}_\Sigma$ is called a *regular algebra*. By Theorem 1.1, the closed μ-regular expressions define exactly the context-free languages. Let $\mathcal{CFL}_\Sigma$ be the class of context-free languages over Σ, and let *μ-regular algebra* (over Σ) be the equational theory of $\mathcal{CFL}_\Sigma$. We are interested in subtheories of μ-regular algebra that can be axiomatized in the language of μ-regular expressions. In the rest of this section, we will consider a very basic theory of this kind, fixing μ to be a least fixed point operator.

We want to look at some ways to define * in terms of μ, and for that reason *we will not use * in μ-regular expressions henceforth*, but only $+$, $\cdot$, 0, 1, and μ. The models of our theory are certain structures $\mathcal{M} = (M, +, 0, \cdot, 1, \mu)$, where $(M, +, 0, \cdot, 1)$ is a first-order structure, and μ a functional on M that associates to every definable function $f : M \to M$ an element $\mu(f)$ of M, the least pre-fixed point of f.

Definition 2.2 The theory of *Kleene Algebra with least fixed points*, *KAF*, consists of (i) the theory of idempotent semirings (as above), and (ii) the following schemata of least pre-fixed points: for all μ-regular expressions r,

$$r[\mu x.r/x] \leq \mu x.r, \quad (6) \qquad \text{and} \qquad \forall x(r \leq x \to \mu x.r \leq x). \quad (7)$$

A more accurate name would be *idempotent semiring with least fixed points*, in particular since we dropped *. The expected monotonicity and fixed-point properties follow easily from the axioms for μ:

Proposition 2.3 *For all μ-regular expressions r,s, and variables $y \not\equiv x \not\equiv z$,*

$$KAF \vdash \forall x(s \leq r) \to \mu x.s \leq \mu x.r, \quad (8)$$
$$KAF \vdash \forall y \forall z\, (y \leq z \to \mu x.r[y/v] \leq \mu x.r[z/v]), \quad (9)$$
$$KAF \vdash \mu x.r = r[\mu x.r/x]. \quad (10)$$

Proof To show (8), from $\forall x(s \leq r)$ we first get $s[\mu x.r/x] \leq r[\mu x.r/x] \leq \mu x.r$, using (6) for the second step. With (7), this then gives $\mu x.s \leq \mu x.r$. To show (9) for $r(x, v)$, note that by induction we may assume $y \leq z \to \forall x(r(x, y) \leq r(x, z))$, using monotonicity of the regular operations for μ-free expressions r. By (8), we get $\mu x.r(x, y) \leq \mu x.r(x, z)$. Finally, for (10), from axiom (6) we get $r[r[\mu x.r/x]/x] \leq r[\mu x.r/x]$ by monotonicity, and by axiom (7) this implies $\mu x.r \leq r[\mu x.r/x]$. □

Let us now see whether we can define Kleene's iteration * in *KAF*. At first sight, Kozen's axioms for * are instances of our μ-axioms for $r(a, b, x) := (b + ax)$,

$$b + a \cdot \mu x(b + ax) \leq \mu x(b + ax), \qquad \text{and} \qquad \forall x(b + ax \leq x \to \mu x(b + ax) \leq x).$$

Writing a^*b for $\mu x(b + ax)$ these are just the first (with $b = 1$) and second part of (4). Similarly, we obtain (5) by taking $\mu x(b + xa)$ for ba^*.

However, this does not give *one* definition for a^*, by taking $b = 1$, but two of them: the *right iteration* $\mu x(1 + ax)$ and the *left iteration* $\mu x(1 + xa)$ of a, respectively. Since $\cdot$ need not be commutative, in some models of *KAF* it can make a difference whether we iterate a to the left or to the right. But in the standard interpretations of the Introduction, both iterations of a coincide, and in fact are equal not only to the *both-sided iteration* $\mu x(1 + ax + xa)$ of a, but also to $\mu x(1 + a + xx)$, the *reflexive transitive closure* of a. It seems that in *KAF*, we can only prove part of this:

Proposition 2.4 *KAF* $\vdash \mu x(1+ax)+\mu x(1+xa) \leq \mu x(1 + ax + xa) \leq \mu x(1 + a + xx)$

Proof The first inequation is obvious. For the second, let $\bar{x}$ be $\mu x(1 + a + xx)$. Then (i) $1 \leq \bar{x}$, (ii) $a \leq \bar{x}$, and (iii) $\bar{x}\bar{x} \leq \bar{x}$. By monotonicity of $+$ and $\cdot$, we get

$$1 + a\bar{x} + \bar{x}a \leq_{(ii)} 1 + \bar{x}\bar{x} + \bar{x}\bar{x} \leq_{(iii)} 1 + \bar{x} + \bar{x} \leq_{(i)} \bar{x} + \bar{x} + \bar{x} = \bar{x}.$$

By the minimality axiom (7), this implies $\mu x(1 + ax + xa) \leq \bar{x}$. □

We note that Conway has given a finite nonstandard model of regular algebra where a^* is *not* the reflexive transitive closure of a. (Such models are excluded by *ACT* and *KA*.) To understand better why the various definitions of a^* coincide in the language and in the relation interpretation, we now look at a specific class of models of *KAF*.

3 Continuous models of KAF

As is well-known, an n-ary function $f : M^n \to M$ on a complete partial order $(M, 0, \leq)$ is *Scott-continuous* if and only if for all i and all parameters $b_1, \ldots, b_{i-1}, b_{i+1}, \ldots, b_n$,

$$f(b_1, \ldots, b_{i-1}, \bigsqcup D, b_{i+1}, \ldots, b_n) = \bigsqcup_{d \in D} f(b_1, \ldots, b_{i-1}, d, b_{i+1}, \ldots, b_n),$$

for each directed set $D \subseteq M$. Moreover, every continuous function $f : M \to M$ has a least fixed-point $\mu x.f(x)$, which, by Kleene's fixed point theorem, is

$$\mu x.f(x) := \bigsqcup_{n \in \omega} f_n(0), \qquad \text{where } f_0(x) = 0,\ f_{k+1}(x) := f(f_k(x)). \tag{11}$$

Definition 3.1 An idempotent semiring $(M, +, 0, \cdot, 1)$ is *continuous*, if the partial ordering $\leq$ induced by $+$ makes $(M, \leq, 0)$ a complete partial order, and $+$ and $\cdot$ are continuous functions with respect to the Scott topology on $(M, 0, \leq)$ and its cartesian product. A *continuous model of KAF* is a model $\mathcal{M}$ of *KAF* whose underlying semiring is continuous.

Proposition 3.2 *The family $\mathcal{L}_\Sigma$ of all languages over an alphabet Σ and the family $\mathcal{R}_K$ of all binary relations over a set K each form a continuous model of KAF.*

The proof is obvious and hence omitted. But note that although the class of context-free languages has enough fixed points to form a model of *KAF*, it is *not* a continuous model, because its partial ordering is incomplete (in fact, not even closed under unions of increasing chains).

The next lemma shows that the various definitions for a^* coincide not only in the standard interpretations, but in all (substructures of) continuous models of *KAF*.

Lemma 3.3 *If $\mathcal{M} \models KAF$ is continuous, $\mathcal{M} \models \forall a.\ \mu x(1+ax) = \mu x(1+a+xx)$.*

Proof By Proposition 2.4, it is sufficient to prove $\mathcal{M} \models \forall a.\ \mu x(1+a+xx) \leq \mu x(1+ax)$. Let $\bar{x}$ be $\mu x(1+ax)$, whence we have $1+a\bar{x} \leq \bar{x}$. From $1+a\bar{x} \leq \bar{x}$ we get (i) $1 \leq \bar{x}$, and then (ii) $a \leq \bar{x}$ using $a = a \cdot 1 \leq a \cdot \bar{x} \leq \bar{x}$. To show (iii) $\bar{x}\bar{x} \leq \bar{x}$, define $x_0 := 0$, $x_{n+1} := r^{\mathcal{M}}[x_n/x]$, where $r = (1+ax)$. By monotonicity of $+$ and $\cdot$, $x_n \leq x_{n+1}$, and by completeness of the partial order, there is a least upper bound $\sqcup_{n\in\omega} x_n \in M$ for $\{x_n \mid n \in \omega\}$. Since $\lambda x.r$ is continuous on M by the continuity of $+$ and $\cdot$, from Kleene's fixed-point theorem (11) we get $\mu x.r = \sqcup_{n\in\omega} x_n$. Finally, we need $x_n\bar{x} \leq \bar{x}$ for all n: this is clear for $n=0$, and by induction,

$$x_{n+1}\bar{x} = (1+ax_n)\bar{x} = \bar{x} + ax_n\bar{x} \leq \bar{x} + a\bar{x} \leq \bar{x},$$

using $a\bar{x} \leq 1 + a\bar{x} \leq \bar{x}$ in the last step. Hence, (iii) holds, since

$$\bar{x}\bar{x} = (\sqcup_{n\in\omega} x_n)\bar{x} = \sqcup_{n\in\omega}(x_n\bar{x}) \leq \sqcup_{n\in\omega}\bar{x} = \bar{x}.$$

From (i) - (iii) it follows that $\mu x(1+a+xx) \leq \bar{x}$, by minimality. □

Theorem 3.4 *Every continuous idempotent semiring $(M,+,0,\cdot,1)$ can uniquely be expanded to a continuous model $\mathcal{M} = (M,+,0,\cdot,1,\mu)$ of KAF.*

To show this, note that least fixed points of continuous functions exist by (11), and so it remains to note that nesting least fixed-points does not take us beyond the continuous functions:

Lemma 3.5 *Let $\mathcal{M}$ be a continuous idempotent semiring, and $r(x, y_1, \ldots, y_n)$ a continuous function from M^{n+1} to M. The function*

$$\lambda(a_1,\ldots,a_n)\ \mu b.r(b,a_1,\ldots,a_n)\ :\ M^n \to M$$

that picks the least fixed point of $\lambda b.r(b,a_1,\ldots,a_n)$ according to (11), is a continuous function on M^n.

Proof It is sufficient to show continuity in each dimension, so let r be $r(x,y)$, and $Y \subseteq M$ a directed set. To show

$$\mu x.r(x, \bigsqcup Y) = \bigsqcup_{y\in Y} \mu x.r(x,y),$$

we write $\bar{x} := \mu x.r(x, \bigsqcup Y)$, $x_y := \mu x.r(x,y)$ for $y \in Y$, and $x_Y := \bigsqcup_{y \in Y} x_y$.

Claim 1: $\bar{x} \geq x_Y$. By definition of $\bar{x}$ and monotonicity of r in y,

$$\bar{x} \geq r(\bar{x}, \bigsqcup Y) \geq \bigsqcup_{y \in Y} r(\bar{x}, y) \geq r(\bar{x}, y)$$

for each $y \in Y$, and hence $\bar{x} \geq \mu x.r(x,y) = x_y$ for each $y \in Y$. Since Y is directed, so is $\{r(\bar{x}, y) \mid y \in Y\}$, and taking the sup gives $\bar{x} \geq \bigsqcup_{y \in Y} x_y = x_Y$.

Claim 2: $\bar{x} \leq x_Y$. Note that

$$\begin{array}{rclcl} r(x_Y, \bigsqcup Y) & = & \bigsqcup_{y \in Y} r(x_Y, y) & = & \bigsqcup_{y \in Y} \bigsqcup_{z \in Y} r(x_z, y) \\ & = & \bigsqcup_{y \in Y, z \in Y} r(x_z, y) & = & \bigsqcup_{y \in Y} r(x_y, y) \\ & \leq & \bigsqcup_{y \in Y} x_y & = & x_Y, \end{array}$$

and hence $\bar{x} \leq x_Y$ by minimization. □

Conway[2] has studied several notions of Kleene algebra, and we next relate continuous models of *KAF* to his Standard Kleene Algebras, which are defined in terms of an infinitary summation Σ.

Definition 3.6 (Conway[2], Chapter 3) A *Standard Kleene Algebra*, or *$\mathcal{S}$-algebra*, is a structure $(M, \sum, 0, \cdot, 1, ^*)$ satisfying the following conditions, for arbitrary index sets I, J, J_i ($i \in I$) and elements $e_i \in M$, using $e^0 = 1$, $e^{n+1} = e^n \cdot e$:

$$\begin{array}{rclrcl} \sum_{i \in \emptyset} e_i & = & 0 & e \cdot 1 = & e & = 1 \cdot e \\ \sum_{i \in I}(\sum_{j \in J_i} e_j) & = & \sum_{j \in \cup_{i \in I} J_i} e_j & (e_1 \cdot e_2) \cdot e_3 & = & e_1 \cdot (e_2 \cdot e_3) \\ (\sum_{i \in I} e_i) \cdot (\sum_{j \in J} e_j) & = & \sum_{(i,j) \in I \times J} e_i \cdot e_j & e^* & = & \sum_{n \in \omega} e^n. \end{array}$$

The binary addition is defined by $e_1 + e_2 := \sum_{i \in \{1,2\}} e_i$. The properties of $\sum$ make $(M, +, 0, \cdot, 1)$ an idempotent semiring, and hence $e_1 \leq e_2 :\leftrightarrow e_1 + e_2 = e_2$ induces a partial order $\leq$ on M. According to Kozen, "$\mathcal{S}$-algebras are defined in terms of an infinitary summation operator $\sum$, whose sole purpose, it seems, is to define *". Actually, there is more to $\sum$ than defining *: by the properties of $\sum$, M is a complete semi-lattice with respect to $\leq$, and the operations $+$ and $\cdot$ are Scott-continuous with respect to $\leq$.

Proposition 3.7 *Every $\mathcal{S}$-algebra $\mathcal{M}$ can be expanded to a continuous model of KAF, by defining a μ-operator via*

$$(\mu x.r)^{\mathcal{M}} := \sum_{n \in \omega} r_n^{\mathcal{M}} \qquad \textit{with } r_0^{\mathcal{M}} := 0, \; r_{n+1}^{\mathcal{M}} := r^{\mathcal{M}}[r_n^{\mathcal{M}}/x].$$

The structure $(\mathcal{M}, \mu)$ satisfies $\forall a(\mu x(1 + ax) = a^*)$.

Proof By Kleene's fixed point theorem (11), the defined μ makes $\mu x.r$ the least fixed point of $\lambda x.r$ on $\mathcal{M}$. This implies that the μ-axioms of *KAF* are satisfied. Continuity and the semiring properties follow from Conway's axioms about $\sum$. Finally, $a^* = \sum_{n \in \omega} a^n = \mu x(1 + ax)^{\mathcal{M}}$. □

Conway[2] (Chapter 4, Theorem 6) shows that $\mathcal{L}_\Sigma$, the family of all formal languages over the alphabet Σ, is the free $\mathcal{S}$-algebra in the generators Σ. We now give a similar characterization for $\mathcal{CFL}_\Sigma$ and the class of substructures of continuous models of KAF. The construction needs a bit more care than the one for $\mathcal{S}$-algebras, since models of KAF need not even be closed under suprema of monotone sequences, as $\mathcal{CFL}_\Sigma$ shows.

Definition 3.8 A subset $A \subseteq M$ of a structure $\mathcal{M}$ for the language $\{+, 0, \cdot, 1, \mu\}$ is a *set of generators for* M if every element of M is the value $r^{\mathcal{M}}[a_1, \ldots, a_n]$ of a pure μ-regular expression $r(x_1, \ldots, x_n)$ with parameters $a_1, \ldots, a_n$ from A. If $\mathcal{C}$ is a class of structures, $\mathcal{M} \in \mathcal{C}$ is *free for* $\mathcal{C}$ in its set $A \subseteq M$ of generators, if for all elements $a_1, \ldots, a_n$ of A and all μ-regular expressions s and t,

$$\mathcal{M} \models (s = t)[a_1/x_1, \ldots, a_n/x_n] \quad \Rightarrow \quad \mathcal{C} \models \forall x_1 \ldots \forall x_n (s = t).$$

This means that in a free structure for a class of models only those relations between generators hold that are universally valid in the class.

Theorem 3.9 *The algebra $\mathcal{CFL}_\Sigma$ of all context-free languages over the finite alphabet Σ is free for the class of substructures of continuous models of KAF with $|\Sigma|$ many generators.*

Proof By Theorem 1.1, $\mathcal{CFL}_\Sigma$ is just the substructure of $\mathcal{L}_\Sigma$ that consists of the μ-regularly definable languages (that is, definable without parameters other than the $a \in \Sigma$). As $\mathcal{L}_\Sigma$ is an $\mathcal{S}$-algebra, by Proposition 3.7 we know that $\mathcal{CFL}_\Sigma$ is a substructure of a continuous model of KAF. We assume $\Sigma = \{a_1, \ldots, a_n\}$ and write $\mathcal{CF}$ instead of $\mathcal{CFL}_\Sigma$.

Let r and s be pure regular expressions such that $\mathcal{CF} \models (r = s)[\{a_1\}, \ldots, \{a_n\}]$, and $\mathcal{M} = (M, +, 0, \cdot, 1, \mu)$ be a continuous model of KAF. To show $\mathcal{M} \models \forall x_1 \ldots x_n . r = s$, let $b_1, \ldots, b_n$ be elements of $\mathcal{M}$. It is sufficient to present a homomorphism from $\mathcal{CF}$ to $\mathcal{M}$, mapping the atoms $\{a_i\}$ to the b_i.

Define $^{-} : \mathcal{CF} \to \mathcal{M}$ by putting $\overline{L} := \bigsqcup\{\sum_{w \in E} \overline{w} \mid E \subseteq L \text{ is finite}\}$, for context-free languages $L \subseteq \Sigma^*$, where $\overline{v \cdot w} := \overline{v} \cdot \overline{w}$, $\overline{a_i} := b_i$, and $\overline{\epsilon} := 1$. We leave it to the reader to check that for all languages L_1, L_2,
Claim 1: $\overline{L_1 + L_2} = \overline{L_1} + \overline{L_2}$, $\quad \overline{L_1 \cdot L_2} = \overline{L_1} \cdot \overline{L_2}$, $\quad$ and $\quad \overline{\emptyset} = 0^{\mathcal{M}}$, $\quad \overline{\{\epsilon\}} = 1^{\mathcal{M}}$.

To see that $^{-}$ is a homomorphism, we show the following
Claim 2: For every μ-regular expression $r(x_0, \ldots, x_n)$ and all languages $L_0, \ldots, L_n$ in $\mathcal{CFL}_\Sigma$:

(i) $\overline{r^{\mathcal{CF}}[L_0, \ldots, L_n]} = r^{\mathcal{M}}[\overline{L_0}, \ldots, \overline{L_n}]$,

(ii) $\overline{r_k^{\mathcal{CF}}[L_1, \ldots, L_n]} = r_k^{\mathcal{M}}[\overline{L_1}, \ldots, \overline{L_n}]$, for all $k \in \omega$, where for all models $\mathcal{A}$ and $L_i \in A$,

$$r_0^{\mathcal{A}}[L_1, \ldots, L_n] := 0^{\mathcal{A}} \text{ and } r_{k+1}^{\mathcal{A}}[L_1, \ldots, L_n] := r^{\mathcal{A}}[r_k^{\mathcal{A}}[L_1, \ldots, L_n], \ldots, L_1, \ldots, L_n].$$

Proof by induction on the nesting depth of μ's in r. If r has μ-depth 0, (i) is clear by Claim 1, as is (ii) for $k = 0$. By induction, one has

$$\begin{aligned}
\overline{r_{k+1}^{\mathcal{CF}}[L_1,\ldots,L_n]} &= \overline{r^{\mathcal{CF}}[r_k^{\mathcal{CF}}[L_1,\ldots,L_n],L_1,\ldots,L_n]} \\
&= r^{\mathcal{M}}[\overline{r_k^{\mathcal{CF}}[L_1,\ldots,L_n]},\overline{L_1},\ldots,\overline{L_n}] \quad \text{(by (i))} \\
&= r^{\mathcal{M}}[r_k^{\mathcal{M}}[\overline{L_1},\ldots,\overline{L_n}],\overline{L_1},\ldots,\overline{L_n}] \quad \text{(by induction)} \\
&= r_{k+1}^{\mathcal{M}}[\overline{L_1},\ldots,\overline{L_n}],
\end{aligned}$$

and hence (ii) is shown. For μ-depth $k+1$, consider $\mu x.r$ where $r(x, x_0, \ldots, x_n)$ has μ-depth at most k. To see part (i), we calculate

$$\begin{aligned}
\overline{(\mu x.r)^{\mathcal{CF}}[L_0,\ldots,L_n]} &= \overline{\bigcup_{k\in\omega} r_k^{\mathcal{CF}}[L_0,\ldots,L_n]} \\
&= \bigsqcup\{\textstyle\sum_{w\in E}\overline{w} \mid E \subseteq \bigcup_{k\in\omega} r_k^{\mathcal{CF}}[L_0,\ldots,L_n] \text{ finite}\} \\
&= \bigsqcup\bigcup_{k\in\omega}\{\textstyle\sum_{w\in E}\overline{w} \mid E \subseteq r_k^{\mathcal{CF}}[L_0,\ldots,L_n] \text{ finite}\} \\
(*) \quad &= \bigsqcup_{k\in\omega} \overline{r_k^{\mathcal{CF}}[L_0,\ldots,L_n]} \\
&= \bigsqcup_{k\in\omega}(r_k^{\mathcal{M}}[\overline{L_0},\ldots,\overline{L_n}]) \quad \text{(by (ii) for } r) \\
&= (\mu x.r)^{\mathcal{M}}[\overline{L_0},\ldots,\overline{L_n}],
\end{aligned}$$

using in (*) that $\bigsqcup(\bigcup_{i\in\omega} M_i) = \bigsqcup_{i\in\omega}(\bigsqcup M_i)$ for ascending chains $M_0 \subseteq M_1 \subseteq \ldots \subseteq M$. Part (ii) follows from (i) for $\mu x.r$, exactly as shown for μ-depth 0 above.

By induction on pure μ-regular expressions, these claims yield

$$\overline{r^{\mathcal{CF}}[L_1,\ldots,L_n]} = r^{\mathcal{M}}[\overline{L_1},\ldots,\overline{L_n}] \tag{12}$$

for every $r(x_1,\ldots,x_n)$; using $L_i = \{a_i\}$ shows that $^-$ is a homomorphism.

If $\mathcal{N}$ is the substructure of $\mathcal{M}$ that is μ-regularly generated by the elements b_i, this homomorphism is onto $\mathcal{N}$. If, in addition, $\mathcal{N}$ is a free structure for the class of substructures of continuous models of *KAF*, then $^-$ is an isomorphism. □

4 Kleene algebras with recursion

In Section 2 we have seen that Kozen's axiom for * could *almost* be seen as instances of the least fixed point properties of *KAF*. The problem was that $\mu x(b + ax)$ and $\mu x(b + xa)$, intended to represent $a^* \cdot b$ and $b \cdot a^*$, do not agree for $b = 1$ to give one candidate for a^*.

Reading the $\mu x(b + xa)$ as a recursive program, it is clear that in terminating executions essentially we *first* do b *and then* repeatedly do a, i.e. the finite behaviour of recursive programs satisfies $\mu x(b + xa) = b \cdot \mu x(1 + ax)$. The following makes this link between recursion and iteration explicit.

Definition 4.1 The theory of *Kleene Algebra with recursion*, *KAR*, is the extension of the above theory *KAF* by the following assumptions, for all a and b:

$$b \cdot \mu x(1 + ax) \leq \mu x(b + xa) \tag{13}$$

$$\mu x(1 + xa) \cdot b \leq \mu x(b + ax) \tag{14}$$

It seems plausible that these axioms are independent of KAF and of each other, because the minimality conditions in axioms (4) and (5) are independent over KA, as shown by Kozen[7].

Proposition 4.2 *In KAF, (13) and (14) together are equivalent to*

$$\mu x(b+xa) = b\cdot\mu x(1+ax) \tag{15}$$
$$\mu x(b+ax) = \mu x(1+xa)\cdot b \tag{16}$$

Proof To see that (15) and (16) follow from the axioms, first note that by taking $b=1$ in (13) and (14), we obtain

$$KAR \vdash \mu x(1+ax)=\mu x(1+xa). \tag{17}$$

The missing inequations can now be obtained by substituting (17) into the following:

$$KAF \vdash \mu x(b+xa) \leq b\cdot\mu x(1+xa) \tag{18}$$
$$KAF \vdash \mu x(b+ax) \leq \mu x(1+ax)\cdot b \tag{19}$$

To show (19), we use $\bar{x} := \mu x(1+ax)$. By definition, we have $1\leq\bar{x}$ and $a\bar{x}\leq\bar{x}$, and thus

$$b+a\bar{x}b\leq b+\bar{x}b\leq(1+\bar{x})b\leq\bar{x}b.$$

This implies $\mu x(b+ax)\leq\bar{x}b$ by minimality. By symmetry, we also have (18). □

We are now ready to see that in KAR all of the previously discussed iterations of a coincide.

Lemma 4.3 $KAR\vdash \mu x(1+ax)=\mu x(1+xa)=\mu x(1+ax+xa)=\mu x(1+a+xx)$.

Proof The claim reduces by (17) and Proposition 2.4 to $\mu x(1+a+xx)\leq\mu x(1+ax)$. Since obviously $1+a\leq\mu x(1+ax)$, we only have to show that

$$\mu x(1+ax)\cdot\mu x(1+ax)\leq\mu x(1+ax), \tag{20}$$

and then use minimality of $\mu x(1+a+xx)$. We can now mimick an argument of Pratt[12], abbreviating $\mu x(b+ax)$ by $[a^*,b]$ for clarity. By the μ-axioms of KAF, we have

$$b+a[a^*,b] \leq [a^*,b], \tag{21}$$
$$b+ax\leq x \rightarrow [a^*,b]\leq x. \tag{22}$$

From (21) and $[a^*,b]\leq[a^*,b]$ we get $[a^*,b]+a[a^*,b]\leq[a^*,b]$, hence $[a^*,[a^*,b]]\leq[a^*,b]$ by (22). Equations (17) and (14) yield $[a^*,1]\cdot b\leq[a^*,b]$, and by instantiating b to $[a^*,b]$ we get

$$[a^*,1]\cdot[a^*,b]\leq[a^*,[a^*,b]]\leq[a^*,b].$$

Taking $b=1$ here gives transitivity $[a^*,1]\cdot[a^*,1]\leq[a^*,1]$ of $[a^*,1]$, which is (20). □

From equations (16) and (17) and the discussion of Kozen's axioms in Section 2, we conclude that all the μ-definable iterations above have the properties of Kleene's iteration operator *:

Corollary 4.4 *Under the translation $a^* := \mu x(1 + ax)$, KA is a subtheory of KAR.*

By Kozen's completeness theorem for *KA*, it follows that *every* equation between regular expressions that is valid in the language interpretation can be proven in *KAR*.

It is natural to ask whether *KAR* is complete for the equational theory of context-free languages, i.e. whether every equation between *μ-regular* expressions valid in the language interpretation is provable in *KAR*. This is not the case: the equivalence between context-free grammars is not recursively enumerable (combine Corollary 1 with the proof of Theorem 8.11 of Hopcroft/Ullman[5], p. 192 and p. 203). Since there is an effective translation between context-free grammars and μ-regular expressions (in a proof of Theorem 1.1), the equational theory of context-free languages in terms of μ-regular expressions is not axiomatizable at all.

Clearly, the standard interpretations give models not only of *KAF*, but also of *KAR*. It follows from Theorem 5.3 below that every continuous model of *KAF* is a model of *KAR*.

5 Elimination of left-recursion

The axioms of *KAR* are special instances of S. Greibach's trick for eliminating left-recursive grammar rules by right-recursive ones and vice versa. We will now express the core of Greibach's method as an equation between μ-regular expressions that appears to be independent of *KAR*, but is valid in all continuous models of *KAF*.

Definition 5.1 The theory of *Kleene Algebra with Greibach's inequations for recursion*, *KAG*, is the extension of the theory *KAF* by the following axiom schemata: for all μ-regular expressions r and s with y not free in r, add the universal closures of

$$\mu x(s \cdot \mu y(1 + ry)) \leq \mu x(s + xr) \tag{23}$$

$$\mu x(\mu y(1 + yr) \cdot s) \leq \mu x(s + rx) \tag{24}$$

Note that *KAG* extends *KAR*: if x is not free in s and r, the μx on the left hand side of (23) and (24) can be dropped; for variables r and s this gives axioms (15) and (16) of *KAR*.

Proposition 5.2 *In KAF, the schemata (23) and (24) together are equivalent to the following two schemata for μ-regular expressions r and s with y not free in r:*

$$\mu x(s \cdot \mu y(1 + ry)) = \mu x(s + xr), \tag{25}$$

$$\mu x(\mu y(1 + yr) \cdot s) = \mu x(s + rx). \tag{26}$$

Proof By the above remark, it is sufficient to show

$$KAR \vdash \mu x(xr + s) \leq \mu x(s \cdot \mu y(1 + ry)). \tag{27}$$

Let $\bar{x}$ be $\mu x(s(x) \cdot \mu y(1 + r(x) \cdot y))$, suppressing other free variables in the notation. We prove

$$\bar{x} \cdot r(\bar{x}) + s(\bar{x}) \leq \bar{x}, \tag{28}$$

which implies (27) by minimization. Let $\bar{y}$ be $\mu y(1+r(\bar{x})\cdot y)$. By properties of $\bar{x}$ and $\bar{y}$, we get $s(\bar{x})\cdot(1+r(\bar{x})\cdot\bar{y}) \leq s(\bar{x})\cdot\bar{y} \leq \bar{x}$, hence $s(\bar{x}) \leq \bar{x}$ and $s(\bar{x})\cdot r(\bar{x})\cdot\bar{y} \leq \bar{x}$. It remains to show $\bar{x}\cdot r(\bar{x}) \leq \bar{x}$. From (15), (16) and (17) we have

$$KAR \vdash \forall a.\ \mu y(1+ay)\cdot a = \mu y(a+ay) = a\cdot\mu y(1+ay).$$

Thus, $\bar{y}\cdot r(\bar{x}) = \mu y(1+r(\bar{x})\cdot y)\cdot r(\bar{x}) = r(\bar{x})\cdot\mu y(1+r(\bar{x})\cdot y) = r(\bar{x})\cdot\bar{y}$, and hence (28) follows by $\bar{x}\cdot r(\bar{x}) = s(\bar{x})\cdot\bar{y}\cdot r(\bar{x}) = s(\bar{x})\cdot r(\bar{x})\cdot\bar{y} \leq \bar{x}$. □

To explain why we call (23) and (24) *Greibach's axioms*, let us recall S. Greibach's[3] way to eliminate left recursive rules from a context-free grammar (c.f. [5, 4]). Suppose

$$A = Av_1 + \cdots + Av_n + w_1 + \cdots + w_m$$

combines all the A-rules of the grammar, where the v_i and w_j are concatenations of grammar symbols, and no w_j begins with the variable A. Using a fresh variable B, the rules corresponding to the above equation are replaced by those corresponding to the equations

$$\begin{aligned} A &= w_1 + \cdots + w_m + w_1B + \cdots + w_mB, \\ B &= v_1 + \cdots + v_n + v_1B + \cdots + v_nB. \end{aligned}$$

Writing r for $(v_1+\cdots+v_m)$ and s for $(w_1+\cdots+w_m)$, we just have replaced the equation $A = Ar + s$ by the equations $A = s + sB$ and $B = r + rB$. Since we are dealing with *least* solutions, we have in fact replaced $\mu A(Ar+s)$ by $\mu A(s + s\cdot\mu B(r+rB))$, i.e. we used

$$\mu A(s + Ar) = \mu A(s + s\cdot\mu B(r+rB)). \tag{29}$$

As B was fresh, in KAR we can reformulate the right hand side as

$$\begin{aligned} \mu A(s + s\cdot\mu B(r+rB)) &= \mu A(s\cdot(1+\mu B(r+rB))) \\ &= \mu A(s\cdot(1+r\cdot\mu B(1+rB))) \\ &= \mu A(s\cdot\mu B(1+rB)), \end{aligned}$$

and (29) becomes $\mu A(s+Ar) = \mu A(s\cdot\mu B(1+rB))$, an instance of (25)[3]. Of course, for suitable s only, in particular those corresponding to the above constraint that no w_j begins with A, can we conclude that $\mu A(s\cdot\mu B(1+rB))$ is not left-recursive (with respect to μA).

It is well known that for *fixed* sets r and s of words, the least solution of $x = xr+s$ is sr^*. This is just the content of axiom (15) of KAR, using the right-recursive definition $\mu x(1+ax)$ for the iteration a^*.[4]

[3]This corresponds to the replacement of the original grammar rules for A by the new equations

$$A = w_1B + \cdots + w_mB, \qquad B = 1 + v_1B + \cdots + v_nB.$$

These are simpler than the standard ones given above, but contain an undesired 'ϵ-rule' in $B = 1+rB$.

[4]Note that $\mu x(b+ax) = a^*\cdot b$ can be read as *'tail recursion is* implementable by *iteration'*.

Greibach's 'elimination of left recursion' is based on a more general fact, expressed in (25): the least solution of $x = xr + s$ is *the least solution of* $x = sr^*$, even if r and s μ-regularly *depend* on x. We finally prove that this is a true identity not only in the language interpretation, but in all continuous models of *KAF*:

Theorem 5.3 *Every continuous model of KAF is a model of KAG.*

Proof Let $\mathcal{M}$ be a continuous model of *KAF*. By symmetry, it is sufficient to show that $\mathcal{M}$ satisfies (23). Let $\bar{x}$ be $\mu x(x \cdot r(x) + s(x))$ in $\mathcal{M}$, under a given assignment for free variables that are not explicitly mentioned. We will show that $s(\bar{x}) \cdot \mu y(1 + r(\bar{x})y) \leq \bar{x}$, which gives (23) by minimization. Let $\bar{y} := \mu y(1 + r(\bar{x})y)$ in $\mathcal{M}$. Since $\bar{x} \cdot r(\bar{x}) \leq \bar{x}$ and $s(\bar{x}) \leq \bar{x}$ by the choice of $\bar{x}$, we get $s(\bar{x}) \cdot \bar{y} \leq \bar{x}\bar{y}$. To show the remaining inequation $\bar{x}\bar{y} \leq \bar{x}$, note that since $\mathcal{M}$ is continuous, $\bar{y} = \bigsqcup_{n \in \omega} y_n$ with $y_0 := 1 + r(\bar{x}) \cdot 0 = 1$, $y_{n+1} := 1 + r(\bar{x}) \cdot y_n$. From $\bar{x}y_0 \leq \bar{x}$ we get $\bar{x}y_{n+1} = \bar{x} + \bar{x} \cdot r(\bar{x}) \cdot y_n \leq \bar{x} + \bar{x}y_n \leq \bar{x}$ by induction, so $\bar{x}\bar{y} = \bigsqcup_{n \in \omega} \bar{x}y_n \leq \bar{x}$ by continuity. □

6 Open problems

We have demonstrated that μ-regular expressions can be useful to study questions about equality and subsumption between context-free languages in an algebraic and logical manner. The two main arguments in this exercise have been (i) minimality of least fixed points on a continuous idempotent semiring and (ii) induction on the number of iterations of a definable monotone function on this ring.

The axiom systems $KAF \subseteq KAR \subseteq KAG$ separate the basic properties of least fixed points from aspects of continuity that can be expressed as equations between μ-regular expressions. But we have left open all non-trivial questions, for example:

(i) Is *KAF* strictly weaker than *KAR*, and *KAR* strictly weaker than *KAG*?

(ii) Is *KAF*, or *KAG* relative to *KAF*, finitely axiomatizable?

(iii) Let $\mathcal{K}$ be a model of *KAR*. Can the algebra $\mathcal{K}_n$ of $n \times n$-matrices over $\mathcal{K}$ be expanded to a model of *KAR* by adding an appropriate μ? Is the same true for *KAG*? (C.f. Conway[2] for $\mathcal{S}$-algebras and Kozen[8] for models of *KA*.)

(iv) Are there natural equations between μ-regular expressions that are valid in all continuous models of *KAF*, but go beyond *KAG*? Good candidates are those equations that arise by transforming a simultaneously regular definition into different μ-regular ones.

(v) Is the equational theory of *linear* languages, i.e. linear μ-regular expressions, more tractable than the general case?

Note that in order to show strictness in (i), we need non-continuous models in which the definable functions have least fixed points.

References

[1] J. R. Büchi. *Finite Automata, Their Algebras and Grammars.* Springer Verlag, New York, 1988.

[2] J. H. Conway. *Regular Algebra and Finite Machines.* Chapman and Hall, London, 1971.

[3] S. A. Greibach. A new normal-form theorem for context-free, phrase-structure grammars. *Journal of the Association of Computing Machinery*, 12:42–52, 1965.

[4] M. Harrison. *Introduction to Formal Languages.* Addison Wesley, Reading, Mass., 1978.

[5] J. Hopcroft and J. D. Ullman. *Formal Languages and their Relation to Automata.* Addison Wesley, Reading, Mass., 1979.

[6] S. C. Kleene. Representation of events in nerve nets and finite automata. In C. E. Shannon and J. McCarthy, editors, *Automata Studies*, pages 3–42. Princeton University Press, 1956.

[7] D. Kozen. On Kleene Algebras and Closed Semirings. In B. Rovan, editor, *15th Int. Symp. on Mathematical Foundations of Computer Science, Banská Bystrica, 1990*, LNCS 452, pages 26–47. Springer Verlag, 1990.

[8] D. Kozen. A completeness theorem for Kleene algebras and the algebra of regular events. In *6th Annual Symposium on Logic in Computer Science, July 15–18, 1991, Amsterdam.* Computer Society Press, Los Alamitos, CA, 1991.

[9] H. R. Lewis and C. H. Papadimitriou. *Elements of The Theory of Computation.* Prentice-Hall International, Inc., London, 1981.

[10] D. Niwinski. Fixed point characterizations of context-free ∞-languages. *Information and Computation*, 63:247 – 276, 1984.

[11] V. Pratt. Dynamic Algebras as a well-behaved fragment of Relation Algebras. In C.H.Bergmann, R.D.Maddux, and D.L.Pigozzi, editors, *Algebraic Logic and Universal Algebra in Computer Science*, volume 425 of *LNCS.* Springer Verlag, 1990.

[12] V. Pratt. Action logic and pure induction. In *Logic in AI. Proceedings JELIA'90, Univ. of Amsterdam*, volume 478 of *Lecture Notes in Artificial Intelligence, Subseries of LNCS*, pages 97–120. Springer Verlag, 1991.

[13] V. N. Redko. On defining relations for the algebra of regular events (russian). *Ukrain. Mat. Z.*, 16:120–126, 1964.

[14] A. Salomaa. Two complete axiom systems for the algebra of regular events. *Journal of the ACM*, 13:158–169, 1966.

[15] A. Salomaa. *Formal Languages.* Academic Press, Inc., New York and London, 1973.

EQUATIONAL SPECIFICATION OF ABSTRACT TYPES AND COMBINATORS

Karl Meinke
Department of Mathematics and Computer Science,
University College Swansea,
Singleton Park,
Swansea SA2 8PP,
Great Britain.

Abstract We introduce an algebraic framework for the equational specification of algebras of types and combinators. A categorical semantics for type specifications is given based on cofibrations of categories of algebras. It is shown that each equational type specification admits an initial model semantics, and we present complete inference systems for type assignments and equations.

0. INTRODUCTION

The algebraic theory of typed combinator systems originates independently in several places, for example Poigné [1986], Möller [1987] and Meinke [1990], [1992]. Common to most approaches is the idea of encoding type structure in the sort structure of a many-sorted universal algebra. In attempting to generalise the algebraic approach to types and combinators beyond simple propositional connectives such as $\times$ and $\rightarrow$ we are forced to take a more algebraic view of the technique of encoding types as sorts.

To illustrate the problems encountered with more complex type constructions, consider the construction of a cartesian product $\Pi_{i\in I} A_{f(i)}$, where I is some (possibly infinite) indexing set, $f : I \rightarrow \mathit{TYPE}$ is some indexing function, TYPE is a set of type names and for each $i \in I$, $A_{f(i)}$ is some set named by $f(i)$. Even if the construction is assumed to be predicative, i.e. $\Pi_{i\in I} A_{f(i)}$ is not one of the sets $A_{f(i)}$, the encoding of this set as a new carrier set of some sort within an algebra of types is not straightforward. In encoding the name of the new type in a sort set we are no longer dealing with a concrete syntactic object (a freely generated type expression) but with an abstract object, in this case involving an indexing function of the form $f : I \rightarrow \mathit{TYPE}$.

Thus from considerations about sorts interpreted as types, we are led to a theory of many-sorted universal algebras in which the sort sets themselves are sets of abstract objects having an algebraic structure. This structure can be syntactically fixed by a signature, semantically explicated by some class of algebras and even axiomatically specified, for example by means of equations.

In this paper we introduce a theory of abstract algebras of types, although we attempt to motivate our methods and results with examples of familiar types and combinators. Section 1 introduces the essential algebraic and category theoretic preliminaries. In Section 2 we formalise the syntax and semantics of abstract algebras of types using simple algebraic and categorical concepts. We consider equational specifications of algebras of types and show that every equational specification admits an initial model semantics. In Section 3 we prove the completeness of a type assignment calculus. In Section 4 we prove the completeness of an equational calculus for type specifications and give a concrete construction of the initial model of an equational type specification.

Our construction of a cofibration semantics for equational type specifications in Section 2 is the result of collaborative research with E.G. Wagner and further results on the semantics and specification of abstract types and combinators will appear in Meinke and Wagner [1992].

1. ALGEBRAIC AND CATEGORICAL PRELIMINARIES

We will not assume any specific knowledge of types or combinators. The novice may consult Hindley and Seldin [1986], more advanced works are Curry and Feys [1958] and Curry et al [1972].

The notation for many–sorted universal algebra that we use is taken from Meinke and Tucker [1992] where most of the basic technical definitions and results on algebra that we require may be found.

Our use of category theory is restricted to one essential construction, the Grothendieck cofibration construction. We define this construction together with some basic categories and functors that will be used later.

1.1. Definition. The category *Sig* of all signatures is defined as follows. An object of *Sig* is a many–sorted signature. A morphism $\theta : (S, \Sigma) \to (S', \Sigma')$ in *Sig* is a *signature morphism*, i.e. a pair $\theta = (\theta_S, \theta_\Sigma)$ where $\theta_S : S \to S'$ is any mapping and θ_Σ is an $S^* \times S$ indexed family of mappings

$$\theta_\Sigma = \langle\, \theta_\Sigma^{s_1 \ldots s_n,\, s_0} \;:\Sigma_{s_1 \ldots s_n,\, s_0} \;\to \Sigma'_{\theta_S(s_1) \ldots \theta_S(s_n),\, \theta_S(s_0)} \;\mid s_0, \ldots, s_n \in S \,\rangle.$$

Given signature morphisms $\phi : (S, \Sigma) \to (S', \Sigma')$, $\psi : (S', \Sigma') \to (S'', \Sigma'')$ we define the composition

$$\theta = \psi\,\phi : (S, \Sigma) \to (S'', \Sigma'')$$

by $\theta = (\theta_S, \theta_\Sigma)$, where $\theta_S : S \to S''$ is the composition $\theta_S = \psi_{S'}\,\phi_S$ in *Set*, and

$$\theta_\Sigma = \langle\, \theta_\Sigma^{s_1 \ldots s_n,\, s_0} \;:\Sigma_{s_1 \ldots s_n,\, s_0} \;\to \Sigma''_{\theta_S(s_1) \ldots \theta_S(s_n),\, \theta_S(s_0)} \;\mid s_0, \ldots, s_n \in S \,\rangle$$

where $\theta_\Sigma^{s_1 \ldots s_n,\, s_0}$ is the composition

$$\theta_\Sigma^{s_1 \ldots s_n,\, s_0} = \psi_{\Sigma'}^{\phi_S(s_1) \ldots \phi_S(s_n),\, \phi_S(s_0)}\;\phi_\Sigma^{s_1 \ldots s_n,\, s_0}$$

in *Set*.

The passage from an S–sorted signature Σ to the associated category $Alg(S, \Sigma)$ of all S–sorted Σ algebras and Σ homomorphisms is captured by the following functor

1.2. Definition. Define the functor $\mu : Sig^{op} \to Cat$ as follows. For any signature $(S, \Sigma) \in Sig$ define

$$\mu(S, \Sigma) = Alg(S, \Sigma).$$

Given a signature morphism $\theta : (S, \Sigma) \to (S', \Sigma')$ in *Sig* where $\theta = (\theta_S, \theta_\Sigma)$, we define the functor $\mu(\theta) : Alg(S', \Sigma') \to Alg(S, \Sigma)$.

For any algebra $A \in Alg(S', \Sigma')$ define $\mu(\theta)A \in Alg(S, \Sigma)$ as follows. For each sort $s \in S$, define the carrier set

$$(\mu(\theta)A)_s = A_{\theta_S(s)}.$$

For each sort $s \in S$ and any constant symbol $c \in \Sigma_{\lambda, s}$ define the constant

$$c_{\mu(\theta)A} = \theta_\Sigma^{\lambda, s}(c)_A.$$

For any $n \geq 1$, any sorts $s_0, \ldots, s_n \in S$ and any function symbol $f \in \Sigma_{s_1 \ldots s_n, s_0}$ define the function

$$f_{\mu(\theta)A} = \theta_\Sigma^{s_1 \ldots s_n, s_0}(f)_A.$$

For any Σ' homomorphism $\phi : A \to B$ in $Alg(S', \Sigma')$ define $\mu(\theta)\phi : \mu(\theta)A \to \mu(\theta)B$ in $Alg(S, \Sigma)$ by

$$(\mu(\theta)\phi)_s = \phi_{\theta_S(s)}.$$

In order to put together classes of algebras of different signature in a uniform way, in Section 2, we require the following construction from category theory known as a cofibration.

1.3. Definition. Let C be a category and *Cat* be the category of all small categories. Let $F : C^{op} \to Cat$ be any functor. We can form the category $Gr(F)$ termed the *(Grothendieck) cofibration of F*. The class of objects of $Gr(F)$ is the coproduct (disjoint sum) of the classes $|F(c)|$ for all $c \in C$, i.e.

$$|Gr(F)| = \coprod_{c \in C} |F(c)| = \{ \ (c, a) \ : \ c \in |C| \text{ and } a \in |F(c)| \ \}.$$

The morphisms of $Gr(F)$ are all pairs $(f, g) : (c, a) \to (c', a')$ such that $f : c \to c'$ is a morphism in C and $g : a \to (Ff)a'$ is a morphism in Fc.

Composition is defined for morphisms

$$(f_0, g_0) : (c_0, a_0) \to (c_1, a_1), \quad (f_1, g_1) : (c_1, a_1) \to (c_2, a_2)$$

as

$$(f_2, g_2) = (f_1, g_1)(f_0, g_0) : (c_0, a_0) \to (c_2, a_2)$$

where $f_2 : c_0 \to c_2$, $f_2 = f_1 f_0$, and $g_2 : a_0 \to (Ff_2)a_2$ in Fc_0 is given by $g_2 = (Ff_0)g_1 g_0$.

We require one simple result about initial objects under cofibrations.

1.4. Proposition. *Let C be any category, Cat be the category of all small categories and $F : C^{op} \to Cat$ be any functor. If C has an initial object and for each object $c \in |C|$ the category $F(c)$ has an initial object then the cofibration $Gr(F)$ has an initial object.*

Proof. It is easy to check that if $I \in |C|$ is initial in C and $i \in |F(I)|$ is initial in $F(I)$ then the pair $(I, i) \in |Gr(F)|$ is initial in $Gr(F)$. □

Further results on cofibrations may be found in Barr and Wells [1990].

2. EQUATIONAL TYPE SPECIFICATIONS

In this section we introduce the syntax and semantics of equational type specifications. An equational type specification consists of two components: (i) an equational specification of types; and, (ii) an equational specification of typed combinators over the type specification of (i). Each component has its own formal syntax given by a signature, and semantics given by a category of algebras. The semantics of a combinator specification will be given by a cofibered category of algebras indexed by the algebras which are possible semantics of the type specification. We begin by formalising the syntax of types within a signature.

2.1. Definition. By a *type signature* we mean a K-sorted signature Γ for some non-empty set K. In this context we term K a set of ***kinds*** and Γ a ***K-kinded type signature.***

2.2. Examples. (i) Let $K = \{ \ type \ \}$ and let Γ be defined by

$$\Gamma_{\lambda,\, type} = \{ \ NAT \ \}, \quad \Gamma_{type\ type,\, type} = \{ \ \times, \rightarrow \ \}$$

and $\Gamma_{w,\,s} = \emptyset$ for all other $w \in K^*$ and $s \in K$. In Γ we have a notation for types constructed from the single type constant NAT using just the propositional connectives $\times$ and $\rightarrow$.

(ii) Let

$$K = \{ \ nat,\ type^i,\ (nat \Rightarrow type^i) \ | \ i \in \mathbf{N} \ \}$$

and let Γ be the signature defined by

$$\Gamma_{\lambda,\, type^0} = \{ \ NAT \ \}, \quad \Gamma_{type^i\ type^i,\, type^i} = \{ \ \times^i, \rightarrow^i \ \},$$

$$\Gamma_{(nat \Rightarrow type^i),\, type^{i+1}} = \{ \ \Pi^i \ \}, \quad \Gamma_{(nat \Rightarrow type^i)\ nat,\, type^i} = \{ \ eval^{(nat \Rightarrow type^i)} \ \},$$

$$\Gamma_{\lambda,\, nat} = \{ \ 0 \ \}, \quad \Gamma_{nat,\, nat} = \{ \ +1 \ \}, \quad \Gamma_{\lambda,\, (nat \Rightarrow type^i)} = \{ \ f^i_j \ | \ j \in \mathbf{N} \ \}$$

for all $i \in \mathbf{N}$, and let $\Gamma_{w,\,s} = \emptyset$ for all other $w \in K^*$ and $s \in K$. Notice that nat occurs as a kind while NAT is a constant symbol of kind $type^0$. In Γ we have a notation for the construction of predicative cartesian product types of the kind discussed in Section 1.

From the point of view of algebra, we are only concerned with the semantics of types up to isomorphism. Thus, in principle we should allow *any* Γ algebra G as a possible semantics of a type signature Γ.

2.3. Definition. Let Γ be a K-kinded type signature. By a ***type algebra*** G we mean a Γ algebra $G \in Alg(\Gamma)$.

Next we introduce the notion of a (typed) ***combinator signature*** Σ over a type signature Γ and a family Y of sets of type variables. To allow for the possibility that not all the kinds in K are necessarily kinds of types (for example the kind nat in Example 2.2.(ii) simply provides a copy of arithmetic) we will allow an arbitrary subset $K' \subseteq K$ of kinds to be the only source of type names in Σ.

2.4. Definition. Let Γ be a K-kinded type signature, $Y = \langle Y_k \ | \ k \in K \rangle$ be any K-indexed family of sets of variables and let $K' \subseteq K$. By a ***combinator signature*** Σ over Γ, Y and K' we mean a $\bigcup_{k \in K'} T(\Gamma, Y)_k$-sorted signature Σ.

2.5. Examples. (*i*) Let Γ and K be as in Example 2.2.(*i*). Let Y be a K-indexed family of infinite sets of variables. Define Σ over Γ, Y and K by

$$\Sigma_{(\sigma\to\tau)\,\sigma,\,\tau} = \{\ eval^{(\sigma\to\tau)}\ \}$$

$$\Sigma_{(\sigma\times\tau),\,\sigma} = \{\ proj^{(\sigma\times\tau),\,\sigma}\ \},\quad \Sigma_{(\sigma\times\tau),\,\tau} = \{\ proj^{(\sigma\times\tau),\,\tau}\ \}$$

for all $\sigma,\,\tau \in T(\Gamma,\,Y)_{type}$ and $\Sigma_{w,\,s} = \emptyset$ for all other $w \in K^*$ and $s \in K$.

(*ii*) Let Γ and K be as in Example 2.2.(*ii*), let Y be a K-indexed family of infinite sets of variables, and let $K' = \{\ type^i \mid i \in \mathbf{N}\ \}$. Define Σ over Γ, Y and K' by

$$\Sigma_{(\sigma\to^i\tau)\,\sigma,\,\tau} = \{\ eval^{(\sigma\to^i\tau)}\ \}$$

$$\Sigma_{(\sigma\times^i\tau),\,\sigma} = \{\ proj^{(\sigma\times^i\tau),\,\sigma}\ \},\quad \Sigma_{(\sigma\times^i\tau),\,\tau} = \{\ proj^{(\sigma\times^i\tau),\,\tau}\ \}$$

for each $i \in \mathbf{N}$ and $\sigma,\,\tau \in T(\Gamma,\,Y)_{type^i}$. Let

$$\Sigma_{\Pi^i(\tau),\ eval\,(\tau,\,n)} = \{\ proj^{\Pi^i(\tau),\,n}\ \}$$

for each $i \in \mathbf{N}$, each $\tau \in T(\Gamma,\,Y)_{(nat\Rightarrow type^i)}$ and each $n \in T(\Gamma,\,Y)_{nat}$. Define $\Sigma_{w,\,s} = \emptyset$ for all other $w \in K'^*$ and $s \in K'$.

In the sequel we consider a fixed, but arbitrarily chosen K-kinded type signature Γ, K-indexed family Y of sets of variables, subset $K' \subseteq K$ and combinator signature Σ over Γ, Y and K'.

We must provide a semantical account of the parameterisation of the combinator signature Σ by the type signature Γ that is achieved by taking $\cup_{k\in K'} T(\Gamma,\,Y)_k$ to be the sort set for Σ. Since any type algebra $G \in Alg(\Gamma)$ can provide a semantical interpretation of type terms, for each such semantics G we obtain a derived combinator signature $\Sigma(G)$ in which classes of combinators of semantically identical type in G are collected together. The uniform passage from a Γ algebra G to a derived combinator signature $\Sigma(G)$ is given by a functor.

2.6. Definition. Let Γ be a K-kinded type signature, Y be any K-indexed family of sets of variables, $K' \subseteq K$ and Σ be a combinator signature over Γ, Y and K'. Then Σ, Γ and K' uniquely determine a functor $\Sigma : Alg(\Gamma) \to Sig$ where for any $G \in Alg(\Gamma)$ the signature $\Sigma(G)$ is the $\coprod_{k\in K'} G_k$-sorted signature defined as follows.

For any $n \geq 0$, any kinds $k_0, \ldots, k_n \in K'$ and any $g_i \in G_{k_i}$ for $1 \leq i \leq n$,

$$\Sigma(G)_{(k_1,\,g_1)\ldots(k_n,\,g_n),\,(k_0,\,g_0)} = \bigcup_{\gamma:Y\to G} \Sigma(G,\,\gamma)_{(k_1,\,g_1)\ldots(k_n,\,g_n),\,(k_0,\,g_0)}$$

where for any variable assignment $\gamma : Y \to G$,

$$\Sigma(G,\,\gamma)_{(k_1,\,g_1)\ldots(k_n,\,g_n),\,(k_0,\,g_0)} =$$

$$\bigcup\{\ \Sigma_{\tau_1\ldots\tau_n,\,\tau_0} \mid \text{ for each } 1 \leq i \leq n,\ \ \tau_i \in T(\Gamma,\,Y)_{k_i} \text{ and } \overline{\gamma_{k_i}}(\tau_i) = g_i\ \}.$$

For any algebras $G,\,G' \in Alg(\Gamma)$ and Γ homomorphism $\phi : G \to G'$, define the signature morphism $\Sigma(\phi) : \Sigma(G) \to \Sigma(G')$ where $\Sigma(\phi) = (\Sigma(\phi)_S,\,\Sigma(\phi)_\Sigma)$ and $\Sigma(\phi)_S : \coprod_{k\in K'} G_k \to \coprod_{k\in K'} G'_k$ is given by

$$\Sigma(\phi)_S(k,\,g) = (k,\,\phi_k(g))$$

for any $k \in K'$ and $g \in G_k$. Also for any $n \geq 0$, any $k_0, \ldots, k_n \in K'$ and any $g_i \in G_{k_i}$ for $1 \leq i \leq n$ define $\Sigma(\phi)_\Sigma^{(k_1, g_1) \ldots (k_n, g_n), (k_0, g_0)}$ to be the inclusion mapping.

Composing the functors $\mu : Sig^{op} \to Cat$ and $\Sigma : Alg(\Gamma) \to Sig$ gives a functor $F = \mu\Sigma : Alg(\Gamma) \to Cat$. Thus we may form the cofibration $Gr(F)$.

The introduction of signatures for both types and combinators allows us to write down equational axiomatisations of each. Let Γ be a K-kinded type signature and $Y = \langle\, Y_k \mid k \in K \,\rangle$ be a family of sets of variable symbols. We have the usual notion of *term algebra* $T(\Gamma, Y)$ over Γ and Y and we may write down *equations* over Γ and Y as formulas of the form $\tau = \tau'$ where $\tau, \tau' \in T(\Gamma, Y)_k$ are type terms of the same kind $k \in K$. Furthermore, given a set ε of equations over Γ and Y we have the usual many-sorted equational calculus which is a sound and complete inference system for equational consequences of ε.

However, the notions of term, equation and validity for a combinator signature Σ over a type signature Γ are complicated by the parametric role of Γ. In the sequel we consider some fixed, but arbitrarily chosen family $X = \langle\, X_\tau \mid k \in K'$ and $\tau \in T(\Gamma, Y)_k \,\rangle$ of disjoint sets of combinator variables. It is clear that the standard definition of terms over Σ and X is too restrictive for our needs. Thus we weaken the notion of term to a notion of *preterm*. Then some well formed preterms may be incorrectly typed in the context of a particular semantical interpretation G of the type signature Γ. (This approach is applied elsewhere in type theory, see for example the survey of Mitchell [1990].)

2.7. Definition. Define the K' indexed family $T_{pre}(\Sigma, X) = \langle\, T_{pre}(\Sigma, X)_k \mid k \in K' \,\rangle$ of sets $T_{pre}(\Sigma, X)_k$ of *preterms of kind* k over Σ and X, and for each preterm $t \in T_{pre}(\Sigma, X)_k$ its *nominal type* $\tau_t \in T(\Gamma, Y)_k$, inductively.

(i) For any kind $k \in K'$, any types $\sigma, \tau \in T(\Gamma, Y)_k$ and any variable $x \in X_\sigma$,

$$x^\tau \in T_{pre}(\Sigma, X)_k,$$

and has nominal type τ.

(ii) For any kind $k \in K'$, any types $\sigma, \tau \in T(\Gamma, Y)_k$ and any constant symbol $c \in \Sigma_{\lambda, \sigma}$,

$$c^\tau \in T_{pre}(\Sigma, X)_k,$$

and has nominal type τ

(iii) For any $n \geq 1$, any kinds $k_0, \ldots, k_n \in K'$, any types $\tau_i \in T(\Gamma, Y)_{k_i}$ for $1 \leq i \leq n$ and $\sigma, \tau \in T(\Gamma, Y)_{k_0}$, any function symbol $f \in \Sigma_{\tau_1 \ldots \tau_n, \sigma}$ and any preterms $t_i \in T_{pre}(\Sigma, X)_{k_i}$ for $1 \leq i \leq n$,

$$f^\tau(t_1, \ldots, t_n) \in T_{pre}(\Sigma, X)_{k_0}.$$

and has nominal type τ.

To check whether a preterm t is well typed in the context of particular type algebra $G \in Alg(\Gamma)$ and variable assignment $\gamma : Y \to G$, and to calculate the value of the type of t, we make the the following definition.

2.8. Definition. Consider any algebra $G \in Alg(\Gamma)$ and variable assignment $\gamma : Y \to G$. Define the K'-indexed family $Type^{G,\gamma} = \langle\, Type_k^{G,\gamma} : T_{pre}(\Sigma, X)_k \rightsquigarrow G_k \mid k \in K' \,\rangle$ of partial functions $Type_k^{G,\gamma}$ inductively.

(i) For any kind $k \in K'$, any types $\sigma, \tau \in T(\Gamma, Y)_k$ and any variable $x \in X_\sigma$,

$$Type_k^{G,\gamma}(x^\tau) = \begin{cases} \overline{\gamma_k}(\tau), & \text{if } \overline{\gamma_k}(\tau) = \overline{\gamma_k}(\sigma); \\ \uparrow, & \text{otherwise.} \end{cases}$$

(ii) For any kind $k \in K'$, any types $\sigma, \tau \in T(\Gamma, Y)_k$ and any constant symbol $c \in \Sigma_{\lambda,\sigma}$,

$$Type_k^{G,\gamma}(c^\tau) = \begin{cases} \overline{\gamma_k}(\tau), & \text{if } \overline{\gamma_k}(\tau) = \overline{\gamma_k}(\sigma); \\ \uparrow, & \text{otherwise.} \end{cases}$$

(iii) For any $n \geq 1$, any kinds $k_0, \ldots, k_n \in K'$, any types $\tau_i \in T(\Gamma, Y)_{k_i}$ for $1 \leq i \leq n$ and $\sigma, \tau \in T(\Gamma, Y)_{k_0}$, any function symbol $f \in \Sigma_{\tau_1 \ldots \tau_n, \sigma}$ and any preterms $t_i \in T_{pre}(\Sigma, X)_{k_i}$ for $1 \leq i \leq n$,

$$Type_k^{G,\gamma}(f^\tau(t_1, \ldots, t_n)) =$$

$$\begin{cases} \overline{\gamma_{k_0}}(\tau_0), & \text{if } \overline{\gamma_{k_0}}(\sigma) = \overline{\gamma_{k_0}}(\tau) \text{ and for each } 1 \leq i \leq n,\ Type_{k_i}^{G,\gamma}(t_i) = \overline{\gamma_{k_i}}(\tau_i); \\ \uparrow, & \text{otherwise.} \end{cases}$$

We say that a preterm $t \in T_{pre}(\Sigma, X)_k$ is *well typed* with respect to G and γ if, and only if, $Type_k^{G,\gamma}(t)$ is defined.

In Section 3 we present a sound and complete inference system for calculating the type of a preterm. Given a Γ algebra G and an assignment $\gamma : Y \to G$, the preterms which are well typed with respect to G and γ are precisely the preterms which can be translated into well formed terms over the derived combinator signature $\Sigma(G)$ and a derived family $X(G)$ of sets of variables.

2.9. Definition. Let $G \in Alg(\Gamma)$ any type algebra. Define the derived family

$$X(G) = \langle\, X(G)_{(k,g)} \mid k \in K' \text{ and } g \in G_k \,\rangle$$

of sets of variables, where for any $k \in K'$ and any $g \in G_k$ we have $X(G)_{(k,g)} = \bigcup_{\gamma : Y \to G} X(G, \gamma)_{(k,g)}$ and for any variable assignment $\gamma : Y \to G$,

$$X(G, \gamma)_{(k,g)} = \bigcup\{\, X_\tau \mid \tau \in T(\Gamma, Y)_k \text{ and } \overline{\gamma_k}(\tau) = g \,\}.$$

For any variable assignment $\gamma : Y \to G$, define the K'-indexed family $Trans^{G,\gamma} = \langle\, Trans_k^{G,\gamma} : T_{pre}(\Sigma, X)_k \rightsquigarrow \bigcup_{g \in G_k} T(\Sigma(G), X(G))_{(k,g)} \,\rangle$ of (partial) *translation mappings* inductively.

(i) For any kind $k \in K'$, any types $\sigma, \tau \in T(\Gamma, Y)_k$ and any variable $x \in X_\sigma$, define

$$Trans_k^{G,\gamma}(x^\tau) = \begin{cases} x^{(k, \overline{\gamma_k}(\tau))}, & \text{if } Type_k^{G,\gamma}(x^\tau)\downarrow; \\ \uparrow, & \text{otherwise.} \end{cases}$$

(ii) For any kind $k \in K'$, any types $\sigma, \tau \in T(\Gamma, Y)_k$ and any constant symbol $c \in \Sigma_{\lambda,\sigma}$, define

$$Trans_k^{G,\gamma}(c^\tau) = \begin{cases} c^{(k, \overline{\gamma_k}(\tau))}, & \text{if } Type_k^{G,\gamma}(c^\tau)\downarrow; \\ \uparrow, & \text{otherwise.} \end{cases}$$

(iii) For any $n \geq 1$, any kinds $k_0, \ldots, k_n \in K'$, any types $\tau_i \in T(\Gamma, Y)_{k_i}$ for $1 \leq i \leq n$ and $\sigma, \tau \in T(\Gamma, Y)_{k_0}$, any function symbol $f \in \Sigma_{\tau_1 \ldots \tau_n, \sigma}$ and any preterms $t_i \in T_{pre}(\Sigma, X)_{k_i}$ for $1 \leq i \leq n$, define

$$Trans_k^{G,\gamma}(\, f^{\tau}(t_1, \ldots, t_n)\,) =$$

$$\begin{cases} f^{(k,\, \overline{\gamma_{k_0}}(\tau))}(\, Trans_{k_1}^{G,\gamma}(t_1), \ldots, Trans_{k_n}^{G,\gamma}(t_n)\,), & \text{if } Type_{k_0}^{G,\gamma}(\, f^{\tau_0}(t_1, \ldots, t_n)\,)\downarrow; \\ \uparrow, & \text{otherwise.} \end{cases}$$

2.10. Proposition. *For any type algebra $G \in Alg(\Gamma)$, any assignment $\gamma : Y \to G$, any kind $k \in K'$ and any preterm $t \in T_{pre}(\Sigma, X)_k$, if $Type_k^{G,\gamma}(t) \downarrow = g$ then $Trans_k^{G,\gamma}(t) \downarrow$ and*

$$Trans_k^{G,\gamma}(t) \in T(\, \Sigma(G),\, X(G)\,)_{(k,\, g)}\, .$$

Proof. By induction on the complexity of t. □

We now use the semantical apparatus introduced for types to make precise the concepts of *well typedness* and *validity* for an *equation* between preterms with respect to a Γ type algebra G.

2.11. Definition.

(i) By an *equation in preterms* over Σ and X of kind $k \in K'$ we mean an expression of the form

$$t = t'$$

where $t,\, t' \in T_{pre}(\Sigma, X)_k$ are preterms of the same kind k.

(ii) For any type algebra $G \in Alg(\Gamma)$ and assignment $\gamma : Y \to G$, we say that $t = t'$ is *well typed* with respect to G and γ if, and only if, $Type_k^{G,\gamma}(t) \downarrow$ and $Type_k^{G,\gamma}(t') \downarrow$ and

$$Type_k^{G,\gamma}(t) = Type_k^{G,\gamma}(t').$$

If $t = t'$ is well typed with respect to G and γ for every $\gamma : Y \to G$ we say that $t = t'$ is well typed with respect to G. If $C \subseteq Alg(\Gamma)$ is any class of algebras then we say that $t = t'$ is well typed with respect to C if, and only if, $t = t'$ is well typed with respect to every $G \in C$.

(iii) For any pair of algebras $(G, A) \in |Gr(\Gamma, \Sigma)|$ and any pair of assignments $(\gamma : Y \to G,\ \alpha : X(G) \to A)$ we say that $t = t'$ is *valid* in (G, A) under (γ, α) and write

$$(\,(G, A), (\gamma, \alpha)\,) \models t = t'$$

if, and only if, $t = t'$ is well typed with respect to G and γ and

$$Val_{(k,\, g)}^{A,\alpha}(Trans_k^{G,\gamma}(t)) = Val_{(k,\, g)}^{A,\alpha}(Trans_k^{G,\gamma}(t'))$$

(where $g = Type_k^{G,\gamma}(t) = Type_k^{G,\gamma}(t')$ and $Val^{A,\alpha} : T(\Sigma(G), X(G)) \to A$ is the term evaluation mapping in A under assignment α).

If $t = t'$ is valid with respect to (G, A) and (γ, α) for all $(\gamma : Y \to G,\ \alpha : X(G) \to A)$ then we say that $t = t'$ is valid in (G, A) and write $(G, A) \models t = t'$.

If $C \subseteq |Gr(\Gamma, \Sigma)|$ is any subclass then we say that $t = t'$ is valid in C and write $C \models t = t'$ if, and only if, $(G, A) \models t = t'$ for every pair $(G, A) \in C$.

(iv) If E is any set of equations over Σ and X then we say that E is valid in $(G, A) \in |Gr(\Gamma, \Sigma)|$ (respectively $C \subseteq |Gr(\Gamma, \Sigma)|$) if and only if $(G, A) \models e$ (respectively $C \models e$) for every equation $e \in E$.

(v) Let $Mod((\Gamma, \varepsilon), (\Sigma, E))$ denote the class

$$\{\ (G, A) \in |Gr(\Gamma, \Sigma)| \ :\ G \models \varepsilon \text{ and } (G, A) \models E\ \}.$$

We next make precise the syntax and initial model semantics of an equational type specification.

2.12. Definition. By an *equational type specification* we mean a pair

$$((\Gamma, \varepsilon), (\Sigma, E))$$

where Γ is a K-kinded type signature, ε is an equational theory over Γ and some family Y of sets of variables, Σ is a combinator signature over Γ, Y and $K' \subseteq K$, and E is a set of equations in preterms over Σ and some family X of sets of variables.

To give an initial model semantics to a specification $((\Gamma, \varepsilon), (\Sigma, E))$ we structure the class $Mod((\Gamma, \varepsilon), (\Sigma, E))$ as a cofibered category.

2.13. Definition. Let $Spec = ((\Gamma, \varepsilon), (\Sigma, E))$ be an equational type specification. Define the functor $Spec : Alg(\Gamma, \varepsilon)^{op} \to Cat$ by

$$Spec(G) = Alg(\Sigma(G), E(G))$$

for each $G \in Alg(\Gamma, \varepsilon)$, where

$$E(G) = \{\ Trans_k^{G,\gamma}(t) = Trans_k^{G,\gamma}(t') \mid k \in K' \text{ and } t, t' \in T_{pre}(\Sigma, X)_k$$

$$\text{and } t = t' \in E \text{ and } \gamma : Y \to G\ \}.$$

For any $G, G' \in Alg(\Gamma, \varepsilon)$ and Γ homomorphism $\phi : G \to G'$ the functor

$$Spec(\phi) : Alg(\Sigma(G'), E(G')) \to Alg(\Sigma(G), E(G))$$

is the restriction of $\mu(\phi)$ (c.f. Definition 1.2) to $Alg(\Sigma(G'), E(G'))$. By the cofibration construction of Definition 1.3 we obtain a category $Gr(Spec)$.

2.14. Proposition. *Let $Spec = ((\Gamma, \varepsilon), (\Sigma, E))$ be an equational type specification. Then the category $Gr(Spec)$ admits an initial object $I(Spec)$.*

Proof. Immediate from Proposition 1.4 and the definition of $Gr(Spec)$. □

Thus each equational type specification $Spec$ admits an initial model semantics $I(Spec)$. In Section 4 we give an explicit construction of the initial model $I(Spec)$. We also present a sound and complete equation calculus for equational type specifications.

3. A CALCULUS OF TYPE ASSIGNMENTS

In Section 2 we introduced the notion of preterms over a combinator signature Σ and family X of variables with respect to a type signature Γ. Simultaneously with each preterm $t \in T_{pre}(\Sigma, X)_k$ over Σ and X of kind k we introduced its nominal type $\tau_t \in T(\Gamma, Y)_k$. In the context of a particular type algebra $G \in Alg(\Gamma)$ and variable assignment $\gamma : Y \to G$ interpreting the type terms of $T(\Gamma, Y)$ a preterm t may or may not be well typed. In this section we will show that for any any set ε of equations over Γ and Y a preterm t is well typed over all type algebras $G \in Alg(\Gamma, \varepsilon)$ and assignments $\gamma : Y \to G$ precisely when t can be assigned its nominal type τ_t under a simple type assignment calculus.

3.1. Definition. By a *type assignment* of kind $k \in K'$ we mean an expression of the form

$$t : \tau$$

where $t \in T_{pre}(\Sigma, X)_k$ is a preterm of kind k and $\tau \in T(\Gamma, Y)_k$ is a type of kind k.

Given an algebra $G \in Alg(\Gamma)$ and an assignment $\gamma : Y \to G$, we say that the type assignment $t : \tau$ is *valid* in G under γ and write $(G, \gamma) \models t : \tau$ if, and only if,

$$Type_k^{G,\gamma}(t)\downarrow \quad \text{and} \quad Type_k^{G,\gamma}(t) = \overline{\gamma_k}(\tau).$$

If $t : \tau$ is valid in G under γ for every assignment $\gamma : Y \to G$ we simply say that $t : \tau$ is valid in G and write $G \models t : \tau$. If $K \subseteq Alg(\Gamma)$ is any class of Γ algebras and $t : \tau$ is valid in G for every $G \in K$ we say that $t : \tau$ is valid in K and write $K \models t : \tau$.

3.2. Definition. Let ε be an equational theory over Γ and Y. Define the inference rules of the *type assignment calculus.*

(i) For any kind $k \in K'$, any types $\sigma, \tau \in T(\Gamma, Y)_k$ and any variable $x \in X_\sigma$,

$$\frac{\varepsilon \vdash \tau = \sigma}{\varepsilon \vdash x^\tau : \tau}$$

(ii) For any kind $k \in K'$, any types $\sigma, \tau \in T(\Gamma, Y)_k$ and any constant symbol $c \in \Sigma_{\lambda, \sigma}$,

$$\frac{\varepsilon \vdash \tau = \sigma}{\varepsilon \vdash c^\tau : \tau}$$

(iii) For any $n \geq 1$, any kinds $k_0, \ldots, k_n \in K'$, any types $\tau_i \in T(\Gamma, Y)_{k_i}$ for $1 \leq i \leq n$ and $\sigma, \tau \in T(\Gamma, Y)_{k_0}$, any function symbol $f \in \Sigma_{\tau_1 \ldots \tau_n, \sigma}$ and any preterms $t_i \in T_{pre}(\Sigma, X)_{k_i}$ for $1 \leq i \leq n$,

$$\frac{\varepsilon \vdash \tau = \sigma, \quad \varepsilon \vdash t_1 : \tau_1 \ , \ldots, \ \varepsilon \vdash t_n : \tau_n}{\varepsilon \vdash f^\tau(t_1, \ldots, t_n) : \tau}$$

(iv) For any kind $k \in K'$, any types $\tau, \tau' \in T(\Gamma, X)_k$ and any preterm $t \in T_{pre}(\Sigma, X)_k$,

$$\frac{\varepsilon \vdash t : \tau, \quad \varepsilon \vdash \tau = \tau'}{\varepsilon \vdash t : \tau'}$$

3.3. Completeness Theorem. *For any equational theory* ε *over* Γ *and a family* Y *of infinite sets of variables, any kind* $k \in K'$, *any type* $\tau \in T(\Gamma, Y)_k$ *and any preterm* $t \in T_{pre}(\Sigma, X)_k$,

$$\varepsilon \vdash t : \tau \Leftrightarrow Alg(\Gamma, \varepsilon) \models t : \tau.$$

Proof. $\Rightarrow$ (Soundness.) By induction on the complexity of derivations.
$\Leftarrow$ (Completeness.) By induction on the complexity of t. □

4. A CALCULUS OF EQUATIONS

In Section 2 we introduced the notion of an equational type specification $Spec = ((\Gamma, \varepsilon), (\Sigma, E))$ where ε is a set of equations over a type signature Γ and E a set of equations in preterms over a combinator signature Σ. As well as formally deriving equations of the form $\tau = \tau'$ between type terms $\tau, \tau' \in T(\Gamma, Y)_k$ from the theory ε, we wish to be able to formally derive well typed equations of the form $t = t'$ between preterms $t, t' \in T_{pre}(\Sigma, X)_k$ from the theory E. For this we must both modify and extend the rules of the many-sorted equational calculus to include both type checking and type manipulation based on the theory ε.

In this section we present a sound and complete calculus for deriving well typed equations in preterms from an equational type specification *Spec*. The proof of completeness, when restricted to ground (variable free) preterms, yields a concrete construction of the initial model of *Spec*.

We begin by formalising the notion of substitution for combinator variables in a preterm.

4.1. Definition. Let $\alpha = \langle \alpha_k : X_k \to T_{pre}(\Sigma, X)_k \mid k \in K' \rangle$ be any family of assignments (where $X_k = \bigcup_{\tau \in T(\Gamma, Y)_k} X_\tau$). Define the family

$$\overline{\alpha} = \langle \overline{\alpha_k} : T_{pre}(\Sigma, X)_k \to T_{pre}(\Sigma, X)_k \mid k \in K' \rangle$$

of substitution mappings $\overline{\alpha_k}$ inductively.

(i) For each $k \in K'$ and $\sigma, \tau \in T(\Gamma, Y)_k$ and constant symbol $c \in \Sigma_{\lambda, \sigma}$ define

$$\overline{\alpha_k}(c^\tau) = c^\tau.$$

(ii) For each $k \in K'$ and $\sigma, \tau \in T(\Gamma, Y)_k$ and variable $x \in X_\sigma$ define

$$\overline{\alpha_k}(x^\tau) = \alpha_k(x).$$

(iii) For any $n \geq 1$, any $k_0, \ldots, k_n \in K'$, any $\tau_i \in T(\Gamma, Y)_{k_i}$ for $1 \leq i \leq n$ and $\sigma, \tau \in T(\Gamma, Y)_{k_0}$, and any $f \in \Sigma_{\tau_1, \ldots, \tau_n, \sigma}$ and any preterms $t_i \in T_{pre}(\Sigma, X)_{k_i}$ for $1 \leq i \leq n$ define

$$\overline{\alpha_k}(f^\tau(t_1, \ldots, t_n)) = f^\tau(\overline{\alpha_{k_1}}(t_1), \ldots, \overline{\alpha_{k_n}}(t_n)).$$

For any $k \in K'$, any type term $\tau \in T(\Gamma, Y)_k$, any variable $x \in X_\tau$ and any preterm $t \in T_{pre}(\Sigma, X)_k$ we let $[x/t]$ denote the family of assignments such that for any $k' \in K'$, any $\tau' \in T(\Gamma, Y)'_k$ and any variable $y \in X_{\tau'}$,

$$[x/t]_{k'}(y) = \begin{cases} t, & \text{if } k' = k \text{ and } \tau' = \tau \text{ and } y = x; \\ y^{\tau'}, & \text{otherwise.} \end{cases}$$

As usual, for any $k' \in K'$ and preterm $t' \in T_{pre}(\Sigma, X)_{k'}$ we write $t'[x/t]$ for $\overline{[x/t]_k}(t)$.

As well as combinator variables, a preterm $t \in T_{pre}(\Sigma, X)_k$ may have free type variables. We also allow substitution for type variables.

4.2. Definition. Let $\gamma = \langle \gamma_k : Y_k \to T(\Gamma, Y)_k \mid k \in K \rangle$ be any family of assignments. Define the family $\overline{\gamma} = \langle \overline{\gamma_k} : T_{pre}(\Sigma, X)_k \to T_{pre}(\Sigma, X)_k \mid k \in K' \rangle$ of substitution mappings $\overline{\gamma_k}$ inductively.

(i) For each $k \in K'$ and $\sigma, \tau \in T(\Gamma, Y)_k$ and constant symbol $c \in \Sigma_{\lambda, \sigma}$ define

$$\overline{\gamma_k}(c^\tau) = c^{\overline{\gamma_k}(\tau)}.$$

(ii) For each $k \in K'$ and $\sigma, \tau \in T(\Gamma, Y)_k$ and variable $x \in X_\sigma$ define

$$\overline{\gamma_k}(x^\tau) = x^{\overline{\gamma_k}(\tau)}.$$

(iii) For any $n \geq 1$, any $k_0, \ldots, k_n \in K'$, any $\tau_i \in T(\Gamma, Y)_{k_i}$ for $1 \leq i \leq n$ and $\sigma, \tau \in T(\Gamma, Y)_{k_0}$, and any $f \in \Sigma_{\tau_1, \ldots, \tau_n, \sigma}$ and any preterms $t_i \in T_{pre}(\Sigma, X)_{k_i}$ for $1 \leq i \leq n$ define

$$\overline{\gamma_{k_0}}(\, f^\tau(t_1, \ldots, t_n)\,) = f^{\overline{\gamma_{k_0}}(\tau)}(\,\overline{\gamma_{k_1}}(t_1), \ldots, \overline{\gamma_{k_n}}(t_n)\,).$$

For any $k \in K'$ and type variable $y \in Y_k$ and type term $\tau \in T(\Gamma, Y)_k$ we let $[y/\tau]$ denote the family of assignments such that for any $k' \in K$ and $y' \in Y_{k'}$

$$[y/\tau](y') = \begin{cases} \tau, & \text{if } k' = k \text{ and } y' = y; \\ y', & \text{otherwise.} \end{cases}$$

Again, for any $k' \in K'$ and preterm $t \in T_{pre}(\Sigma, X)_{k'}$ we write $t[y/\tau]$ for $\overline{[y/\tau]_k}(t)$.

Next we present inference rules for equations in combinator preterms.

4.3. Definition. Let ε be any set of equations over Γ and Y and E be any set of equations over Σ and X. Define the *rules of equational deduction* over ε and E.

(i) For any equation $e \in E$,

$$\frac{}{E \vdash e}$$

is an *axiom introduction rule.*

(ii) For any kind $k \in K'$ and any preterm $t \in T_{pre}(\Sigma, X)_k$,

$$\frac{\varepsilon \vdash t : \tau_t}{E \vdash t = t}$$

is a *reflexivity rule*, where $\tau_t \in T(\Gamma, Y)_k$ is the nominal type of t.

(iii) For any kind $k \in K'$ and any preterms $t_1, t_2 \in T_{pre}(\Sigma, X)_k$,

$$\frac{E \vdash t_1 = t_2}{E \vdash t_2 = t_1}$$

is a *symmetry rule.*

(iv) For any kind $k \in K'$ and any preterms $t_1, t_2, t_3 \in T_{pre}(\Sigma, X)_k$,

$$\frac{E \vdash t_1 = t_2, \quad E \vdash t_2 = t_3}{E \vdash t_1 = t_3}$$

is a *transitivity rule.*

(v) For any kind $k \in K'$, any preterms $t,\ t' \in T_{pre}(\Sigma,\ X)_k$, any kind $k' \in K'$, any preterms $t_1,\ t_2 \in T_{pre}(\Sigma,\ X)_{k'}$, any type $\tau \in T(\Gamma,\ Y)_{k'}$ and any variable $x \in X_\tau$,

$$\frac{E \vdash t = t', \quad E \vdash t_1 = t_2, \quad \varepsilon \vdash t_1 : \tau}{E \vdash t[x/t_1] = t'[x/t_2]}$$

is a *substitution rule for combinators.*

(vi) For any kinds $k,\ k' \in K'$, any preterms $t,\ t' \in T_{pre}(\Sigma,\ X)_k$, any type variable $y \in Y_k$ and any types $\tau,\ \sigma \in T(\Gamma,\ Y)_k$,

$$\frac{E \vdash t = t', \quad \varepsilon \vdash \tau = \sigma \quad \varepsilon \vdash t[y/\tau] : \tau'[y/\tau]}{E \vdash t[y/\tau] = t'[y/\sigma]}$$

is a *substitution rule for types,* where τ' is the nominal type of t.

(vii.a) For any $n \geq 0$ any kinds $k_0, \ldots, k_n \in K'$, any types $\tau_i \in T(\Gamma,\ Y)_{k_i}$ for $1 \leq i \leq n$ and $\sigma,\ \tau \in T(\Gamma,\ Y)_{k_0}$, any operation symbol $f \in \Sigma_{\tau_1 \ldots \tau_n,\ \tau}$, and any preterms $t_i \in T_{pre}(\Sigma,\ X)_{k_i}$ for $1 \leq i \leq n$

$$\frac{\varepsilon \vdash \tau = \sigma}{E \vdash f^\tau(t_1, \ldots, t_n) = f^\sigma(t_1, \ldots, t_n)}$$

is a *type replacement rule* (for constants and operations).

(vii.b) For any kind $k \in K'$, any types $\sigma,\ \tau \in T(\Gamma,\ Y)_k$, any variable $x \in X_\tau$

$$\frac{\varepsilon \vdash \tau = \sigma}{E \vdash x^\tau = x^\sigma}$$

is a *type replacement rule* (for variables).

4.4. Completeness Theorem. *Let ε be any set of equations over Γ and a family Y of infinite sets of variables. Let E be any set of equations in preterms over Σ and a family X of infinite sets of variables. Suppose that each equation $e \in E$ is well typed with respect to $Alg(\Gamma,\ \varepsilon)$.*

For any $k \in K'$ and preterms $t,\ t' \in T_{pre}(\Sigma,\ X)_k$,

$$E \vdash t = t' \quad \Leftrightarrow \quad Mod(\,(\Gamma,\ \varepsilon),\ (\Sigma,\ E)\,) \models t = t'.$$

Proof. $\Rightarrow$ (Soundness) By induction on the complexity of proofs.

$\Leftarrow$ (Completeness) Define the family

$$\equiv^\varepsilon = \langle \equiv^\varepsilon_k \ \mid k \in K' \rangle$$

of binary relations $\equiv^\varepsilon_k$ on $T(\Gamma,\ Y)_k$ by

$$\tau \equiv^\varepsilon_k \tau' \quad \Leftrightarrow \quad \varepsilon \vdash \tau = \tau'.$$

Clearly $\equiv^\varepsilon$ is a congruence on $T(\Gamma,\ Y)$. We let $[\tau]$ denote the equivalence class of any type term $\tau \in T(\Gamma,\ Y)_k$, for any $k \in K$ and we let $T_\varepsilon(\Gamma,\ Y) = T(\Gamma,\ Y)/\equiv^\varepsilon$.

Define the family

$$T_{typed}(\Sigma, X) = \langle T_{typed}(\Sigma, X)_{k,[\tau]} \mid k \in K' \text{ and } \tau \in T(\Gamma, Y)_k \rangle$$

where

$$T_{typed}(\Sigma, X)_{k,[\tau]} = \{ t \in T_{pre}(\Sigma, X)_k \mid \varepsilon \vdash t : \tau \}.$$

Then define the family

$$\equiv^E = \langle \equiv^E_{k,[\tau]} \mid k \in K' \text{ and } \tau \in T(\Gamma, Y)_k \rangle$$

of binary relations $\equiv^E_{k,[\tau]}$ on $T(\Sigma, X)_{k,[\tau]}$ by

$$t \equiv^E_{k,[\tau]} t' \Leftrightarrow E \vdash t = t'$$

for any $k \in K'$, $\tau \in T(\Gamma, Y)_k$ and $t, t' \in T_{typed}(\Sigma, X)_{k,[\tau]}$. It is easily shown that $\equiv^E$ forms a family of equivalence relations on $T_{typed}(\Sigma, X)$. Thus we may define the family

$$T_{\varepsilon, E}(\Sigma, X) = \langle T_{\varepsilon, E}(\Sigma, X)_{k, [\tau]} \mid k \in K' \text{ and } \tau \in T(\Gamma, Y)_k \rangle$$

where

$$T_{\varepsilon, E}(\Sigma, X)_{k, [\tau]} = T_{typed}(\Sigma, X)_{k,[\tau]} / \equiv^E_{k,[\tau]}.$$

Now $T_{\varepsilon,E}(\Sigma, X)$ forms a $\Sigma(T_\varepsilon(\Gamma, Y))$ algebra under the obvious interpretation of constants and function symbols. Furthermore, for any $k \in K'$ and preterms $t, t' \in T_{pre}(\Sigma, X)_k$ we have

$$(T_\varepsilon(\Gamma, Y), T_{\varepsilon,E}(\Sigma, X)) \models t = t' \Leftrightarrow E \vdash t = t'.$$

Thus $(T_\varepsilon(\Gamma, Y), T_{\varepsilon,E}(\Sigma, X)) \models E$ and for any $k \in K'$ and preterms $t, t' \in T_{pre}(\Sigma, X)_k$, if $E \not\vdash t = t'$ then $(T_\varepsilon(\Gamma, Y), T_{\varepsilon,E}(\Sigma, X)) \not\models t = t'$. □

A special case of the proof of the Completeness Theorem 4.4 for equations in ground preterms yields a concrete construction of the initial model of an equational type specification.

4.5. Initiality Theorem. *Let ε be any set of equations over Γ and Y. Let E be any set of equations over Σ and X which are well typed with respect to $Alg(\Gamma, \varepsilon)$. For any $k \in K'$ and ground preterms $t, t' \in T_{pre}(\Sigma, X)_k$,*

$$E \vdash t = t' \Leftrightarrow I((\Gamma, \varepsilon), (\Sigma, E)) \models t = t'.$$

Proof. Take X and Y to be the families of empty sets of variables and consider the pair of type and combinator algebras

$$(T_\varepsilon(\Gamma, Y), T_{\varepsilon,E}(\Sigma, X))$$

defined in the proof of Theorem 4.4 above. By above, we know that for any $k \in K'$ and ground preterms $t, t' \in T_{pre}(\Sigma, X)_k$ we have

$$E \vdash t = t' \Leftrightarrow (T_\varepsilon(\Gamma, Y), T_{\varepsilon,E}(\Sigma, X)) \models t = t'.$$

A lengthy but routine calculation shows that

$$I(\,(\Gamma,\,\varepsilon),\,(\Sigma,\,E)\,) \cong (\,T_{\varepsilon}(\Gamma,\,Y),\,T_{\varepsilon,E}(\Sigma,\,X)\,)$$

and hence the result follows. □

It is a pleasure to thank J.R. Hindley, J.V. Tucker and E.G. Wagner for helpful comments on this work. We also acknowledge the financial support of the Science and Engineering Research Council, the Nuffield Foundation and IBM T.J. Watson Research Center.

REFERENCES

M. Barr and C. Wells, Category Theory for Computing Science, Prentice Hall, Englewood Cliffs, 1990.

H.B. Curry and R. Feys, Combinatory Logic, Vol. I, North Holland, Amsterdam, 1958.

H.B. Curry, J.R. Hindley and J.P. Seldin, Combinatory Logic, Vol. II, North Holland, Amsterdam, 1972.

J-Y. Girard, Y. Lafont and P. Taylor, Proofs and Types, Cambridge University Press, Cambridge, 1989.

J.R. Hindley and J.P. Seldin, Introduction to Combinators and λ–Calculus, Cambridge University Press, Cambridge, 1986.

K. Meinke, Universal algebra in higher types, Report CSR 12–90, Dept. of Computer Science, University College Swansea, to appear in Theoretical Computer Science, Volume 99, 1990.

K. Meinke, Subdirect representation of higher type algebras, to appear in K. Meinke and J.V. Tucker (eds), Many–Sorted Logic and its Applications, John Wiley, 1992.

K. Meinke, A recursive second order initial algebra specification of primitive recursion, Report CSR 8–91, Department of Computer Science, University College of Swansea,1991.

K. Meinke and J.V. Tucker, Universal algebra, to appear in: S. Abramsky, D. Gabbay and T.S.E. Maibaum, (eds) Handbook of Logic in Computer Science, Oxford University Press, Oxford, 1992.

K. Meinke and E. Wagner, Algebraic specification of types and combinators, IBM research report, in preparation, 1992.

J.C. Mitchell, Type systems for programming languages, in: J. van Leeuwen (ed), Handbook of Theoretical Computer Science, Volume B, Elsevier, Amsterdam, 1990.

B. Möller, Higher–order algebraic specifications, Facultät für Mathematik und Informatik, Technische Universität München, Habilitationsschrift, 1987b.

M. Nivat, J. Reynolds (eds), Algebraic Methods in Semantics, Cambridge University Press, Cambridge, 1985.

A. Poigné, On specifications, theories and models with higher types, Information and Control 68, (1986) 1–46.

Normal Forms in infinite-valued Logic: the Case of one Variable

Daniele Mundici

Department of Computer Science, University of Milan
via Comelico 39/41, 20135 Milan, Italy
mundici@imiucca.csi.unimi.it

Abstract Let [0,1] be the real unit interval. A Schauder hat is a Λ-shaped function $h:[0,1] \to [0,1]$ whose four pieces are given by linear polynomials with integral coefficients. Rose and Rosser gave an effective method to represent every Schauder hat by a sentence in the infinite-valued calculus of Lukasiewicz. We give an effective method to reduce every sentence ψ with one variable, to an equivalent sentence ϕ which is a disjunction of Schauder hat sentences. Since the equivalence between ψ and ϕ holds in all n-valued calculi, our normal form reduction may be used for a uniform (i.e., n-free) treatment of deduction in these calculi. For the case under consideration, our methods already yield a self-contained and *constructive* proof of McNaughton's theorem stating that in the infinite-valued calculus every piecewise linear function with integral coefficients is representable by some sentence.

1 Introduction

Automated deduction in many-valued logic has been studied with increasing attention in recent years [2], [6], [7], [8], [10]. Following our approach in [5], rather than developing ad hoc techniques for each n-valued calculus, we aim at a uniform theory, where n is regarded as a parameter, and as much as possible of the deductive machinery is independent of n. In this paper we consider normal form reductions in the infinite-valued sentential calculus of Lukasiewicz [11, § 4], [12, § 4.3]. Our reductions automatically hold for all n-valued calculi.

The set S of sentences in the infinite-valued calculus, as well as in every n-valued calculus, is the same as for the two-valued case. Each sentence ψ whose variables are among $X_1,\ldots, X_n$ determines a function $f_\psi\colon [0,1]^n \to [0,1]$ according to the following stipulations:

$$f_{X_i} = x_i, \quad \text{the canonical projection on the } i\text{th axis}$$
$$f_{\text{not}\,\phi} = (f_\phi)^* = 1 - f_\phi$$
$$f_{\phi \text{ or } \chi} = f_\phi \oplus f_\chi = \min(1, f_\phi + f_\chi)$$
$$f_{\phi \text{ and } \chi} = f_\phi \cdot f_\chi = \max(0, f_\phi + f_\chi - 1).$$

Two sentences ϕ and χ are *equivalent*, in symbols $\phi \equiv \chi$, iff $f_\phi = f_\chi$. From the definition it follows that a function $f\colon [0,1]^n \to [0,1]$ coincides with f_ψ for some $\psi \in S$ iff f belongs to the smallest set M of functions containing each canonical projection, and closed under the pointwise operations of *negation* $*$, and truncated addition (or, *disjunction*) $\oplus$.

2 Proposition *The set* M *is closed under the pointwise lattice operations* $g \wedge h = \min(g, h)$ *and* $g \vee h = \max(g, h)$, *as well as under the* conjunction *operation* $g \bullet h = \max(0, g+h-1)$.
Proof. $g \vee h = (g^* \oplus h)^* \oplus h$; $g \wedge h = (g^* \vee h^*)^*$; $g \bullet h = (g^* \oplus h^*)^*$.
QED

3 Definition A function $f: [0,1]^n \to [0,1]$ is *piecewise linear with integral coefficients*, for short $f \in P$, iff f obeys the following two conditions:
(i) f is continuous with respect to the natural topology of the n-cube $[0,1]^n$, and
(ii) there are linear polynomials $p_1, \ldots, p_k$ with integral coefficients, say $p_i(x_1,\ldots,x_n) = a_{i1}x_1 + \ldots + a_{in}x_n + b_i$, with $a_{i1},\ldots,a_{in}, b_i \in \mathbf{Z}$, such that for each $x = (x_1,\ldots,x_n) \in [0,1]^n$ there is an index $j \in \{1,\ldots, k\}$ with $f(x) = p_j(x)$.

Since each projection $x_i: [0,1]^n \to [0,1]$ belongs to P, and P is closed under negation and disjunction, it follows that $M \subseteq P$. Using a nonconstructive argument, McNaughton [4] proved that, conversely, $P \subseteq M$. The following is a constructive proof of McNaughton's theorem for functions of one variable:

4 Theorem
(i) *If* $f: [0,1] \to [0,1]$ *is a piecewise linear function with integral coefficients, then there is a sentence* $\phi \in S$ *such that* $f = f_\phi$.
(ii) *Suppose that, in addition,* f *is concretely specified by the following data:*

—*linear polynomials* $p_1,\ldots, p_k$, *say* $p_i(x) = c_i x + d_i$, *with integral coefficients* c_i *and* d_i, *and*
—*for each rational* $r \in [0, 1]$ *of denominator* $\leq 4 \max(|c_1|,\ldots,|c_k|)$ *an index* $j(r) \in \{1,\ldots,k\}$ *such that* $f(r) = p_{j(r)}(r)$.
Then ϕ *can be effectively computed by a Turing machine.*

Proof. (i) There is a finite set R of rational numbers $0 = m_1/n_1 < m_2/n_2 < \ldots < m_{t-1}/n_{t-1} < m_t/n_t = 1$ such that f is linear over each interval $[m_i/n_i, m_{i+1}/n_{i+1}]$. Assume each fraction m_i/n_i is in irreducible form, i.e., $\gcd(m_i, n_i) = 1$, $n_i > 0$. Then by our assumption about f, for each $i = 1,\ldots,t$, $f(m_i/n_i)$ is an integral multiple of $1/n_i$. Let $d = \max(n_1,\ldots,n_t)$. Then all the elements of R will occur in the dth Farey series F_d, where F_d is the ascending sequence of all irreducible fractions between 0 and 1 whose denominator is $\leq d$. Let us display F_d as follows: $0 = a_1/b_1 < a_2/b_2 < \ldots < a_{u-1}/b_{u-1} < a_u/b_u = 1$. Since F_d contains R, for every $j = 1,2,\ldots, u-1$ the function f will be linear over the interval $[a_j/b_j, a_{j+1}/b_{j+1}]$, and for every $l = 1,2,\ldots,u$ there will exist an integer $k_l \geq 0$ such that

$$f(a_l/b_l) = k_l/b_l. \tag{1}$$

As proved by Cauchy [1] (see [3, Theorem 28]), F_d obeys the *unimodularity* law, to the effect that any two consecutive fractions a_j / b_j and a_{j+1} / b_{j+1} satisfy the identity

$$\det\begin{pmatrix} a_{j+1} & a_j \\ b_{j+1} & b_j \end{pmatrix} = 1$$

Let $mx + q$ be the line joining the two points $(a_j / b_j;\ 0)$ and $(a_{j+1}/b_{j+1};\ 1/b_{j+1})$. An elementary computation shows that both m and q are integers, indeed $m = b_j$ and $q = -a_j$. Similarly, letting $nx + r$ be the line through points $(a_{j+1}/b_{j+1};\ 1/b_{j+1})$ and $(a_{j+2}/b_{j+2};\ 0)$, the unimodularity law again implies that both n and r are integers. Let h_{j+1} be the *Schauder hat* of F_d of height $1/b_{j+1}$ at point a_{j+1}/b_{j+1}, i.e., let h_{j+1} be given by

(2) $$h_{j+1}(x) = \max(0, \min(nx + r, mx + q)), \quad \text{for all } x \in [0,1].$$

The graph of h_{j+1} is as follows:

(3)

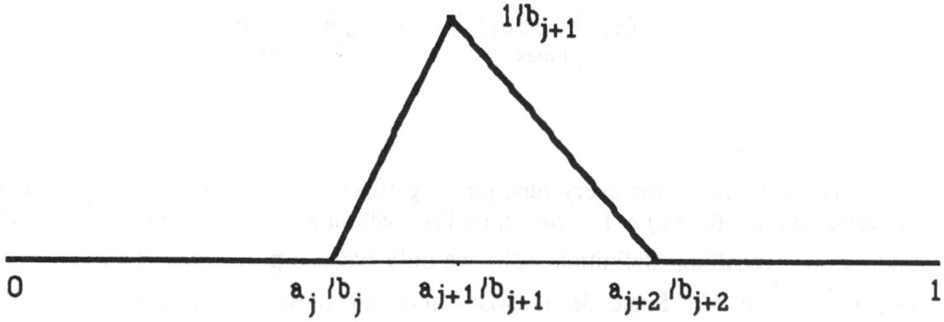

The graphs of the extremal hats h_1 and h_u are as follows:

(3')

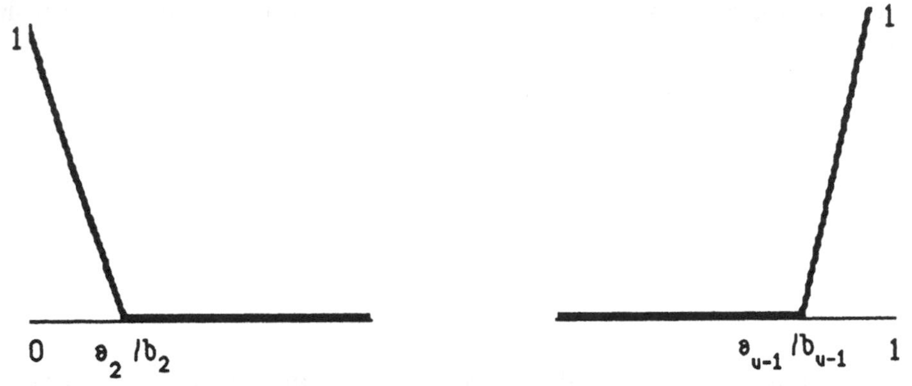

Since n, r, m and q are integers, each hat $h_1, h_2, \ldots, h_u$ is a piecewise linear function with integral coefficients. We claim that

(4) $$f = k_1 h_1 + k_2 h_2 + \ldots + k_u h_u \,, \quad \text{for all} \quad x \in [0,1].$$

As a matter of fact, by (1) and (2) the identity holds at each point of F_d. On the other hand, both f and all hats are linear over each interval $[a_i / b_i, \; a_{i+1} / b_{i+1}]$, and this is sufficient to establish identity (4) over all [0,1]. Replacing throughout in (4) addition by truncated addition $\oplus$, we see that f can be written as a truncated sum of Schauder hats as follows:

(5) $$f = \underbrace{(h_1 \oplus \ldots \oplus h_1)}_{k_1 \text{ times}} \oplus \ldots \oplus \underbrace{(h_u \oplus \ldots \oplus h_u)}_{k_u \text{ times}}.$$

Claim. Each hat is in M.

As a matter of fact, for every function $g{:}[0,1] \to \mathbf{R}$ let us write $g^{=}$ as an abbreviation of $(0 \vee g) \wedge 1$. Since, by Proposition 2, M is closed under max and min, recalling definition (2) we only have to prove that $(mx+q)^{=}$ and $(nx+r)^{=}$ are in M. Since M is closed under negation, it suffices to prove that $(mx+q)^{=}$ is in M. Following Rose and Rosser [9, § 12], we shall argue by induction on m. The basis is trivial. For the induction step, let $z = z(x) = (m-1)x + q$. By induction hypothesis, both $z^{=}$ and $(z+1)^{=}$ are in M. Then for the proof of the claim it is sufficient to establish the following identity

(6) $$(z+x)^{=} = (z^{=} \oplus x) \cdot (z+1)^{=} \,,$$

where • is conjunction. The cases $z(x) > 1$ and $z(x) < -1$ are trivial. If $z(x) \in [0,1]$ then $z^{=} = z$, $(z+1)^{=} = 1$, and (6) boils down to $(z+x)^{=} = z \oplus x$, which certainly holds, because $x \geq 0$. Finally, if $z \in [-1,0]$, then $z^{=} = 0$, $(z+1)^{=} = z+1$, and (6) becomes $(z+x)^{=} = x \bullet (z+1)$, i.e., $\max(0, z+x) = \max(0, x+z+1-1)$. The claim is proved.

The proof of the claim yields, via (6), sentences $\eta_1,\dots,\eta_u$ such that $h_j = f_{\eta_j}$ for each $j = 1,\dots,u$. Let ϕ_j be the disjunction of k_j copies of η_j. Let $\phi = \phi_1$ or ϕ_2 or,..., or ϕ_u be the disjunction of the ϕ_j' s. Then, recalling (5), we have $f = f_\phi$, as required to complete the proof of the first part.

(ii) Let I denote the concrete specification of ϕ. Thus, I is the sequence of integers $c_1, d_1,\dots, c_k, d_k$, followed by the sequence of pairs $(r, j(r))$ for each rational number $r \in [0,1]$ of the Farey series F_ν, where $\nu = 4 \max_i |c_i|$. We will describe a Turing machine T yielding, over input I, a sentence ϕ such that $f = f_\phi$. Without loss of generality, $(c_i, d_i) \neq (c_j, d_j)$ whenever $i \neq j$. For any two indices $i < j$ in $\{1,\dots,k\}$, T first computes the (rational) abscissa r_{ij} of the intersection of the lines represented by p_i and p_j—if these are not parallel. Among the r_{ij} ' s, T only chooses those in the unit interval, and writes them in irreducible form, as follows: $0 = m_1/n_1 < m_2/n_2 < \dots < m_{t-1}/n_{t-1} < m_t/n_t = 1$. Observe that

(7) $$\max(n_1,\dots,n_t) \leq 2 \max_i |c_i|.$$

Let $d = \max(n_1,\dots,n_t)$ and let F_d be the dth Farey series, which we display as $0 = a_1/b_1 < a_2/b_2 < \dots < a_{u-1}/b_{u-1} < a_u/b_u = 1$. Since F_d includes each m_j/n_j, f is linear over each interval $I_i = [a_i / b_i, a_{i+1} / b_{i+1}]$. As proved by Cauchy, the mediant $w_i = (a_{i+1} + a_i) / (b_{i+1} + b_i)$ is in irreducible form and falls in the interval I_i (see [3, 3.1]). By (7), the denominator $b_{i+1} + b_i$ of w_i is $\leq 4 \max_i |c_i| = \nu$. Therefore, by definition of I, T can compute the following numbers:

—the index $j = j(w_i)$ such that $f(w_i) = p_j(w_i)$; then f coincides with p_j over the whole I_i;

—the integer $k_i = f(a_i/b_i) / (1/b_i) = p_j(a_i/b_i) / (1/b_i)$.

Using the inductive procedure of the above claim, T now proceeds to write down sentences $\eta_1,\dots,\eta_u$ such that each hat h_i of F_d has the form $h_i = f_{\eta_i}$. As in the final part of (i), T finally writes down the disjunction ϕ_i of k_i copies of η_i, $(i = 1,\dots,u)$, and the disjunction ϕ of the ϕ_i' s. The proof of (i) shows that $f = f_\phi$.

QED

As in the proof of (i), we say that a piecewise linear function $h : [0,1] \to [0,1]$ with integral coefficients is a Schauder hat iff the graph of h is of the form (3) or (3'). We also say that $\phi \in S$ is a *Schauder hat sentence* iff f_ϕ is a Schauder hat.

5 Corollary (DNF Reduction)

(i) *In the infinite-valued calculus, every sentence* ψ *of one variable can be effectively reduced to an equivalent sentence* ϕ *that is a disjunction of Schauder hat sentences;*
(ii) *The equivalence* $\phi \equiv \psi$ *also holds in each n-valued calculus.*

Proof. (i) We can give an account of a Turing machine U that performs the reduction of ψ to ϕ as follows: Proceeding by induction on the number of connectives occurring in ψ, U first computes linear polynomials $p_1,..., p_k$ with integral coefficients, together with an indexing function $j(r)$ for f_ψ as in statement (ii) of Theorem 5. U now simulates the Turing machine T of the proof of the theorem, thus finally obtaining the required sentence ϕ.
(ii) Every tautology (and a fortiori, every equivalence) in the infinite-valued calculus is also a tautology in each n-valued calculus [11], [12]. QED

References

[1] A. L. CAUCHY, Démonstration d'un théorème curieux sur les nombres, *Bull. Sc. Soc. Philomatique Paris,* (3) **3** (1816) 133-135. Reproduced in *Exercices Math.*, **1** (1826) 114-116, and in *Oeuvres,* (2) **6** (1887) 146-148.

[2] R. HÄHNLE, Uniform notation tableau rules for multiple-valued logic, *Proc. 21th Int. Symp. on Multiple-Valued Logic,* Victoria, B.C., Canada, IEEE Press, 1991, pp. 238-245.

[3] G. H. HARDY, E. M. WRIGHT, "An Introduction to the Theory of Numbers", Fifth Edition, Oxford University Press, London, 1979.

[4] R. MCNAUGHTON, A theorem about infinite-valued sentential logic, *Journal of Symbolic Logic,* **16** (1951) 1-13.

[5] D. MUNDICI, Satisfiability in many-valued sentential logic is *NP*-complete, *Theoretical Computer Science,* **52** (1987) 145-153.

[6] N. V. MURRAY, E. ROSENTHAL, Improving tableau deduction in multiple-valued logic, *Proc. 21th Int. Symp. on Multiple-Valued Logic,* Victoria, B.C., Canada, IEEE, 1991, pp. 230-237.

[7] A. MYCROFT, Logic programs and many-valued logic, *Lecture Notes in Computer Science,* **166** (1984) 274-286.

[8] P. O'HEARN, Z. STACHNIAK, Resolution framework for finitely-valued first-order logic, *Journal of Symbolic Computation,* to appear.

[9] A. ROSE, J. B. ROSSER, Fragments of many-valued statement calculi, *Trans. Amer. Math. Soc.*, **87** (1958) 1-53.

[10] P. SCHMITT, Computational aspects of three-valued logic, *Lecture Notes in Computer Science,* **230** (1986) 190-198.

[11] A. TARSKI, J. LUKASIEWICZ, Investigations into the Sentential Calculi, In: "Logic, Semantics, Metamathematics", Oxford University Press, 1956, pp. 38-59. Reprinted by Hackett Publishing Company, 1983.

[12] R. WOJCICKI, "Theory of Logical Calculi", Kluwer Academic Publishers, Dordrecht, 1988.

A Fragment of First Order Logic Adequate for Observation Equivalence

Halit Oğuztüzün*

Department of Computer Science, University of Iowa
Iowa City, IA 52242

Abstract. We present a logical characterization of the Milner's notion of observation equivalence of processes ("at most one observable action at a time" variant) by using a restricted class of first order formulas. We use the game technique due to Ehrenfeucht as a means to achieve this characterization. First we extend the Ehrenfeucht game by introducing a pair of compatibility relations as a parameter to the game so that we can restrict the moves of the players on the basis of the previous moves. We then define the logic which corresponds to the extended game. Second we characterize the observation equivalence on a restricted class of labelled transition systems (with τ moves), called trace-unique labelled transition systems (t-τ*lts*'s), as the equivalence induced by the games played on certain reducts of the given t-τ*lts*'s where the compatibility relations are defined in terms of bounded reachability in the t-τ*lts*'s. Combining these two characterizations we get our main result.

1 Introduction

The question of when two processes should be considered equivalent plays a crucial role in the study of concurrent systems. One widely accepted notion of equivalence is defined by Milner for his *Calculus of Communicating Systems* [4]. The general idea is that two processes are identified if an observer cannot distinguish them by finite observation. In this paper we present a logical characterization of the Milner's notion of observation equivalence of processes ("at most one observable action at a time" variant) using a restricted class of first order formulas. In Section 2 we extend the Ehrenfeucht game [2, 8] by introducing a pair of *compatibility* relations as a parameter to the game. We then define the logic $\mathcal{L}_\Sigma$, where Σ is some finite signature, which corresponds to the games played under compatibility so that the logical equivalence in $\mathcal{L}_\Sigma$ coincides with the equivalence induced by the extended game. We adopt labelled transition systems (*lts*'s) as models of processes, and we view an *lts* over an alphabet Δ (including the special symbol τ) as a relational Δ-structure. In Section 3 we first define the class of *lts's with silent τ moves* (τ*lts*'s) and the class of *trace-unique τlts's* (*t*-τ*lts*'s) as models of certain axioms. Next we characterize the observation equivalence on *t*-τ*lts*'s as the equivalence induced by the games, where the compatibility relations are defined in terms of bounded

* Author's present address is Ministry of National Education Data Processing Department, Bakanlıklar, 06648 Ankara, Turkey

reachability in the t-τlts's and the games are played on the Λ-reducts of the t-τlts's ($\Lambda=\Delta-\{\tau\}$). In Section 4, by specializing $\mathcal{L}_\Sigma$ and expanding this language with a special constant symbol, we obtain the logic $\mathcal{L}^+_\Lambda$. Combining these two characterizations we get the logical characterization in $\mathcal{L}^+_\Lambda$ of the observation equivalence on t-τlts's. Such a logic $\mathcal{L}^+_\Lambda$ is said to be *adequate* for the observation equivalence. The modal logic developed by Hennessy and Milner is also adequate for observation equivalence [1, 3]. Further, two τlts's are observation equivalent if and only if their unfoldings are $\mathcal{L}^+_\Lambda$-equivalent.

The idea of extending the Ehrenfeucht games with compatibility and characterizing the observation equivalence by the extended games is introduced by the author, in a preliminary form, in [6]. This paper is based on the author's thesis [7].

2 An Extended Ehrenfeucht Game and Its Associated Logic

We extend the Ehrenfeucht games by introducing the notion of compatibility, and define the logic associated with these games. Our line of development in this section is adapted from [8,§6.1,13.1].

Preliminaries

A *signature* Σ is a pair $\Sigma=\langle R,\sigma\rangle$, where R is a set (of symbols) and σ is a function mapping R into ω-{0}. We shall use the symbol Σ in place of both R and σ. For a signature Σ, a *(relational) Σ-structure* $\mathcal{A}$ is a pair $\mathcal{A}=\langle A,r\rangle$, where A is a nonempty set, called the *domain* of the structure, and r is a function mapping each $\rho\in\Sigma$ to some $\Sigma(\rho)$-ary relation $\rho^{\mathcal{A}}$ on A. For the rest of this section, $\mathcal{A}$ and $\mathcal{B}$ denote the Σ-structures $\mathcal{A}=\langle A,\{\rho^{\mathcal{A}}:\rho\in\Sigma\}\rangle$ and $\mathcal{B}=\langle B,\{\rho^{\mathcal{B}}:\rho\in\Sigma\}\rangle$.

Notations for Sequences: The concatenation of two sequences is denoted by their juxtaposition. When there is no danger of confusion, we write a single element sequence $\langle a\rangle$ simply as a. We use both s_i and $s(i)$ to denote the i'th element of the sequence s. By convention, for $m=0$, the sequence $\langle d_1,\ldots, d_m\rangle$ is $\langle\rangle$, the empty sequence. The $rg(s)$ denotes the set of elements that occur in s.

An Ehrenfeucht Game with Compatibility

The Ehrenfeucht game is played by two players, Player I and Player II, on two given Σ-structures, $\mathcal{A}$ and $\mathcal{B}$. Let n be a fixed natural number. With $\mathcal{A}$ [resp. $\mathcal{B}$] we associate a reflexive binary relation $C^{\mathcal{A}}_n$ on A [resp. $C^{\mathcal{B}}_n$ on B], called a *compatibility relation.* We then denote the Ehrenfeucht game with compatibility by $G_n(\mathcal{A},C^{\mathcal{A}}_n,\mathcal{B},C^{\mathcal{B}}_n)$, or simply by G_n if the structures and compatibility relations involved are understood. A play of this game consists of n rounds, each of which is played as follows. First, Player I chooses either structure $\mathcal{A}$ or structure $\mathcal{B}$, and picks an element from the domain of the structure he has chosen. Then Player II picks an element from the domain of the other structure. This completes one round. In any round, the choice of an element is subject to the following condition, called the *compatibility condition:* An element chosen from the domain of $\mathcal{A}$ [resp. $\mathcal{B}$] must be (pairwise) "compatible" with all the elements chosen from the domain of $\mathcal{A}$ [resp. $\mathcal{B}$] in the preceding rounds of the play. Player II *wins* a particular play of the game if the correspondence by rounds between the sequence of elements chosen from $\mathcal{A}$ and the sequence of elements chosen from $\mathcal{B}$ preserves the relations of Σ; otherwise Player I wins this play. We make these notions more precise shortly.

Example 2.1. Consider the trees $\mathcal{A}$ and $\mathcal{B}$ in figure 1 below. In the original Ehrenfeucht game, Player I wins in three rounds by choosing b_0, b_2 and b_3 regardless of what Player II chooses. We observe that the winning power of Player I comes from his ability to pick elements that lie on different paths of the same tree. Indeed, under the compatibility relations which require that choices must be on the same directed path of the same tree, Player II has an obvious winning strategy. Such a compatibility relation can be defined by the predicate Cxy= "there is a directed path of length at most 2 either from x to y or from y to x". Note that we do not want to distinguish between the trees in figure 1. ▯

Example 2.2. Consider the trees of figure 2 below. Now, Player I has a winning strategy in the three round game under the compatibility relations defined by the predicate C, given in the previous example. We indicate the choice in round 1 of Player I with "I,1", and so on. Player II's move in the last round (not shown) is immaterial. In fact, Player I has a winning strategy even in the two round game. Note that we want to distinguish between the trees in figure 2.▯

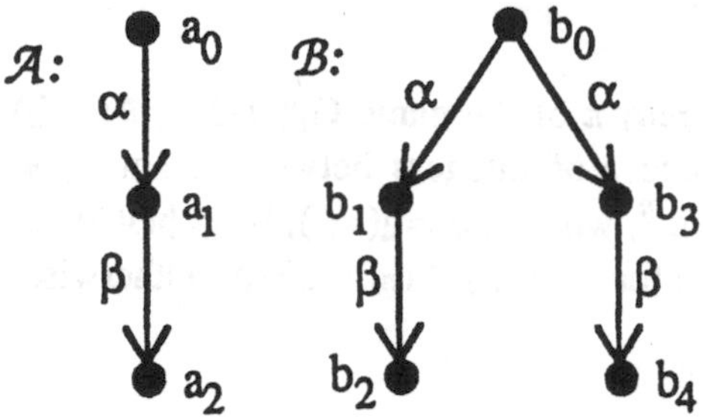

Figure 1. Trees for example 2.1.

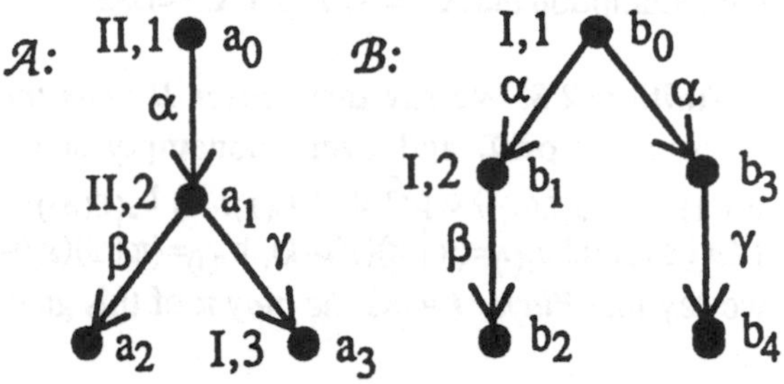

Figure 2. Trees for example 2.2.

Definition 2.3. Let $\mathcal{A}$ be a structure with the compatibility relation $C^{\mathcal{A}}$. Let $A' \subseteq A$. An element $a \in A$ is said to be *compatible* with A' (w.r.t. $C^{\mathcal{A}}$) if for every $a' \in A'$, $\langle a',a \rangle \in C^{\mathcal{A}}$. The set of elements of A that are compatible with A' is denoted by $Com(\mathcal{A},A')$. We simply write $Com(\mathcal{A},s)$ as an abbreviation of $Com(\mathcal{A},rg(s))$ for a sequence s over A. A sequence

$\langle a_1,\ldots, a_m\rangle$, where $m\in\omega$, is said to be *admissible* (w.r.t. $C^{\mathcal{A}}$) if $a_j\in Com(\mathcal{A},\langle a_1,\ldots,a_{j-1}\rangle$, for $1\le j\le m$. The set of all admissible sequences over A of length i is denoted by $Adm(\mathcal{A},i)$.

We may assume that the first k rounds have already been played, and we let the players play n more rounds. For fixed $n,k\in\omega$, we denote such a game as $G^k_n(\mathcal{A},C^{\mathcal{A}}_n,\underline{a},\mathcal{B},C^{\mathcal{B}}_n,\underline{b})$, where $\underline{a}\in Adm(\mathcal{A},k)$ and $\underline{b}\in Adm(\mathcal{B},k)$. To keep our notations simple, from this point on we assume that in a game the domains of the two structures are disjoint.

Notation: Let $\mathcal{A}$ and $\mathcal{B}$ be two structures. For a sequence p over $A\cup B$, $p\,|\,\mathcal{A}$ [resp. $p\,|\,\mathcal{B}$] denotes the sequence obtained by deleting from p exactly those elements that are not from the domain of $\mathcal{A}$ [resp. $\mathcal{B}$].

Definition 2.4. A *play* π of the game $G^k_n(\mathcal{A},C^{\mathcal{A}},\underline{a},\mathcal{B},C^{\mathcal{B}},\underline{b})$ is a sequence $\pi=\langle c_{I,1},c_{II,1},\ldots,c_{I,n},c_{II,n}\rangle$ over $A\cup B$ such that the following two conditions are satisfied:

- either $c_{I,i}\in A$ and $c_{II,i}\in B$, or $c_{I,i}\in B$ and $c_{II,i}\in A$, for $1\le i\le n$;
- $\underline{a}\pi\,|\,\mathcal{A}$ and $\underline{b}\pi\,|\,\mathcal{B}$ are both admissible (the *compatibility condition*).

In an n-round play π, where $n\ge 1$, $(\pi\,|\,\mathcal{A})(i)$ [resp. $(\pi\,|\,\mathcal{B})(i)$] is the element of A [resp. B] selected in round i, for $1\le i\le n$. Note that for any prefix p of the play π, $\underline{a}p\,|\,\mathcal{A}$ and $\underline{b}p\,|\,\mathcal{B}$ are both admissible sequences. These two sequences together can be thought of as the *configuration* of the play after some number of moves. The set $Com(\mathcal{A}, \underline{a}p\,|\,\mathcal{A})$ [resp. $Com(\mathcal{B}, \underline{b}p\,|\,\mathcal{B})$] is the set of possible choices from $\mathcal{A}$ [resp. $\mathcal{B}$] for the next move. The set $Com(\ldots)$ is never empty, because of our stipulation that the compatibility relations be reflexive.

The game $G^k_n(\mathcal{A},C^{\mathcal{A}},\underline{a},\mathcal{B},C^{\mathcal{B}},\underline{b})$ reduces to the original Ehrenfeucht game $G^k_n(\mathcal{A},\underline{a},\mathcal{B},\underline{b})$ under the assumption that $C^{\mathcal{A}}=A\times A$ and $C^{\mathcal{B}}=B\times B$.

Definition 2.5. We say that *Player II wins the play* π of the game $G^k_n(\mathcal{A},C^{\mathcal{A}},\underline{a},\mathcal{B},C^{\mathcal{B}},\underline{b})$ if for every $\rho\in\Sigma$ and every nonempty sequence s of integers between 1 and $k+n$, $\langle a_{s(1)},\ldots, a_{s(\Sigma(\rho))}\rangle\in\rho^{\mathcal{A}}$ iff $\langle b_{s(1)},\ldots, b_{s(\Sigma(\rho))}\rangle\in\rho^{\mathcal{B}}$, where $a_{s(i)}=\underline{a}(s(i))$, $b_{s(i)}=\underline{b}(s(i))$ for $1\le s(i)\le k$, and $a_{s(i)}=(\pi\,|\,\mathcal{A})(s(i)-k)$, $b_{s(i)}=(\pi\,|\,\mathcal{B})(s(i)-k)$ for $k+1\le s(i)\le k+n, 1\le i\le\Sigma(\rho)$. Otherwise, we say that *Player I wins the play* π of this game.

A strategy, intuitively speaking, is a method by which a player decides on his next move in a given configuration during the play of the game. A winning strategy is the one that enables the player to win any play of the game no matter which strategy his opponent is using. We do not bother to give formal definitions for these concepts here; the reader may consult [7]. *Notation:* We write $G^k_n\in$ II as an abbreviation of the statement "Player II has a winning strategy in game G^k_n". Similarly, we write $G^k_n\in$ I for "Player I has a winning strategy in G^k_n". We know from game theory that $G^k_n\in$ II iff $G^k_n\notin$ I.

We would like to have a necessary and sufficient condition for Player II to have a winning strategy in an $(n+1)$-round game in terms of his having a winning strategy in n-round games, for a fixed pair of compatibility relations. This result, stated as lemma 2.6 below, will be helpful in carrying out inductive arguments. (This is analogous to theorem 13.4 in [8].)

Lemma 2.6. $G^{k}_{n+1}(\mathcal{A},C^{\mathcal{A}},\underline{a},\mathcal{B},C^{\mathcal{B}},\underline{b}) \in$ II iff (i) for every $a \in Com(\mathcal{A},\underline{a})$ there is $b \in Com(\mathcal{B},\underline{b})$ such that $G^{k+1}_{n}(\mathcal{A},C^{\mathcal{A}},\underline{a}a,\mathcal{B},C^{\mathcal{B}},\underline{b}b) \in$ II, and (ii) for every $b \in Com(\mathcal{B},\underline{b})$ there is $a \in Com(\mathcal{A},\underline{a})$ such that $G^{k+1}_{n}(\mathcal{A},C^{\mathcal{A}},\underline{a}a,\mathcal{B},C^{\mathcal{B}},\underline{b}b) \in$ II.

We now define the induced equivalence relation on Σ-structures by the existence of a winning strategy for Player II.

Definition 2.7. For fixed $n \in \omega$, we define $\mathcal{A} \sim_n \mathcal{B}$ (w.r.t. $C^{\mathcal{A}}$ and $C^{\mathcal{B}}$) if $G_n(\mathcal{A},C^{\mathcal{A}},\mathcal{B},C^{\mathcal{B}}) \in$ II. Further, $\mathcal{A} \sim \mathcal{B}$ (w.r.t. the sequence of the pairs of compatibility relations, $\langle C^{\mathcal{A}}_n, C^{\mathcal{B}}_n \rangle_{n \in \omega}$) if $\mathcal{A} \sim_n \mathcal{B}$ (w.r.t. $C^{\mathcal{A}}_n$ and $C^{\mathcal{B}}_n$) for all $n \in \omega$. We call the relation $\sim$ *the game equivalence*. It is straightforward to check that $\sim_n$ and $\sim$ are indeed equivalence relations.

Game Equivalence as Logical Equivalence

We define a fragment of the first order logic, denoted $\mathcal{L}_\Sigma$, for a given finite signature Σ, and then we show that $\mathcal{L}_\Sigma$ is the logic that corresponds to our games. Our approach consists in defining $\mathcal{L}_\Sigma$ as a "first order like" formal language and giving a translation of the $\mathcal{L}_\Sigma$ formulas to the well-formed formulas of the first order logic. First we describe $\mathcal{L}_\Sigma$, and then we proceed with defining its formulas inductively.

The predicate symbols of $\mathcal{L}_\Sigma$ are the elements of $\Sigma \cup \{C_m : m \in \omega\}$. For each $\rho \in \Sigma$, ρ is a $\Sigma(\rho)$-place predicate symbol and each C_m, $m \in \omega$, is a binary predicate symbol. There are no function symbols or equality symbol in $\mathcal{L}_\Sigma$. There are infinitely many individual variables $v_1, v_2, \ldots, v_j, \ldots$. $\mathcal{L}_\Sigma$ has the usual logical connectives, $\neg$, $\rightarrow$, $\vee$, $\wedge$ and $\leftrightarrow$; we take $\neg$ and $\rightarrow$ as the basis. Instead of the usual first order quantifiers, $\exists$ and $\forall$, $\mathcal{L}_\Sigma$ has the *pseudo-quantifiers* ${}^{m}\exists$ and ${}^{m}\forall$, for each $m \in \omega$. The pseudo-quantifiers ${}^{m}\exists$ and ${}^{m}\forall$ are defined in terms of $\exists$ and $\forall$, and the predicate symbols C_m. Hence, a pseudo-quantifier is merely a syntactic device to enforce the restriction we need on the usual first order formulas. We shall see how this is done when we define the translation (2.11).

Given a Σ-structure $\mathcal{A}$, the predicate symbols in Σ are to be interpreted as the relations of $\mathcal{A}$, and the predicate symbols C_m are to be interpreted as the compatibility relations, $C^{\mathcal{A}}_m$, $m \in \omega$, associated with $\mathcal{A}$. As far as the interpretation of the first order formulas are concerned, we regard a compatibility relation as a part of the Σ-structure like other relations of Σ, although we do not augment the signature Σ explicitly. We may write $\mathcal{L}$ instead of $\mathcal{L}_\Sigma$ when Σ is understood.

Definition 2.8. We define the set of *expressions* of $\mathcal{L}_\Sigma$ as the set of finite strings over the set of symbols $\{\neg, \rightarrow, (,)\} \cup \{v_i : i \in \omega - \{0\}\} \cup \{{}^m\exists v_i : m \in \omega, i \in \omega - \{0\}\} \cup \Sigma$. We define the operations $F_\neg$, $F_\rightarrow$, and ${}^mE_i$, for $m \in \omega$, $i \in \omega - \{0\}$, on expressions as follows:

(1) $F_\neg(e) = \neg e$; (2) $F_\rightarrow(e_1, e_2) = (e_1 \rightarrow e_2)$; (3) ${}^mE_i(e) = {}^m\exists v_i e$.

An expression of the form $\rho v_{i_1} \ldots v_{i_{\Sigma(\rho)}}$, where $\rho \in \Sigma$, is called an *atomic formula*.

Definition 2.9. For each $m \in \omega$, we define the set ${}^m\mathcal{P}$ as the set generated from the set of atomic formulas by the operations $F_\neg$, $F_\rightarrow$, and ${}^mE_i$, for all $i \in \omega - \{0\}$. Then the set $\mathcal{P}$ of the *preformulas* is defined as $\mathcal{P} = \bigcup_{m \in \omega} {}^m\mathcal{P}$.

The notions of a free variable and a bound variable in a preformula and the quantifier depth of a preformula are carried over naturally from the first order logic. (*Notation:* $\mathcal{F}ree(\varphi)$ denotes the set of free variables in φ.) Other connectives are defined in terms of $\neg$ and $\rightarrow$ in the usual way. The pseudo-quantifier ${}^m\forall$ is the dual of ${}^m\exists$, i.e. ${}^m\forall v_i \varphi = \neg {}^m\exists v_i \neg \varphi$, for $\varphi \in {}^m\mathcal{P}$.

Definition 2.10. First we introduce some notations. For fixed $k \in \omega$, let $t = \langle t_1, \ldots, t_l \rangle$, $0 \leq l \leq k$, such that $1 \leq t_1 < \ldots < t_l \leq k$. (We may regard t as an ordered set.) We write $j \in t$ as an abbreviation for $j \in rg(t)$, $t' \subseteq t$ for a sequence t' obtained from t by deleting zero or more elements from t, and $t \cup j$ for the sequence obtained by inserting j into t at the proper position ($t \cup j$ is the same as t if $j \in t$). Now let $\mathcal{P}_0$ be the set generated from atomic formulas by $F_\neg$ and $F_\rightarrow$, and let $\mathcal{P}_0^t = \{\varphi \in \mathcal{P}_0 : \mathcal{F}ree(\varphi) = \{v_{t_1}, \ldots, v_{t_l}\}\}$ and for all $m \in \omega$ ${}^m\mathcal{P}_0^t = \mathcal{P}_0^t$. Define ${}^m\mathcal{P}_{n+1}^t$ as the set generated from ${}^m\mathcal{U}_{n+1}^t$ by $F_\neg$ and $F_\rightarrow$ where ${}^m\mathcal{U}_{n+1}^t = {}^m\mathcal{P}_n^t \cup_{j \notin t} {}^mE'_j({}^m\mathcal{P}_n^{t \cup j}) \cup_{j \notin t} {}^mE_j({}^m\mathcal{P}_n^t)$. Here ${}^mE'_j(\ldots)$ denotes the alphabetic variants of the preformulas in ${}^mE_j(\ldots)$ under the change of bound variable v_j. Then we define the set of *formulas* of $\mathcal{L}$ as $\bigcup {}^{k+n}\mathcal{P}_n^t$ (union over all $k, n \in \omega$ and $t \subseteq \langle 1, \ldots, k \rangle$). A formula with no free variables is called a *sentence*. Hence, the set of sentences of $\mathcal{L}$ is given by $\bigcup_{n \in \omega} {}^n\mathcal{P}_n^{\langle\rangle}$.

Definition 2.11. A preformula φ of $\mathcal{L}$ is translated into the formula φ^* of the first order logic as follows.

(1) If φ is an atomic formula, then $\varphi^* = \varphi$;
(2) If φ is $\neg\psi$, then $\varphi^* = \neg\psi^*$;
(3) If φ is $(\psi_1 \rightarrow \psi_2)$, then $\varphi^* = (\psi_1^* \rightarrow \psi_2^*)$;
(4) If φ is ${}^m\exists v_i \psi$ with $\mathcal{F}ree(\varphi) = \{v_{t_1}, \ldots, v_{t_l}\}$, then $\varphi^* = \exists v_i((\bigwedge_{1 \leq j \leq l} C_m v_{t_j} v_i) \wedge \psi^*)$.

Regarding ${}^m\forall$ we have the dual of (4):

(4′) If φ is ${}^m\forall v_i \psi$ with $\mathcal{F}ree(\varphi) = \{v_{t_1}, \ldots, v_{t_l}\}$, then $\varphi^* = \forall v_i((\bigwedge_{1 \leq j \leq l} C_m v_{t_j} v_i) \rightarrow \psi^*)$.

Example 2.12. Consider example 2.1. There are sentences in the first order logic whose quantifier depth is 3 to distinguish the trees $\mathcal{A}$ and $\mathcal{B}$ in figure 1. Indeed, the following sentence is satisfied by $\mathcal{A}$ but not by $\mathcal{B}$: $\forall v_1 (\exists v_3\, \alpha v_3 v_1 \rightarrow \forall v_2 \forall v_4 (\beta v_2 v_4 \rightarrow \beta v_1 v_4))$. If we used pseudo-quantifiers (of index at least 3) uniformly in this sentence in place of the usual quantifiers we would have the sentence σ of $\mathcal{L}$: ${}^3\forall v_1\, ({}^3\exists v_3\, \alpha v_3 v_1 \rightarrow {}^3\forall v_2 {}^3\forall v_4\, (\beta v_2 v_4 \rightarrow \beta v_1 v_4))$. We apply the translation defined in 2.11 to σ to get the first order sentence $\sigma^* = \forall v_1\, (\exists v_3\, (C_3 v_1 v_3 \wedge \alpha v_3 v_1) \rightarrow \forall v_2\, (C_3 v_1 v_2 \rightarrow \forall v_4\, ((C_3 v_1 v_4 \wedge C_3 v_2 v_4) \rightarrow (\beta v_2 v_4 \rightarrow \beta v_1 v_4))))$. The predicate C_3 can be interpreted as the compatibility relation C defined in example 2.1. Observe now that $\mathcal{A}$ and $\mathcal{B}$ both satisfy σ^*. In fact, it will follow from the main result of this section that they satisfy exactly the same sentences of $\mathcal{L}$ with quantifier depth at most 3 since Player II has a winning strategy in the three round game under this compatibility. ▯

Example 2.13. Consider example 2.2. Recall that Player I has a winning strategy in the two round game. The following sentence of $\mathcal{L}$ distinguishes the trees $\mathcal{A}$ and $\mathcal{B}$ given in figure 2: ${}^2\exists v_1({}^2\exists v_2\, \beta v_1 v_2 \wedge {}^2\exists v_2\, \gamma v_1 v_2)$. ▯

For a given Σ-structure $\mathcal{A}$, a preformula $\varphi(v_1,\ldots,v_k)$ and a sequence $\underline{a}$ of length k over A, for some $k \in \omega$, $\varphi(v_1,\ldots,v_k)$ $[\underline{a}]$ denotes the assignment of values to the free variables of φ such that v_j, if free, is assigned to $\underline{a}(j)$, for $1 \le j \le k$. (The notation $\varphi(v_1,\ldots,v_k)$ indicates that the free variables of the preformula φ are among $v_1,\ldots,v_k$.) The inductive definition of satisfaction is given in the usual way. We show only the pseudo-quantifier case below.

$\mathcal{A} \models {}^m\exists v_{k+1}\varphi\, (v_1,\ldots,v_k)\, [a_1,\ldots,a_k]$, if $\mathcal{A} \models \varphi(v_1,\ldots,v_{k+1})\, [a_1,\ldots,a_k, a_{k+1}]$ for some $a_{k+1} \in Com(\mathcal{A}, \{a_i: v_i \in Free(\varphi), 1 \le i \le k\})$ (w.r.t. $C_m^{\mathcal{A}}$).

Regarding ${}^m\forall$ we have the following, which is easily provable:

$\mathcal{A} \models {}^m\forall v_{k+1}\varphi\, (v_1,\ldots,v_k)\, [a_1,\ldots,a_k]$, if $\mathcal{A} \models \varphi(v_1,\ldots,v_{k+1})\, [a_1,\ldots,a_k, a_{k+1}]$ for all $a_{k+1} \in Com(\mathcal{A}, \{a_i: v_i \in Free(\varphi), 1 \le i \le k\})$ (w.r.t. $C_m^{\mathcal{A}}$).

Definition 2.14. Two Σ-structures $\mathcal{A}$ and $\mathcal{B}$ are *$\mathcal{L}_\Sigma$-equivalent*, denoted $\mathcal{A} \equiv_{\mathcal{L}_\Sigma} \mathcal{B}$, if for any sentence σ of $\mathcal{L}_\Sigma$, $\mathcal{A} \models \sigma$ iff $\mathcal{B} \models \sigma$.

The following lemma, whose counterpart for the first order logic appears as lemma 13.10 in [8], is crucial for the proof of the theorem.*

Lemma 2.15. For every $k,n \in \omega$, $t \subseteq \langle 1,\ldots,k\rangle$ there is a finite subset ${}^m\Phi_n^t$ of ${}^m\mathcal{P}_n^t$ such that every preformula of ${}^m\mathcal{P}_n^t$ is logically equivalent to some preformula of ${}^m\Phi_n^t$.

We are now ready to prove the main result of this section.

Theorem 2.16. Let Σ be a finite signature, and $\mathcal{A}$ and $\mathcal{B}$ be two Σ-structures. Then for all $n \in \omega$, $\mathcal{A} \sim_n \mathcal{B}$ (w.r.t. $C_n^{\mathcal{A}}$ and $C_n^{\mathcal{B}}$) iff for any sentence σ of $\mathcal{L}_\Sigma$ of quantifier depth at most n, $\mathcal{A} \models \sigma$ iff $\mathcal{B} \models \sigma$.

Proof: First we introduce some notations. Let $\underline{a}$ be a sequence of elements of A of length k. Let $t=\langle t_1,\ldots,t_l\rangle$, $0 \le l \le k$, such that $1 \le t_1 < \ldots < t_l \le k$ (see definition 2.10). Then $\underline{a}|t$ denotes the sequence of length l such that $(\underline{a}|t)(j)=\underline{a}(t_j), 1 \le j \le l$. In particular, $\underline{a}|t=\langle\rangle$ if $t=\langle\rangle$. Clearly, it is sufficient to show that for any $n,k \in \omega$, sequences $\underline{a}$ and $\underline{b}$ of length k over A and B, respectively, and a sequence t of length l defined as above such that $\underline{a}|t \in Adm(\mathcal{A},l)$ and $\underline{b}|t \in Adm(\mathcal{B},l)$, $G_n^l(\mathcal{A},C_{k+n}^{\mathcal{A}},\underline{a}|t,\mathcal{B},C_{k+n}^{\mathcal{B}},\underline{b}|t) \in$ II iff for any $\varphi \in {}^{k+n}\mathcal{P}_n^t$, $\mathcal{A} \models \varphi\, [\underline{a}]$ iff $\mathcal{B} \models \varphi$

[$\underline{b}$]. The "only if" part is by induction on n. Here we discuss only the pseudo-quantifier case of the inductive step. Consider the formula ${}^{k+n+1}\exists v_{k+1}\psi$, where $\psi \in {}^{k+n+1}\mathcal{P}_n^{t\cup k+1}$. From the assumption that $G^{\ell}_{n+1}(\mathcal{A}, C^{\mathcal{A}}_{k+n+1}, \underline{a}|t, \mathcal{B}, C^{\mathcal{B}}_{k+n+1}, \underline{b}|t) \in \text{II}$ by lemma 2.6 we have (i) for every $a \in Com(\mathcal{A}, \underline{a}|t)$ there is $b \in Com(\mathcal{B}, \underline{b}|t)$ such that $G^{\ell+1}_n(\mathcal{A}, C^{\mathcal{A}}_{k+n+1}, (\underline{a}|t)a, \mathcal{B}, C^{\mathcal{B}}_{k+n+1}, (\underline{b}|t)b) \in \text{II}$, and (ii) symmetric to (i). But $(\underline{a}|t)a = \underline{a}a|t \cup k+1$ and $(\underline{b}|t)b = \underline{b}b|t \cup k+1$. Hence by inductive hypothesis we have (i′) for every $a \in Com(\mathcal{A}, \underline{a}|t)$ there is $b \in Com(\mathcal{B}, \underline{b}|t)$ such that for any $\psi \in {}^{k+n+1}\mathcal{P}_n^{t\cup k+1}$ $\mathcal{A} \models \psi\,[\underline{a}a]$ iff $\mathcal{B} \models \psi\,[\underline{b}b]$, and (ii′) symmetric to (i′). From (i′), for some $a \in Com(\mathcal{A}, \underline{a}|t)$ $\mathcal{A} \models \psi\,[\underline{a}a]$ implies for some $b \in Com(\mathcal{B}, \underline{b}|t)$ $\mathcal{B} \models \psi\,[\underline{b}b]$. Hence $\mathcal{A} \models {}^{k+n+1}\exists v_{k+1}\psi\,[\underline{a}]$ implies $\mathcal{B} \models {}^{k+n+1}\exists v_{k+1}\psi\,[\underline{b}]$. We get the reverse direction from (ii′). For the "if" part of the theorem we argue as in the proof of theorem 13.11 in [8] using lemma 2.15. ▯

Corollary. Let Σ be a finite signature, and $\mathcal{A}$ and $\mathcal{B}$ be two Σ-structures. Then, $\mathcal{A} \sim \mathcal{B}$ (w.r.t. the sequence $\langle C^{\mathcal{A}}_n, C^{\mathcal{B}}_n \rangle_{n \in \omega}$) iff $\mathcal{A} \equiv_{\mathcal{L}_\Sigma} \mathcal{B}$.

By choosing the compatibility relation as the universal relation in theorem 2.16, we obtain the Ehrenfeucht's characterization. This theorem does not hold in case the predicate symbols are infinitely many. An interested reader may consult [2] for a counter-example.

Given two Σ-structures $\mathcal{A}$ and $\mathcal{B}$, we have considered the game $G^k_n(\mathcal{A}, C^{\mathcal{A}}, \underline{a}, \mathcal{B}, C^{\mathcal{B}}, \underline{b})$ – call this game Γ. Now let $\mathcal{A}'$ and $\mathcal{B}'$ be the Σ'-reducts of $\mathcal{A}$ and $\mathcal{B}$, respectively, and consider the game $G^k_n(\mathcal{A}', C^{\mathcal{A}}, \underline{a}, \mathcal{B}', C^{\mathcal{B}}, \underline{b})$ – call this game Γ'. Any play of Γ' is also a play of Γ, and vice versa. The winning of a play π, when viewed as a play of Γ', is determined by the relations of Σ', and as a play of Γ, by the relations of Σ (cf. definition 2.5). As an immediate consequence, if Player II has a winning strategy in Γ, then the same winning strategy works for Γ'; similarly, if Player I has a winning strategy in Γ', then this strategy works for Γ. By the corollary to theorem 2.16, Player II has a winning strategy in $G_n(\mathcal{A}', C^{\mathcal{A}}_n, \mathcal{B}', C^{\mathcal{B}}_n)$ for every $n \in \omega$ iff $\mathcal{A}' \equiv_{\mathcal{L}_{\Sigma'}} \mathcal{B}'$ iff $\mathcal{A} \equiv_{\mathcal{L}_{\Sigma'}} \mathcal{B}$. The latter "iff" is from the fact that for any formula φ of $\mathcal{L}_{\Sigma'}$, $\mathcal{A}' \models \varphi\,[\underline{a}]$ iff $\mathcal{A} \models \varphi\,[\underline{a}]$, because φ does not involve any predicate symbols from the set $\Sigma - \Sigma'$.

3 Observation Equivalence as Game Equivalence

We establish the connection between our games and the observation equivalence on a particular class of labelled transition systems, called the trace-unique labelled transition systems with silent τ-moves (t-τlts's). Roughly speaking, by defining the compatibility relations in terms of bounded reachability in the given t-τlts's and playing the game on their Λ-reducts, the game equivalence defined in the preceding section coincides with the bounded variant of Milner's observation equivalence [4] on this class of lts's.

Preliminaries

By the term *alphabet* we mean a nonempty set; we refer to the elements of an alphabet as *symbols* or *labels.* We use Δ to denote an alphabet including the special symbol τ. We use Λ to denote the set $\Delta-\{\tau\}$. For an alphabet Δ, Δ^* [resp. Δ^+], denotes the set of all [resp. nonempty] finite strings over Δ. The length of a string u is denoted $lh(u)$. The symbol ε denotes the empty string, the string of length zero.

Definition 3.1. A *labelled transition system* (*lts*) $\mathcal{A}$ with alphabet Δ is a Δ-structure $\mathcal{A}=\langle A, \{\overset{\alpha}{\Rightarrow}^{\mathcal{A}}: \alpha\in\Delta\}\rangle$, where the relations of $\mathcal{A}$ are all binary. We call each relation $\overset{\alpha}{\Rightarrow}^{\mathcal{A}}$ a *transition relation.* In the sequel, we simply write $\overset{\alpha}{\Rightarrow}$ instead of $\overset{\alpha}{\Rightarrow}^{\mathcal{A}}$ whenever $\mathcal{A}$ is understood.

We may think of the elements of the domain A as the states of the machine $\mathcal{A}$ under consideration. For each label $\alpha\in\Delta$, $a_1 \overset{\alpha}{\Rightarrow} a_2$ (i.e. $(a_1,a_2)\in\overset{\alpha}{\Rightarrow}$) indicates a transition from state a_1 upon performing the action α into state a_2. We sometimes distinguish $a_0\in A$ as the *initial* state. Then we use the pair $\langle\mathcal{A},a_0\rangle$ to denote an *lts* $\mathcal{A}$ with initial state a_0.

Definition 3.2. An *lts* $\mathcal{A}=\langle A, \{\overset{\alpha}{\Rightarrow}: \alpha\in\Delta\}\rangle$ satisfying the following axiom scheme is called a *labelled transition system with silent τ-moves* (τ*lts*).

(TC) For each $\alpha\in\Delta$,

$$\forall x_1\forall x_2\forall x_3\,((x_1 \overset{\alpha}{\Rightarrow} x_2 \wedge x_2 \overset{\tau}{\Rightarrow} x_3) \rightarrow x_1 \overset{\alpha}{\Rightarrow} x_3)$$
$$\forall x_1\forall x_2\forall x_3\,((x_1 \overset{\tau}{\Rightarrow} x_2 \wedge x_2 \overset{\alpha}{\Rightarrow} x_3) \rightarrow x_1 \overset{\alpha}{\Rightarrow} x_3).$$

Intuitively speaking, (TC) reflects the "silent" nature of τ. In the sequel, $\mathcal{A}$ and $\mathcal{B}$ denote τ*lts*'s with alphabet Δ. We are ***not*** assuming that our τ*lts*'s are ***image-finite***, i.e. we allow the set $\{a'\in A: a \overset{\alpha}{\Rightarrow} a'\}$ to have infinitely many elements, for $\alpha\in\Delta$, $a\in A$.

Definition 3.3. Given a τ*lts* $\mathcal{A}$, for each $u\in\Lambda^*$, define the *extended transition relation* $\overset{u}{\Rightarrow}$ as follows:

$$\overset{u}{\Rightarrow} = \overset{\tau}{\Rightarrow} \cup \mathrm{Id}_A \text{ if } u=\varepsilon \text{ (}\mathrm{Id}_A\text{ denotes the identity relation on A);}$$
$$\overset{u}{\Rightarrow} = \overset{\lambda}{\Rightarrow} \circ \overset{v}{\Rightarrow} \text{ if } u=\lambda v \text{ for some } \lambda\in\Lambda,\ v\in\Lambda^*.$$

If $\overset{u}{\Rightarrow}$ is not the empty relation, we call u a *trace* of $\mathcal{A}$, i.e. a trace u is a (possibly empty) sequence of observable actions that can take machine $\mathcal{A}$ from one state to another.

Definition 3.4. A τ*lts* $\mathcal{A}$ is called a *trace-unique* τ*lts* (t-τ*lts*) if it satisfies the following axiom scheme:

(TU) For all distinct $u,v\in\Lambda^*$, $\forall x_1\forall x_2\,(x_1\overset{u}{\Rightarrow}x_2 \rightarrow x_1\overset{v}{\not\Rightarrow}x_2)$.

By (TU), in a t-τ*lts* all the directed paths between any two nodes "spell" the same string of labels when τ's are ignored. To put another way, if there is a trace between two nodes, this trace is unique (thus the term "trace-unique").

An immediate consequence of (TU) is that in a ι-τ*lts* $\mathcal{A}$ a $\overset{u}{\nRightarrow}$ a for any a∈ A, u∈ Λ^+, i.e. "no cycles with a non-ε trace". On the other hand, for two distinct elements a_1 and a_2, it is possible to have $a_1 \overset{\varepsilon}{\Rightarrow} a_2$ and $a_2 \overset{\varepsilon}{\Rightarrow} a_1$. Then, it is impossible to separate a_1 and a_2 by using the relations of $\mathcal{A}$ only. Hence, the result of a play will not be affected by choosing a_1 instead of a_2, and vice versa, provided they behave the same way w.r.t. the compatibility relation as well.

Observe that the models of the axioms (TC) and (TU), in general, are not connected (when viewed as graphs). It is known that connectedness cannot be enforced in a first order way. In our later treatment, though, the setup will be such that we deal only with the connected component including the initial state.

We study the behavior of an *lts* by means of a tree representing its behavior. We obtain such a tree by "unfolding" the *lts*. The resulting tree is called a *labelled transition tree* (*ltt*). The synchronization trees of Milner [4] can be viewed as *ltt*'s. An *ltt* is in fact a special kind of *lts*. It has no cycles (even undirected ones); hence it satisfies the axiom (TU). By adding the edges (and only those edges) required by (TC) we get the corresponding τ*lts*. The τ*lts* thus obtained is a ι-τ*lts*.

Observation Equivalence and Its Game Characterization

Milner defined the observation equivalence of processes by means of an ω-indexed chain of equivalence relations starting with the universal relation and obtaining the n+1'st relation as a refinement of the n'th one [4]. The intersection of the whole sequence gives the desired equivalence. Here we present our version of the his weaker variant, called the (bounded) observation equivalence, denoted ≈.

Notation: For each $n \in \omega$, ${}^n\Lambda$ [resp. ${}^{<n}\Lambda$] denotes the set of strings of Λ^* whose length is at most [resp. less than] 2^n.

Definition 3.5. Let $\mathcal{A}$ and $\mathcal{B}$ be two τ*lts*'s. We first define a sequence of relations, $\langle \approx_n \rangle_{n\in\omega}$, between A and B. Let a∈ A and b∈ B.
(1) a $\approx_0$ b.
(2) a $\approx_{n+1}$ b iff for any $u \in {}^n\Lambda$
  (i) for any a′, a $\overset{u}{\Rightarrow}$ a′ implies that there is b′ such that b $\overset{u}{\Rightarrow}$ b′ and a′ $\approx_n$ b′, and
  (ii) for any b′ , b $\overset{u}{\Rightarrow}$ b′ implies that there is a′ such that a $\overset{u}{\Rightarrow}$ a′ and a′ $\approx_n$ b′.

We then define the *(bounded) observation equivalence*, denoted ≈, as follows:

$$a \approx b \text{ iff } a \approx_n b \text{ for all } n \in \omega.$$

By replacing ${}^n\Lambda$ by Λ^* in the above definition we obtain the definition of the unbounded observation equivalence ("an arbitrary sequence of observable actions at a time"), and by replacing ${}^n\Lambda$ by ${}^0\Lambda$ we obtain that of the weak observation equivalence ("at most one observable action at a time"). The weak observation equivalence is indeed weaker than the unbounded observation equivalence, although they coincide on image-finite systems [1]. Note that in this paper the term "observation equivalence" is synonymous with "bounded observation equivalence".

It is easy to prove that the (bounded) observation equivalence coincides with weak observation equivalence, denoted $\approx^W$. In fact, for all $n \in \omega$, $\approx_n \subseteq \approx^W_n$ and $\approx^W_{2^n-1} \subseteq \approx_n$. Also, with an arbitrary bound on the length of strings, "bounded" observation equivalence coincides with $\approx^W$. We choose the binary exponential bound, because in our characterizations the correlation between the number of rounds in the game (and quantifier depth in sentences) and the "depth" of equivalence comes out sharper this way.

Definition 3.6. Given a τlts $\mathcal{A}$, we define the *n-reachability* relation for $\mathcal{A}$, denoted $\Rightarrow_n$, as follows: for all $a, a' \in A$, $a \Rightarrow_n a'$ iff $a \overset{u}{\Rightarrow} a'$ for some $u \in {}^{<n}\Lambda$. Further, we define the *n-history* relation, denoted $\Leftrightarrow_n$, as the symmetric closure of $\Leftrightarrow_n$. For fixed $a \in A$, we define the *n-history relation with starting point a* as $\Leftrightarrow^a_n = \Leftrightarrow_n - \{(a, a'): a \not\Rightarrow_n a', a' \in A\}$.

We adopt $\Leftrightarrow^a_n$ as a compatibility relation, that is, we consider two elements compatible if one of them is reachable from the other within at most 2^n-1 observable steps, and to be compatible with the starting point a, an element must be reachable from a.

Given two t-τlts's $\mathcal{A}$ and $\mathcal{B}$, we shall consider games played on their Λ-reducts, denoted $\mathcal{A}^{-\tau}$ and $\mathcal{B}^{-\tau}$, respectively. The compatibility relations will still be defined in terms of bounded reachability in $\mathcal{A}$ and $\mathcal{B}$. Any play on $\mathcal{A}$ and $\mathcal{B}$ is also a play on $\mathcal{A}^{-\tau}$ and $\mathcal{B}^{-\tau}$, and vice versa, as long as they are played under the same pair of compatibility relations. At the end of a play, however, winning is determined with respect to the relations of $\mathcal{A}^{-\tau}$ and $\mathcal{B}^{-\tau}$, i.e. the $\overset{\varepsilon}{\Rightarrow}$ relations of $\mathcal{A}$ and $\mathcal{B}$ are ignored in determining the outcome of a play. Later we describe some winning strategies for Player II so that he wins with respect to the extended transition relations $\overset{u}{\Rightarrow}$, $u \in \Lambda^+$. We indicate the existence of a winning strategy for Player II in this stronger sense by using the phrase "w.r.t. Λ^+". In the discussion below we use the following terminology. Given a t-τlts $\mathcal{A}$, we say that an element a_3 is *between* a_1 and a_2, if either $a_1 \overset{u_1}{\Rightarrow} a_3 \overset{u_2}{\Rightarrow} a_2$ or $a_2 \overset{u_1}{\Rightarrow} a_3 \overset{u_2}{\Rightarrow} a_1$ for some u_1, $u_2 \in \Lambda^*$. If $a_1 \overset{u}{\Rightarrow} a_2$ for some $u \in \Lambda^*$, we say that a_2 is *downstream* from a_1, and a_1 is *upstream* from a_2.

Suppose Player I moves downstream from a previously chosen element, say, in $\mathcal{A}$. Now Player II must also move downstream from the corresponding previously chosen element in $\mathcal{B}$. Otherwise, Player I can easily exploit the disparity in directions by using what we call his "middlemost strategy", which involves choosing a middlemost point between two appropriate previously chosen points. Part (1) of lemma 3.7 below makes this point precise. Furthermore, suppose elements a_1 and a_2 from $\mathcal{A}$, and b_1 and b_2 from $\mathcal{B}$ are already picked in a play. If the trace from a_1 to a_2 is different from the one from b_1 to b_2, then Player I can exploit this difference to win within a sufficiently large number of rounds, again using his middlemost strategy. This idea is formulated as the part (2) of lemma 3.7 below. On the other hand, if the traces are the same, Player II always wins, provided the play is restricted to the regions between a_1 and a_2 in $\mathcal{A}$, and between b_1 and b_2 in $\mathcal{B}$. Note that due to the existence of the τ-transitions, these regions need not be isomorphic. Yet Player II can always keep the traces between corresponding selections the same. We call this strategy the "trace matching" strategy of Player II. Lemma 3.8 below makes this second point precise.

Lemma 3.7. Let $\mathcal{A}$ and $\mathcal{B}$ be two ι-τlts's. Suppose $a_1 \overset{u}{\Rightarrow} a_2, b_1 \overset{v}{\Rightarrow} b_2$ for some $u\in \Lambda^+$, $v\in \Lambda^*$. Let n and m be integers such that $lh(u)\leq 2^n$ and $lh(u), lh(v)\leq 2^m-1$. Then,
(1) $G_n^2(\mathcal{A}^{-\tau}, \Leftrightarrow_m, a_1a_2, \mathcal{B}^{-\tau}, \Leftrightarrow_m, b_2b_1)\in$ I;
(2) $G_n^2(\mathcal{A}^{-\tau}, \Leftrightarrow_m, a_1a_2, \mathcal{B}^{-\tau}, \Leftrightarrow_m, b_1b_2)\in$ I if $u\neq v$.
Moreover, in both (1) and (2), Player I needs to select only those elements from $\mathcal{A}$ that are between a_1 and a_2.

Lemma 3.8. Let $\mathcal{A}$ and $\mathcal{B}$ be two ι-τlts's. Suppose $a_1 \overset{u}{\Rightarrow} a_2, b_1 \overset{u}{\Rightarrow} b_2$ for some $u\in \Lambda^*$. Then, $G_n^2(\mathcal{A}^{-\tau}, [a_1,a_2], a_1a_2, \mathcal{B}^{-\tau}, [b_1,b_2], b_1b_2)\in$ II w.r.t. Λ^+, for all $n\in\omega$, where the compatibility relation $[a_1,a_2]$ [resp. $[b_1,b_2]$] is defined such that two elements are compatible with each other if both are between a_1 and a_2 (in $\mathcal{A}$) [resp. b_1 and b_2 (in $\mathcal{B}$)].

The following is the main result of this section.

Theorem 3.9. Let $\mathcal{A}$ and $\mathcal{B}$ be two ι-τlts's. For all $n,m\in\omega$, $a\in A$, $b\in B$,
(1) $a \approx_{m+n-1} b$ and $m\geq n$ implies $G_n^1(\mathcal{A}^{-\tau}, \Leftrightarrow_m^a, a, \mathcal{B}^{-\tau}, \Leftrightarrow_m^b, b)\in$ II, w.r.t. Λ^+;
(2) $G_n^1(\mathcal{A}^{-\tau}, \Leftrightarrow_n^a, a, \mathcal{B}^{-\tau}, \Leftrightarrow_n^b, b)\in$ II implies $a \approx_n b$.

Proof: We first prove part (1). As a special case, we assume n=0, and easily see that the consequence holds. Proceed with induction on n. For the basis (for n=1), assume $a \approx_m b$, for some $m\geq 1$. Suppose Player I picks some a′ from $\mathcal{A}$ such that $a \overset{u}{\Rightarrow} a'$, $u\in {}^{<m}\Lambda$. There are $a_1,a_2,\dots,a_m$ (not necessarily distinct) such that $a \overset{u_1}{\Rightarrow} a_1 \overset{u_2}{\Rightarrow}\dots\overset{u_m}{\Rightarrow} a_m = a'$, for some $u_i\in {}^{m-i}\Lambda$, $1\leq i\leq m$, where $u_1\dots u_m=u$. Then, by definition 3.5, there are $b_1,b_2,\dots,b_m$ such that $b \overset{u_1}{\Rightarrow} b_1 \overset{u_2}{\Rightarrow} \dots \overset{u_m}{\Rightarrow} b_m = b'$ and $a_i \approx_{m-i} b_i$, $1\leq i\leq m$. Hence $b \overset{u}{\Rightarrow} b'$, so we have $G_0^2(\mathcal{A}^{-\tau}, \Leftrightarrow_m^a, aa', \mathcal{B}^{-\tau}, \Leftrightarrow_m^b, bb')\in$ II, w.r.t. Λ^+. By symmetry and by lemma 2.6, $G_1^1(\mathcal{A}^{-\tau}, \Leftrightarrow_m^a, a, \mathcal{B}^{-\tau}, \Leftrightarrow_m^b, b)\in$ II. Now let $n\geq 1$, and assume (1) for n. Pick $m\geq n+1$, and suppose $a \approx_{m+n} b$. We shall show that there is a winning strategy (w.r.t. Λ^+) for Player II in the game $G_{n+1}^1(\mathcal{A}^{-\tau}, \Leftrightarrow_m^a, a, \mathcal{B}^{-\tau}, \Leftrightarrow_m^b, b)$. By definition 3.5, (i) for any $a'\in A$, $u\in {}^{m+n-1}\Lambda$, $a \overset{u}{\Rightarrow} a'$ implies that there is $b'\in B$ such that $b \overset{u}{\Rightarrow} b'$ and $a' \approx_{m+n-1} b'$, and (ii) symmetric to (i). In the first round, suppose Player I picks some $a' \in Com(\mathcal{A}^{-\tau}, a)$. By the definition (3.6) of the compatibility relation, $a \overset{u}{\Rightarrow} a'$ for some $u\in {}^{<m}\Lambda$ (recall that the history starts at a). Then Player II picks b′ as given by (i). Also, $b' \in Com(\mathcal{B}^{-\tau}, b)$. Hence, after the first round, we have $a \overset{u}{\Rightarrow} a'$, $b \overset{u}{\Rightarrow} b'$, and $a' \approx_{m+n-1} b'$. By the inductive hypothesis, $G_n^1(\mathcal{A}^{-\tau}, \Leftrightarrow_m^{a'}, a', \mathcal{B}^{-\tau}, \Leftrightarrow_m^{b'}, b')\in$ II, w.r.t. Λ^+. We claim $G_n^2(\mathcal{A}^{-\tau}, \Leftrightarrow_m^a, aa', \mathcal{B}^{-\tau}, \Leftrightarrow_m^b, bb')\in$ II, w.r.t. Λ^+. The idea is that whenever Player I picks some a″ such that $a' \Leftrightarrow_m^a a''$ – we call this a "forward" move, Player II picks b″ by his strategy in the game G_n^1, i.e. he

plays as dictated by the observation equivalence; whenever Player I picks a″ such that $a \overset{u_1}{\Rightarrow} a'' \overset{u_2}{\Rightarrow} a'$ where $u_1 u_2 = u$ – we call this an "inside" move, Player II responds according to his trace matching strategy as in the game $G^2_n(\mathcal{A}^{-\tau}, [a,a'], aa', \mathcal{B}^{-\tau}, [b,b'], bb')$ in lemma 3.8, i.e. he chooses some b″ such that $b \overset{u_1}{\Rightarrow} b'' \overset{u_2}{\Rightarrow} b'$. The situation is of course symmetric if Player I makes a move on $\mathcal{B}$. To make sure that this is indeed a winning strategy for Player II we observe that the interaction of the forward moves with the inside moves does not affect the winning of Player II adversely. By symmetry and by lemma 2.6, $G^1_{n+1}(\mathcal{A}^{-\tau}, \Leftrightarrow^a_m, a, \mathcal{B}^{-\tau}, \Leftrightarrow^b_m b) \in II$.

Proof of part (2): We establish by induction that for every $n \in \omega$, (*) $a \not\approx_n b$ implies $G^1_n(\mathcal{A}^{\tau}, \Leftrightarrow^a_n, a, \mathcal{B}^{\tau}, \Leftrightarrow^b_n, b) \in I$ for $a \in A$, $b \in B$. This is vacuously true for $n=0$. Let $n \in \omega$, and assume (*) for n. Suppose $a \not\approx_{n+1} b$. Then, by definition 3.5, (i) there is $a' \in A$ and $u \in {}^n\Lambda$ such that $a \overset{u}{\Rightarrow} a'$, but there is no b′ with $b \overset{u}{\Rightarrow} b'$ and $a' \approx_n b'$, or (ii) symmetric to (i). In the first round, if (i) holds, Player I chooses a′ from $\mathcal{A}$ as given by (i); otherwise he chooses b′ from $\mathcal{B}$ as given by (ii). In other words, he plays as dictated by the inequivalence. Without loss of generality, assume the former case, and consider the possible responses of Player II.

Case 1: Player II picks b′ with $b \overset{u}{\Rightarrow} b'$. Then $a' \not\approx_n b'$. By the inductive hypothesis, $G^1_n(\mathcal{A}^{\tau}, \Leftrightarrow^{a'}_n, a', \mathcal{B}^{\tau}, \Leftrightarrow^{b'}_n, b') \in I$. We claim $G^2_n(\mathcal{A}^{\tau}, \Leftrightarrow^a_{n+1}, aa', \mathcal{B}^{\tau}, \Leftrightarrow^b_{n+1}, bb') \in I$. In this game, Player I employs his strategy in G^1_n. As long as Player II moves within the region defined by $\Leftrightarrow^{a'}_n$ and a′ in $\mathcal{A}$ [resp. $\Leftrightarrow^{b'}_n$ and b′ in $\mathcal{B}$] – call this region the "$\not\approx_n$-region", Player I wins in G^2_n. In game G^2_n, though, Player II can either move upstream or far downstream from the $\not\approx_n$-region. In the former case Player I proceeds with his middlemost strategy as in lemma 3.7(1), and in the latter case (cf. Case 2 below) he proceeds with his middlemost strategy as in lemma 3.7(2). We omit further details of this analysis.

Case 2: Player II picks b′ such that $b \overset{v}{\Rightarrow} b'$, $u \neq v$. By lemma 3.7(2), Player I uses his middlemost strategy between a and a′ in game $G^2_n(\mathcal{A}^{-\tau}, \Leftrightarrow_{n+1}, aa', \mathcal{B}^{-\tau}, \Leftrightarrow_{n+1}, bb')$. Hence, Player I wins in any case. By lemma 2.6, $G^1_{n+1}(\mathcal{A}^{-\tau}, \Leftrightarrow^a_m, a, \mathcal{B}^{-\tau}, \Leftrightarrow^b_m, b) \in I$. []

Corollary. Let $\langle \mathcal{A}, a_0 \rangle$ and $\langle \mathcal{B}, b_0 \rangle$ be two *t-τlts*'s.
Then, $a_0 \approx b_0$ iff $G^1_n(\mathcal{A}^{-\tau}, \Leftrightarrow^{a_0}_n, a_0, \mathcal{B}^{-\tau}, \Leftrightarrow^{b_0}_n, b_0) \in II$ for all $n \in \omega$.

We use the trees depicted in figure 3 (taken from [1]) to argue that our game characterization cannot be extended for unbounded observation equivalence. $\mathcal{A}$ and $\mathcal{B}$ are (bounded) observation equivalent, but they are not unbounded observation equivalent. Furthermore, they are elementarily equivalent. Hence a logical characterization of the unbounded variant of observation equivalence cannot be obtained within the first order framework.

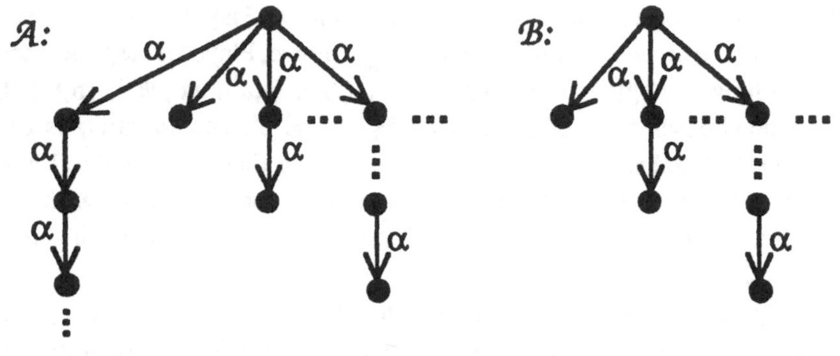

Figure 3. Two observation equivalent $\tau\ell ts$'s which are not image-finite.

4 An Adequate Logic for Observation Equivalence

By combining the results of the two preceding sections we now obtain a logical characterization of the (bounded) observation equivalence of the ι-$\tau\ell ts$'s.

Consider the logic $\mathcal{L}_\Sigma$ defined in Section 2. Given the set Λ of labels (not including the special symbol τ), the signature Σ is chosen such that its set of symbols is $\Lambda \cup \{\Leftrightarrow_m : m \in \omega\}$. We expand $\mathcal{L}_\Sigma$ thus defined, by adding a constant symbol r to be interpreted as the initial state (root) of the ι-$\tau\ell ts$. The compatibility predicate symbols $\Leftrightarrow_m$, $m \in \omega$, will be first order definable in terms of the relations of the $\tau\ell ts$'s. We call the new language $\mathcal{L}^+_\Lambda$. The interpretation of the pseudo-quantifiers ${}^m\exists$ and ${}^m\forall$ of $\mathcal{L}^+_\Lambda$ in the first order logic has to be slightly different so as to take r into account; we modify the translation defined in 2.11 accordingly. For a preformula φ of $\mathcal{L}^+_\Lambda$ with $\mathcal{F}ree(\varphi) - \{v_i\} = \{v_{t_1}, \ldots, v_{t_l}\}$, we now have

$$({}^m\exists v_i \varphi)^* = \exists v_i((\wedge_{1\le j\le l}\, v_{t_j} \Leftrightarrow_m v_i) \wedge r \Leftrightarrow_m v_i \wedge \varphi^*)$$
$$({}^m\forall v_i \varphi)^* = \forall v_i(((\wedge_{1\le j\le l}\, v_{t_j} \Leftrightarrow_m v_i) \wedge r \Leftrightarrow_m v_i) \to \varphi^*).$$

Lemma 4.1. Given two ι-$\tau\ell ts$'s, $\langle \mathcal{A}, r^{\mathcal{A}}\rangle$ and $\langle \mathcal{B}, r^{\mathcal{B}}\rangle$ with a finite alphabet Δ, let $\langle \Leftrightarrow^{\mathcal{A}}_n \rangle_{n\in\omega}$ [resp. $\langle \Leftrightarrow^{\mathcal{B}}_n \rangle_{n\in\omega}$] be the sequence of n-history relations with starting point $r^{\mathcal{A}}$ [resp. $r^{\mathcal{B}}$], defined as in definition 3.6. Then for all $n \in \omega$, $G^1_n(\mathcal{A}^{-\tau}, \Leftrightarrow^{\mathcal{A}}_n, r^{\mathcal{A}}, \mathcal{B}^{-\tau}, \Leftrightarrow^{\mathcal{B}}_n, r^{\mathcal{B}}) \in \mathrm{II}$ iff for any sentence σ of $\mathcal{L}^+_\Lambda$ of quantifier depth at most n, $\mathcal{A} \models \sigma$ iff $\mathcal{B} \models \sigma$.
Proof: Essentially the same as the proof of theorem 2.16. ▯

Corollary. Given two ι-$\tau\ell ts$'s, $\langle \mathcal{A}, r^{\mathcal{A}}\rangle$ and $\langle \mathcal{B}, r^{\mathcal{B}}\rangle$ with a finite alphabet Δ. Then for all $n \in \omega$ $G^1_n(\mathcal{A}, \Leftrightarrow^{\mathcal{A}}_n, r^{\mathcal{A}}, \mathcal{B}, \Leftrightarrow^{\mathcal{B}}_n, r^{\mathcal{B}}) \in \mathrm{II}$ iff $\mathcal{A} \equiv_{\mathcal{L}^+_\Lambda} \mathcal{B}$.

Now we are ready to connect the observation equivalence and $\mathcal{L}^+_\Lambda$-equivalence through the idea of game equivalence. Combining the above corollary with the corollary to theorem 3.9 we directly get the following main result.

Theorem 4.2. For any two ι-$\tau\ell ts$'s, $\langle \mathcal{A}, r^{\mathcal{A}}\rangle$ and $\langle \mathcal{B}, r^{\mathcal{B}}\rangle$, $r^{\mathcal{A}} \approx r^{\mathcal{B}}$ iff $\mathcal{A} \equiv_{\mathcal{L}^+_\Lambda} \mathcal{B}$.

We have developed the game characterization of observation equivalence by playing our game on the Λ-reducts of t-$\tau\ell ts$'s. We cannot get the analogous result for arbitrary $\tau\ell ts$'s by playing the same game. Yet it is desirable to relate the logic $\mathcal{L}_{\Lambda}^{+}$ with arbitrary $\tau\ell ts$'s. To this end, we unfold a given $\tau\ell ts$ to a t-$\tau\ell ts$, which is observation equivalent to the given $\tau\ell ts$. (An automata theoretic construction for unfolding is presented in [7,§ 4.2]. For an axiomatic treatment of unfolding, see [5].) Then we see that two $\tau\ell ts$'s are observation equivalent iff their respective t-$\tau\ell ts$'s are $\mathcal{L}_{\Lambda}^{+}$-equivalent.

Acknowledgments. The help received from George C. Nelson and M.N. Muralidharan is deeply appreciated.

References

1. S.D. Brookes, W.C. Rounds: Behavioral equivalence relations induced by programming logics. Lecture Notes in Computer Science 154 (Proc. ICALP), Berlin: Springer-Verlag, 1983, pp. 97-108
2. A. Ehrenfeucht: An application of games to the completeness problem for formalized theories. Fundamenta Mathematicae XLIX, 1961, pp.129-141
3. M. Hennessy, R. Milner: Algebraic laws for nondeterminism and concurrency. Journal of the ACM 32, no. 1, January 1985, pp. 137-161
4. R. Milner: A calculus of communicating systems. Lecture Notes in Computer Science 154, Berlin: Springer-Verlag, 1980
5. G.C. Nelson, H. Oğuztüzün: Characterization of finitely axiomatizable labelled transition trees. Submitted for publication
6. H. Oğuztüzün: A game characterization of the observational equivalence of processes (extended abstract). First Int. Conf. on Algebraic Methodology and Software Technology (AMAST), Iowa City, Iowa, May 22-24, 1989, pp. 195-196
7. H. Oğuztüzün: A logical characterization of the observation equivalence of processes. Ph.D Thesis, University of Iowa, Iowa City, Iowa, August 1991
8. J.G. Rosenstein: Linear Orderings. New York: Academic Press, 1982

Ordinal Processes in Comparative Concurrency Semantics

S. Pinchinat*

Laboratoire d'Informatique Fondamentale
et d'Intelligence Artificielle,
Institut Imag - CNRS,
Grenoble - FRANCE

Abstract

When comparing concurrency semantics, one of the main difficulties is to find counter-examples to show that two equivalences do not coincide, which takes on considerable proportions if these equivalences coincide in the framework of finitely branching programs. The study of Ordinal Processes of Klop provides us with a wide family of simple counter-examples. The considered semantics are induced by temporal logics CTL, CTL^* and L_μ. We pursue this approach by, in addition, considering CCS-like contexts built up from the *tyft* operators of Groote and Vaandrager.

1 Introduction

There exist several ways to define the semantics of processes. Main cases, among others are through a behavioural equivalence (e.g. bisimulation [Par81] and its variants) or through a logical language (e.g. through modal logics such as HML [HM85]). Each one of these frameworks has its own advantages and finding connections between them guarantees a better understanding of the domain.

A comparison of two different views of processes appears in the so-called *Adequacy* result of Hennessy and Milner [HM85]. They have shown that bisimulation coincides with the equivalence generated by the modal logic HML. However, the result only applies to finitely branching systems, i.e. systems where in any given state, there is only a finite number of possible choices.

There exist, in the literature, several comparisons between different concurrency semantics [HM85, BR83, Mil80, Klo88, SP90, Her87]. In order to show that two semantics differ, one of the main difficulties is to generate appropriate counter-examples. This difficulty takes on considerable proportions when the compared equivalences coincide in the framework of non-deterministic programs with the *finitely branching*

*LIFIA-IMAG, 46 Av. Félix Viallet,38031 GRENOBLE Cedex, FRANCE.E-mail:sp@lifia.imag.fr *restriction* since the counter-examples have to be infinitely branching systems.

In [Klo88], in which bisimulation and variants are compared, a family of models, called *ordinal processes*, is introduced and seems to be a pleasant tool for generating counter-examples. This idea is used again in [SP90] for the comparison between some variants of bisimulation and the semantics induced by the branching-time temporal logic CTL (where to processes are equivalent if they satisfy the same formulae). These study is pursued in the general framework of *infinitely branching* graphs, which reveals to be very interesting as the adequacy results proved to be different.

Continuing work initiated in [SP90], we study the cases of the logics CTL, CTL^* and L_μ. Moreover, we extend the investigation when in addition contexts are introduced: two processes are equivalent if embedded in any same context, they satisfy the same formulae. This topic is very close to the full abstraction problem in the algebraic semantics of processes.

In this paper, instead of considering contexts defined by a fixed set of combinators as is usual, we allow all the *tyft* operators of [GV88] (bisimulation is a congruence w.r.t. all of them). We also propose a new class, so-called tyft *sans copy* (this class contains all the CCS combinators).

In section 2, we present the framework: labeled graphs, temporal logics, etc. Section 3 introduces the notion of distinguishing power and studies the different semantics. Section 4 contains the presentation of tyft operators and tyft "sans" copy, as well as the study of equivalences induced by logic CTL and contexts built up from these operators. We conclude in section 5 with suggestions for future work.

2 Concurrency Semantics

The semantics we shall present are based on equivalences between processes.

2.1 Labeled transition systems

We assume a nonempty set $A = \{a, \ldots\}$ of *action names* or *labels*. A (*labeled*) *transition system* (or *rooted labeled graph*) over A is a tuple $G = (Q, A, (\xrightarrow{a})_{a \in A}, r)$, where

- $Q = \{q, \ldots\}$ is any set of states,
- r is a particular element of Q, called the root (of G),
- $\xrightarrow{a} \subseteq Q \times Q$, for any $a \in A$

In the following and when it is clear from the context, we often identify the transition system with its root: we refer to r instead of G itself. And in general, a state q denotes the subgraph rooted at q.

A tuple (q, q') of $\xrightarrow{a}$ is called an *a-step* or an *a-transition*. We write $q \xrightarrow{a} q'$ when $(q, q') \in \xrightarrow{a}$ and $q \rightarrow q'$ if $q \xrightarrow{a} q'$ for some a. We then say that q' is a *successor* of q. A state with no successor is *blocked*.

A *path* in G is a maximal sequence σ of the form $q_0 \xrightarrow{a_1} q_1 \xrightarrow{a_2} \ldots q_n \ldots$ where $q_{i-1} \xrightarrow{a_i} q_i$ is a step of G. It may be finite (ending with a blocked state) or infinite. $|\sigma|$ denotes the length of σ, i.e. the number of steps. For all paths $\sigma = q_0 \xrightarrow{a_1} q_1 \xrightarrow{a_2} \ldots$ and $i \leq |\sigma|$, we define $\sigma(i)$ as the state q_i and σ^i as the suffix path $q_i \xrightarrow{a_{i+1}} q_{i+1} \xrightarrow{a_{i+2}} \ldots$. We write $Ex(q) = \{\sigma, \pi, \ldots\}$ for the set of all paths starting from q.

We say that a graph is *finitely-branching* (f.b.) (or *image finite*) if for all $q \in Q$, for all $a \in A$, the set $\{q' \in Q \mid q \xrightarrow{a} q'\}$ is finite. Otherwise the graph is not f.b. also called *infinitely branching (i.b.)*. We explicitly do not restrict to f.b. graphs.

When the graph is labeled by the same action, we write $G = (Q, \rightarrow, r)$, where the singleton set A of action names and the label of $\rightarrow$ are omitted.

In the following we present several equivalences over states of the same transition systems. This is no loss of generality compared to equivalences between graphs since any state can be seen as the rooted sub-graph it canonically induces.

2.2 Behavioral equivalences

Given $G = (Q, A, (\xrightarrow{a})_{a\in A}, r)$, we define *bisimulation* as the largest symmetric relation $\underline{\leftrightarrow}$ over Q for which, whenever $q_1 \underline{\leftrightarrow} q_2$ and $q_1 \xrightarrow{a} q_1'$, there exists q_2' s.t. $q_2 \xrightarrow{a} q_2'$ and $q_1' \underline{\leftrightarrow} q_2'$. This definition is due to Park. It characterizes bisimulation as the greatest fixed point of some monotonic mapping $\mathcal{F}$ over the complete lattice $2^{Q\times Q}$ of relations over Q. It is well known (e.g. [Mil90]) that bisimulation can be defined iteratively by using the mapping $\mathcal{F}$. Given a relation R over Q, $\mathcal{F}(R)$ is defined by: for all $q_1, q_2 \in Q, q_1\mathcal{F}(R)q_2$ iff for all steps $q_1 \xrightarrow{a} q_1'$ there is a $q_2 \xrightarrow{a} q_2'$ with $q_1' R q_2'$ and reciprocally. We define the approximations of $\underline{\leftrightarrow}$ by

$$\equiv_0 \stackrel{\text{def}}{=} Q \times Q \qquad \equiv_{n+1} \stackrel{\text{def}}{=} \mathcal{F}(\equiv_n) \qquad \equiv_\omega \stackrel{\text{def}}{=} \bigcap_n \equiv_n$$

Then, if G is finitely branching, $\mathcal{F}$ is anticontinuous, implying that $\underline{\leftrightarrow} = \equiv_\omega$. In the general case of i.b. graphs the approximation $\equiv_\omega$ is larger than $\underline{\leftrightarrow}$, does not "distinguish termination" and satisfies the approximation induction principle of [BBK87].

2.3 Temporal logics

Another way to define the semantics of processes is through a logical language. Pnueli and Lamport initiated the use of Temporal Logics for specifying and proving the correctness of parallel programs. These logics belong to the class of Modal Logics originally developed by philosophers to study different "modes" of truth. Temporal Logic provides a formal system for describing and reasoning about how the truth values of assertions vary with time. Such a system is equipped with various temporal operators. Typical temporal operators are *sometime* and *always*. *sometimes* A is true now if there is a future moment at which A becomes true and *always* B is true if B is true at all future moments.

In this section, we present three branching-time temporal logics, respectively called CTL, CTL^* and L_μ, with increasing expressive power. These logics are interpreted in transition systems and the formulae of the language denote properties of states in the graph.

2.3.1 *CTL* logic

CTL [CE81] is a paradigmatic example. It is possible to express the *sometimes* and *always* modalities. The syntax of *CTL* is given by

$$(CTL \ni)\ f, g ::= \top \mid f \wedge g \mid \neg f \mid \mathbf{EX} f \mid \mathbf{E} f \mathbf{U} g \mid \mathbf{A} f \mathbf{U} g$$

Informally, in a given graph G, the formula $\top$ ("true") holds at any state. A formula of the form $f \wedge g$ holds at state q if f and g hold at q. $\neg f$ holds at q if f does not hold at q. $\mathbf{EX} f$ can be read as "there exists a next state satisfying f", $\mathbf{E} f \mathbf{U} g$ as " there exists a path (starting from here) satisfying f all along until g is eventually satisfied", and $\mathbf{A} f \mathbf{U} g$ as "all paths (starting from here) satisfy f all along until g is eventually satisfied". For instance, the assertion "*sometimes*" f is definable in *CTL* by $\mathbf{E} \top \mathbf{U} f$. The formal semantics of the formulae is given below, where $q \models f$ means that "f holds at q" or equivalently that "q satisfies f". We introduce a few practical abbreviations: $f \vee g = \neg(\neg f \wedge \neg g)$, $\mathbf{AX} f = \neg \mathbf{EX} \neg f$.

The semantics of the formulae is defined by induction over their structure: For all $q, q_0 \in Q$ and $f, g \in CTL$,

- $q \models \top, q \models f \wedge g, q \models \neg f$ have their obvious definition.
- $q \models \mathbf{EX} f$ iff there exists some $q \to q'$ s.t. $q' \models f$
- $q_0 \models \mathbf{E} f \mathbf{U} g$ iff there exists a path $q_0 \to q_1 \to q_2 \to \ldots$ and some $n \in \mathbf{N}$ s.t. $q_n \models g$ and $q_i \models f$ for all $i < n$.
- $q_0 \models \mathbf{A} f \mathbf{U} g$ iff for all path $q_0 \to q_1 \to q_2 \to \ldots$ there is some $n \in \mathbf{N}$ s.t. $q_n \models g$ and $q_i \models f$ for all $i < n$.

The fragment of *CTL* which contains only the operators $\top$, $\wedge$, $\neg$, $\mathbf{EX}$ is the so-called *HML* logic of [HM85].

2.3.2 *CTL** logic

The syntax of CTL^* (from [EH86]) is given by

$$(CTL^* \ni)\ f, g ::= \top \mid f \wedge g \mid \neg f \mid \mathbf{X} f \mid f \mathbf{U} g \mid \forall f$$

Formulae of CTL^* hold of paths. We write $\pi \models f$ with the following definition:

- $\top$, $f \wedge g$, $\neg f$ have their obvious definitions,
- $\pi \models \mathbf{X} f$ iff $\pi^1 \models f$,

- $\pi \models f \mathbf{U} g$ iff there exists some n s.t. $\pi^n \models g$ and for all $i < n$, $\pi^i \models f$,
- $\pi \models \forall f$ iff for all $\sigma \in Ex(\pi(0))$, $\sigma \models f$.

We then extend the semantics over states by: $q \models f \overset{\text{def}}{\Leftrightarrow}$ for all $\pi \in Ex(q), \pi \models f$
Writing $\exists f$ for $\neg\forall\neg f$, it is easy to check that CTL^* includes CTL: for instance, the CTL formulae $\mathbf{E}\, f \,\mathbf{U}\, g$ and $\mathbf{A}\, f \,\mathbf{U}\, g$ correspond respectively to the CTL^* formulae $\exists(f \mathbf{U}\, g)$ and $\forall(f \,\mathbf{U}\, g)$.

2.3.3 L_μ logic

The μ-calculus, L_μ, of [Koz83] has the following syntax

$$(L_\mu \ni) \;\; f, g ::= \top \mid f \wedge g \mid \neg f \mid \mathbf{EX} f \mid X \mid \mu X.f(X)$$

where X belongs to a set V of variables. $\mu X.f(X)$ denotes the least solution of $X \equiv f(X)$. Analogously, μ behaves as a quantifier in First Order Logic and induces the usual notions of free and bounds occurrences of variables, closed formulae,.... In L_μ, we only allow closed formulae, where for all subformulae $\mu X.f(X)$, any free occurrence of X in $f(X)$ is

- *guarded*, i.e. under the scope of at least one $\mathbf{EX}$,
- *positive*, i.e. under the scope of an even number of $\neg$. (This ensures that $f(X)$ is a monotonic function of X and then that the least fixpoint of $X \longmapsto f(X)$ exists).

It is known that CTL combinators can be defined in L_μ, where CTL modalities are characterized in fixpoint style [EC80]: $\mathbf{E}\, f \,\mathbf{U}\, g = \mu X.(g \vee f \wedge \mathbf{EX}\, X)$ and $\mathbf{A}\, f \,\mathbf{U}\, g = \mu X.(g \vee f \wedge \mathbf{EX} \top \wedge \mathbf{AX}\, X)$. L_μ can encode CTL^* (see [Dam90]) and allows expression of extended modalities such as "B is true at all even moments along all futures", which is captured by $\nu X.(B \wedge \mathbf{AX}\,\mathbf{AX}\, X)$. We refer to [Koz83] for the formal semantics of L_μ.

2.4 Logical equivalences

Once a logical language L is defined, the underlying semantics of processes is easely definable: we say that two states q and q' are L-equivalent, written $q \sim_L q'$, if they satisfy the same formulae of L. Formally, for all states q and q', $q \sim_L q' \overset{\text{def}}{\Leftrightarrow}$ (for all $f \in L, q \models f$ iff $q' \models f$).

3 Comparative Concurrency semantics

3.1 Ordinal Processes and Distinguishing Power

The comparison method is based on the use of particular non-f.b. graphs, called *Ordinal Processes*, [Klo88]. We recall their definition, their main properties and we use them in order to formalize the discriminating power of the semantics equivalences, called the *Distinguishing Power*.

Given an ordinal λ, the associated ordinal graph is $G_\lambda = (\lambda + 1, \rightarrow, \lambda)$, where we have a transition $\beta \rightarrow \alpha$ iff $\alpha < \beta \leq \lambda$. A basic property is that G_λ has no infinite path (as $>$ is well-founded) and that $\rightarrow$ is transitive. Note that G_λ is non-f.b. as soon as $\lambda \geq \omega$.

Given a semantic equivalence $\sim$ over graphs, we define the *distinguishing power* of $\sim$, written $dp(\sim)$, as the least ordinal γ s.t. for all ordinals α and β [1] $\alpha \sim \beta$ whenever $\alpha, \beta \geq \gamma$. Note that this ordinal needs not always exist. Clearly, over G_λ, $\underline{\leftrightarrow}$ reduces to equality [SP90]. Indeed, suppose $\alpha_0 \underline{\leftrightarrow} \alpha_1$ with $\alpha_1 < \alpha_0 \leq \lambda$. Then $\alpha_0 \rightarrow \alpha_1$, and $\alpha_0 \underline{\leftrightarrow} \alpha_1$ implies that there exists $\alpha_1 \rightarrow \alpha_2$ s.t. $\alpha_1 \underline{\leftrightarrow} \alpha_2$. Carrying on, we shall build an infinite path $\alpha_0 \rightarrow \alpha_1 \rightarrow \alpha_2 \rightarrow \ldots$ contradicting the well-foundedness of $>$ over ordinals.

Thus, if the d.p. of an equivalence exists, this equivalence is strictly larger than bisimulation on ordinal graphs. Moreover, if $dp(\sim) = \gamma < \gamma' = dp(\sim')$ then $\sim \not\Rightarrow \sim'$ with the immediate counter-examples γ and γ' ($\gamma \sim \gamma'$ but $\gamma \not\sim' \gamma'$).

3.2 The distinguishing power of Temporal Logics

Let f be a formula of some logic. We define a relation $\sim_f$ between states by $q \sim_f q'$ iff ($q \models f \Leftrightarrow q' \models f$) and we call the *height* of f, written $h_L(f)$, the least ordinal γ s.t. $\alpha \sim_f \beta$ whenever $\alpha, \beta \geq \gamma$.

Now, for a logic L, it is easy to check that $dp(\sim_L)$ (written $dp(L)$ in the following) is the least upper bound of the $h_L(f)$ for all $f \in L$.

In this section, we prove, that the height of CTL, CTL^* and L_μ formulae is an integer.

3.2.1 *CTL* logic

Concerning the CTL logic, the proof is straightforward. We define for each $f \in CTL$ its *depth* (also called *modal height*), corresponding to the maximum number of nested operators **EX** and show $d(f)$ is a bound on $h_{CTL}(f)$. Formally we define $d : CTL \rightarrow$ N by induction over the structure of the formula:

$$d(\mathrm{T}) = 0 \qquad d(\mathbf{EX}\, f) = d(f) + 1$$
$$d(f \wedge g) = max(d(f), d(g)) \qquad d(\mathbf{E}\, f\, \mathbf{U}\, g) = max(d(f), d(g))$$
$$d(\neg f) = d(f) \qquad d(\mathbf{A}\, f\, \mathbf{U}\, g) = max(d(f), d(g))$$

Lemma 1 [SP90] *for all $f \in CTL$, $\alpha, \beta \geq d(f)$ implies $\alpha \sim_f \beta$*

Proof We have to show that for all $\alpha, \beta \geq d(f), \alpha \sim_f \beta$. We prove by induction over the structure of f that $\alpha \sim_f d(f)$.

---

[1] α and β are seen as the ordinal sub-graph they canonically induce

- Assume $f = \mathbf{EX}\, g$ and $n = d(g)$, and consider $\alpha \geq n+1$. If $n+1 \models \mathbf{EX}\, f$ there exists $\beta < n+1 \leq \alpha$ s.t. $\beta \models g$. Then $\alpha \models f$. If $n+1 \not\models f$, then $k \not\models g$, for any $k \leq n$, which implies $n \not\models g$. By ind. hyp., $\beta \not\models g$ for any $\beta \geq n$ because $n = d(g)$. Finally $\alpha \not\models f$.

- Assume $f = \mathbf{E}\, g\, \mathbf{U}\, h$, and $n = d(f)$, and consider $\alpha \geq n$. First suppose that $n \models f$. Then $n \models g$ or $n \models h$. If $n \models h$, then by ind. hyp., $\alpha \models h$ and then $\alpha \models f$. Otherwise $n \models g$, and there exists $n \to k$ s.t. $k \models h$. Now $\alpha \models g$ (by ind. hyp.) and as $\alpha \to k$, we have $\alpha \models f$. The other possibility is $n \not\models f$, and then $n \not\models h$, implying $\alpha \not\models h$ (by ind. hyp.). If now $\alpha \models f$ then $\alpha \models g$ (and then $n \models g$) and there exists $\alpha \to \beta$ with $\beta \models h$. As β must be less than n (or else we would have $n \models h$), we have $n \to \beta$, contradicting $n \not\models f$.

A similar proof applies to the $\mathbf{A}\, g\, \mathbf{U}\, h$ case, and the remaining cases are obvious. □

Corollary 2 *$h_{CTL}(f) \leq d(f)$ and $dp(CTL) \leq \omega$*

Remark 1 *The bound is reached: consider formulae of the form* $\mathbf{EX} \ldots \mathbf{EX}\, \top$ *with n nested* $\mathbf{EX}$ *operators, also written* $\mathbf{EX}^n\top$. *Then $h_{CTL}(\mathbf{EX}^n\top) = n$. As $dp(CTL)$ bounds $h_{CTL}(f)$, for all $f \in CTL$, $dp(CTL)$ is greater than any integer n, and then cannot be less than ω.*

3.2.2 L_μ logic

We prove below that the height of a L_μ formula f is bounded by the number of subformulae of f, of the form $\mathbf{EX}\, g$. As fixed point definitions of the formulae are used in the L_μ, we capture the notion of (closed) subformulae by the Fischer-Ladner closure.

Given $f \in L_\mu$, we define $FL(f)$, the *Fischer-Ladner closure* of f as the least set satisfying:

$$FL(\top) = \{\top\} \qquad FL(\neg f) = \{\neg f\} \cup FL(f)$$
$$FL(f \wedge g) = \{f \wedge g\} \cup FL(f) \cup FL(g) \qquad FL(\mathbf{EX}\, f) = \{\mathbf{EX}\, f\} \cup FL(f)$$
$$FL(\mu X.f(X)) = \{\mu X.f(X)\} \cup FL(f(\mu X.f(X)))$$

Proposition 3 *For any formula $f_0 \in L_\mu$, $h_{L_\mu}(f_0) \leq |\{\mathbf{EX}\, f \in FL(f_0)\}|$*

Example 1 *For the L_μ formula $f_0 \equiv \mu Y.(\neg\mathbf{EX}\,\neg(Y \wedge \mathbf{EX}\, Y))$, the set $\{\mathbf{EX}\, f \in FL(f_0)\} = \{\mathbf{EX}\,\neg(f_0 \wedge \mathbf{EX}\, f_0), \mathbf{EX}\, f_0\}$ and its cardinal is 2.*

Proof The proof is given in [Pin91]. □

Remark 2 *In [Her89], a similar study is pursued in the framework of formulae in guarded positive normal form and uses results on L_μ and games to show that $h_{L_\mu}(f_0)$ is bounded by the maximum of $Card\{\mathbf{EX}\, f \in FL(f_0)\}$ and $Card\{\mathbf{AX}\, f \in FL(f_0)\}$.*

3.2.3 CTL^* logic

In order to simplify the exposition, we fix some notations. We define $SF(F)$, for a set $F \subseteq CTL^*$ of formulae, as the least set containing F and which is closed by subformulae; we define $BC(F)$ as the least set (modulo semantical equivalence over the formulae) containing F, closed by subformulae and boolean combinations.

Proposition 4 *For any formula $f_0 \in CTL^*$, $h_{CTL^*}(f_0) \leq |BC(f_0)|$*

Proof It is similar to Proposition 3, but is more complicated, e.g. for CTL^* formulae, $q \sim_f q'$ does not imply $q \sim_{\neg f} q'$. We refer to [Pin91] for the complete proof. □

Remark 3
- *The question whether the bound $|BC(f_0)|$ is reached is not yet solved ($|BC(\mathbf{EX}^n\top)| >> n$).*
- *As formulae of the form $\mathbf{EX}^n\top$ belong to CTL^* and L_μ, we can state that dp(CTL^*) and dp(L_μ) are at least equal to ω.*

Theorem 5 *$dp(CTL) = dp(HML) = dp(L_\mu) = dp(CTL^*) = \omega$*

Proof These d.p.'s are clearly less than or equal to ω by Propositions 2, 3 and 4 which guarantee that the formulae can distinguish only up to an integer. Next, by Remark 1 and 3, it is clear that these d.p.'s are at least equal to ω. □

4 Adding contexts

4.1 Definitions of the "tyft" contexts

In the following, we assume a countably infinite set X of *variables* with typical elements $x, y, x_1, y_1, \ldots$, and a (single sorted) signature $\Sigma = (F, r)$, where F is a set of *function names* disjoint from X, and $r : F \to \mathbf{N}$ gives the arity of a function name. T_Σ denotes the set of all closed terms over the signature Σ, and we write $T_\Sigma(X)$ for the set $T(\Sigma \bigcup X)$, i.e. terms over Σ that may contain variables from X. $Var(t) \subseteq X$ denotes the set of variables occurring in term t.

A *Transition System Specification (TSS)* is a triple $P = (\Sigma, A, R)$ where Σ is a signature, A is a set of actions and R is a set of rules of the form $\dfrac{\{t_i \xrightarrow{a_i} y_i \mid i \in I\}}{t \xrightarrow{a} t'}$ where I is a finite set, $t_i, t'_i, t, t' \in T_\Sigma(X)$, $a, a_i \in A$.

The elements of $\{t_i \xrightarrow{a_i} y_i \mid i \in I\}$ are called the *premises* of the rule, and $t \xrightarrow{a} t'$ is its *conclusion*. Rules without any premise are called *axioms*.

We assume the reader knows what a proof of a transition $t \xrightarrow{a} t'$ from P is (where t and t' are closed terms): the proof is built up from rules and an instantiation of the variables occurring in these rules. We write $P \vdash t \xrightarrow{a} t'$ if a proof exists. We extend naturally this notion to any sequence of transitions $t \xrightarrow{a_1} t_1 \xrightarrow{a_2} \ldots$.

We are now able to define the transition system specified by P:

Let P be a TSS. Given $t_0 \in T_\Sigma$, the transition system, or labeled graph, specified by P at t_0 is $TS(t_0) = (T_\Sigma, A, (\xrightarrow{a}_P)_{a \in A}, t_0)$ where $\xrightarrow{a}_P$ is defined by $t \xrightarrow{a}_P t'$ iff $P \vdash t \xrightarrow{a} t'$. We always drop the P subscript when it is clear from the context.

The TSS's of [GV88] can be understood as the specification of operators in a natural way : the operators are simply named by the function symbols of the signature Σ and we get their operational definitions by looking at the rules in R. We say that a rule r *describes* an operator f if the left hand side of its conclusion has the form $f(\ldots)$.

Example 2 *The rules* $\dfrac{x \xrightarrow{a} x'}{x + y \xrightarrow{a} x'}$ $\dfrac{y \xrightarrow{a} y'}{x + y \xrightarrow{a} y'}$ *provide a description of "+", the non-deterministic choice operator of CCS.*

A rule r is in *tyft format* if it has the form $\dfrac{\{t_i \xrightarrow{a_i} y_i \mid i \in I\}}{f(x_1, \ldots, x_n) \xrightarrow{a} t}$ where I is a finite set, $x_j (1 \leq j \leq n), y_i (i \in I)$ are all different variables, $a, a_i \in A$, $t_i, t \in T_\Sigma(X)$.

Rules containing a circular reference in their variables give rise to difficulties, e.g. regarding to bisimulation. This is the reason why, as in [GV88], we exclude them.

Given a rule $r \equiv \dfrac{\{t_i \xrightarrow{a_i} y_i \mid i \in I\}}{f(x_1, \ldots, x_n) \xrightarrow{a} t}$, we write $Var(r)$ for the set of variables occurring in r. A variable in $Var(r)$ is *free* if it does not occur in the left hand side of the conclusion or in the right hand side of a premise. A rule r is *pure* if it is non circular (see [GV88] for the formal definition) and contains no free variables.

If all the rules of a TSS P are in pure tyft format, we say that P is a tyft TSS.

In the following, we explicitly restrict to tyft TSS's. This is not a very strong restriction as all the CCS operators are (usually) defined in this format, see example 2.

We propose now a subclass of tyft TSS's. We call them tyft "*sans copy*" TSS's. The rules' format also gives the ability to define all the CCS combinators. Informally, the rules cannot produce copies of the processes. In order to formalize this notion, we first define, for each rule r, a relation of "dependency" between its variables.

Let r be a (pure tyft) rule of the form $\dfrac{\{t_i \xrightarrow{a_i} y_i \mid i \in I\}}{f(x_1, \ldots, x_n) \xrightarrow{a} t}$. If y is an occurrence of a variable in term t, we say that y *depends on* variable x_k if there exist premisses $p_1, \ldots, p_m$ $(m \geq))$ in the rule r of the form $t_{i_1} \xrightarrow{a_{i_1}} y_{i_1}, \ldots, t_{i_m} \xrightarrow{a_{i_m}} y_{i_m}$ s.t. $x_k \in Var(t_{i_1})$, $y_{i_m} = y$, and for all $1 \leq j \leq m-1$, $y_{i_j} \in Var(t_{i_{j+1}})$.

A (pure tyft) rule r is "*sans copy*" if for all x_k, $k = 1, \ldots, n$, there exists at most one occurrence of a variable in term t which depends on x_k. When all the rules of a TSS P are sans copy we talk about a *sans copy* TSS.

Example 3 *The rule* $\frac{x_0 \to x_1 \;\; x_1 \to x_2 \;\; y_0 \to y_1}{g(x_0, y_0) \to g(x_2, y_1)}$ *is sans copy while* $\frac{x \to y}{f(x) \to g(y, x)}$ *is not, as the variables* y *and* x *occurring in* t *(i.e.* $g(y, x)$*) both depend on* x.

Given a TSS $P = (\Sigma, A, R)$, we define an *n-place context* as a term of T_Σ with n holes at some positions. Syntactically, a context is written $C[.,\ldots,.]$ where "$[.,\ldots,.]$" symbolizes the holes. We talk about *sans copy contexts* if the TSS P is sans copy.

We consider now the semantics generated by the logical language CTL when in addition 1-place contexts are allowed. Then, the defined equivalence, i.e. obtained by allowing tyft (resp. tyft sans copy) contexts corresponds to the greatest congruence w.r.t. tyft (resp. tyft sans copy) combinators contained in the CTL-equivalence.

Let G be a labeled graph over A, and $q, q' \in Q$. We define two equivalence relations over Q as:

$$q \simeq_{CTL} q' \overset{\text{def}}{\Leftrightarrow} \text{ for all contexts } C[.], C[q] \sim_{CTL} C[q']$$

$$q \cong_{CTL} q' \overset{\text{def}}{\Leftrightarrow} \text{ for all sans copy contexts } C[.], C[q] \sim_{CTL} C[q']$$

Clearly adding contexts may increase the discriminating power of the semantics.

4.2 The distinguishing powers

We investigate the d.p. of $\simeq_{CTL}$ and $\cong_{CTL}$ defined in section 4.1. This study is relevant since CTL-equivalence is not a congruence w.r.t. parallel combinators, see [Sch90a]. Before carrying on with the proofs, let us first be acquainted with contexts and behaviors of an ordinal graph in these contexts.

Example 4 *Consider the rules*

$$\frac{x \to y}{g(x) \to f(x, y)} \qquad \frac{x \to y}{f(x, z) \to f(x, y)} \qquad \frac{x \to y}{f(x, z) \to f(f(x, z), y)} \qquad f(x, y) \to y$$

In the context $g(.)$*, the ordinal* ω *can perform the infinite behavior* $g(\omega) \to f(\omega, 0) \to f(f(\omega, 0), 0) \to f(f(f(\omega, 0), 0), 0) \ldots$*. We can build a proof of this sequence by using on the one hand the rules and on the other hand the valid transition* $\omega \to 0$ *in the graph* ω.

Intuitively, given a TSS $P = (\Sigma, A, R, t_0)$, a context $C[.,\ldots,.]$ and ordinals $\alpha_1, \ldots, \alpha_n$, the term $C[\alpha_1, \ldots, \alpha_n]$ denotes a state in the transition system specified by extending P: we add to Σ all the ordinal constants (and to R all the axioms $\beta \to \gamma$, where $\beta > \gamma$). We write $\hat{\Sigma}$ for the extended Σ.
We call a *behavior* of a term $t_0 \in T_{\hat{\Sigma}}$ any possible maximal sequence (possibly infinite) of transitions $t_0 \overset{a_1}{\to} t_1 \overset{a_2}{\to} t_2 \ldots$. This behavior is nothing else than a path in the graph rooted at t_0. For this reason, we use the notations of section 2.1 and write $\pi, \pi', \ldots$ for the behaviors.

Note that contexts may give rise to infinite behaviors (think of $g(\omega)$ in example 4). Moreover, the number of occurrences of ordinals in the terms can grow along the behavior. However, when restricted to tyft sans copy rules, a transition of the form $C[\alpha] \rightarrow t'$ implies that t' contains at most one occurrence of an ordinal $\alpha' \leq \alpha$ (this property can be generalized to n-place contexts but is technically difficult to state). In contrast, if the contexts are not sans copy, as in Example 4, the number and the value of ordinals may increase.

Given an ordinal γ, we define a relation, $\overset{\gamma}{\smile}$, over ordinals by $\alpha \overset{\gamma}{\smile} \beta$ iff $\alpha = \beta$ or $\alpha, \beta \geq \gamma$. Clearly, $\gamma \geq \gamma'$ and $\alpha \overset{\gamma}{\smile} \beta$ implies $\alpha \overset{\gamma'}{\smile} \beta$.

We extend this equivalence relation over terms of $T_{\hat{\Sigma}}$ by $f(u_1, \ldots, u_n) \overset{\gamma}{\smile} g(v_1, \ldots, v_m)$ iff $f = g$ and $u_i \overset{\gamma}{\smile} v_i$, $i = 1, \ldots, n$, and over behaviors (of same length) by $\pi \overset{\gamma}{\smile} \pi'$ iff $\pi(i) \overset{\gamma}{\smile} \pi'(i)$, for all $i = 0, \ldots$

Lemma 6 *Let $C[.,\ldots,.]$ be a n-place context and γ be an ordinal,*

1. *If $t_0 = C[\alpha_1, \ldots, \alpha_n] \overset{\gamma+\omega_1}{\smile} C[\alpha'_1, \ldots \alpha'_n] = t'_0$, then for all $\pi \in Ex(t_0)$, there exists a behavior π' of t'_0 s.t. $\pi \overset{\gamma}{\smile} \pi'$. ($\omega_1$ is the first uncountable ordinal)*

2. *When only sans copy contexts are allowed, ω_1 can be replaced by ω if the behavior π is finite and by $\omega * 2$ otherwise.*

Proof The proof is in [Pin91]. □

In order to compare the CTL properties of different ordinals in contexts, we define, by induction over the structure of a CTL formula f, two notions of depth, written $d'(f)$ and $d''(f)$, different from the one of section 3.2:

$d'(\top) = 0$	$d''(\top) = 0$
$d'(f \wedge g) = max(d'(f), d'(g))$	$d''(f \wedge g) = max(d''(f), d''(g))$
$d'(\neg f) = d'(f)$	$d''(\neg f) = d''(f)$
$d'(\mathbf{EX}\, f) = d'(f) + \omega$	$d''(\mathbf{EX}\, f) = d''(f) + \omega$
$d'(\mathbf{E}\, f\, \mathbf{U}\, g) = max(d'(f), d'(g)) + \omega$	$d''(\mathbf{E}\, f\, \mathbf{U}\, g) = max(d''(f), d''(g)) + \omega_1$
$d'(\mathbf{A}\, f\, \mathbf{U}\, g) = max(d'(f), d'(g)) + \omega * 2$	$d''(\mathbf{A}\, f\, \mathbf{U}\, g) = max(d''(f), d''(g)) + \omega_1$

Note that for all $f \in CTL$, $d'(f) < \omega^2 (= \bigcup_{n \in \mathbf{N}} \omega * n)$ and $d''(f) < \omega_1 * \omega$. $d'(f)$ has been defined for the first time in [Sch90b] for a subclass of our contexts. The proofs below are a generalization.

Lemma 7 *1. For all $f \in CTL$, for all n-place ($n \geq 0$) context $C[.,\ldots,.]$ and all $\alpha_i \overset{d''(f)}{\smile} \alpha'_i$, $i = 1, \ldots, n$, $C[\alpha_1, \ldots, \alpha_n] \sim_f C[\alpha'_1, \ldots, \alpha'_n]$*

2. $d''(f)$ can be replaced by $d'(f)$ when only sans copy contexts are allowed.

Proof The proofs for $d'(f)$ and $d''(f)$ are similar. They are made by induction over the structure of f using the results of Proposition 2 and Lemma 6; for details see [Pin91]. □

Theorem 8 *$dp(\simeq_{CTL}) = \omega_1 * \omega$ and $dp(\cong_{CTL}) = \omega^2$*

Proof Lemma 7 entails that the d.p.'s have at least these values. These bounds are reached, see [Pin91]. □

5 Conclusion

This paper shows that the use of Ordinal Processes in Comparative Concurrency Semantics is a very fruitful idea: it provides us with a unified method of comparison, based on the determination of the distinguishing power (in [Pin91], other semantics are studied, e.g. trace congruences). Several remarks rise up from this work:

- Beyond the computation of the d.p. of Temporal Logics and through the introduction of the height of the formulae is emphasized a discriminating comparison of the logics.

- As other logics may have other d.p.'s (e.g. it can be shown that the logic *RTL* of [BR83] has $dp(RTL) = \omega^2$), the determination of the d.p.'s can be used in the classification of Temporal Logics w.r.t. the semantics they induce. This classification is different from the usual one which is based on the expressiveness of the logical languages. However, the two notions are related: if a logic L is more expressive than a logic L', then $dp(L) \geq dp(L')$. But the reciprocal does not hold.

- The investigation on the d.p. of tyft combinators (with or without the "sans copy" restriction) enables us to assert that there exists a fundamental difference of expressiveness between the two formats. It may be interesting to pursue this study by allowing negative premisses in the rules (ntyft format of Groote), or by considering other formats, e.g. the GSOS format of Bloom.

6 Acknowledgements

We thank Philippe Schnoebelen for his helpful comments and suggestions.

References

[BBK87] J. C. M. Baeten, J. A. Bergstra, and J. W. Klop. On the consistency of Koomen's fair abstraction rule. *Theoretical Computer Science*, 51(1):129–176, 1987.

[BR83] S. D. Brookes and W. C. Rounds. Behavioural equivalence relations induced by programming logics. In *Proc. 10th ICALP, Barcelona, LNCS 154*, pages 97–108. Springer-Verlag, July 1983.

[CE81] E. M. Clarke and E. A. Emerson. Design and synthesis of synchronization skeletons using branching time temporal logic. In *Proc. Logics of Programs Workshop, Yorktown Heights, LNCS 131*, pages 52–71. Springer-Verlag, May 1981.

[Dam90] M. Dam. Translating CTL* into the modal μ-calculus. Research Report ECS-LFCS-90-123, Lab. for Foundations of Computer Science, Edinburgh, November 1990.

[EC80] E. A. Emerson and E. M. Clarke. Characterizing correctness properties of parallel programs using fixpoints. In *Proc. 7th ICALP, Noordwijkerhout, LNCS 85*, pages 169–181. Springer-Verlag, July 1980.

[EH86] E. A. Emerson and J. Y. Halpern. "Sometimes" and "Not Never" revisited: On branching versus linear time temporal logic. *Journal of the ACM*, 33(1):151–178, 1986.

[GV88] J. F. Groote and F. W. Vaandrager. Structured operational semantics and bisimulation as a congruence. Research Report CS-R8845, CWI, November 1988.

[Her87] B. Herwig. Equivalence in *PDL* and $L\mu$. In *Proc. 5th Easter Conf. Model Theory, Berlin*, pages 46–56, April 1987.

[Her89] B. Herwig. *Zur Modelltheorie von* $L\mu$. PhD thesis, Albert-Ludwigs Univ., Freiburg, September 1989.

[HM85] M. Hennessy and R. Milner. Algebraic laws for nondeterminism and concurrency. *Journal of the ACM*, 32(1):137–161, January 1985.

[Klo88] J. W. Klop. *Bisimulation Semantics*. Lectures given at the REX School/-Workshop, Noordwijkerhout, NL, May 1988.

[Koz83] D. Kozen. Results on the propositional μ-calculus. *Theoretical Computer Science*, 27(3):333–354, 1983.

[Mil80] R. Milner. *A Calculus of Communicating Systems*, volume 92 of *Lecture Notes in Computer Science*. Springer-Verlag, 1980.

[Mil90] R. Milner. Functions as processes. Research Report 1154, INRIA, February 1990.

[Par81] D. Park. Concurrency and automata on infinite sequences. In *Proc. 5th GI Conf. on Th. Comp. Sci., LNCS 104*, pages 167–183. Springer-Verlag, March 1981.

[Pin91] S. Pinchinat. Ordinal processes in comparative concurrency semantics. Research report, LIFIA-IMAG, Grenoble, 1991. To appear.

[Sch90a] Ph. Schnoebelen. Congruence properties of the process equivalence induced by temporal logic. Research Report 831-I, LIFIA-IMAG, Grenoble, October 1990.

[Sch90b] Ph. Schnoebelen. *Sémantique du parallélisme et logique temporelle. Application au langage FP2*. Thèse de Doctorat, I.N.P. de Grenoble, France, June 1990.

[SP90] Ph. Schnoebelen and S. Pinchinat. On the weak adequacy of branching-time temporal logic. In *Proc. ESOP'90, Copenhagen, LNCS 432*, pages 377–388. Springer-Verlag, May 1990.

Logical Semantics of Modularisation

Gerard R. Renardel de Lavalette

University of Groningen, Department of Computing Science,
P.O.Box 800, 9700 AV Groningen, the Netherlands

Abstract. An algebra of theories, signatures, renamings and the operations import and export is investigated. A normal form theorem for terms of this algebra is proved. Another algebraic approach and the relation with a fragment of second order logic are also considered.

1 Introduction

Modularisation is (together with parametrisation) a key feature in order to describe and design complex objects in a manageable and comprehensible way. In this paper we study the logical aspects of modularisation in formal (programming or specification) languages. This is done by investigating some natural and useful operations on *theories*, the objects that express the logical semantics of such languages. The usual names for these operations in the jargon of computer science are import (in logical terms: combination of theories), export (restricting the signature of a theory) and renaming. The results are presented in an algebraic fashion.

1.1 Relation with Other Work

Operators on modules and their semantics have been studied in e.g. [1] (in the context of CLEAR), [5] (in the context of PLUSS), [3] and [4] (using category theory), [16] (using model class semantics). Our main source of inspiration has been [2], where the approach is similar to Wirsing's in [16], extended to theory semantics and countable model semantics. The role of the interpolation theorem for the theory semantics of import and export has been pointed out in [7]. Besides giving a survey of logical aspects of modularisation, this paper contains, to the best of our knowledge, the following new points:

- investigation of the behaviour of import and export in combination with non-bijective renamings on theories;
- a normal form theorem for theory terms constructed with these operators;
- a (trivial) counterexample for interpolation in conditional equational logic;
- definition of import and export on theories using two orthogonal closure properties;
- relation with the $[\&, \exists]$-fragment of second order logic.

1.2 Survey of the Rest of the Paper

In Sect. 2 we introduce signatures, theories, renamings and operations defined on them. Axioms for these operations are given in Sect. 3, where also the relation of one of these properties with the interpolation theorem is considered, as well as some results on interpolation in (conditional) equational logic. Section 4 is about normal forms of so-called theory terms. In the last two sections we sketch some related ideas: reducing the theory operations of Sect. 2 to two orthogonal closure operators, and a theory semantics for the [&, ∃]-fragment of second order logic.

1.3 Acknowledgements

The research reported here has been performed while the author was at the Department of Philosophy of the University of Utrecht. Partial support has been received from the European Communities under ESPRIT projects METEOR and ATMOSPHERE through Philips Research Laboratories Eindhoven and SERC (Software Engineering Research Centre, Utrecht), respectively. Part of the work reported here is also in [11]. Discussion and collaboration with Jan Bergstra, Loe Feijs, Hans Jonkers, Karst Koymans and Piet Rodenburg were influential and are gratefully acknowledged. Finally I thank Maureen Reyers and Peter de Bruin for converting an earlier version of this paper to LaTeX.

2 Signatures, Renamings and Theories

We assume some logical language L with a derivability relation $\vdash$. L contains signature elements, e.g. sorts, functions, predicates. Signature elements can have a type: an arity (required number of arguments) or a sort type (a list of input and output sorts). We assume that, for every type, there are infinitely many signature elements having that type: this will allow us to apply the *fresh signature element principle* (see the end of this section). A signature is a finite set of signature elements; it is called *closed* if it contains all sorts occurring in the types of its elements (observe that closedness is preserved under union and intersection). The closure $c(\Sigma)$ of a signature is the least closed signature containing Σ. If X is a (collection of) expression(s) in the language of L then $S(X)$ is the closure of the collection of all signature elements occurring in (elements of) X.

From now on, we adopt the default convention that Σ and Π range over closed signatures.

Let Γ be a collection of sentences of L, Σ a signature, then the *closure of Γ in Σ* is defined by

$$Cl(\Sigma, \Gamma) =_{\mathrm{def}} \{A \mid \Gamma \vdash A \text{ and } S(A) \subseteq \Sigma\} .$$

These closures are called *theories*. The *union* of two theories is the smallest theory containing them, defined by

$$T + U =_{\mathrm{def}} Cl(S(T \cup U), T \cup U) .$$

It is obvious that $+$ is commutative, associative and idempotent. The *restriction* of a theory to a signature is defined as

$$\Sigma \sqcap T =_{\mathrm{def}} Cl(\Sigma \cap S(T), T) \; (= \{A \mid A \in T \text{ and } S(A) \subseteq \Sigma\}) .$$

We also define the trivial theory over a signature:

$$T(\Sigma) =_{\text{def}} Cl(\Sigma, \emptyset) \; (= \{A \mid \vdash A \text{ and } S(A) \subseteq \Sigma\}) \; .$$

Renamings are finitely generated mappings defined on expressions of L, changing only signature elements and commuting with taking types (i.e. the type of a renamed signature element is the renaming of the type of that signature element). Observe that we are liberal in the definition of renamings in the sense that they need not be bijective, so e.g. $[P := R, Q := R]$ is a correct renaming (if P, Q and R have the same type); this is in contrast to [2], where renamings are bijective and even involutive (i.e. if $\rho(P) = Q$ then $\rho(Q) = P$). We define domain and range of a renaming by

$$\begin{aligned} dom(\rho) &=_{\text{def}} \{I \mid \rho(I) \neq I\} \\ rg(\rho) &=_{\text{def}} \{\rho(I) \mid \rho(I) \neq I\} \; (= \rho(dom(\rho))) \end{aligned}$$

All renamings are finitely generated, so domain and range of a renaming are finite and hence signatures. We shall use some injectivity properties of renamings, defined by (here Σ and Π are arbitrary, i.e. not necessarily closed)

$$\begin{aligned} inj(\rho, \Sigma, \Pi) &=_{\text{def}} \text{for all } I \in \Sigma, J \in \Pi, \text{ if } \rho(I) = \rho(J) \text{ then } I = J \\ inj(\rho, \Sigma) &=_{\text{def}} inj(\rho, \Sigma, \Sigma) \\ inj(\rho) &=_{\text{def}} inj(\rho, dom(\rho)) \end{aligned}$$

We also put, for arbitrary Σ, Π:

ρ *renames* Σ *outside* Π iff $dom(\rho) = \Sigma \cap \Pi$, $rg(\rho) \cap (\Sigma \cup \Pi) = \emptyset$ and $inj(\rho, \Sigma)$.

We shall use the *good renaming property*:

for arbitrary Σ, Π there is a renaming $\gamma(\Sigma, \Pi)$ renaming Σ outside Π,

which follows directly from the finiteness of signatures and the fresh signature element principle. The application of renamings on theories is defined by

$$\rho(T) =_{\text{def}} Cl(\rho(S(T)), \{\rho(A) \mid A \in T\}) \; .$$

Observe that $\rho(T) = \{\rho(A) \mid A \in T\}$ only if $inj(\rho, S(T))$, e.g. if $\rho = [P := Q]$ and $T = Cl(\{P, Q\}, \{P, \neg Q\})$ then $\bot \in \rho(T)$, $\bot \notin \{\rho(A) \mid A \in T\}$.

3 Properties

We now list some properties of the operations defined above. Completeness is not our aim and we restrict ourselves to the properties used in the proof of the normal form theorem in the next section; most of them deal with permutation of two operators. We let T, U, V range over theories, Σ and Π over (closed) signatures, ρ and σ over renamings.

(Sr)	$S(\rho(T)) = \rho(S(T))$
$(S+)$	$S(T+U) = S(T) \cup S(U)$
$(S\square)$	$S(\Sigma \square T) = \Sigma \cap S(T)$
(rr)	$\rho(\sigma(T)) = (\rho \cdot \sigma)(T)$
$(r+)$	$\rho(T+U) = \rho(T) + \rho(U)$
$(r\square)$	$inj(\rho, S(T) - \Sigma, S(T) \cup \Sigma) \Rightarrow \rho(\Sigma \square T) = \rho(\Sigma) \square \rho(T)$
$(\square r)$	$dom(\rho) \cap \Sigma \cap S(T) = \emptyset \ \& \ \rho(S(T) - \Sigma) \cap \Sigma = \emptyset \ \& \ inj(\rho, S(T))$ $\Rightarrow \Sigma \square T = \Sigma \square \rho(T)$
$(\square S)$	$T = S(T) \square T$
$(\square\square)$	$\Sigma \square (\Pi \square T) = (\Sigma \cap \Pi) \square T$
$(\square+)$	$S(T) \cap S(U) \subseteq \Sigma \Rightarrow \Sigma \square (T+U) = \Sigma \square T + \Sigma \square U$
(γ)	$dom(\gamma(\Sigma, \Pi)) = \Sigma \cap \Pi \ \& \ rg(\gamma(\Sigma, \Pi)) \cap (\Sigma \cup \Pi) = \emptyset \ \& \ inj(\gamma(\Sigma, \Pi), \Sigma)$

The axioms (Sr), $(S+)$, $(r+)$, $(\square S)$, $(\square\square)$ and $(\square+)$ are also found in [2]. The condition in $(r\square)$ is required to prevent new identifications of names in $\rho(T)$ and between names in $\rho(\Sigma)$ and $\rho(T)$ not present in $\rho(\Sigma \square T)$; the condition in $(\square r)$ guarantees that ρ does not affect names in $\Sigma \square T$, neither introduces new identifications in T. Most of the properties listed have a straightforward proof; only the proof of $(\square+)$ is more involved and follows now.

Lemma 1. *Property $(\square+)$ holds iff L satisfies the following version of the Interpolation Theorem:*

(*) if $\Gamma \cup \Delta \vdash A$ then there is an interpolant I with $\Gamma \vdash I$, $\Delta \cup \{I\} \vdash A$ and $S(I) \subseteq S(\Gamma) \cap (S(\Delta) \cup S(A))$.

Proof sketch of the 'if' part (the other part is trivial; see [2] for more details). Assume $S(T) \cap S(U) \subseteq \Sigma$ and let $A \in \Sigma \square (T+U)$, i.e. $T \cup U \vdash A$ and $S(A) \subseteq \Sigma \cap (S(T) \cup S(U))$. Now use the interpolant I with $T \vdash I, U \cup \{I\} \vdash A$ to see (after some rewriting of parameter conditions) that $Cl(\Sigma, T) \vdash I$ and $Cl(\Sigma, U) \vdash I \rightarrow A$, hence $Cl(\Sigma, T) \cup Cl(\Sigma, U) \vdash A$. Conclusion: $A \in \Sigma \square T + \Sigma \square U$. This proves the $\subseteq$-part of the equality, the other inclusion is trivial. □

Remark 1. There are several variants of the axiom $(\square+)$:

$(\square+1)$	$\Sigma \square (T \cup U) = \Sigma \square ((\Sigma \cup S(U)) \square T + (\Sigma \cup S(T)) \square U)$
$(\square+2)$	$(S(T) \cap S(U)) \square (T+U) = S(U) \square T + S(T) \square U$
$(\square+3)$	$S(T) \square (T+U) = T + S(T) \square U$
$(\square+4)$	$\Sigma \square (T(\Sigma) + T) = T(\Sigma) + \Sigma \square T$
$(\square+5)$	$\Sigma \square (T(\Pi) + T) = T(\Sigma \cap \Pi) + \Sigma \square T$

These equations have in common that they all require some version of the interpolation theorem for the underlying logic (see below). We have the following relations:

$$(\square+) \Leftrightarrow (\square+1) \Leftrightarrow (\square+2) \Leftrightarrow (\square+3) \Rightarrow (\square+4) \Leftrightarrow (\square+5)\ .$$

For proofs, we refer to [2] for $(\square+2) \Rightarrow (\square+)$, [12] for $(\square+3) \Rightarrow (\square+2)$ and [15] and [14] for $(\square+4) \Rightarrow (\square+5)$; the other implications are fairly easy (for $(\square+) \Rightarrow (\square+1)$, use that $\Sigma \square (T \cup U) = \Sigma \square (\Sigma \cup (S(T) \cap S(U)) \square (T+U))$.

Remark 2. Interpolation as formulated in (*) holds for predicate logic, but not for equational logic (EL) neither for conditional equational logic (CEL). CEL has the easiest counterexample:

$$\Gamma = \{fc = c\},\ \Delta = \{fx = x \rightarrow a = b\},\ A = (a = b)$$

If I were an interpolant then $S(I) \subseteq \{f\}$, $fc = c \vdash I$ and we claim $I \vdash \exists x(fx = x)$; this would lead to definability of existential quantification in CEL, *quod non*. So it remains to prove the claim: observe $I \vdash (\forall x(fx = x) \rightarrow a = b) \rightarrow a = b$, hence $I \vdash \exists x(fx = x) \vee a = b$; if M is a model for the signature $\{f\}$ and $M \models I$, then either M has only one element and $M \models \exists x(fx = x)$, or M has at least two elements and we can expand M to a model M' for the signature $\{f, a, b\}$ with $M' \models a \neq b$, so $M' \models \exists x(fx = x)$ and hence $M \models \exists x(fx = x)$. This proves the claim. This counterexample to (*) in CEL can be transferred to EL using the embedding of CEL into EL given in [2, 5.2.1]. The essence of this embedding is given by

$s = s' \rightarrow t = t'$ becomes $\{g(x, x, y) = y, g(s, s', t) = g(s, s', t')\}$
(g a fresh function);

it leads to the counterexample $\Gamma = \{fc = c\}$, $\Delta = \{g(x, x, y) = y, g(x, fx, a) = g(x, fx, b)\}$, $A = (a = b)$ that is given in [2] and [10]. Another counterexample is given in [7]: $\Gamma = \{f(gx) = x\}$, $\Delta = \{f(fx) = fx\}$, $A = (fx = x)$: none of the possible interpolants (e.g. $\forall x \exists y(fy = x)$ or $f(fx) = fx \rightarrow fx = x$) is in EL.

Remark 3. Logics EL and CEL do satisfy a weaker interpolation theorem, viz. (*) with $\Delta = \emptyset$ and I a finite conjunction of formulas. This has been proved for EL by Rodenburg and Van Glabbeek in [15] (see also [8] and [13]) and for CEL by Rodenburg in [14]. It turns out that this weaker version of interpolation corresponds exactly with the axiom ($\Box$+4) (or, equivalently, ($\Box$+5)), as is proved in [15] and [14].

Remark 4. If one replaces the theory semantics by the model class semantics, the axiom ($\Box$+) follows directly and the interpolation theorem is not required. See [2].

4 Normal Forms

Now we consider normal forms of expressions involving theories, $\Box$, + and renamings. Define the language TL of *theory terms* by

$$T ::= Th \mid \rho(T) \mid \Sigma \Box T \mid T + T\ ,$$

where Th denotes constants $Th_1, Th_2, \ldots$ referring to specific theories. We shall show the following Normal Form Theorem: any expression of TL is equivalent to a normal form expression

$$\Sigma \Box (\rho_1(Th_1) + \cdots + \rho_n(Th_n))\ .$$

To prove this, it suffices to show that the collection of normal forms contains the constants Th_i and is closed under renamings, $\Box$ and +. Now $Th_i = S(Th_i) \Box \rho_i(Th_i)$ which is in normal form, and closure under $\Box$ follows from ($\Box\Box$); for closure under renamings and + we use the next lemma.

Lemma 2. *i) For any Σ, T, ρ there is a renaming σ with*

$$\rho(\Sigma \,\Box\, T) = \rho(\Sigma) \,\Box\, \sigma(T)\,.$$

ii) For any Σ, Π, T, U there are renamings ρ, σ with

$$\Sigma \,\Box\, T + \Pi \,\Box\, U = (\Sigma \cup \Pi) \,\Box\, (\rho(T) + \sigma(U))\,.$$

Proof. i) Property $(r\Box)$ cannot be applied directly, therefore we put $\tau = \gamma(S(T) - \Sigma, S(T) \cup \Sigma \cup dom(\rho) \cup rg(\rho))$, so

$$\begin{aligned}&dom(\tau) = S(T) - \Sigma,\\ &rg(\tau) \cap (S(T) \cup \Sigma \cup dom(\rho) \cup rg(\rho)) = \emptyset,\\ &inj(\tau, S(T) - \Sigma)\end{aligned}$$

and one straightforwardly verifies the conditions of $(\Box r)$ (reading τ for ρ), so we have $\Sigma \,\Box\, T = \Sigma \,\Box\, \tau(T)$, hence $\rho(\Sigma \,\Box\, T) = \rho(\Sigma \,\Box\, \tau(T))$. Now the result follows if we can apply $(r\Box)$ to this last term, so we have to check $inj(\rho, S(\tau(T)) - \Sigma, S(\tau(T)) \cup \Sigma)$, i.e. (using the fact that $rg(\tau) = S(\tau(T)) - \Sigma$) $inj(\rho, rg(\tau), rg(\tau) \cup \Sigma)$. To see this, assume $I \in rg(\tau)$, $J \in rg(\tau) \cup \Sigma$, $\rho(I) = \rho(J)$; since $rg(\tau) \cap (dom(\rho) \cup rg(\rho)) = \emptyset$, we have $\rho(I) = I \notin rg(\rho)$ and ($\rho(J) = J$ or $J \in dom(\rho)$); the second alternative yields contradiction with $\rho(I) = \rho(J)$ and $\rho(I) \notin rg(\rho)$, so we conclude $I = J$ and the injectivity condition is proved. So we have $\rho(\Sigma \,\Box\, T) = \rho(\Sigma) \,\Box\, \sigma(T)$ with $\sigma = \rho \cdot \tau$.

ii) Let $\rho = \gamma(S(T) \cap \Pi - \Sigma, S(T) \cup \Sigma \cup \Pi)$, then $dom(\rho) = S(T) \cap \Pi - \Sigma$, $rg(\rho) \cap (S(T) \cup \Sigma \cup \Pi) = \emptyset$ and $inj(\rho, S(T) \cap \Pi - \Sigma)$, so

$$\begin{aligned}\Sigma \,\Box\, T &= \Sigma \,\Box\, \rho(T) && (\text{ by } (\Box r))\\ &= \Sigma \cap S(\rho(T)) \,\Box\, \rho(T) && (\text{ by } (\Box S) \text{ and } (\Box\Box))\\ &= (\Sigma \cup \Pi) \cup S(\rho(T)) \,\Box\, \rho(T) && (\Pi \cap S(\rho(T)) \subseteq \Sigma)\\ &= (\Sigma \cup \Pi) \,\Box\, \rho(T) && (\text{ by } (\Box S) \text{ and } (\Box\Box))\end{aligned}$$

Analogously: if $\rho' = \gamma(S(U) \cap \Sigma - \Pi, S(U) \cup \Sigma \cup \Pi)$, then $\Pi \,\Box\, U = (\Sigma \cup \Pi) \,\Box\, \rho'(U)$. Now put $\tau = \gamma(S(\rho(T)) \cap S(\rho'(U)) - (\Sigma \cup \Pi), S(\rho(T)) \cup S(\rho'(U)) \cup \Sigma \cup \Pi)$, then

$$\begin{aligned}&dom(\tau) = S(\rho(T)) \cap S(\rho'(U)) - (\Sigma \cup \Pi),\\ &rg(\tau) \cap (S(\rho(T)) \cup S(\rho'(U)) \cup \Sigma \cup \Pi) = \emptyset,\\ &inj(\tau, S(\rho(T)) \cap S(\rho'(U)) - (\Sigma \cup \Pi)).\end{aligned}$$

After some calculation we see $dom(\tau) \cap (\Sigma \cup \Pi) \cap S(\rho'(U)) = \emptyset$, $\tau(S(\rho'(U) - (\Sigma \cup \Pi)) \cap (\Sigma \cup \Pi) = \emptyset$ and $inj(\tau, S(\rho'(U)))$. Now

$$\begin{aligned}\Sigma \,\Box\, T + \Pi \,\Box\, U &= (\Sigma \cup \Pi) \,\Box\, \rho(T) + (\Sigma \cup \Pi) \,\Box\, \rho'(U)\\ &= (\Sigma \cup \Pi) \,\Box\, \rho(T) + (\Sigma \cup \Pi) \,\Box\, \tau \cdot \rho'(U) && (\text{ by } (\Box r) \text{ and } (rr))\\ &= (\Sigma \cup \Pi) \,\Box\, (\rho(T) + \tau \cdot \rho'(U)) && (\text{ by } (\Box +))\\ &= (\Sigma \cup \Pi) \,\Box\, (\rho(T) + \sigma(U))\end{aligned}$$

where $\sigma = \tau \cdot \rho'$. □

Now we can finish the proof of the Normal Form Theorem. Closure of the collection of normal forms under renamings follows from Lemma 2(i), $(r+)$ and (rr); closure under $+$ follows from Lemma 2(ii), $(r+)$, (rr) and the properties of $+$.

5 Two Orthogonal Closures

It is possible to define the operations on theories presented above (except renaming) in terms of two orthogonal closure operators: signature closure s and theory closure t. They are defined on the collection of *all* sets of sentences (i.e. not only sets closed w.r.t. derivability, as in Sect. 2). The definitions read (X, Y, Z range over sets of sentences, i.e. subsets of L):

$$sX =_{\mathrm{def}} \{A \mid S(A) \subseteq S(X)\}$$
$$tX =_{\mathrm{def}} \{A \mid X \vdash A\}$$

Lemma 3. *The following hold:*

(s1) $X \subseteq sX$
(s2) $ssX = sX$
(s3) $X \subseteq Y \Rightarrow sX \subseteq sY$
(s4) $s(sX \cap sY) = sX \cap sY$
(s5) $s(X \cup sY) = s(X \cup Y)$

(t1) $X \subseteq tX$
(t2) $ttX = tX$
(t3) $X \subseteq Y \Rightarrow tX \subseteq tY$
(t4) $t(tX \cap tY) = tX \cap tY$
(t5) $t(X \cup tY) = t(X \cup Y)$

(st1) $stX = tsX = L$
(st2) $s(sX \cap tY) = sX$
(st3) $t(sX \cap tX) = tX$
(st4) $sX \cap t(sX \cap tY) = sX \cap tY$
(st5) $sX \cap (tY \cup tZ) = sX \cap t((tY \cap sY \cap s(X \cup Z)) \cup (tZ \cap sZ \cap s(X \cup Y)))$

Proof (sketch). Verifying ($s1-3$) is easy; moreover, they imply ($s4-5$). Similarly for ($t1-5$).
($st1$): use $\{\varphi \to \varphi \mid \varphi \in L\} \subseteq t\emptyset$ and $\bot \in s\emptyset$.
($st2$): use $\{\varphi \to \varphi \mid \varphi \in sX\} \subseteq t\emptyset$.
($st3$): by ($s1$), ($t1-3$) we have $tX \subseteq t(sX \cap tX) \subseteq ttX = tX$.
($st4$): by ($t1-3$) we have $sX \cap tY \subseteq sX \cap t(sX \cap tY) \subseteq sX \cap ttY = sX \cap tY$.
($st5$): analogous to the proof of ($\Box$+1), requiring the Interpolation Theorem. □

By the definition of Cl in Sect. 2 it is obvious that X is a theory iff it satisfies $X = sX \cap tX$. The definition of the operations on theories introduced in Sect. 2 now reads:

$$T(X) =_{\mathrm{def}} sX \cap t\emptyset$$
$$X \Box Y =_{\mathrm{def}} sX \cap Y$$
$$X + Y =_{\mathrm{def}} s(X \cup Y) \cap t(X \cup Y)$$

Now (S+), ($S\Box$), ($\Box S$), ($\Box\Box$) and ($\Box$+) follow from the previous Lemma; for ($\Box$+), use the instance of ($st5$) obtained by taking $Z = \emptyset$.

6 First-Order Semantics for a Second-Order Fragment

A natural logical interpretation of the notions of combination and hiding is: conjunction and existential quantification. So let us consider the collection $L2$ of second-order formulae, characterised by

$$A ::= \varphi \mid A\&A \mid \exists f A$$

where φ stands for first-order formulae (elements of L) and f for signature elements. We consider the following interpretation:

$$\begin{aligned}
[\![\varphi]\!] &=_{\text{def}} \{\psi \mid \varphi \vdash \psi \text{ and } S(\psi) \subseteq S(\varphi)\} \\
[\![A\&B]\!] &=_{\text{def}} \{\varphi \mid [\![A]\!] \cup [\![B]\!] \vdash \varphi \text{ and } S(\varphi) \subseteq S(A) \cup S(B)\} \\
[\![\exists f A]\!] &=_{\text{def}} \{\varphi \mid \varphi \in [\![A]\!] \text{ and } f \notin S(\varphi)\}
\end{aligned}$$

Validity is defined by

$$\begin{aligned}
A \models B &=_{\text{def}} [\![A]\!] \supseteq [\![B]\!] \\
A \equiv B &=_{\text{def}} A \models B \text{ and } B \models A
\end{aligned}$$

It is obvious that $[\![A]\!]$ is always a theory (i.e. of the form $Cl(\Sigma, X)$) and that

$$\begin{aligned}
[\![A\&B]\!] &= [\![A]\!] + [\![B]\!] \\
[\![\exists f A]\!] &= (S[\![A]\!] - \{f\}) \,\square\, [\![A]\!]
\end{aligned}$$

Moreover, assuming that L has the substitution property, i.e.

> if $\Gamma \vdash \varphi$ then $\Gamma[f := \lambda x.t] \vdash \varphi[f := \lambda x.t]$
> (for unary function symbols f; analogously for other signature elements),

we also have

$$[\![A[f := \lambda x.t]]\!] \supseteq [\![A]\!][f := \lambda x.t] \, .$$

Lemma 4. *The following hold:*

- (&L) $A\&B \models A, B$
- (&R) *if* $A \models B$ *and* $A \models C$ *then* $A \models B\&C$
- ($\exists R$) $A[f := \lambda x.t] \models \exists f A$
- ($\exists L$) *if* $A \models B$ *and* f *not free in* B *then* $\exists f A \models B$
- (&$\exists$) *if* f *not free in* A *then* $A\&\exists f B \equiv \exists f(A\&B)$

Proof (sketch). (&L): easy, use $S[\![A]\!] = S(A)$.

(&R): if $\varphi \in [\![B\&C]\!]$ then there are $\psi \in [\![B]\!], \chi \in [\![C]\!]$ with $\psi \wedge \chi \vdash \varphi$. Hence $\psi, \chi \in [\![A]\!]$, but $[\![A]\!]$ is closed, so $\varphi \in [\![A]\!]$.

($\exists R$): we only have to show $[\![\exists f A]\!] \subseteq [\![A]\!][f := \lambda x.t]$, which follows directly from the substitution property for L.

($\exists L$): straightforward.

(&$\exists$): follows from ($\square$+). As a consequence, (&$\exists$) requires the Interpolation Theorem for L. $\square$

6.1 Remark

A natural question is: can the collection of formulae used for the definition of the interpretation of $L2$ not be replaced by a single formula? This is what Pitts actually does in [9] for second-order intuitionistic propositional logic. In order to do so, he proves a uniform interpolation theorem for first-order intuitionistic propositional logic (IpL), viz.

> if A is a formula of IpL and $\Sigma \subseteq S(A)$, then there is an $I = I(A, \Sigma)$ satisfying $A \vdash I$, $S(I) \subseteq \Sigma$ and ($A \vdash B$ and $\Sigma \subseteq S(A) \cap S(B)$ imply $I \vdash B$).

However, uniform interpolation does not hold in predicate logic, as has been observed by Henkin in [6]. As a consequence, it remains a nontrivial question how to extend the interpretation $[\![\cdot]\!]$ to negation and, via negation, to the other logical operations.

References

1. R.M. Burstall and J.A. Goguen, The semantics of CLEAR, a specification language, in: Abstract software specifications, LNCS 86, Springer-Verlag (1980) 292–332
2. J.A. Bergstra, J. Heering, P. Klint, Module algebra, Journal of the ACM **37** (1990) 335–372
3. H. Ehrig and B. Mahr, Fundamentals of algebraic specification 1, Equations and initial semantics, Springer-Verlag, 1985
4. H. Ehrig and B. Mahr, Fundamentals of algebraic specifications 2, Module specifications and constraints, Springer-Verlag, 1990
5. M.-C. Gaudel, Toward structured algebraic specifications, in: Esprit '85: Status report of continuing work, vol. 1, North-Holland (1986) 493–510
6. L. Henkin, An extension of the Craig-Lyndon interpolation theorem, The Journal of Symbolic Logic **28** (1963) 201–216
7. T.S.E. Maibaum and M.R. Sadler, Axiomatising specification theory, in: H.-J. Kreowski (ed.), Recent trends in data type specification, 3rd Workshop on Theory and Applications of Abstract Data Types, Informatik-Fachberichte 116, Springer-Verlag (1985) 171–177
8. D. Pigozzi, The join of equational theories, Colloquium Mathematicum **30** (1974), 15–25
9. A.M. Pitts, On an interpretation of second order quantification in first order intuitionistic propositional logic, to appear in The Journal of Symbolic Logic
10. G.R. Renardel de Lavalette, Modularisation, parametrisation, interpolation, Journal of Information Processing and Cybernetics EIK **25** (1989) 283–292
11. G.R. Renardel de Lavalette, COLD-K^2, the Static Kernel of COLD-K, Report RP/mod-89/8, Software Engineering Research Centre, Utrecht, 1989
12. P.H. Rodenburg, On the ω-completeness of a fragment of module algebra, manuscript (1990)
13. P.H. Rodenburg, A simple algebraic proof of the equational interpolation theorem, Algebra Universalis **28** (1991) 48–51
14. P.H. Rodenburg, Interpolation in conditional equational logic, Fundamenta Informaticae **15** (1991) 80–85
15. P.H. Rodenburg, Interpolation in equational logic, Report P9201, Programming Research Group, University of Amsterdam, 1992

16. M. Wirsing, Structured algebraic specifications: a kernel language, Habilitationsschrift, Institut für Informatik, Technische Universität, München, 1983

This article was processed using the LaTeX macro package with LLNCS style

A Cut-Elimination Procedure Designed for Evaluating Proofs as Programs

Ulf R. Schmerl

Fakultät für Informatik, Universität der Bundeswehr München, D-8014 Neubiberg

Abstract. This paper describes a method which can be used to determine output values from given input values of programs taking the form of formal proofs. The proofs are constructed in a logic calculus which may include user-defined non-logical inference rules. This calculus is complete in the sense that each computable function can be computed by means of a proof in this calculus. The evaluation method is based mainly on Gentzen's cut-elimination procedure. However the presence of non-logical inference rules requires additional steps to eliminate these rules. Termination of the method is proved by assigning constructive ordinal numbers.

1. Introduction. It is well-known that formal mathematical proofs can be used as programs. In fact, given a constructive proof P of an existential statement $\exists y\varphi(y)$, an effective procedure can be applied to extract from P a term t such that $\varphi(t)$, too, is provable. Thus t can be regarded as the output of P. More generally, if P is a proof of a formula of the form $\forall x\exists y\psi(x,y)$, then, for each input e, P can be specialized to a proof P_e of the existential formula $\exists y\psi(e,y)$ providing an output term t_e. Meanwhile, the method of programming by means of proofs, called "proofs as programs", has been systematically analyzed and implemented. Systems particularly suited for this purpose are described e.g. by *Martin-Löf* [13], *Manna und Waldinger* [12], *Hayashi* [11], *Schwichtenberg* [16], or *Alexi* [1]. Implemented systems supporting the method of using proofs as programs have been developed e.g. by *Constable* et al. [3,4,5,7,8] and *Hayashi* [11].

Different methods can be applied to extract a term t from the proof of an existential statement. Alexi, for example, uses normalization in natural deductions. In this paper we consider a method based on cut-elimination in a sequent calculus which we will introduce first. Then we describe the evaluation procedure for proofs in this calculus and show that all recursive functions can be computed by such proofs.

2. The sequent calculus $\mathbb{K}$. The formula ψ in a proof P of $\forall x\exists y\psi(x,y)$ represents the specification of the program represented by P. In general ψ will be a rather long formula thus making the proof of $\forall x\exists y\psi(x,y)$ accordingly complex. Furthermore, many steps in this proof are usually irrelevant for computing the output values (of course, these steps are relevant to make sure that the values computed will satisfy the specification). For this reason we suppose that besides the specification ψ we have some idea of an algorithm that can be used to compute the output values. Using this idea, we formally prove that for each input value some output value can be obtained, i.e. we set up a formal proof of a formula $\forall x\exists yA(x,y)$ where A is only a *name* for what is computed. This proof will in general be much simpler than the one of $\forall x\exists y\psi(x,y)$. In particular, it contains only proof steps which have a precise computational meaning. In other words, proofs which are to be evaluated as programs will be formulated in a very restricted formalism. Clearly, the proof of $\forall x\exists yA(x,y)$ only ensures that the algorithm terminates, but it does not guarantee that the computed output will satisfy the specification ψ. For that purpose one has to establish the partial correctness $A(x,y)\rightarrow\psi(x,y)$, the proof of which, however, need not be formalized.

Our formalism, called $\mathbb{K}$, is formulated as a sequent calculus in minimal logic (see *Gentzen* [10] or *Gallier* [9] for a description of sequent calculi). Thus proofs in $\mathbb{K}$ consist of sequents of the form $\varphi_1,..,\varphi_n \Rightarrow \psi$ where $\varphi_1,..,\varphi_n$ and ψ are $\forall\exists$-quantified prime formulae or related subformulae. The logical meaning of a sequent is: If $\varphi_1,..,\varphi_n$ hold true, then ψ holds true. In computational terms the meaning is as follows: The function represented by ψ can be computed using a proof of this sequent; the computation needs auxiliary functions described by $\varphi_1,..,\varphi_n$. A proof to be used as a program ends with a sequent of the form $\Rightarrow\forall x\exists y A(x,y)$ where x and y represent finite sequences of variables and A is an atom.

$\mathbb{K}$ allows user-defined non-logical rules. However, these rules must be "solvable" in the following sense: when applying the rule to a closed term of the language then it must be possible to replace this rule by steps in the pure predicate logic. Here we consider non-logical rules for case distinctions, induction and the μ-operator on natural numbers. The solvability of these rules will be shown explicitely below.

In the following, upper case letters A, B, C, .. are used for atomic formulae which represent functions to be computed. Lower case letters f, g, h, .. denote "built-in" functions which can immediately be evaluated by the computer. Accordingly, p, q, r, .. denote built-in predicates. For simplicity, typing of variables, functions, and predicates will be omitted. Formulae are denoted by small Greek letters such as φ, ψ, χ, .. and lists of formulae by capital letters such as Γ, Δ, Π, .. .

Definition: Calculus $\mathbb{K}$

Axioms have the form $\varphi_1,..., \varphi_n \Rightarrow \psi$, where $\varphi_1,..., \varphi_n, \psi$ are prime formulae. The axioms should be valid in some intended interpretation M of the language.

Inference rules:

• Structural rule: (S) $\dfrac{\Gamma\Rightarrow\varphi}{\Pi\Rightarrow\varphi}$,

if Π is obtained from Γ by interchanging formulae, inserting new formulae, and contracting multiple occurrences of formulae.

• Cut rule: (C) $\dfrac{\Gamma\Rightarrow\varphi \quad \Pi\Rightarrow\psi}{\Gamma,\Pi^*\Rightarrow\psi}$,

where Π^* results from Π by deleting all occurrences of φ; Π contains at least one occurrence of φ.

• Quantifier rules:

$$(\exists\Rightarrow) \quad \frac{\Gamma,\varphi(x),\Pi\Rightarrow\psi}{\Gamma,\exists y\varphi(y),\Pi\Rightarrow\psi}\ !, \qquad \varphi \text{ existential formula}$$

where ! denotes the usual variable condition. x is called eigen variable of this rule.

$$(\Rightarrow\exists) \quad \frac{\Gamma\Rightarrow\varphi(t)}{\Gamma\Rightarrow\exists y\varphi(y)} \qquad \varphi \text{ existential formula, t arbitrary term}$$

$$(\forall\Rightarrow) \quad \frac{\Gamma,\psi(t),\Pi\Rightarrow\varphi}{\Gamma,\forall x\psi(x),\Pi\Rightarrow\varphi} \qquad \text{in } \psi \text{ all } \forall \text{ preceding } \exists\text{, t arbitrary term}$$

$$(\Rightarrow\exists) \quad \frac{\Gamma\Rightarrow\varphi(x)}{\Gamma\Rightarrow\forall y\varphi(y)}\ !, \qquad \text{in } \varphi \text{ all } \forall \text{ preceding } \exists;$$

"!" denotes the variable condition, x is the eigenvariable of this rule.

• **Case distinction by built-in predicates:**

$$(\vee)\quad \frac{\Gamma, p_1(\mathbf{t_1}) \Rightarrow \varphi \quad .. \quad \Gamma, p_n(\mathbf{t_n}) \Rightarrow \varphi}{\Gamma \Rightarrow \varphi}$$

if $p_1,..,p_n$ are built-in predicates, $\mathbf{t_1},..,\mathbf{t_n}$ are sequences of terms corresponding to the arity of the p_i, and $p_1(\mathbf{t_1}) \vee .. \vee p_n(\mathbf{t_n})$ is valid in M.

• **Induction on natural numbers:**

$$(I)\quad \frac{\Gamma \Rightarrow \varphi(0) \quad \Gamma, \varphi(x) \Rightarrow \varphi(x+1)}{\Gamma \Rightarrow \forall x \varphi(x)}\ ! \qquad \text{in } \varphi \text{ all } \forall \text{ preceding } \exists,$$

"!" denotes the variable condition, x is the eigenvariable of this rule.

• **μ-Rule:**

$$(\mu)\quad \frac{\Rightarrow \forall x \forall y \exists z A(x,y,z) \quad \Gamma, \forall x \exists y A(x,y,0) \Rightarrow \varphi}{\Gamma \Rightarrow \varphi},$$

if z is of type "natural number" and the formula $\forall x \exists y A(x,y,0)$ is valid in M. •

3. Evaluation of proofs in 𝕂. In this section, we construct an algorithm for evaluating proofs in 𝕂 as programs. This algorithm is mainly based on Gentzen's cut-elimination procedure. The presence of non-logical inference rules, however, require additional steps to eliminate these rules too. We begin by defining when a proof is reducible or irreducible.

Definition: (i) Occurrences of axioms, quantifier rules, structural rules and inductions are always irreducible.
(ii) An occurrence of the cut-rule is irreducible if it occurs in the following context:

$$\frac{\dfrac{\Gamma \Rightarrow \varphi(0) \quad \Gamma, \varphi(x) \Rightarrow \varphi(x+1)}{\Gamma \Rightarrow \forall x \varphi(x)} \qquad \dfrac{\Pi, \varphi(t) \Rightarrow \psi}{\Pi, \forall x \varphi(x) \Rightarrow \psi}}{\Gamma, \Pi \Rightarrow \psi},$$

where the formula $\forall x \varphi(x)$ does not occur in Π and t is a term which cannot be evaluated to a natural number as it contains at least one variable. All other occurrences of the cut-rule are reducible.
(iii) Occurrences of the case distinction rule ($\vee$) are irreducible if the disjunction $p_1(\mathbf{t_1}) \vee .. \vee p_n(\mathbf{t_n})$ is not decidable because at least one of the p_i contains a variable (remember that the p_i are built-in predicates and can thus be evaluated for arguments without variables. Hence, if all the $\mathbf{t_i}$ are constant and the distinction $p_1(\mathbf{t_1}) \vee .. \vee p_n(\mathbf{t_n})$ is valid, it can effectively be determined for which i, $1 \leq i \leq n$, $p_i(\mathbf{t_i})$ is true.) All other occurrences of the rule ($\vee$) are reducible.
(iv) An occurrence of the μ-rule is irreducible if it occurs in the following context:

$$\frac{\Rightarrow \forall x \forall \mathbf{x} \forall y \exists z A(x,\mathbf{x},y,z) \qquad \dfrac{\Gamma, \forall \mathbf{x} \exists y A(t,\mathbf{x},y,0) \Rightarrow \varphi}{\Gamma, \forall x \forall \mathbf{x} \exists y A(x,\mathbf{x},y,0) \Rightarrow \varphi}}{\Gamma \Rightarrow \varphi}$$

where $M \models \forall x \forall \mathbf{x} \exists y A(x,\mathbf{x},y,0)$ and the term t contains at least one variable. All other occurrences of the μ-rule are reducible.
(v) A proof is irreducible if it contains no reducible occurrence of an inference rule. Otherwise the proof is reducible. •

In the following, we consider only proofs which satisfy certain variable conditions:

Definition: (i) A proof is called *regular* if all eigenvariables occurring in it are pairwise disjoint and if each eigenvariable occurs only in the subtree above the rule where it is bound.
(ii) A proof is called *strongly regular* if it is regular and if each free variable occurring in it is an eigenvariable or occurs in the end-sequent of that proof. •

It is a rather trivial observation that each proof can be transformed into a strongly regular one by simply renaming variables and by substituting default constants. Therefore we adopt the convention that *all proofs considered below are strongly regular*. For strongly regular proofs of existential formulae we can now show:

Theorem: Let P be an irreducible proof.
(i) If the end sequent of P has the form $\Rightarrow A$, where A is a closed atom, then the entire proof P consists only of the axiom $\Rightarrow A$.
(ii) If the end sequent of P is closed and has the form $\Rightarrow \exists y_1..\exists y_n A(y_1..y_n)$, then P takes the following form:

$$\cfrac{\cfrac{\Rightarrow A(t_1,..,t_n)}{\Rightarrow \exists y_n A(t_1,..,t_{n-1},y_n)}}{\cfrac{\vdots}{\Rightarrow \exists y_1..\exists y_n A(y_1,..,y_n)}}\ .$$

where $t_1,..,t_n$ are closed terms.
Proof: (i) In strongly regular and irreducible proofs a sequent of this form cannot be obtained by applying any inference rule.
(ii) Induction on n. In accordance with the assumptions on P the last application of a rule can only be $(\Rightarrow\exists)$. •

We will now describe how to transform any strongly regular proof into an irreducible one. The transformation procedure relies on two operators, $\mathbb{R}$ und $\mathbb{S}$. Operator $\mathbb{R}$ performs one-step transformations. $\mathbb{S}$ iterates $\mathbb{R}$ until an irreducible proof is obtained. To assure termination of the transformation process the sequents are assigned constructive ordinal numbers which will decrease when applying $\mathbb{R}$ and $\mathbb{S}$. Due to the well-foundedness of the ordering of ordinal numbers, the termination procedure will eventually terminate after finitely many steps. These ordinal numbers can also be used to define the operators $\mathbb{R}$ und $\mathbb{S}$ by ordinal recursion.

The ordinals we use are represented by notations from the Veblen hierarchy up to $\varphi_\omega 0$ (see *Bachmann* [2] or *Schütte* [15] for these notations). Strictly increasing ordinal functions φ_n, $n \in \mathbb{N}$, are used which are defined as follows: $\varphi_0(\alpha)=\omega^\alpha$, φ_{n+1} is the enumeration function of the fixed points of φ_n. Furthermore, $\alpha\#\beta$ denotes the natural sum of α and β.

Definition: Ordinal assignment to the sequents of a proof
By $\Gamma\Rightarrow\varphi\{\alpha\}$ we denote the fact that the ordinal α is assigned to the sequent $\Gamma\Rightarrow\varphi$.
(i) The ordinal 0 is assigned to initial sequents.
(ii) The ordinal assigned to the conclusions of all rules, except for cut and the μ-rule, is the supremum of the ordinals assigned to the premises of this rule; hence

$$\frac{\Gamma_1\Rightarrow\varphi_1\{\alpha_1\}\ ..\ \Gamma_n\Rightarrow\varphi_n\{\alpha_n\}}{\Gamma\Rightarrow\varphi\{\max(\alpha_1,..,\alpha_n)+1\}}\ .$$

(iii) Let α and β be the ordinals assigned to the premises of an application of the cut or μ-rule. Further, let n be the number of logical symbols of the principal formula ψ of the rule considered; this is denoted by $\#(\psi)=n$. Then the ordinal $\varphi_n(\alpha\#\beta)+1$ is assigned to the conclusion.

(iv) The ordinal assigned to the end sequent of a proof P is called the ordinal of P and is denoted by $|P|$. •

Definition and Theorem: There are effective operators $\mathbb{R}$ and $\mathbb{S}$ for which the following holds:

(i) If P is a proof which ends with an application of the cut or μ-rule, then either $|\mathbb{R}(P)|<|P|$ or $\mathbb{R}(P)$ ends with an irreducible cut or μ-rule, possibly followed by an application of the structural rule, and the ordinals assigned to the premises of the irreducible cut are less than $|P|$.

(ii) $\mathbb{S}$ will transform any proof into irreducible form.

Proof: $\mathbb{R}$ and $\mathbb{S}$ are simultaneously defined by recursion on ordinal numbers. At the same time, statements (i) and (ii) are proved by transfinite induction.

(i) **Case A:** P ends with an application of the cut rule:

$$\frac{P_0\colon \Gamma\Rightarrow\varphi\{\alpha\} \quad P_1\colon \Pi\Rightarrow\psi\{\beta\}}{P\colon \Gamma,\Pi^*\Rightarrow\psi}$$

With $\#(\varphi)=n$ we obtain $|P|=\varphi_n(\alpha\#\beta)+1$. A case distinction is made according to the last inference rules of P_0 und P_1:

• Case A.1: The last rules of P_0 and P_1 generate the cut formula φ and Π contains only this occurrence of φ.

- Subcase A.1.1: $P_1\colon \Pi\Rightarrow\psi\{\beta\}$ is an initial sequent. Then ψ and φ must be prime formulae. As φ is also produced at the end of P_0, $P_0\colon \Gamma\Rightarrow\varphi$ must also be an initial sequent. Hence $\alpha=\beta=0$, $|P|=\varphi_0(0)+1=2$. Let $\mathbb{R}(P)$ be the initial sequent $\Gamma,\Pi^*\Rightarrow\psi\{0\}$; then obviously $|\mathbb{R}(P)|<|P|$.

- Subcase A.1.2: P_1 ends with an application of the structural rule with the single occurrence of φ being introduced in Π. Then the cut can be replaced by applying the structural rule to P_1, and $|\mathbb{R}(P)|<|P|$.

- Subcase A.1.3: P_1 ends with an application of $(\exists\Rightarrow)$:

$$\frac{\dfrac{P_{00}\colon \Gamma\Rightarrow\varphi(t)\{\alpha\}}{P_0\colon \Gamma\Rightarrow\exists y\varphi\{\alpha+1\}} \qquad \dfrac{P_{10}\colon \Pi,\ \varphi(y)\Rightarrow\psi\{\beta\}}{P_1\colon \Pi,\exists y\varphi\Rightarrow\psi\{\beta+1\}}}{P\colon \Gamma,\Pi\Rightarrow\psi}$$

With $\#(\exists y\varphi)=n+1$, we obtain $|P|=\varphi_{n+1}(\alpha+1\#\beta+1)+1$. Now let $\mathbb{R}(P)$ be the following proof:

$$\frac{\dfrac{P_{00}\colon \Gamma\Rightarrow\varphi(t) \quad P_{10}(t)\colon \Pi,\varphi(t)\Rightarrow\psi}{P_0\colon \Gamma,\Pi^0\Rightarrow\psi}}{P\colon \Gamma,\Pi\Rightarrow\psi}$$

Then $|\mathbb{R}(P)|<|P|$.

- Subcase A.1.4: P_1 ends with an application of $(\forall\Rightarrow)$ in the cut formula. Then, P_0 ends with either $(\Rightarrow\forall)$ or (I). In the former case, $\mathbb{R}(P)$ is defined in a way similar to A.1.3, so that $|\mathbb{R}(P)|<|P|$. In the latter case, P takes the following form:

$$\frac{\dfrac{P_{00}\colon \Gamma\Rightarrow\varphi(0)\{\alpha_0\} \ P_{01}(x)\colon \Gamma,\varphi(x)\Rightarrow\varphi(x+1)\{\alpha_1\}}{P_0\colon \Gamma\Rightarrow\forall x\varphi\{\alpha\}} \quad \dfrac{P_{10}(t)\colon \Pi,\varphi(t)\Rightarrow\psi\{\beta\}}{P_1\colon \Pi,\forall x\varphi\Rightarrow\psi\{\beta+1\}}}{P\colon \Gamma,\Pi\Rightarrow\psi}$$

where $\alpha=\max(\alpha_0,\alpha_1)+1$, and with $\#(\forall x\varphi)=n+1$, we obtain $|P|=\varphi_{n+1}(\alpha\#\beta+1)+1$. Moreover, the following distinctions are made:

- If t can be evaluated to a natural number, then let $\mathbb{R}(P)$ be the following proof:

$$\cfrac{P_{01}(t\text{-}2):\Gamma,\varphi(t\text{-}2)\Rightarrow\varphi(t\text{-}1)\{\alpha_1\} \quad \cfrac{P_{01}(t\text{-}1):\Gamma,\varphi(t\text{-}1)\Rightarrow\varphi(t)\{\alpha_1\} \quad P_{10}(t):\Pi,\varphi(t)\Rightarrow\psi\{\beta\}}{\Gamma,\varphi(t\text{-}1),\Pi^0\Rightarrow\psi}}{\vdots}$$

$$\cfrac{\cfrac{\cfrac{P_{00}:\Gamma\Rightarrow\varphi(0)\{\alpha_0\} \quad \Gamma,..\varphi(0),..\Rightarrow\psi}{\Gamma,\Gamma^0,..\Rightarrow\psi}}{\Gamma,\Pi\Rightarrow\psi}}{}$$

Then $|\mathbb{R}(P)|=\varphi_n(\alpha_0\#\varphi_n(\alpha_1\#\varphi_n(..\#\varphi_n(\alpha_1\#\beta)+1)+1)..)+2 < |P|$.

- if t contains a variable, the cut is irreducible. We set $\mathbb{R}(P):=P$. Now, $\mathbb{R}(P)$ ends with an irreducible cut, and the ordinals of the premises of this cut are less than $|P|$.

• Case A.2: Case A.1 is not applicable. A distinction is made according to the last inference rules of P_0 and P_1. For the sake of brevity we will discuss only the most important steps.

- Subcase A.2.1: P_0 or P_1 end with an inference rule which is not a cut or the μ-rule and does not produce an occurrence of the cut formula. In this case, the application of the cut rule is exchanged with the last inference rule applied in P_0 or P_1 yielding a smaller ordinal.

- Subcase A.2.2: P_0 and P_1 end with an inference rule which produces an occurrence of the cut formula. The most complex situation results if P_0 ends with an induction and P_1 ends with a rule $(\forall\Rightarrow)$, producing an occurrence of the cut formula. Unlike case A.1.4, however, multiple occurrences of the cut formula may be found in Π. Nevertheless the definition of $\mathbb{R}(P)$ is similar to case A.1.4. Initially, the additional occurrences of the cut formula are eliminated by a cut with P_0 and P_{10}. Then, again $|\mathbb{R}(P)|<|P|$.

- Subcase A.2.3: The subcases discussed above are not applicable. Then P_0 or P_1 ends with a cut or μ-rule. Let us look at the example where P_0 ends with a cut rule:

$$\cfrac{\cfrac{P_{00}:\Gamma_0\Rightarrow\chi\{\gamma\} \quad P_{01}(x):\Gamma_1\Rightarrow\varphi\{\delta\}}{P_0:\Gamma_0,\Gamma_1{}^0\Rightarrow\varphi\{\alpha\}} \quad \Pi\Rightarrow\psi\{\beta\}}{P:\Gamma_0,\Gamma_1{}^0,\Pi^*\Rightarrow\psi}$$

With $\#(\chi)=k$, $\#(\varphi)=n$, we have $\alpha=\varphi_k(\gamma\#\delta)+1$ and $|P|=\varphi_n(\alpha\#\beta)+1$. Since $|P_0|<|P|$, $\mathbb{R}(P_0)$ is already defined. By induction hypothesis, either $|\mathbb{R}(P_0)|<|P_0|=\alpha$ or $\mathbb{R}(P_0)$ ends with an irreducible cut or μ-rule, possibly succeeded by (S), and $|\mathbb{R}(P_0)_i|<|P_0|$ or $|\mathbb{R}(P_0)_{0i}|<|P_0|$ for $i=0,1$. Moreover, the following distinctions are made:

-- If $\xi:=|\mathbb{R}(P_0)|<|P_0|$, then let $\mathbb{R}(P)$ be the following proof:

$$\cfrac{\mathbb{R}(P_0):\Gamma_0,\Gamma_1{}^0\Rightarrow\varphi\{\xi\} \quad P_1:\Pi\Rightarrow\psi\{\beta\}}{P:\Gamma_0,\Gamma_1{}^0,\Pi^*\Rightarrow\psi}$$

Then $|\mathbb{R}(P)|=\varphi_n(\xi\#\beta)+1<|P|$, as $\xi<\alpha$.

-- Otherwise $\mathbb{R}(P_0)$ ends with an irreducible cut or μ-rule, possibly succeeded by (S). As an example, we consider the situation where $\mathbb{R}(P_0)$ ends with an irreducible μ-rule succeeded by (S):

$$\dfrac{\dfrac{\mathbb{R}(P_0)_{00}: \Rightarrow\forall x\forall x\forall y\exists z A(x,x,y,z)\{\mu\} \qquad \dfrac{\mathbb{R}(P_0)_{010}: \Gamma,\forall x\exists y A(t,x,y,0)\Rightarrow\varphi\{\nu\}}{\mathbb{R}(P_0)_{01}: \Gamma,\forall x\forall x\exists y A(x,x,y,0)\Rightarrow\varphi\{\nu+1\}}}{\mathbb{R}(P_0)_0: \Gamma\Rightarrow\varphi}}{\mathbb{R}(P_0): \Gamma_0,\Gamma_1{}^{\circ}\Rightarrow\varphi}$$

with $M \models \forall x\forall x\exists y A(x,x,y,0)$ and $\mu,\ \nu+1< |P_0|=\alpha$. Then let $\mathbb{R}(P)$ be the following proof:

$$\dfrac{\dfrac{\mathbb{R}(P_0)_{00}:\Rightarrow\forall x\forall x\forall y\exists z A(x,x,y,z)\{\mu\} \qquad \dfrac{\dfrac{\mathbb{R}(P_0)_{010}: \Gamma,\forall x\exists y A(t,x,y,0)\Rightarrow\varphi\{\nu\} \quad P_1:\Pi\Rightarrow\psi\{\beta\}}{\Gamma,\forall x\exists y A(t,x,y,0),\Pi^*\Rightarrow\psi}}{\Gamma,\forall x\forall x\exists y A(x,x,y,0),\Pi^*\Rightarrow\psi}}{\Gamma,\Pi^*\Rightarrow\psi}}{\Gamma_0,\Gamma_1{}^{\circ},\Pi^*\Rightarrow\psi}$$

$\mathbb{R}(P)$ now ends with an irreducible μ-rule, and $|\mathbb{R}(P)_{00}| = |\mathbb{R}(P_0)_{00}| = \mu<|P|$ and $|\mathbb{R}(P)_{01}|=\varphi_n(\nu\#\beta)+2<\varphi_n(\alpha\#\beta)+1=|P|$.

Case B: P ends with an application of the μ-rule:

$$\dfrac{P_0: \Rightarrow\forall x\forall y\exists z A(x,y,z)\{\alpha\} \quad P_1: \Gamma,\forall x\exists y A(x,y,0)\Rightarrow\psi\{\beta\}}{P: \Gamma^*\Rightarrow\psi},$$

where $M \models \forall x\exists y A(x,y,0)$. With $\#(\forall x\exists y A(x,y,0))=n+1$, $|P|=\varphi_{n+1}(\alpha\#\beta)+1$. $\mathbb{R}(P)$ is defined in terms of a case distinction by the last inference rule of P_1 and is largely similar to case A. Therefore, only two relevant cases are outlined below:

• Case B.1: P_1 ends with $(\forall\Rightarrow)$ in the μ-formula and this formula occurs in Π only once:

$$\dfrac{P_0: \Rightarrow\forall x\forall x\forall y\exists z A(x,x,y,z)\{\alpha\} \quad \dfrac{P_{10}: \Gamma,\forall x\exists y A(t,x,y,0)\Rightarrow\psi\{\beta'\}}{P_1: \Gamma,\forall x\forall x\exists y A(x,x,y,0)\Rightarrow\psi\{\beta'+1\}}}{P: \Gamma\Rightarrow\psi}$$

with $|P|=\varphi_{n+1}(\alpha\#\beta+1)+1$. The following distinctions are made:

- if t is not free of variables, then the occurrence of the μ-rule is irreducible and we set $\mathbb{R}(P):=P$;

- if t is variablefree, we transform P_0 into a specialized proof P' of the sequent $\Rightarrow\forall x\forall y\exists z A(t,x,y,z)\{\gamma\}$ where $\gamma<\varphi_{n+1}(\alpha+1)$. From $M \models \forall x\forall x\exists y A(x,x,y,0)$ we arrive at the trivial conclusion that $M \models \forall x\exists y A(t,x,y,0)$. Therefore, let $\mathbb{R}(P)$ be the proof:

$$\dfrac{P': \Rightarrow\forall x\forall y\exists z A(t,x,y,z)\{\gamma\} \quad P_{10}: \Gamma,\forall x\exists y A(t,x,y,0)\Rightarrow\psi\{\beta\}}{P: \Gamma\Rightarrow\psi}$$

Then $|\mathbb{R}(P)|=\varphi_n(\gamma\#\beta)+1<\varphi_{n+1}(\alpha\#\beta+1)+1=|P|$.

Case B.2: P_1 ends with $(\exists\Rightarrow)$ in the μ-formula and this formula occurs in Π only once:

$$\dfrac{P_0: \Rightarrow\forall y\exists z A(y,z)\{\alpha\} \quad \dfrac{P_{10}: \Gamma,A(y,0)\Rightarrow\psi\{\beta'\}}{P_1: \Gamma,\exists y A(y,0)\Rightarrow\psi\{\beta'+1\}}}{P: \Gamma\Rightarrow\psi}$$

where $M \models \exists y A(y,0)$ (*). With $\#(\exists y A(y,0))=2$, we obtain $|P|=\varphi_2(\alpha\#\beta+1)+1$. Let f be the enumeration function of the type of y. For n=0,1,2, ..the proof P_n can be transformed into a proof $P_n': \Rightarrow\exists z A(f(n),z)\{\gamma_n\}$, where $\gamma_n<\varphi_2(\alpha+1)$. Operator $\mathbb{S}$ which is already defined by recursion can now be applied to P_n' so as to determine natural numbers k_n so that $\Rightarrow A(f(n),k_n)\{0\}$, hence $M \models A(f(n),k_n)$. In this way we find the smallest n_0, such that $k_{n_0}=0$ (this exists due to (*)). Now let $\mathbb{R}(P)$ be the following proof:

$$\frac{\Rightarrow A(f(n_0),0)\{0\} \quad P_{10}(f(n_0)): \Gamma, A(f(n_0),0)\Rightarrow\psi\{\beta\}}{P: \Gamma\Rightarrow\psi}.$$

Then $|\mathbb{R}(P)|=\varphi_0(\beta)+2<\varphi_2(\alpha\#\beta+1)+1=|P|$.

(ii) In order to define $\mathbb{S}$, we distinguish between the following cases:

- **Case 1:** $P: \Gamma\Rightarrow\varphi\{\alpha\}$ is an initial sequent. We set $\mathbb{S}(P):=P$ so that $\mathbb{S}(P)$ is irreducible.
- **Case 2:** P ends with an irreducible application of an inference rule

$$\frac{P_1: \Gamma_1\Rightarrow\varphi_1\{\alpha_1\} \;..\; P_n: \Gamma_n\Rightarrow\varphi_n\{\alpha_n\}}{P: \Gamma\Rightarrow\varphi}$$

with $\alpha_1,..,\alpha_n<|P|$. Then let $\mathbb{S}(P)$ be the following proof:

$$\frac{\mathbb{S}(P_1): \Gamma_1\Rightarrow\varphi_1 \;..\; \mathbb{S}(P_n): \Gamma_n\Rightarrow\varphi_n}{\mathbb{S}(P): \Gamma\Rightarrow\varphi}.$$

Since $\alpha_1,...,\alpha_n<|P|$, the proofs $\mathbb{S}(P_1),..., \mathbb{S}(P_n)$ are defined and by induction hypothesis they are irreducible. Hence $\mathbb{S}(P)$, too, is irreducible.

- **Case 3:** P ends with a reducible occurrence of an inference rule.

- Subcase 3.1: P ends with a $(\vee)$–rule

$$\frac{P_1: \Gamma_1,p_1\Rightarrow\varphi_1\{\alpha_1\} \quad P_n: \Gamma_n,p_n\Rightarrow\varphi_n\{\alpha_n\}}{P: \Gamma\Rightarrow\varphi}$$

and for some i_0, $1\leq i_0\leq n$, $M \models p_{i_0}$. Then we set $\mathbb{S}(P):=\mathbb{S}(P_{i_0}')$, where P_{i_0}' results from P_{i_0} by deleting p_{i_0} in the antecedent of all sequents. Since $|P_{i_0}'|=\alpha_{i_0}<|P|$, $\mathbb{S}(P)$ is defined and irreducible.

- Subcase 3.2: P ends with a cut or μ-rule. Consider $\mathbb{R}(P)$. According to (i), $|\mathbb{R}(P)|<|P|$ or $\mathbb{R}(P)$ ends with an irreducible cut or μ-rule so that $|\mathbb{R}(P)_i|<|P|$ for i=0,1. If $|\mathbb{R}(P)|<|P|$, we set $\mathbb{S}(P):=\mathbb{S}(\mathbb{R}(P))$. Otherwise, let $\mathbb{S}(P)$ be the proof resulting from $\mathbb{S}(\mathbb{R}(P)_0)$ and $\mathbb{S}(\mathbb{R}(P)_1)$ by applying an irreducible cut or μ-rule. •

Remarks: (i) The reduction operator $\mathbb{S}$ transforms a proof P from the root to the leaves of the proof tree. The irreducible form of P is uniquely determined.

(ii) The ordinal assignment as employed here is not optimal in the sense that the ordinals used are minimal. Proof-theoretic considerations show that ordinal notations for ordinals up to ε_0 $(=\varphi_1 0)$ will suffice. However, the assignment employed here has the advantage of being convenient and notationally simple.

Now the evaluation procedure for proofs as programs can easily be defined as follows: Given a proof P in $\mathbb{K}$ of a sequent $\Rightarrow\forall x\exists y A(x,y)$ and input values **e**, we first specialize P in the following way to a proof $P_{\mathbf{e}}$ using the input values **e**:

$$P: \Rightarrow \forall x \exists y A(x,y) \quad \dfrac{\dfrac{A(e,y) \Rightarrow A(e,y)}{\vdots}}{\forall x \exists y A(x,y) \Rightarrow \exists y A(e,y)}$$
$$\overline{\quad \Rightarrow \exists y A(e,y) \quad}$$

Then we apply the operator $\mathbb{S}$ to P_e in order to obtain the irreducible proof:

$$\dfrac{\dfrac{\Rightarrow A(e,t_1(e),..,t_n(e))}{\vdots}}{\Rightarrow \exists y A(e,y)} \cdot$$

The output values are the values of the terms $t_1(e),..,t_n(e)$. •

This procedure has been implemented by *C. Bornschein* [6] and *F.Regensburger* [14] in their thesis at the Technical University of Munic; their program was successfully applied to numerous examples.

4. Completeness of $\mathbb{K}$. We now prove that all computable functions can be computed by means of a proof in $\mathbb{K}$.

Definition: (i) Let L be a first-order language with the following symbols: a constant symbol 0, a unary function symbol N, and a n+1-ary predicate symbol A_f for each n-ary recursive function f. For each natural number n, let K_n be the numeral N(N(..N(O)..)) with n occurrences of N.
(ii) Let M be the following interpretation compatible with L: The universe of M is N, $0^M := 0 \in N$, N^M: $N \to N$ with $a \to a+1$, and $A_f^M(a_1,..,a_n,b)$ iff. $f(a_1,..,a_n)=b$.•

Theorem: For each n-ary recursive function f, there is a proof in $\mathbb{K}$ of the sequent $\Rightarrow \forall x_1..\forall x_n \exists y A_f(x_1,..,x_n,y)$, so that the axioms used in that proof are valid in the interpretation M.

Remarks: (i) The above discussion clearly shows that such a proof ensures that f is correctly computed.

(ii) Obviously, our calculus $\mathbb{K}$ is not an axiomatized logical theory. Since all valid prime sequents of a given language can potentially be axioms in a proof in $\mathbb{K}$, the property of being an axiom of a proof in $\mathbb{K}$ is in general not decidable.

Proof: (i) For the basic functions C^n_a, I^n_i, N the following holds:

$$\dfrac{\dfrac{\Rightarrow A_{C^n_a}(x,K_a)}{\vdots}}{\Rightarrow \forall x \exists y A_{C^n_a}(x,y)} \qquad \dfrac{\dfrac{\Rightarrow A_{I^n_i}(x,x_i)}{\vdots}}{\Rightarrow \forall x \exists y A_{I^n_i}(x,y)} \qquad \dfrac{\dfrac{\Rightarrow A_N(x,N(x))}{\vdots}}{\Rightarrow \forall x \exists y A_N(x,y)} \cdot$$

(ii) Let $f(\mathbf{x})=h(g_1(\mathbf{x}),..,g_m(\mathbf{x}))$ with recursive functions $g_1,..,g_m,h$. By induction hypothesis, proofs exist for the following sequents:

(*) $\quad \Rightarrow \forall x \exists y A_{g_1}(x,y), .., \ \Rightarrow \forall x \exists y A_{g_m}(x,y), \ \Rightarrow \forall x_1..\forall x_m \exists y A_h(x_1,..,x_m,y).$

According to the definition of f, the sequent

$$A_{g_1}(\mathbf{x},y_1), .., A_{g_m}(\mathbf{x},y_m), A_h(y_1,..,y_m,z) \Rightarrow A_f(\mathbf{x},z)$$

is valid in M. By introducing quantifiers and performing cuts with the sequents (*), we obtain a proof of $\Rightarrow \forall x \exists y A_f(x,y)$.

(iii) Let $f(0,\mathbf{x}) = g(\mathbf{x})$, $f(y+1,\mathbf{x}) = h(y,\mathbf{x},f(y,\mathbf{x}))$. As in (ii), proofs of $\Rightarrow\forall\mathbf{x}\exists y A_g(\mathbf{x},y)$ and $\Rightarrow\forall y\forall\mathbf{x}\forall z\exists u A_h(y,\mathbf{x},z,u)$ with sequents

$$A_g(\mathbf{x},u) \Rightarrow A_f(0,\mathbf{x},u) \text{ und } A_f(y,\mathbf{x},v), A_h(y,\mathbf{x},v,w) \Rightarrow A_f(N(y),\mathbf{x},w)$$

yield proofs of $\Rightarrow\forall\mathbf{x}\exists z A_f(0,\mathbf{x},z)$ and $\forall\mathbf{x}\exists z A_f(y,\mathbf{x},z)\Rightarrow\forall\mathbf{x}\exists z A_f(y+1,\mathbf{x},z)$. By applying the rule of induction, we obtain a proof of $\Rightarrow\forall y\forall\mathbf{x}\exists z A_f(y,\mathbf{x},z)$.

(iv) Let $f(\mathbf{x}) = \mu y.\ g(\mathbf{x},y)=0$, with $\forall\mathbf{x}\exists y\ g(\mathbf{x},y)=0$. By induction hypothesis, there is a proof of (*) $\Rightarrow\forall\mathbf{x}\forall y\exists z Ag(\mathbf{x},y,z)$, and $M \models \forall\mathbf{x}\exists y Ag(\mathbf{x},y,0)$. Let Πg be the function defined by $\Pi g(\mathbf{x},y) = \prod_{z<y} g(\mathbf{x},z)$ and g^* the function defined by $g^*(\mathbf{x},y) = sg'(\Pi g(\mathbf{x},y))+sg(g(\mathbf{x},y))$. According to the definition of g^* we obviously obtain: $g^*(\mathbf{x},y)=0$ iff. $f(\mathbf{x})=y$. Therefore the axiom $A_{g^*}(\mathbf{x},y,0) \Rightarrow A_f(\mathbf{x},y)$ is valid in M. By quantifier introduction we obtain $\forall\mathbf{x}\exists y A_{g^*}(\mathbf{x},y,0) \Rightarrow\forall\mathbf{x}\exists y A_f(\mathbf{x},y)$. Since g^* is a primitive recursive function in g, the existence of a proof of $\forall\mathbf{x}\forall y\exists z A_g(\mathbf{x},y,z) \Rightarrow\forall\mathbf{x}\forall y\exists z A_{g^*}(\mathbf{x},y,z)$ is implied by (i) to (iii). A cut with (*) then yields $\Rightarrow\forall\mathbf{x}\forall y\exists z A_{g^*}(\mathbf{x},y,z)$. By applying the μ-rule we obtain a proof $\Rightarrow\forall\mathbf{x}\exists y A_f(\mathbf{x},y)$ •

References

[1] Alexi, M., Extraction and verification of programs by analysis of formal proofs, TCS 61 (1988) 225-258

[2] Bachmann, H., Transfinite Zahlen, Springer Verlag, Berlin 1955

[3] Bates, J.L., A logic for correct program development, Cornell Univ. 1979

[4] Bates, J.L. a. Constable, R.L., Definition of Micro-PRL, Tech. Rp. 82-492, Cornell Univ. 1981

[5] Bates, J.L. a. Constable, R.L., Proofs as programs, ACM Trans. on Progr. Lang. a. Syst. 7,1 (1985) 113-136

[6] Bornschein, C., Implementierung eines Verfahrens zur Herleitungstransformation, Diplomarbeit, Technische Universität München 1989

[7] Constable, R.L., Proofs as Programs, Inf. Process. Lett. 16,3 (1983) 105-112

[8] Constable, R.L. a. Bates, J.L., The nearly ultimate PRL, Techn. Rep. 83-551, Cornell Univ. 1983

[9] Gallier, J.H., Logic for Computer Science, New York 1986

[10] Gentzen, G., Untersuchungen über das logische Schließen, Math. Zeitschrift 39 (1935) 176-210, 405-431

[11] Hayashi, S., PX: A System extracting Programs from Proofs

[12] Manna, Z. a. Waldinger, R., A deductive approach to program synthesis, ACM Trans. on Progr. Lang. a. Syst. 2,1 (1980) 92-121

[13] Martin-Löf, P., Constructive mathematics and computer programming, 6th. Intern. Congr. Logic, Method. a. Philos. of Science, Hannover 1979

[14] Regensburger, F., Implementierung eines Schnitteliminationsverfahrens, Diplomarbeit, Technische Universität München 1989

[15] Schütte, K., Proof theory, Springer Verlag, Berlin 1977

[16] Schwichtenberg, H., LCF with realizing terms: a framework for the development and verification of programs, 1988

Minimal from Classical Proofs

Helmut Schwichtenberg

Mathematisches Institut, Universität München
Theresienstr. 39, D–8000 München 2

We consider the $\to\forall$–fragment of first order logic with a distinguished predicate symbol $\bot$ (for falsity); as usual we write $\neg\varphi$ for $\varphi \to \bot$. Gentzen's natural deduction system for minimal logic in this language consists just of introduction and elimination rules for $\to$ and $\forall$. Hence any proof in this system gives rise to a type–free λ–term, possibly with assumption variables. If in addition a proof (and hence also its associated λ–term) is normal, then from its context, i.e. the assignment of assumption formulas to its assumption variables, and from its endformula we can recover all formulas in the proof. This representation of formal proofs seems to be useful: for instance it allows an efficient implementation of normalization by evaluation (cf. [1], [2]).

It is well known that any proof can be transformed into a unique normal form with respect to β–conversion. Using η–expansion we can then construct the *long* normal form, where all minimal formulas are atomic.

We are interested in the problem of how to find proofs in minimal logic, from a somewhat practical point of view.* In particular we want to make use of existing theorem provers based on classical logic. So our problem is to review under what circumstances a classical proof can be converted into a proof in minimal logic, and moreover to describe reasonable algorithms which do this conversion. A good survey of the subject can be found in [3, Chapter 2.3]. Here we add a new result.

Note first that a convenient way to represent classical logic in our setting is to add stability assumptions of the form

$$\text{stab}_P : \forall\vec{x}.\neg\neg P\vec{x} \to P\vec{x}$$

for all predicate symbols P. For then we can easily derive $\neg\neg\varphi \to \varphi$ for an arbitrary formula φ, using

$$(\neg\neg\psi \to \psi) \to \neg\neg(\varphi \to \psi) \to \varphi \to \psi$$

$$(\forall x.\neg\neg\varphi \to \varphi) \to \neg\neg\forall x\varphi \to \forall x\varphi,$$

which are derivable in our $\to\forall$–fragment of minimal logic. Hence by a classical proof of ψ from assumptions $\varphi_1, \dots, \varphi_n$ we mean a proof in minimal logic using stability assumptions in addition to the given assumptions $\varphi_1, \dots, \varphi_n$.

A formula is called *Horn formula* if it has the form $\forall x_1, \dots, x_n.A_1 \to \dots \to A_m \to B$ with A_i and B atomic. It is called ***definite Horn formula*** if in addition we have $B \neq \bot$. If instead of atomic A_i we allow universally quantified atomic formulas, the result is called a ***generalized (definite) Horn formula.***

---

* At the conference I gave a more general lecture on "Proofs and Programs". Since most of what I have said is already published (in [1] and [2]), this note only elaborates one part of the lecture dealing with a very special aspect of the field.

Theorem 1. *Let $\varphi_1, \ldots, \varphi_n$ be generalized Horn formulas. We have a quadratic algorithm transforming a classical proof in long normal form of $\bot$ from $\varphi_1, \ldots, \varphi_n$ into a proof in minimal logic of $\bot$ from the same assumptions.*

The proof is by induction on the total number of stability axioms used. Note first that bound assumption variables u in the given normal proof can only occur in the context

$$\mathrm{stab}_{P}\vec{r}(\lambda u d)$$

with u of type $\neg P\vec{r}$ and d of type $\bot$. The reason for this is that all top formulas different from stability axioms are generalized Horn formulas which never have an implication in the premise of another implication.

Case 1. There is at least one occurrence of a bound assumption variable in the proof. Since we assume our proof to be in long normal form, any of the occurrences of an assumption variable u of type $\neg P\vec{r}$ must be the main premise of an $\to$-elimination, i.e. must be in a context ud_1 where u derives $P\vec{r}$. Now choose an uppermost occurrence of a bound assumption variable, i.e. a subderivation ud_1 where d_1 does not contain an occurrence of any bound assumption variable. Since d_1 derives $P\vec{r}$, we can replace the whole subderivation $\mathrm{stab}_{P}\vec{r}(\lambda u d)$ of $P\vec{r}$ (the one where u is bound) by d_1. Hence we have removed one occurrence of a stability axiom.

Case 2. Otherwise. If there are no more stability axioms in the proof, we are done. If not, choose an uppermost occurrence of a stability axiom, i.e. a subderivation $\mathrm{stab}_{P}\vec{r}(\lambda u d)$ where d does not contain stability axioms. Since we are in case 2 here d also cannot contain free assumption variables which are bound elsewhere in the proof. But since d derives $\bot$, we can replace the whole proof (which also has $\bot$ as its end formula) by d and hence we are done again.

Note that Theorem 1 is best possible in the sense that it becomes false if we allow an implication in the body of one of the Horn formulas. A counterexample (due to U. Berger) is

$$((P \to Q) \to \bot) \to (P \to \bot) \to \bot,$$

which is provable in classical but not in minimal logic. For if it were, we could replace $\bot$ in this proof (which in minimal logic is just another propositional variable) by P, and hence we would obtain a proof in minimal logic of the Peirce formula

$$((P \to Q) \to P) \to P,$$

which is known to be underivable.

By essentially the same argument we obtain the following variant of Theorem 1 for generalized *definite* Horn formulas:

Theorem 2. *Let $\varphi_1, \ldots, \varphi_n$ be generalized definite Horn formulas. We have a quadratic algorithm transforming a classical proof in long normal form of an atomic formula B from $\varphi_1, \ldots, \varphi_n$ into a proof in minimal logic of B from the same assumptions.*

The proof is by a simple modification of the argument for Theorem 1. Note that in case 2 it cannot happen that stability axioms occur in the proof since then we would have a derivation d of $\bot$ from definite Horn formulas, which is clearly impossible.

References

1. U. Berger, H. Schwichtenberg: An inverse of the evaluation functional for typed lambda calculus. Proc. 6th IEEE Symp. on Logic in Computer Science, (ed.: R. Vemuri), Los Alamitos: IEEE Computer Society Press, pp. 203–211, 1991
2. H. Schwichtenberg: Proofs as Programs. To appear in: Leeds Proof Theory '90 (ed.: P. Aczel, H. Simmons, S.S. Wainer), Cambridge: University Press, 1992
3. A. Troelstra, D. van Dalen: Constructivism in Mathematics. Vol. I, Amsterdam: North–Holland, 1988

Quantifier Hierarchies over Word Relations

Sebastian Seibert*
Institut für Informatik und Praktische Mathematik
Christian-Albrechts-Universität zu Kiel
Olshausenstr. 40, DW-2300 Kiel 1
e-mail ss@informatik.uni-kiel.dbp.de

Abstract

We consider analogues of the arithmetical hierarchy over word relations, obtained by replacing the class of recursive relations with some other classes which are defined by various types of finite and pushdown automata or by concatenation formulas.

Most of the new hierarchies turn out to be downward prolongations of the arithmetical hierarchy: They reach the class of recursive relations at their second level. The lower levels of the different new hierarchies are shown to be incomparable w.r.t. set inclusion for the most part.

1 Introduction

We classify the complexity of relations of finite words using quantifier alternation as a measure, similarly to the arithmetical hierarchy of recursion theory. The approach of the present paper is to substitute the "kernel" of that hierarchy, i.e. the class RCS of recursive relations, by several other classes and to study the structure of the resulting hierarchies. For specifying new kernels, we will use different models of finite and pushdown automata (specialising Turing machines which determine the class RCS). We will look additionally at the class CON, defined in terms of equations where each side is a concatenation of words and variables ranging over words.

For a given class C, representing such a kernel, we consider the question whether the obtained corresponding "arithmetical hierarchy", called the C-hierarchy, is infinite. More precisely we want to know which of the inclusions between different levels of the hierarchy are strict. In Section 2 a strong connection between these questions is shown, assuming only some closure properties of the kernel.

Relation classes based on multitape automata are defined in Section 3, and there we investigate the infinity of the hierarchies based on these classes and their relation to the arithmetical hierarchy and to each other. In Section 4 the analogue is done for the concatenation defined class CON, and in addition its relation to the automata defined hierarchies is investigated.

*This work is partially supported by ESPRIT Basic Research Action 3166 'Algebraic and Syntactic Methods In Computer Science' (ASMICS)

2 Infinite vs Strict Hierarchies

Definition 2.1 Let C be a class of word relations over a given alphabet $\Sigma \supseteq \{a, b\}$, that is, each relation is a subset of $(\Sigma^*)^k$ for some $k \in \mathbb{N}$. Following [14] we define the classes of relations $\sum_n^0(C)$, $\prod_n^0(C)$ and $\Delta_n^0(C)$ by: $R \subseteq (\Sigma^*)^k$ is in $\sum_n^0(C)$ iff there exists a relation $R' \in C$ s.t. for all $\mathbf{x} \in (\Sigma^*)^k$:

$R\mathbf{x} \Leftrightarrow \exists \mathbf{x}_1 \forall \mathbf{x}_2 \ldots \overset{\exists}{\forall} \mathbf{x}_n R'\mathbf{x}\mathbf{x}_1\mathbf{x}_2 \ldots \mathbf{x}_n$.

(Boldface letters indicate tuples of variables and $R\mathbf{x}$ means $\mathbf{x} \in R$.) Analogously we define $R \in \prod_n^0(C)$ with leading $\forall$, and $\Delta_n^0(C) := \sum_n^0(C) \cap \prod_n^0(C)$.

Since we have allowed empty tuples of variables in this definition, we get, for an arbitrary class C, the following diagram (each line means that the class to its left is included in the one to the right):

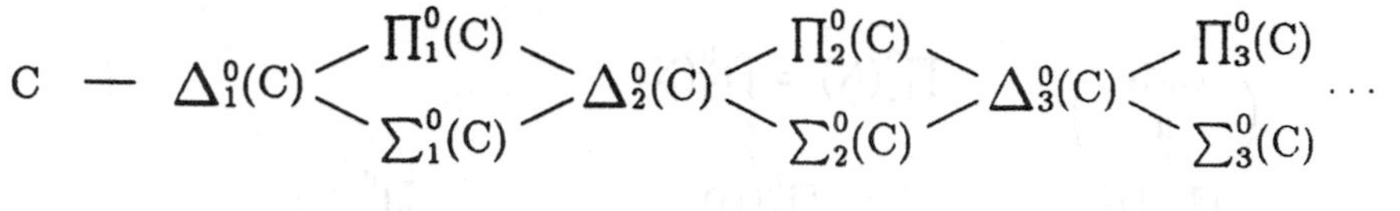

Figure 1: Arithmetical hierarchy over an arbitrary relation class C

This chain of inclusions is called the *arithmetical hierarchy over C*, or the *C-hierarchy*. Given C, the *kernel* of the hierarchy, we can ask which of the inclusions are strict or whether there are infinitely many strict inclusions. We will see a strong relation between these two questions in case C is closed under projection or complementation. For the latter we denote the complement of a k-ary relation R by $\overline{R} = (\Sigma^*)^k \setminus R$. The main result of this section shows the strictness of a hierarchy to be consequence of its infinity and the considered closure properties of its kernel.

All classes in a hierarchy are identical to the kernel if the kernel is closed under projection and complementation. If on the other hand a hierarchy is infinite and its kernel is closed either under projection or under complementation, the strictness of the hierarchy is determined - except for one single inclusion. First we will state this for a pair of kernels, which one should imagine as the classes of all relations accepted by deterministic resp. nondeterministic automata of a given model. This result follows from exchange rules for quantifiers and complementation by easy induction.

Theorem 2.2 *Let D and N be two classes of word relations.*

(a). If D is closed under complementation, a relation is in $\sum_n^0(D)$ iff its complement is in $\prod_n^0(D)$, in short notation:

(i) co-$\sum_n^0(D) = \prod_n^0(D)$

(b). If $N = \sum_1^0(D)$, the N-hierarchy can be extracted from the D-hierarchy in the following sense:

(ii) $N = \sum_1^0(N) = \Delta_1^0(N)$ [$= \sum_1^0(D)$];

and for all $n \in \mathbb{N}$*:*

(iii) $\sum^0_{2n}(N) = \sum^0_{2n+1}(N) = \Delta^0_{2n+1}(N) = \sum^0_{2n+1}(D)$;

(iv) $\prod^0_{2n-1}(N) = \prod^0_{2n}(N) = \Delta^0_{2n}(N) = \sum^0_{2n}(D)$.

(c). If in addition to the assumptions of (a) and (b) the D-hierarchy (or equivalently the N-hierarchy) is infinite, then strictness of inclusions indicated by lines in Figure 2 below holds. In particular, we have for all $n \in \mathbb{N}$*:*

(v) $\Delta^0_n(D) \subsetneq \sum^0_n(D)$;

(vi) $\sum^0_n(D) \subsetneq \Delta^0_{n+1}(D)$

and the same holds for $\prod^0_n(D)$ *instead of* $\sum^0_n(D)$.

(vii) $\sum^0_n(D) \neq \prod^0_n(D)$. □

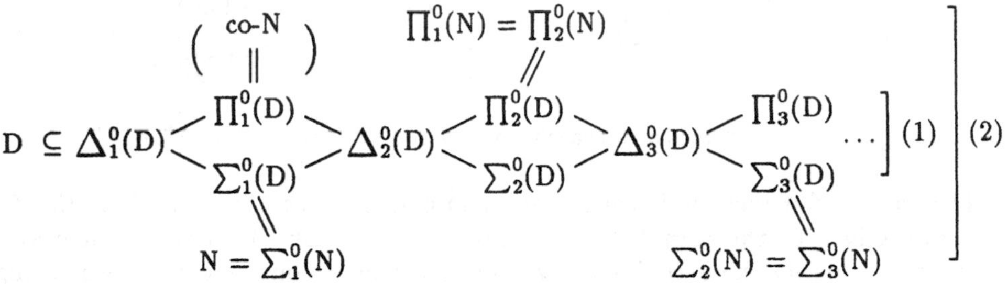

Figure 2: Hierarchies over a pair of kernels. (1) D-hierarchy; (2) N-hierarchy.

The inclusion $D \subseteq \Delta^0_1(D)$ may be strict or may be an equality. We will look in the sequel at several examples of pairs of classes which fulfill the conditions for D and N in Theorem 2.2 and where this inclusion is strict. Only in case of (RCS, R.E.), which can be substituted for (D, N), we have RCS = Δ^0_1(RCS) (R.E. is the class of recursively enumerable relations).

Corollary 2.3 *If a class D of word relations is closed under complementation and defines an infinite arithmetical hierarchy, then parts (a) and (c) of Theorem 2.2 hold for the D-hierarchy.* □

We omit the analogue for kernels closed under projection, and end this section with an observation which allows the transfer of fundamental properties from one hierarchy to another.

Lemma 2.4 *("Hierarchy Lemma") Let A be some relation class included in the arithmetical hierarchy over another class B. This means that there is an* $n \in \mathbb{N}$ *such that* $B \subseteq A \subseteq \Delta^0_n(B)$.

Then both hierarchies exhaust each other and the A-hierarchy is infinite iff the B-hierarchy is infinite. □

3 Automata Definable Relations

Extending the notion of pushdown automata and finite automata to multitape models (in order to handle relations instead of languages) we have to distinguish between synchronous and asynchronous automata: A synchronous automaton is only allowed to make an ε–step on one tape if it does so on all tapes. Otherwise it has to read on each tape a single letter from Σ or the endmarker \$ (which is not in Σ). Asynchronous automata are allowed to read ε on some tape and elements of $\Sigma \dot\cup \{\$\}$ on other tapes, so they are more powerful in general.

Definition 3.1 An *k-tape finite* (resp. *pushdown*) *automaton* $\mathfrak{A}$ is a tuple $(Q, \Sigma, q_0, \Delta, F)$ (resp. $(Q, \Sigma, \Gamma, q_0, Z_0, \Delta, F)$), having the usual components: finite set of states Q, initial state $q_0 \in Q$, final state set $F \subseteq Q$ (and stack alphabet Γ with start symbol Z_0).

The multi-tape extension appears only in the definition of the transition relation Δ, defined in the (more general) *asynchronous* case as

$\Delta \subseteq Q \times [(\Sigma \dot\cup \{\$, \varepsilon\})^k \setminus \{\$\}^k] \times Q$ (resp. $Q \times [(\Sigma \dot\cup \{\$, \varepsilon\})^k \setminus \{\$\}^k] \times \Gamma \times \Gamma^* \times Q$),

and in the *synchronous* case as

$\Delta \subseteq Q \times [\{\varepsilon\}^k \dot\cup (\Sigma \dot\cup \{\$\})^k \setminus \{\$\}^k] \times Q$

(resp. $Q \times [\{\varepsilon\}^k \dot\cup (\Sigma \dot\cup \{\$\})^k \setminus \{\$\}^k] \times \Gamma \times \Gamma^* \times Q$).

For technical reasons two special features are invented, concerned with the endmarker \$. In the above definition we explicitely have forbidden transitions, in which an automaton reads \$ on each tape. Hence our k-tape automata coincide in case "k=1" precisely with the usual automata for language recognition. Moreover the notion of synchronism requires that the endmarker is read, but not passed, so that it may be read an arbitrary number of times.

An input tuple $(x_1, \ldots, x_k) \in (\Sigma^*)^k$ is accepted by a k-tape automaton, if there exists a run from the initial configuration $\kappa_1 = (q_0, x_1\$, \ldots, x_k\$)$ (resp. $(q_0, x_1\$, \ldots, x_k\$, Z_0)$) to a configuration $(q, \$, \ldots, \$)$ (resp. $(q, \$, \ldots, \$, \gamma), \gamma \in \Gamma^*$) where $q \in F$, and κ_{i+1} results from κ_i according to Δ, for $i = 1, \ldots, r-1$.

In case of synchronous finite automata the deterministic ones are as powerful as nondeterministic automata, so that the whole hierarchy collapses to the class SRC of synchronous recognizable relations (see [14]). Therefore we will concentrate upon the other classes where we have to distinguish between deterministic and nondeterministic automata.

The classes of relations accepted by deterministic resp. nondeterministic asynchronous finite automata are DRAT and RAT ("deterministic rational" resp. "rational" relations, cf. e. g. [2, 5, 10, 14]), and we define DPS, NPS, DPA and NPA to be the classes of all relations accepted by deterministic resp. nondeterministic, synchronous resp. asynchronous pushdown automata.

Essential for our treatment are the following closure properties of these classes:

Lemma 3.2 *(a) DRAT, DPS, and DPA are closed under complementation;*
(b) RAT $= \sum_1^0$*(DRAT), NPS* $= \sum_1^0$*(DPS), NPA* $= \sum_1^0$*(DPA).* □

The main result of this section describes the relation of the automata defined hierarchies to the RCS-hierarchy.

Theorem 3.3 *(a)* $RCS = \Delta_2^0(DRAT) = \Delta_2^0(DPS) = \Delta_2^0(DPA)$;
(b) $R.E. = \sum_2^0(DRAT) = \sum_2^0(DPS) = \sum_2^0(DPA)$.

Proof We prove (b), then (a) follows immediately.

By NPA $\subseteq$ RCS and the closure of RCS under complementation we obtain $\prod_1^0(\text{DPA}) \overset{2.2(a)}{=}$ co-NPA $\subseteq$ RCS and hence $\sum_2^0(\text{DPA}) = \sum_1^0(\prod_1^0(\text{DPA})) \subseteq \sum_1^0(\text{RCS}) = $ R.E. . The analogue inclusions for DPS and DRAT instead of DPA are simple consequences of this by DPS $\subseteq$ DPA and DRAT $\subseteq$ DPA.

Main parts of the proof are the inclusions (I) R.E.$\subseteq\sum_2^0$(DRAT) and (II) R.E.$\subseteq\sum_2^0$(DPS). R.E.$\subseteq\sum_2^0$(DPA) is then again an easy consequence.

To show (I) we describe the existence of a successful run of a Turing machine $\mathfrak{A}$ in terms of finite automata. Given two words y and y' for an input $\mathbf{x}$, we split up the test, whether $y = y'$ and whether y codes a successful run on $\mathbf{x}$, into a conjunction of tests executable by deterministic asynchronous finite automata. Doubling the code allows testing by a finite automaton whether or not a configuration, coded in a certain part of y, results from its predecessor by a transition of $\mathfrak{A}$. Substituting a universal quantifier for the conjunction we obtain a $\sum_2^0$(DRAT)-representation.

Let $R \in (\Sigma^*)^k$ be recursively enumerable, accepted by $\mathfrak{A} = (Q, \Sigma, \Gamma, q_0, \Delta, F)$, using w.l.o.g. only one tape where its input is given as $w_1 \mathfrak{b} w_2 \ldots \mathfrak{b} w_k$, and $\Gamma = \Sigma \dot{\cup} \{\mathfrak{b}\}$. An input is accepted by $\mathfrak{A}$, iff there exists a run $\kappa_1, \ldots, \kappa_r$ such that

(i) $\kappa_1 = \mathfrak{b} \ldots \mathfrak{b} q_0 w_1 \mathfrak{b} w_2 \ldots \mathfrak{b} w_k \mathfrak{b} \ldots \mathfrak{b}$,
(ii) $\kappa_r = \alpha q \beta$, where $\alpha, \beta \in \Gamma^*, q \in F$,
(iii) κ_{i+1} results from κ_i according to Δ, and
(iv) all κ_i have the same length (including an appropriate number of blanks).

The whole run is coded in y as $y = code(\#\alpha_1\#a^{j_1}\#\beta_1\# \ldots \alpha_r\#a^{j_r}\#\beta_r\#)$, where $\kappa_i = (\alpha_i q_{j_i} \beta_i)$. The mapping *code* substitutes $\mathfrak{b}$ by bb, $\#$ by ba, and $c \in \Sigma$ by ac. Then there exists a deterministic finite automaton $\mathfrak{B}$, executing the following test on input $\mathbf{x}, y, y'$ and z:

(i) if $z = \varepsilon$, test whether $y = y'$;
(ii) if $z = a$, test whether y is the code of some run;
(iii) if $z = aa$, test whether a prefix of y codes an initial configuration of $\mathfrak{A}$ on $\mathbf{x}$;
(iv) if $z = aaa$, test whether y ends with an accepting configuration;
(v) if $z = b^i (i > 0)$, test whether the $(i+1)$-st configuration in y (if it exists) results from the i-th in y' according to Δ;
(vi) accept the input in all other cases.

Let R' be the relation accepted by $\mathfrak{B}$, then we have as desired $R\mathbf{x} \Leftrightarrow \exists y y' \forall z R' \mathbf{x} y y' z$.

A similar technique is used to show inclusion (II). Therefore we equivalently assume a pushdown automaton $\mathfrak{A}$ with two stacks (cf. "two stack machine" in [7]), accepting the given recursively enumerable relation, instead of a Turing machine.

Assume a run of $\mathfrak{A}$ is given as sequence of transitions (instead of configurations), coded in an additional input $\mathbf{y}$. Provided with $\mathbf{x}$,$\mathbf{y}$, and a further input z, a deterministic single stack pushdown automaton $\mathfrak{B}$ can test in two runs, whether $\mathbf{y}$ codes a legal and successful computation of $\mathfrak{A}$ on $\mathbf{x}$, using its single stack as the first one of $\mathfrak{A}$ if $z = \varepsilon$, otherwise as the second. Letting R' be the relation accepted by $\mathfrak{B}$, this yields again $R\mathbf{x} \Leftrightarrow \exists \mathbf{y} \forall z R'\mathbf{x}\mathbf{y}z$. □

The infinity of the arithmetical hierarchies over DRAT, DPS, and DPA follows immediately by Hierarchy Lemma 2.4. Consequently Theorem 2.2 can be applied to the pairs (DRAT, RAT), (DPS, NPS), and (DPA, NPA). We show strictness of DRAT $\subset \Delta_1^0$(DRAT), using the suffix relation, defined by $Suf(x,y) :\Leftrightarrow \exists z(y = zx)$, being in Δ_1^0(DRAT) \ DRAT. $Pal := \{ww^R | w \in \Sigma^*\} \in \Delta_1^0$(DPS) \ DPA shows the strictness of DPS $\subset \Delta_1^0$(DPS) and DPA $\subset \Delta_1^0$(DPA). Thus we have proved

Corollary 3.4 *Each of the pairs (DRAT, RAT), (DPS, NPS) and (DPA, NPA) determines a hierarchy as in Figure 2, where the pair in question has to be substituted for (D, N). The leftmost inclusion is strict in each of these cases.* □

It follows from Theorem 3.3 that the classes of the three hierarchies over DRAT, DPS and DPA coincide from $\Delta_2^0()$ upwards with the classes of the arithmetical hierarchy with decremented index, that is for $n \geq 2$:

$\Delta_n^0(\text{DRAT}) = \Delta_n^0(\text{DPS}) = \Delta_n^0(\text{DPA}) = \Delta_{n-1}^0(\text{RCS})$ (analogue for $\sum$ and $\prod$).
However, considering the low levels of these hierarchies we have

Theorem 3.5 *Below RCS all classes of the considered hierarchies are different and no inclusion holds which is not an obvious consequence of the fact that one automaton model is a generalization of another one. So the inclusion diagram of Figure 3 holds where each line stands for a strict inclusion and where each two classes are incomparable (w.r.t. to set inclusion), if there is no inclusion shown via a path in the diagram.*

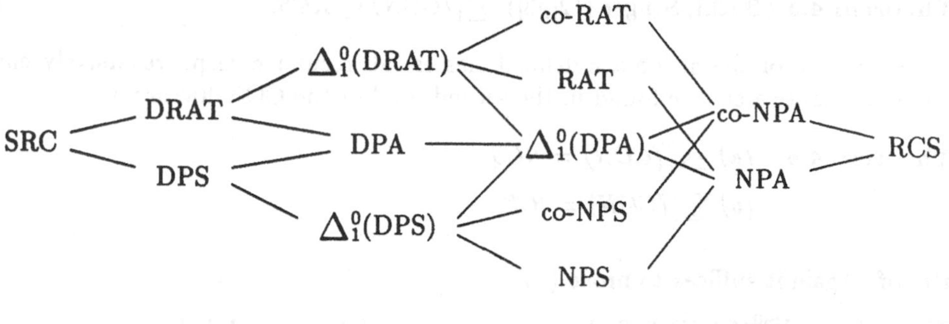

Figure 3: Inclusion diagram of classes below RCS

Proof To show the incomparability of classes defined by synchronous pushdown automata and those defined by asynchronous finite automata, we give example relations, contained in the kernel of one hierarchy but not in a maximal subrecursive class of the other. These examples are $\{a^i b^i | i \in \mathbb{N}\} \in$ DPS \ RAT,

$\{(a^i, a^{2i}, a^{3i}) | i \in \mathbb{N}\} \in$ DRAT $\setminus$ NPS, and their complements. As an easy consequence, the classes defined by asynchronous pushdown automata include the corresponding classes of the other hierarchies strictly.

The remaining incomparabilities, concerning classes of the DPA-hierarchy, are consequences of different closure properties of different relation classes, and of $Suf \notin$ DPA. □

4 The Concatenation Hierarchy

Definition 4.1 An *equation over the concatenation* is of the form $\alpha = \beta$ where the terms α and β are built up by concatenating word variables and words over Σ. A concatenation formula is

- each equation over the concatenation;

- $\varphi_1 \vee \varphi_2$, $\varphi_1 \wedge \varphi_2$, $\neg\varphi$, $\exists x\varphi$, $\forall x\varphi$, if $\varphi, \varphi_1, \varphi_2$ are concatenation formulas;

having the usual semantics (quantification ranges over Σ^*). A relation $R \subseteq (\Sigma^*)^k$ is defined by a concatenation formula φ if at most k variables $x_1, \ldots, x_k$ occur free in φ and $Rx_1, \ldots, x_k \Leftrightarrow \varphi(x_1, \ldots, x_k)$. CON is the class of all word relations over Σ defined by some concatenation formula without quantifier.

Theorem 4.2 (Quine [9]) *Every arithmetical relation is definable by some concatenation formula.* □

It follows immediately that the CON-hierarchy exhausts the RCS-hierarchy. Conversely by CON $\subset$ RCS, every concatenation formula defines an arithmetical relation. For a refinement of this, we will use the following theorem, based on a result of [8] that for every single equation over the concatenation it is decidable whether there exist values for the free variables so that the equation holds.

Theorem 4.3 (Büchi, Senger [12],[3]) *Σ_1^0(CON) $\subseteq$ RCS.* □

As in case of the automata defined classes, all recursive resp. recursively enumerable relations can be found in the second level of the CON-hierarchy.

Theorem 4.4 *(a) Δ_2^0(CON) = RCS;*
(b) Σ_2^0(CON) = R.E. .

Proof Again it suffices to prove (b).

The inclusion Σ_2^0(CON) $\subseteq$ R.E. is a consequence of Theorem 4.3: By closure of RCS under complement and Σ_1^0(CON) = co-Π_1^0(CON), that assertion implies Π_1^0(CON) $\subseteq$ RCS and Σ_2^0(CON) $\subseteq$ Σ_1^0(RCS) = R.E. .

For the converse we assume, a recursively enumerable relation R and a Turing machine $\mathfrak{A} = (Q, \Sigma, \Gamma, q_0, \Delta, F)$ accepting it are given, as in the proof of Theorem 3.3. A successful run of $\mathfrak{A}$ can be represented over the alphabet $\Sigma' := Q \dot\cup \Sigma \dot\cup \{\mathfrak{b}, \#, |\} \supset \Sigma$ as a word $x = \#\#\kappa_1\#|\#\kappa_2\#||\#\kappa_3\#\ldots\#|^{r-1}\#\kappa_r\#|^r\#$.

The i-th symbol in Σ', c_i, is coded as $\tilde{c}_i := aab^iab$. A whole run, given as a word x, is coded as $\tilde{x}$, the concatenated code of the symbols in x. This code has, following [3], the property that any two codes of different symbols do not overlap nor occur inside each other. Hence any word, consisting of symbol codes, can be decomposed uniquely into these.

Using this coding we have the equivalence between the existence of a successful run of $\mathfrak{A}$ on $(w_1, \ldots, w_k)$ and the existence of some $x \in \Sigma^*$ such that

(i) x starts with $\widetilde{\#}\widetilde{\#}\tilde{\not{b}} \ldots \tilde{\not{b}}\tilde{q}_0\tilde{w}_1\tilde{\not{b}}\tilde{w}_2\tilde{\not{b}} \ldots \tilde{w}_k\tilde{\not{b}} \ldots \tilde{\not{b}}\widetilde{\#}\tilde{|}\widetilde{\#}$,
(ii) x consists only of (equal length) configuration codes, separated by marks $\widetilde{\#}\tilde{|}^i\widetilde{\#}$,
(iii) the configuration transitions are according to Δ, and
(iv) x contains some $\tilde{q}$ where $q \in F$.

Note that the unique decomposition property of the code guarantees that, if (i)-(iii) hold, $\tilde{q}$ can occur only as a part of a coded accepting configuration in x, not as a part of some other code words. Therefore it suffices to postulate the existence of $\tilde{q}$ instead of the existence of the whole configuration.

We will now give a concatenation formula $\varphi(w_1, \ldots, w_k)$, describing the existence of a successful run of $\mathfrak{A}$ on the input $(w_1, \ldots, w_k)$, following the above pattern:

$$\varphi(w_1, \ldots, w_k) \equiv \exists x\Big($$

$$\left.\begin{array}{ll} \exists y_1, y_2, y_3, w_1', \ldots, w_k'[& x = \widetilde{\#}\widetilde{\#} y_1 \tilde{\not{b}}\tilde{q}_0 w_1' \tilde{\not{b}} \ldots w_k' \tilde{\not{b}}\tilde{\not{b}} y_2 \widetilde{\#}\tilde{|}\widetilde{\#} y_3 \\ & \wedge y_1 \in \tilde{\not{b}}^* \wedge y_2 \in \tilde{\not{b}}^* \\ & \wedge \bigwedge_{i=1}^{k} w_i' = \tilde{w}_i] \end{array}\right\}\text{(i)}$$

$$\left.\begin{array}{ll} \wedge \forall z_1, z_2, z_3, u, v \bigwedge\limits_{c,c' \in \Gamma} \bigwedge\limits_{q \in Q} & [(x = z_1 \widetilde{\#} z_2 \widetilde{\#} u \tilde{c}' \tilde{q} \tilde{c} v \widetilde{\#} z_2 \tilde{|} \widetilde{\#} z_3 \wedge z_2 \in \tilde{|}^*) \longrightarrow \\ & (z_3 = \varepsilon \\ & \vee [u \neq \varepsilon \wedge \bigvee\limits_{(q,c,c'',l,q') \in \Delta} Pre(u\tilde{q}'\tilde{c}'\tilde{c}'' v \widetilde{\#} z_2 \tilde{|}\tilde{|} \widetilde{\#}, z_3)] \\ & \vee [v \neq \varepsilon \wedge \bigvee\limits_{(q,c,c'',r,q') \in \Delta} Pre(u\tilde{c}'\tilde{c}''\tilde{q}' v \widetilde{\#} z_2 \tilde{|}\tilde{|} \widetilde{\#}, z_3)] \\ & \vee [\bigvee\limits_{(q,c,c'',n,q') \in \Delta} Pre(u\tilde{c}'\tilde{q}'\tilde{c}'' v \widetilde{\#} z_2 \tilde{|}\tilde{|} \widetilde{\#}, z_3)])] \end{array}\right\}\text{(iii)}$$

$$\left.\wedge \exists z, z' \bigvee_{q \in F} x = z\tilde{q}z'\Big)\ \right\}\text{(iv)}$$

Condition (ii) holds by induction, if parts (i) and (iii) of the formula hold. For this induction we postulate in part (iii) the nonemptiness of u and v, and in part (i) the existence of at least one blank to the left and two to the right of the input.

The use of "numbered" markers $\widetilde{\#}\widetilde{\#}, \ldots, \widetilde{\#}\tilde{|}^r\widetilde{\#}$ ensures that u and v can only be parts of one configuration between successive markers and cannot contain larger parts of x.

φ can be obtained as a concatenation formula with $\exists\forall$-prefix, because the prefix relation is in Δ_1^0(CON), "$z \in \tilde{b}^*$" is defined by $z\tilde{b} = \tilde{b}z$ using the special properties of the code (cf. [3]), and the relation $Code, Code(w, w') \Leftrightarrow w' = \tilde{w}$, is in $\sum_2^0$(CON). □

The infinity of the CON-hierarchy follows immediately by Hierarchy Lemma 2.4, and since CON is closed under complement by definition, Corollary 2.3 provides us with the strictness of all inclusions (but the leftmost) of the CON-hierarchy. The leftmost inclusion is shown to be strict by $Pre \in \Delta_1^0(\text{CON}) \setminus \text{CON}$.

As in case of the automata defined classes we have again four classes below RCS, but different concepts of automata and formulas lead to incomparability of those classes, that is

Lemma 4.5 *Below RCS each relation class of the CON-hierarchy is incomparable w.r.t. set inclusion with each of the automata defined relation classes. In particular*
(a) the relation Dub, defined by $Dub(x, y){:}\Leftrightarrow x = yy$, *is in* $CON \setminus NPA$;
(b) the equal length relation Elg, defined by $Elg(x, y){:}\Leftrightarrow |x| = |y|$, *is not in* $\sum_1^0$*(CON)* [4] *but in SRC.* □

We end this section by showing some consequences of our results on the CON-hierarchy in relation to other work.

The equality RCS = Δ_2^0(CON) implies the following asymmetry: Each relation in $\sum_1^0$(CON) can be defined by a formula with a single equation preceded by a block of existential quantifiers ([3]). Hence the use of Boolean combinations (allowed in the definition of CON) is not necessary in case of $\sum_1^0$(CON). Moreover Senger [12] has shown that each relation defined by a single equation together with an $\exists\forall$-prefix is recursive. Since $\sum_2^0$(CON) = R.E., not every concatenation formula with an $\exists\forall$-prefix defines a recursive relation. This implies that for $\sum_2^0$(CON), and consequently also for $\prod_1^0$(CON) and CON itself, Boolean combinations in the definition can't be omitted without changing these classes.

As a last remark we refer to the class of the "rudimentary relations" RUD, defined in [13] as the closure of CON under bounded quantification (in our words). There it is shown that R.E.=$\sum_1^0$(RUD), but RUD$\subsetneq$RCS. This implies the identity of the arithmetical and the RUD-hierarchy already from level one onwards.

5 Conclusion

The results of this paper clarify the relation between Turing machines (defining the class RCS) and several natural restricted machine models with respect to application of quantifiers. It solves some of the problems raised in [14] in case of finitary relations. We have obtained different types of "extensions" of the arithmetical hierarchy below the recursive sets. Full proofs of all results stated here can be found in [11], an extension of this paper.

As a field of future research our results may be extended to the case of infinitary relations. In the line of [14] where it is shown that McNaughton's Theorem implies that the (infinitary) SRC-hierarchy collapses at its third level, one should expect that the infinity of the hierarchies investigated here provides in the infinitary case an infinite chain of acceptance conditions for the underlying automata models.

Acknowledgement: I wish to thank W. Thomas for his constant support of my work. He has suggested the questions which are solved here and supervised my diploma thesis where most results of this paper were originally developed.

References

[1] A.V.Aho, J.D.Ullman: The Theory of Parsing, Translation and Compiling, Vol.1, Prentice-Hall, Englewood Cliffs, N.J. 1972

[2] J.Berstel: Transductions and Context-free Languages, Teubner, Stuttgart 1979

[3] J.R.Büchi, S.Senger: Coding in the Existential Theory of Concatenation, Arch. Math. Logik 26 (1986/87), pp. 101-106

[4] J.R.Büchi, S.Senger: Definability in the Existential Theory of Concatenation and Undecidable Extensions of this Theory, Z. Math. Logik Grundlag. Math. 34 (1988), pp. 337-342

[5] P.C.Fischer, A.L.Rosenberg: Multitape One-Way Nonwriting Automata, J. Comp. Syst. Sci. 2 (1968), pp. 88-101

[6] P.G.Hinman: Recursion Theoretic Hierarchies, Springer, Berlin 1978

[7] J.E.Hopcroft, J.D.Ullman: Introduction to Automata Theory, Languages and Computation, Addison-Wesley, Reading, Mass. 1979

[8] G.S.Makanin: The Problem of Solvability of Equations in a Free Semigroup, Math. USSR-Sb. 32, no. 2 (1977), pp. 129-198

[9] W.V.Quine: Concatenation as a Basis for Arithmetic, J. Symbolic Logic 11 (1946), pp. 105-114

[10] M.O.Rabin, D.Scott: Finite Automata and Their Decision Problems, IBM J. Res. Develop. 3 (1959), pp. 114-125 (reprinted in: Sequential Machines: Selected Papers, (E.Moore, Ed.), Addison-Wesley, Reading, Mass. 1965, pp. 63-91)

[11] S.Seibert: Quantifier Hierarchies over Word Relations, Bericht 9204, Institut f. Informatik u. Prakt. Math., Univ. Kiel 1992

[12] S.Senger: The Existential Theory of Concatenation, Ph. D. Dissertation, Purdue University 1982

[13] R.M.Smullyan: Theory of Formal Systems, Princeton University Press 1961

[14] W.Thomas: Automata and Quantifier Hierarchies, in: Formal Properties of Finite Automata and Applications (J.E. Pin, ed.), Lect. Notes in Comp. Sci. 386, Springer, Berlin 1989, pp. 104-119

Complexity results for the default- and the autoepistemic logic

Eckhard Steffen

Fakultät für Mathematik
Universität Bielefeld
Postfach 8640, D-4800 Bielefeld 1

Abstract. The default logic (Rei 80) and the autoepistemic logic (Mo 85) are non-monotonic logics. In default logic a default theory consists of a set of propositional sentences W and a set of defaults D. A sentence is defined to be derivable from a given default theory if it belongs to an extension of the default theory. A central question is: Given a default theory $\Delta = (D, W)$ and a sentence β, is there an extension of Δ which contains β?

This question is the *extension-membership-problem* for Δ and β (abbr. $EMP(\Delta, \beta)$). We show that this problem, as well as the problem whether a given sentence belongs to every extension of a given default theory (abbr. $AEMP(\Delta, \beta)$), is $\mathcal{NP}$-hard for default theories and for normal default theories.

Let $\Delta = (D, W)$ be a finite normal default theory such that the union of W and the consequents of the elements of D is satisfiable. We show for such theories that the EMP is polynomial time Turing reducible to the satisfiability problem in classical propositional logic and it is solvable in polynomial time, if the elements of W, the prerequesites and consequents of the defaults and $\neg\beta$ are Horn sentences. Moreover it is shown that the problem is $\mathcal{P}$-complete for such theories.

On account of the theorem of Konolige (Ko 88) we get the analogous results for propositional autoepistemic logic.

Furthermore we show the $\mathcal{NP}$-hardness of the EMP and $AEMP$ for the modified default-logic (Lu 88). We show that the EMP for normal theories for the default logic and the EMP for the modified default logic are polynomial time nondeterministic Turing reducible to the satisfiability problem in classical propositional logic.

1 Default Logic

For a detailed presentation of the motivation for non-monotonic logics and especially default logic see (Rei 80) or (Bes 89). The set of natural numbers is denoted by ω. Any ordinary formal language of propositional logic $\mathcal{L}$ is a language

for default logic. We call the elements of $\mathcal{L}$ sentences. Variables for sentences are $\alpha, \alpha_1, \alpha_2, \ldots \beta, \beta_1, \ldots \gamma, \gamma_1, \ldots$ and $1 \in \mathcal{L}$ denotes the always true sentence and $\vdash$ is the derivability relation of classical propositional logic. Let $S \subseteq \mathcal{L}$. Then $Th(S) := \{\alpha | S \vdash \alpha\}$ is called the deductive closure of S. Furthermore SAT denotes the set of satisfiable sentences and $TAUT$ the set of tautologies of $\mathcal{L}$.
A sentence α is in conjunctive normal form iff it is of the form $(\alpha_{11} \vee \alpha_{12} \vee \ldots \vee \alpha_{1n^1}) \wedge \ldots \wedge (\alpha_{m1} \vee \alpha_{m2} \vee \ldots \vee \alpha_{mn^m})$. The set of the sentence in conjunctive normal form is denoted by CNF. The disjunctions of a sentences of CNF are also called clauses and we often write the sentence α as a set of clauses like this: $\{\{\alpha_{11}, \alpha_{12}, \ldots \alpha_{1n^1}\}, \ldots \{\alpha_{m1}, \alpha_{m2}, \ldots \alpha_{mn^m}\}\}$. There is an equivalent sentence in conjunctive normal form for every sentence. Thus it is no restriction to consider only sentences in conjunctive normal form. We introduce two special subsets of CNF. Firstly the set of Horn sentences ($HORN$) which consists of all sentences of CNF for which each disjunction is in the form $(\neg\pi_{i1} \vee \ldots \vee \neg\pi_{in^i} \vee \lambda_i)$ where π_{ij} is an atomic sentence and λ_i is a literal that is an atomic sentence or the negation of an atomic sentence. Secondly the set of Krom sentences ($KROM$), the set of sentences of CNF where each conjunction consists at most of the disjunction of two literals.

Definition 1.1 *A default (abbr. δ) is any expression of the form $\frac{\alpha:\beta_1,\ldots,\beta_n}{\gamma}$, where $\alpha, \beta_1, \ldots, \beta_n, \gamma$ are sentences. α is called the prerequisite, $\beta_1, \ldots, \beta_n$ is called the justification, and γ is called the consequent of the default δ.*

Given a set of defaults D we define $PRE(D)$ to be the set of the prerequisites of the defaults of D, $\neg PRE(D) := \{\neg\alpha | \alpha \in PRE(D)\}$, PRE^D to be the conjunction of the elements of $PRE(D)$, $\neg PRE^D$ to be the conjunction of the elements of $\neg PRE(D)$, $JUS(D)$ to be the set of the justifications of the defaults of D, $CON(D)$ to be the set of the consequents of the defaults of D and CON^D to be the conjunction of the elements of $CON(D)$.

Definition 1.2 *A default theory Δ is a pair (D, W), where W is a set of sentences, axioms of Δ, and D is a set of defaults.*

The application of a default is a central notation in default logic. The modified default logic from Łukaszewicz is different from Reiters default logic in this notion.

Definition 1.3 *Let two sets $S \subseteq \mathcal{L}$, $T \subseteq \mathcal{L}$ be given. A default $\delta = \frac{\alpha:\beta_1,\ldots,\beta_n}{\gamma}$ applies to T with respect to S (noted $\delta\nabla_S T$) iff $\alpha \in T$ together with $\neg\beta_i \notin S$ for all $i \leq n$ implies $\gamma \in T$.*

The extensions of a default theory are defined in (Rei 80) as follows:

Definition 1.4 *Let a default theory $\Delta = (D, W)$ be given. For any set $S \subseteq \mathcal{L}$ let $\Gamma(S)$ be the smallest set satisfying the following three properties:*
i) $W \subseteq \Gamma(S)$
ii) $Th(\Gamma(S)) = \Gamma(S)$
iii) if $\delta \in D$ then $\delta\nabla_S\Gamma(S)$
Then $E \subseteq \mathcal{L}$ is an extension of Δ iff $E = \Gamma(E)$.

Reiter gave another characterisation of extensions.

Theorem 1.1 (Reiter 1980) *Let $E \subseteq \mathcal{L}$ be a set of sentences, and let $\Delta = (D, W)$ be a default theory. Define $E_0 := W$, and for $i \geq 0$:*

$$E_{i+1} := Th(E_i) \cup \{\gamma | \frac{\alpha : \beta_1, \ldots, \beta_n}{\gamma} \in D \text{ where } \alpha \in E_i \text{ and } \neg\beta_j \notin E, \text{ for } 0 \leq j \leq n\}.$$

Then E is an extension of Δ iff $E = \bigcup_{i=0}^{\infty} E_i$.

Reiter (1980) was convinced that the only interesting defaults are normal defaults, where the single justification is equivalent to the consequent.

Definition 1.5 *A normal default is a default of the form $\frac{\alpha:\beta}{\beta}$.*

Definition 1.6 *A default theory (D, W) is a normal default theory iff every default of D is a normal default.*

Normal theories have a lot of nice properties (see (Rei 80)). We only mention:

Theorem 1.2 (Reiter 1980) *Every normal default theory has an extension.*

Theorem 1.3 (Reiter 1980) *Let $\Delta = (D, W)$ be a normal default theory and $W \cup CON(D) \subseteq SAT$. Then Δ has a unique extension.*

To decide the extension membership problem, Reiter developed a proof theory for normal default theories. First we define a default proof.

Definition 1.7 *Let $\Delta = (D, W)$ be a normal default theory and β a sentence. A finite sequence $D_0, D_1, .., D_k$ $(k \in \omega)$ of finite subsets of D is a default proof of β with respect to Δ iff*
P1) $W \cup CON(D_0) \vdash \beta$
P2) *for* $1 \leq i \leq k$: $W \cup CON(D_i) \vdash PRE^{D_{i-1}}$
P3) $D_k = \emptyset$
P4) $W \cup \bigcup_{i=0}^{k} CON(D_i) \subseteq SAT$

Theorem 1.4 (Reiter 1980) *A consistent normal default theory Δ has an extension E such that $\beta \in E$ iff β has a default proof with respect to Δ.*

Obviously the extension membership problem is decidable for finite normal default theories. There are $2^{|D|}$ many subsets of D. Thus you can choose one of the $2^{2^{|D|}}$ possible sequences of subsets of D. If you have a sequence which contains - for example - n subsets of D then there are $n!$ possibilities to order the subsets. If you have an order you check your sequence with the procedure given above, because it is decidable whether a given sentence is tautological or satisfiable or a consequence of a set of sentences. If it is not a default proof try it with the next sequence and so on.

1.1 Łukaszewicz' default logic

There are default theories which have no extensions. For example $\Delta = (\{\frac{1:\neg\alpha}{\alpha}\}, \emptyset)$ is one of these. Suppose there is a set E which is an extension of Δ. We have $E = \Gamma(E)$ and so by the definition of an extension we get $\alpha \in E \iff \alpha \notin E$. This is a contradiction. The problem is the applicability of the default.
Łukaszewicz (Łu 88) gave an alternative formalization of default logic to exclude such cases. Furthermore he devoloped a proof theory for the so called modified default logic.
First we have to define a new version of the notion of application for defaults.

Definition 1.8 *Let S, S', T, T' be subsets of $\mathcal{L}$. A default $\delta = \frac{\alpha:\beta_1,\ldots,\beta_n}{\gamma}$ applies to (S',T') with respect to (S,T) (noted $\delta\nabla_{(S,T)}(S',T')$), iff $\alpha \in S'$ together with $S \cup \{\gamma\} \not\vdash \neg\beta$, for all $\beta \in T \cup \{\beta_1, \ldots, \beta_n\}$, implies $\gamma \in S'$ and $\{\beta_1, \ldots, \beta_n\} \subseteq T'$.*

Definition 1.9 *Let $\Delta = (D, W)$ be a default theory. For any sets $S \subseteq \mathcal{L}$ and $T \subseteq \mathcal{L}$, let $\Gamma^C(S,T)$ and $\Gamma^J(S,T)$ be the smallest sets satisfying the following properties:*
i) $W \subseteq \Gamma^C(S,T)$
ii) $Th(\Gamma^C(S,T)) = \Gamma^C(S,T)$
iii) If $\delta \in D$ then $\delta\nabla_{(S,T)}(\Gamma^C(S,T), \Gamma^J(S,T))$
Then $E \subseteq \mathcal{L}$ is a modified extension of Δ iff there is a set $T \subseteq \mathcal{L}$, called the set of justifications for E, such that $(E,T) = (\Gamma^C(E,T), \Gamma^J(E,T))$.

Theorem 1.5 (Łukaszewicz 1988) *Every default theory has a modified extension.*

For normal default theories there is no difference between modified extensions and extensions. The extension membership problem for the modified default logic (abbr. $EMP_{mod}(\Delta, \beta)$) is the question: Given default theory Δ and a sentence β, is there a modified extension of Δ which contains β?
Łukaszewicz also developed a proof theory to study the extension membership problem for the modified default logic.

Definition 1.10 *Let S be a set of sentences and D be a set of defaults. A default δ is S-applicable with respect to D (noted: $S - Appl(\delta, D)$) iff*
1. $PRE(\{\delta\}) \subset Th(S)$
2. for all $\beta \in JUS(D) \cup JUS(\{\delta\}) : S \cup CON(\{\delta\}) \not\vdash \neg\beta$.

Definition 1.11 *Let S be a set of sentences and D be a set of defaults. A sequence of defaults $(\delta_i)_{i\in\omega}$ is S-applicable iff $(\delta_i)_{i\in\omega}$ is the empty sequence or*
1. $S - Appl(\delta_0, \emptyset)$ and
2. $(S \cup CON(D_i)) - Appl(\delta_i, D_i)$ where $D_i = \{\delta_0, \ldots, \delta_{i-1}\}$ for $i \geq 1$.

Definition 1.12 *Let $\Delta = (D, W)$ be a default theory and let β be a sentence. A finite sequence $\delta_0, \delta_1, \ldots, \delta_k, (k \in \omega)$ consisting of defaults of D is a modified default proof of β with respect to Δ iff*
MP1) $\delta_0, \delta_1, \ldots, \delta_k$ is W-applicable and
MP2) $W \cup CON(\{\delta_0, \delta_1, \ldots, \delta_k\}) \vdash \beta$.

Theorem 1.6 (Łukaszewicz 1988) *A sentence β is in some modified extension of a default theory iff β has a modified default proof with respect to Δ.*

We see that the set of the tautologies is a modified extension of the default theory mentioned at the beginning of this section. There is no modified extension of this theory which contains α or $\neg\alpha$, if α or $\neg\alpha$ is not a tautology. The reason is that the default is not applicable.

2 Proof Algorithm for Normal Theories

A default proof consists of a sequence of finite subsets of the set of defaults. Reiter specified a top down and a bottom up search procedure for verifying such a sequence. The top down algorithm exactly corresponds to the definition of a default proof. We define the bottom up procedure as follows:

Let $\Delta = (D, W)$ and β be given.

1. Define $D_0 := \emptyset$. If $W \vdash \beta$ then goto 3 else goto 2.
2. For $i \geq 1$ let D_{i-1} be given and determine a subset $D_i \subseteq D$ such that (*) $D_{i-1} \subset D_i$ and $W \cup CON(D_{i-1}) \vdash PRE^{D_i}$.
   If $W \cup CON(D_i) \vdash \beta$ then goto 3 else carry out step 2 with (i+1) instead of i.
3. If $W \cup CON(D_i) \subseteq SAT$ then halt in an accepting state else halt in a rejecting state.

If the algorithm halts in the accepting state, $D_0, D_1, \ldots, D_i$ is a (bottom up) default proof of β with respect to Δ. (*) is not required in Reiters original definition. For a default proof of β with respect to $\Delta = (D, W)$ at most all defaults of D are needed and every time, when carrying out step 2, at least one new default is added to the set of used defaults. Thus property (*) guaranties that step 2 must be carried out $|D|$ times at most. To show that the definition is correct we prove:

Theorem 2.1 *Let $\Delta = (D, W)$ be a normal default theory and $\beta \in \mathcal{L}$. There is a (bottom up) default proof $D_0 \subset D_1 \subset \ldots \subset D_k \subseteq D$ of β with respect to Δ iff there is a default proof of β with respect to Δ in D_k.*

Proof: ($\Rightarrow$) Let $D_0 \subset D_1 \subset \ldots \subset D_k \subseteq D$ be the bottom up default proof. Suppose there is no default proof $D^0, \ldots, D^n$ of β with respect to Δ. Then we have to distinguish four cases:
Case 1: There is no subset D' of D such that $W \cup CON(D') \subseteq SAT$. This is a contradiction to $W \cup CON(D_k) \subseteq SAT$.
Case 2: There is no subset D' of D such that $W \vdash PRE^{D'}$. This is a contradiction to $W \vdash D_0$.
Case 3: There is no subset D' of D such that $W \cup CON(D') \vdash \beta$. This is a contradiction to $W \cup CON(D_k) \vdash \beta$.
Case 4: Every time when we check *P2* there is an $s \in \omega$ such that $W \cup CON(D^0) \vdash \beta$, $W \cup CON(D^1) \vdash PRE^{D^0}$, $\ldots$ $W \cup CON(D^s) \vdash PRE^{D^{s-1}}$ and there is no

$D' \subseteq D_k$ such that $W \cup CON(D') \vdash PRE^{D^*}$. Thus there is one default $\delta \in D^*$ at least which prerequisite is not derivable from W and the consequents of a subset of D_k. This is a contradiction to the fact that $W \cup CON(D_k) \vdash PRE^{\delta}$ for every $\delta \in D_k$ which we get by the bottom up procedure. Hence it follows that there must be a default proof of β with respect to Δ.
($\Leftarrow$) Let $D^0, D^1, \ldots, D^n$ be a default proof of β with respect to Δ. Define $D_i := \bigcup_{j=0}^{i} D^{n-j}$. Then we have $D_0 \subset D_1 \subset \ldots \subset D_n \subseteq D$ and $W \cup CON(D_n) \subseteq SAT$ and $W \vdash PRE^{D_0}$, $W \cup CON(D^0) \vdash PRE^{D^1}$, ... $W \cup CON(D^{n-1}) \vdash PRE^{D^n}$, $W \cup CON(D^n) \vdash \beta$. Thus $D_0, D_1, \ldots D_n$ is a (bottom up) default proof of β with respect to Δ. •

Theorem 2.2 *Let $\Delta = (D, W)$ be a finite normal default theory, $\beta \in \mathcal{L}$ and $W \cup CON(D) \subseteq SAT$. If $W \cup CON(D) \cup \neg PRE(D) \cup \{\neg\beta\} \subseteq HORN$ the $EMP(\Delta, \beta)$ is decidable in polynomial time.*

Proof: We know by theorem 1.3 that the given default theory has a unique extension. Now we specify an algorithm which decides the extension membership problem for such default theories. With no loss of generality we assume that $D = \{\delta_0, \delta_1, \ldots \delta_n\}$ is in a fixed order and $W, CON(D), \neg PRE(D)$ are sets of clauses.

Let $\Delta = (D, W)$ and β be given.

1. For every $\delta \in D$ do: If $W \vdash PRE^{\{\delta\}}$ then $\delta \in D_0^*$ else $\delta \notin D_0^*$.
2. For every $\delta \in D \setminus D_i^*$ do: If $W \cup CON(D_i^*) \vdash PRE^{\{\delta\}}$ then $\delta \in D_{i+1}$ else $\delta \notin D_{i+1}$.
   After testing all defaults of $D \setminus D_i^*$ define $D_{i+1}^* := D_i^* \cup D_{i+1}$ and goto 3.
3. If $CON(D_{i+1}^*) = CON(D_i^*)$ then define $D^* := D_{i+1}^*$ and goto 4
   else carry out step 2 with (i+1) instead of i.
4. If $W \cup CON(D^*) \vdash \beta$ then halt in an accepting state
   else halt in a rejecting state.

We prove that the algorithm is effectively computable. The tests whether a sentence α is derivable of a set of sentences S can be done by a linear resolution proof of $\Box$ from the clauses of $\neg\alpha$ and S. Because linear resolution is complete and correct for the set of the contradictions and effectively computable, every test in step 1, 2 or 4 is effectively computable. The test in step 3 is also effectively computable. Thus the set D^* is effectively constructable and the algorithm halts either in the accepting or in the rejecting state. We prove that the algorithm decides the $EMP(\Delta, \beta)$.
Case 1: The algorithm halts in the accepting state. Then there is a set $D^* \subseteq D$ such that $W \cup CON(D^*) \vdash \beta$. Thus there is a $k \in \omega$ such that $D^* = D_k^*$. The sequence $D_0^* \subset D_1^* \subset \ldots \subset D_k^*$ satisfies the condition $W \vdash PRE^{D_0^*}$, $W \cup CON(D_0^*) \vdash PRE^{D_1^*}$, ... $W \cup CON(D_{k-1}^*) \vdash PRE^{D_k^*}$ and $W \cup CON(D_k^*) \vdash \beta$. $W \cup CON(D) \subseteq SAT$ implies that $W \cup CON(D^*) \subseteq SAT$ such that the sequence satisfies the conditions of the bottom up procedure. Hence it follows by the previous

theorem that there is a default proof of β with respect to Δ and by theorem 1.4 this implies that there is an extension of Δ which contains β.

Case 2: The algorithm stops in the rejecting state that is $W \cup CON(D^*) \not\vdash \beta$. For every $\delta \in D^*$ there is an $i \leq k$ such that $\delta \in D_i^*$. Thus $W \cup CON(D_{i-1}^*) \vdash PRE^{D_i^*}$ implies that there is a subset D_δ of D_{i-1}^* such that $W \cup CON(D_\delta) \vdash PRE^{\{\delta\}}$. For each of the elements of D_δ there is a subset of D_{i-2}^* such that the prerequisite of the element is derivable from the union of W and the consequents of the subsets. In the next step we will find a appropriate set in D_{i-3}^* and so on. After at most i steps we have a subset of D_0 which prerequisites are derivable from W. We will say that every element of D^* has a history in D^* with respect to W.

Notice that this property of closure is a consequence of $W \cup CON(D) \subseteq SAT$. If this is not given you can construct a set of defaults whose set of consequents is inconsistent. Then there is a smallest $k \in \omega$ such that $W \cup CON(D_k^*) \not\subseteq SAT$ and $W \cup CON(D_{k-1}^*) \subseteq SAT$. The room for solutions can be too large and it is no longer guaranteed to find a subset of D_{k-1}^* for each default of D_k^* such that the prerequisite of the default is derivable from the union of W and the consequents of the subset.

In a minimal default proof $D_0, D_1, \ldots, D_m$ of β with respect to Δ (that is every default is necessary for the proof), every default has a history in the union of the involved sets of defaults with respect to W. If an involved default has no such history one will come to a default δ whose prerequisites are not derivable from the union of W and the consequents of one of $\emptyset, D_0, D_1, \ldots, D_{m-1}$. Hence it follows that $D_0, D_1, \ldots, D_m$ is no default proof of β with respect to Δ because *P2* is violated.

On the other hand every time when step 1 or 2 is carried out, the set of all defaults which satisfy the test, is selected. Thus D^* contains every default of D which has a history in D with respect to W. The maximality of D^* implies that every (minimal) default proof of β with respect to Δ must consist of a sequence of subsets of D^*. Hence it follows that there is no default proof of β with respect to Δ if $W \cup CON(D^*) \not\vdash \beta$. Theorem 1.4 implies that there is no extension of Δ which contains β.

Complexity: Let $W \cup CON(D) \cup \neg PRE(D) \cup \{\neg\beta\} \subseteq HORN$ and define n to be the maximum of the numbers of clauses of the elements of $\neg PRE(D)$, $CLAUSES(\Delta) := W \cup CON(D)$ and $N := |CLAUSES(\Delta)| + n$.

The tests in step 1 resp. 2 can be carried out by unit resolution proofs of $\Box$ from $W \cup \neg PRE(\{\delta\})$ resp. from $W \cup CON(D_i^*) \cup \neg PRE(\{\delta\})$. Unit resolution is complete and correct for *HORN* (He 74). It is decideable whether there is a resolution proof of $\Box$ from a given set of Horn clauses and the algorithm needs less than m^2 computation steps where m is the number of clauses of the given set of Horn clauses (Jo 77). If $D = \emptyset$ we must test whether $W \vdash \beta$ is satisfied. This can be done by the algorithm given in (Jo 77) in N^2 steps. Now we suppose $D \neq \emptyset$.

In step 1 exactly $|D|$ unit resolution proofs must be carried out. In step 2 less than $|D|$ unit resolution proofs must be carried out. Since $D_0^* \subset \ldots \subset D_n^* \subseteq D$ at most $|D|$ sets must be constructed. Thus the test in step 2 must be carried out $|D|$ times at most and hence the total number of computation steps for carrying out step 1 and 2 is less than $(\Sigma_{i=0}^{|D|-1}(|D| - i))N^2 = (\frac{|D|^2 - |D|}{2})N^2$. For the test in step 3, less than N^2 computation steps are necessary and the test must be carried out $|D|$ times

at most. Thus the total number of computation steps in step 3 is less than $|D|N^2$. For the test in point 4 less than N^2 computation steps are necessary and step 4. must be carried out only once.
Hence it follows that the number of necessary computation steps for the execution of the algorithm is less than $(\frac{|D|^2+|D|}{2})N^2 + N^2 \leq (|D|N)^2$, for $|D| \geq 2$. Thus the algorithm solves the $EMP(\Delta, \beta)$ in $O((|D|N)^2)$–time. •
Remark: There are $O(n)$–time algorithms to decide whether a given Horn sentence α is a consequence of a given set H of Horn sentences, where n is the total number of occurrences of literals in α and the elements of H (Do 84). We define N to be the sum of the maximum of the total numbers of occurrences of literals in elements of $\neg PRE(D)$ and the total number of occurrences of literals in $W \cup CON(D)$. Then the algorithm given in theorem 2.2 solves the $EMP(\Delta, \beta)$ in $O(N|D|^2)$–time.

Theorem 2.3 *Let $\Delta = (D, W)$ be a finite normal default theory, $W \cup CON(D) \subseteq SAT$ and $W \cup CON(D) \cup \neg PRE(D) \cup \{\neg\beta\} \subseteq KROM$ then the $EMP(\Delta, \beta)$ is decidable in polynomial time.*

Proof: The satisfiability problem for Krom sentences is solvable in polynomial time (Co 71). Hence the proposition follows analogously to the proposition of the previous theorem. •

Theorem 2.4 *Let $\Delta = (D, W)$ be a finite normal default theory and $W \cup CON(D) \subseteq SAT$. Then the $EMP(\Delta, \beta)$ is polynomial time Turing reducible to the satisfiability problem in propositional logic.*

Proof: We have to show that there exists a polynomial time oracle machine with oracle SAT which accepts EMP. We use the algorithm given in theorem 2.2. We replace each test in step 1, 2 or 4 of the algorithm by a question to the oracle. Hence it follows that we must put $\frac{|D|^2+|D|}{2} + 1$ questions to the oracle at most and the length of each question is less than N. Each test in step 3 is computable in less than N^2 computation steps. That the machine works well is shown in theorem 2.2. Thus the computation is bounded by a polynomial in $|D|N$. •
As an immediate corollary of the previous theorem we can relate the problems to the polynomial time hierarchy (Ba 88) as follows:

Corollary 2.1 *If $W \cup CON(D) \subseteq SAT$ for a finite normal default theory $\Delta = (D, W)$ then $EMP(\Delta, \beta) \in \Delta_2^p$.*

3 The Hardness of the Extension Membership Problem

We ascertain the hardness of the EMP for the default logic, the modified default logic and the autoepistemic logic. The problem whether there is no extension of Δ which contains β is denoted by $\overline{EMP(\Delta, \beta)}$ and $\overline{AEMP(\Delta, \beta)}$ denotes the problem, whether there is an extension of Δ which does not contain β. The abbreviations $\overline{EMP_{mod}(\Delta, \beta)}$ respectively $\overline{AEMP_{mod}(\Delta, \beta)}$ are analogously defined.

3.1 Normal Default Theories

Lemma 3.1 $\beta \in SAT \iff$ *There is a default proof of* β *with respect to the normal default theory* $\Delta := (\{\frac{1:\beta}{\beta}\}, \emptyset)$.

Proof: ($\Rightarrow$) If $\beta \in SAT$ we have to distinguish two cases:
Case 1: $\beta \in TAUT$. This is equivalent to $\vdash \beta$. Thus $\emptyset$ is a default proof of β with respect to Δ.
Case 2: $\beta \notin TAUT$. Thus we have $\neg\beta \notin TAUT$. Define $D_1 := \{\frac{1:\beta}{\beta}\}$ then we have $W \vdash PRE^{D_1}$, $W \cup CON(D_1) \subset SAT$ and $W \cup CON(D_1) \vdash \beta$. Thus $\emptyset, D_1$ is a (bottom up) default proof of β with respect to Δ.
($\Leftarrow$) If there is a default proof of β with respect to Δ we have to distinguish two cases:
Case 1: $\emptyset$ is such a proof. That is we have $\vdash \beta$. Thus $\beta \in TAUT \subset SAT$.
Case 2: $\{\delta\}$ is a default proof of β with respect to Δ. Then we have $\vdash 1$ and $\emptyset \cup \{\beta\} \subset SAT$. •

Theorem 3.1 $\beta \in SAT \iff$ *There is an unique extension of* $\Delta := (\{\frac{1:\beta}{\beta}\}, \emptyset)$ *which contains* β.

Proof: Because of lemma 3.1 and theorem 1.2 we have: β is satisfiable iff there is a default proof of β with respect to Δ iff there is an extension of Δ which contains β. Theorem 1.3 implies that there is only one extension. •
Thus the $AEMP(\Delta, \beta)$ is as hard as the $EMP(\Delta, \beta)$ for normal default theories Δ.

Corollary 3.1 *Let* $\Delta = (D, W)$ *be a normal default theory and* $\beta \in \mathcal{L}$. *Then the* $EMP(\Delta, \beta)$ *and the* $AEMP(\Delta, \beta)$ *are* $\mathcal{NP}-$ *hard.*

Proof: We show that the translation of propositional formulas into normal default theories is computable in constant time. We write a default $\frac{1:\beta}{\beta}$ in the form $\delta 1, \beta$ and the default theory $\Delta = (\{\frac{1:\beta}{\beta}\}, \emptyset)$ in the form $\emptyset; \delta 1, \beta$. A machine which starts on the left cell next to the input will need four tape cells only to compute the translation independly from the input. Theorem 3.1 completes the proof. •
On account of this and the results of Jones & Laaser (Jo 77) we get the following corollary:

Corollary 3.2 *Let* $\Delta = (D, W)$ *be a normal default theory and* $\beta \in HORN$. *Then the* $EMP(\Delta, \beta)$ *and* $AEMP(\Delta, \beta)$ *are* $\mathcal{P}-$ *hard.*

Because for the translation only constant many computation steps are needed, theorem 2.2, theorem 2.3 and corollary 3.2 imply:

Corollary 3.3 *Let* $\Delta = (D, W)$ *be a finite normal default theory,* $\beta \in \mathcal{L}$ *and* $W \cup CON(D) \subseteq SAT$. *If* $W \cup \neg PRE(D) \cup CON(D) \cup \{\beta, \neg\beta\} \subseteq HORN$ *or* $W \cup \neg PRE(D) \cup CON(D) \subseteq KROM$ *and* $\beta \in HORN$ *contains two clauses at most, then the* $EMP(\Delta, \beta)$ *is* $\mathcal{P}-$ *complete.*

Proof: If $\beta \in CNF$ consists of at most two conjunctions $\neg\beta$ is a Krom sentence.•

Theorem 3.2 *$EMP(\Delta, \beta)$ is polynomial time nondeterministic Turing reducible to the satisfiabilty problem in propositional logic for finite normal default theories.*

Proof: We have to show that there exists a nondeterministic polynomial time oracle machine with oracle SAT which accepts EMP.

Let $\Delta = (D, W)$ and β be given.

1. Choose nondeterministically a sequence $\emptyset = D_0 \subset D_1 \ldots \subset D_k \subseteq D$, ($k \leq |D|$) and goto 2.
2. If $W \cup CON(D_k) \subseteq SAT$ then goto 3 else halt in a rejecting state.
3. If $W \cup CON(D_k) \cup \{\neg\beta\} \subseteq SAT$ then halt in a rejecting state else goto 4.
4. For i = 0 to k-1 do: If $W \cup CON(D_i) \cup \neg PRE(D_{i+1} \setminus D_i) \subseteq SAT$ for an $i \leq k-1$ then halt in a rejecting state
   else halt in an accepting state.

To choose less than $|D|$ subsets of D can be done in polynomial time. Let l_W resp. l_C resp. l_P be the length of the conjunction of the elements of W resp. $CON(D)$ resp. $\neg PRE(D)$ and l_β be the length of $\neg\beta$ then the length of the question in 2 resp. in 3 is less than $l_W + l_C$ resp. less than $l_W + l_C + l_\beta$. The length of every question in 4. is less than $l_W + l_C + l_P$. Less than $|D|$ subsets of D must be chosen such that the machine asks the oracle $|D| + 2$ times at most. The given machine is a nondeterministic machine whose number of computation steps is bounded by a polynomial in the sum of the length of the sentences and the cardinality of D. •
As an immediate corollary of the previous theorem we can relate the problems to the polynomial time hierarchy (Ba 88) as follows:

Corollary 3.4 *$EMP(\Delta, \beta) \in \Sigma_2^p$ and $\overline{EMP(\Delta, \beta)} \in \Pi_2^p$*

3.2 Default Theories

Remark: To analyse the complexity of a translation from propositional formulas into default theories we write the default $\delta = \frac{1:\alpha}{\beta}$ in the form $\delta = \delta 1, \alpha/\beta$. The function f, which maps the formula $\alpha \wedge \beta$ to the default theory $\Delta = \emptyset; \delta$ will be computed by a machine, which starts under the "$\wedge$" of the input, replaces $\wedge$ by /; will go left if it reads a symbol of the alphabet of $\mathcal{L}$; if it reads a "blank" it will write ", 1 δ ; $\emptyset$ " on its tape analogously to the given machine for normal default theories. Obviously the computation is accomplished in polynomial time.

Lemma 3.2 $\alpha \wedge \beta \in SAT \iff \beta \not\vdash \neg\alpha$, *for* $\{\alpha, \beta\} \subset \mathcal{L}$.

Proof: $\beta \vdash \neg\alpha \iff$ (Deduction Theorem) $\vdash \beta \rightarrow \neg\alpha \iff \vdash \neg(\alpha \wedge \beta) \iff$ (Completeness & Soundness of $\vdash$) $\neg(\alpha \wedge \beta) \in TAUT \iff (\alpha \wedge \beta) \notin SAT$ •

Extensions

Theorem 3.3 *Let $\beta \notin TAUT$.*[1] *There is an extention E of the default theory $\Delta := (\{\frac{1:\alpha}{\beta}\}, \emptyset)$ and $\beta \in E$ iff $\alpha \wedge \beta \in SAT$.*

Proof: ($\Rightarrow$) Let E be an extension of Δ and $\beta \in E$. We must have $\delta \nabla_E \Gamma(E)$, because $\delta = \frac{1:\alpha}{\beta}$ is the only default and $\not\vdash \beta$. Thus $1 \in \Gamma(E)$ and $\neg\alpha \notin E$. E is deductively closed and $\beta \in E$, hence it follows that $\beta \not\vdash \neg\alpha$. Lemma 3.2 implies $\alpha \wedge \beta \in SAT$.
($\Leftarrow$) With lemma 3.2 we have $\beta \not\vdash \neg\alpha$ and hence we get $\not\vdash \neg\alpha$ by the monotonicity of $\vdash$. Thus $\neg\alpha \notin Th(\beta)$ always holds. We show: $Th(\beta)$ is an extension of Δ. We "construct" the extension as in theorem 1.1. We get $E_0 = \emptyset$, $E_1 = Th(\emptyset)$, and $E_2 = Th(E_1) \cup \{\beta\}$ since $1 \in Th(\emptyset)$ and $\neg\alpha \notin Th(\beta)$. Obviously we have $Th(E_1) = Th(Th(\emptyset)) = Th(\emptyset)$, thus $E_2 = Th(\emptyset) \cup \{\beta\}$. Since there are no more defaults in D, we get $E_3 = Th(E_2) = Th(Th(\emptyset) \cup \{\beta\})$. A sentence is derivable from β iff it is derivable from the union of β and $Th(\emptyset)$, and thus $E_3 = Th(\{\beta\})$. Since there are no more defaults in D we have $E_{i+1} = E_i = Th(\{\beta\})$, for $i \geq 3$ and hence it follows that $Th(\{\beta\}) = \bigcup_{i=0}^{\infty} E_i$. Theorem 1.1 implies that $Th(\{\beta\})$ is an extension of Δ. •
If E is an extension of $\Delta := (\{\frac{1:\alpha}{\beta}\}, \emptyset)$ and $\beta \notin E$, we have $\neg\alpha \in E$ and therefore $\vdash \neg\alpha$ since $\neg\alpha$ cannot be a consquence of β because $\beta \notin E$. The default is never applicable and $\beta \notin TAUT$. Thus β belongs to no extension of Δ.
On the other hand if there is an extension of Δ which contains β and $\beta \notin TAUT$ then the default δ is applicable. Thus $\not\vdash \neg\alpha$ and $\beta \not\vdash \neg\alpha$. It follows that the default is always applicable. Thus the default theory $\Delta := (\{\frac{1:\alpha}{\beta}\}, \emptyset)$ has a unique extension E and either $\beta \in E$ or $\beta \notin E$.
Hence it follows that the $AEMP(\Delta, \beta)$ is as hard as the $EMP(\Delta, \beta)$ and immediately we get the following corollary.

Corollary 3.5 *Let $\Delta = (D, W)$ be a default theory and $\beta \notin TAUT$. Then the $EMP(\Delta, \beta)$ and the $AEMP(\Delta, \beta)$ are $\mathcal{NP}$–hard.*

Proof: The translation of the sentence, given in theorem 3.3, into a default theory is computable in polynomial time by the previous rema rk. Thus theorem 3.3 implies the assertion. •

Modified Extensions

We now proceed to analyse the modified version of the default logic developed by Lukaszewicz (1988).

Lemma 3.3 *Let $\beta \notin TAUT$: $\alpha \wedge \beta \in SAT \iff$ There is a modified default proof of β with respect to $\Delta := (\{\frac{1:\alpha}{\beta}\}, \emptyset)$.*

Proof:($\Rightarrow$) We prove: 1) δ is $\emptyset$-applicable and 2) $\emptyset \cup CON(\{\delta\}) \vdash \beta$.
Obviously 2) is satisfied since $CON(\{\delta\}) = \{\beta\}$. We have to prove 1): By lemma

[1] On these premises the application of the default is guaranteed. This premise is convenient, because $TAUT$ is a subset of every extension of a default theory.

3.2 $1 \in Th(\emptyset)$ and $\alpha \wedge \beta \in SAT$ imply $\beta \not\vdash \neg\alpha$ for all $\alpha \in JUS(\{\delta\})$ and therefore there is a modified default proof of β with respect to Δ.
($\Leftarrow$) If there is a modified default proof of β with respect to Δ and $\beta \notin TAUT$, β must be infered by the only default $\frac{1:\alpha}{\beta}$. Therefore we have $\beta \not\vdash \neg\alpha$ and it follows by lemma 3.2 that $\alpha \wedge \beta \in SAT$. •
As a consequence of the previous lemma and theorem 1.6 we formulate the following theorem.

Theorem 3.4 *Let $\beta \notin TAUT$: $\alpha \wedge \beta \in SAT \iff$ There is a modified extension of $\Delta := (\{\frac{1:\alpha}{\beta}\}, \emptyset)$ which contains β.*

Analogously to the previous section we can prove that there exists one and only one modified extension of $\Delta := (\{\frac{1:\alpha}{\beta}\}, \emptyset)$ which contains β if $\beta \notin TAUT$ and $\alpha \wedge \beta \in SAT$. Hence it follows that the $AEMP_{mod}(\Delta, \beta)$ is as hard as the $EMP_{mod}(\Delta, \beta)$ and immediately we get the following corollary.

Corollary 3.6 *$EMP_{mod}(\Delta, \beta)$ and $AEMP_{mod}(\Delta, \beta)$ are $\mathcal{NP}-$hard.*

Proof: We saw in the remark in 3.2 that the translation of prositional formulas in default theories is computable in polynomial time. The proposition follows analogously to the proposition of corollary 3.5. •
There is no difference between modified extensions and extensions of normal default theories (Lu 88), thus the corollaries 3.1 - 3.5 are also correct for the EMP_{mod} instead of the EMP.

Theorem 3.5 *EMP_{mod} is nondeterministic polynomial time Turing reducible to the satisfiability problem in propositional logic.*

Proof: We have to show that there exists a nondeterministic polynomial time oracle machine with oracle SAT which accepts EMP_{mod}.

Let $\Delta = (D, W)$ and β be given.

1. Choose nondeterministically a sequence $(\delta_i)_{i=0}^{k}$ $(k \leq |D|)$ of defaults of D.
2. If $W \cup CON(\{\delta_0, \ldots, \delta_k\}) \cup \{\neg\beta\} \subseteq SAT$ then halt in a rejecting state else goto 3.
3. Let $D_0 = \emptyset$ and $D_i := \{\delta_0, \ldots, \delta_{i-1}\}$ for $i = 1$ to $k+1$ do:
   (a) If $W \cup CON(D_{i-1}) \cup \{\neg PRE^{\{\delta_{i-1}\}}\} \subseteq SAT$ then halt in a rejecting state else goto (b).
   (b) If $W \cup CON(D_i) \cup JUS(D_i) \subseteq SAT$ then goto (c) else halt in a rejecting state.
   (c) If $i = k+1$ halt in an accepting state else goto 3.

To choose a sequence which length is less than $|D|$ can be done in polynomial time. Let l_W resp. l_C resp. l_J be the length of the conjunction of the elements of W resp.

$CON(D)$ resp. $JUS(D)$, l_P be the length of $\neg PRE^D$ and l_β be the length of β, then the length of the first question is less than $l_W + l_C + l_\beta$ and the sum of the questions in step 3 is bounded by $2(l_W + l_C) + l_P + l_J$. The machine writes $2|D| + 1$ questions on the oracle tape at most. To ask a question costs constant time and the test in 3.(c) whether $i + 1 = k$ can be done in polynomial time. The machine is nondeterministic and its computation time is bounded by a polynomial in the length of the propositional formulas and the cardinal number of D.
As an immediate corollary of the previous theorem we can relate the problem to the polynomial time hierarchy (Ba 88) as follows:

Corollary 3.7 $EMP_{mod}(\Delta, \beta) \in \Sigma_2^p$ *and* $\overline{EMP_{mod}(\Delta, \beta)} \in \Pi_2^p$

3.3 Autoepistemic Logic

Autoepistemic logic, developed by Moore[2] 1985 is another formalisation of nonmonotonic inference. Autoepistemic logic is formulated in an ordinary formal language $\mathcal{L}_L$ of propositional logic with an added unary modal operator L. If α is a sentence of $\mathcal{L}_L$ then $L\alpha$ also is a sentence of $\mathcal{L}_L$. The equivalents of defaults are represented by modal sentences in the autoepistemic logic. In autoepistemic logic it is possible to infer modal sentences. This is the main difference between the default und the autoepistemic logic. An autoepistemic extension (ae-extension) is defined as follows:

Definition 3.1 *A set T is an ae-extension of $\mathcal{A} \subseteq \mathcal{L}_L$ iff it satisfies the equation $T = \{\alpha | \mathcal{A} \cup LT \cup \neg L\overline{T} \vdash \alpha\}$ where $LT := \{L\alpha | \alpha \in T\}$ and $\neg L\overline{T} := \{\neg L\alpha | \alpha \notin T\}$.*

The extension membership problem for the autoepistemic logic is: Given a set $\mathcal{A} \subseteq \mathcal{L}_L$ and $\beta \in \mathcal{L}$, is there an ae-extension of $\mathcal{A}$ which contains β?
The extension membership problem for the autoepistemic logic is denoted by $EMP_{AEL}(\mathcal{A}, \beta)$ and $\overline{EMP_{AEL}(\mathcal{A}, \beta)}$, $AEMP_{AEL}(\mathcal{A}, \beta)$ and $\overline{AEMP_{AEL}(\mathcal{A}, \beta)}$ are analogously defined to the corresponding problems in default logic. Konolige proved in (Ko 88) the following relation between default and autoepistemic logic.

Theorem 3.6 (Konolige 1988) *Let $\mathcal{A} \subseteq \mathcal{L}_L$ be the AE transform of a default theory $\Delta = (D, W)$. A set $E \subseteq \mathcal{L}$ is a default extension of Δ iff E is the kernel*[3] *of a strongly grounded ae-extension of $\mathcal{A}$.*

To construct an AE transform of a default theory Konolige defined a translation T of defaults in sentences of $\mathcal{L}_L$ as follows:

$$\frac{\alpha : \beta_1, \ldots, \beta_n}{\gamma} \quad \overset{T}{\longmapsto} \quad ((L\alpha \wedge \neg L \neg \beta_1 \wedge \ldots \wedge \neg L \neg \beta_n) \rightarrow \gamma)$$

A special formulation of the theorem of Konolige is the following lemma:

Lemma 3.4 *There is an extension E of the default theory $\Delta = (\{\frac{1:\alpha}{\beta}\}, \emptyset)$ iff E is the kernel of a strongly grounded ae-extension of $\mathcal{A} = \{(L1 \wedge \neg L \neg \alpha) \rightarrow \beta\}$.*

---

[2]You can find a detailed representation of the autoepistemic logic in (Mo 85).
[3]The kernel of a set $\mathcal{A} \subseteq \mathcal{L}_L$ is $\mathcal{A} \cap \mathcal{L}$.

A machine which computes the AE transform of a default theory is essentially a copying machine. Thus the translation is computable in polynomial time. Every extension of $\mathcal{A} = \{(L1 \wedge \neg L\neg\alpha) \rightarrow \beta\}$ which contains β is strongly grounded. Hence it follows that the $EMP(\Delta, \beta)$ is polynomial time reducible to the $EMP_{AEL}(\mathcal{A}, \beta)$ and that the $AEMP(\Delta, \beta)$ is polynomial time reducible to the $AEMP_{AEL}(\mathcal{A}, \beta)$. The conversion of the theorem of Konolige is also true.

Theorem 3.7 (Konolige 1988) *For any set of sentences $\mathcal{A} \subseteq \mathcal{L}_L$ in extended normal form, there is an effectively constructable default theory $\Delta = (D, W)$ such that $E \subseteq \mathcal{L}$ is an extension of Δ iff E is the kernel of a strongly grounded ae-extension of $\mathcal{A}$.*

The crucial point in the proof of this theorem is the fact that every set $\mathcal{A} \subseteq \mathcal{L}_L$ has an autoepistemic equivalent set in which each sentence is of the form $\neg L\alpha \vee L\beta_1 \vee \ldots \vee L\beta_n \vee \gamma \quad (\{\alpha, \beta_1, \ldots, \beta_n, \gamma\} \subset \mathcal{L}, \quad n \in \omega)$. If there are $\neg L\alpha$ or all $L\beta_i$ absent we extend the sentence to the autoepistemic equivalent sentence $\neg L1 \vee L\beta_1 \vee \ldots \vee L\beta_n \vee \gamma$ or to $\neg L\alpha \vee L0 \vee \gamma$. Then the resulting sentence is in extended normalform. Such sentences can be translated into defaults by the following translation T':

$$\neg L\alpha \vee L\beta_1 \vee \ldots \vee L\beta_n \vee \gamma \quad \overset{T'}{\longmapsto} \quad \frac{\alpha : \neg\beta_1, \ldots, \neg\beta_n}{\gamma}$$

Obviously the translation can be computed by a polynomial time bounded machine and hence the theorem of Konolige implies that the $EMP_{AEL}(\mathcal{A}, \beta)$ is polynomial time reducible to the $EMP(\Delta, \beta)$ and that the $AEMP_{AEL}(\mathcal{A}, \beta)$ is polynomial time reducible to the $AEMP(\Delta, \beta)$. Hence it follows that the $EMP(\Delta, \beta)$ and the $EMP_{AEL}(\mathcal{A}, \beta)$ resp. the $AEMP(\Delta, \beta)$ and the $AEMP_{AEL}(\mathcal{A}, \beta)$ are polynomial time equivalent.

References:

(Ba 88) Balcázar, J. L. & Díaz, J. & Gabarró, J.: *Structural Complexity I* Springer-Verlag Berlin Heidelberg New York (1988)

(Bes 89) Besnard, P.: *An Introduction to Default Logic* Springer-Verlag Berlin Heidelberg New York (1989)

(Co 71) Cook, S. A.: *The Complexity of Theorem-Proving Procedures* 3rd ACM Symposium on Theory of Computing, p. 151 - 158

(Do 84) Dowling, W.F. & Gallier, J. H.: *Linear time algorithms for testing the satisfiability of propositional Horn formulae* Journal Logic Programming 1 no. 3 (1984) p. 267 - 284

(He 74) Henschen, L. & Wos, L.: *Unit refutations and Horn sets* Journal of the ACM 21 (1974), p. 590 - 605

(Jo 77) Jones, N.D. & Laaser, W.T.: *Complete problems for deterministic polynomial time* Theor. Comp. Science 3 (1977), p. 105 - 117

(Ko 88) Konolige, K.: *On the relation between default and autoepistemic logic* Artificial Intelligence 35 (1988) p. 343 - 382

(Łu 88) Łukaszewicz, W.: *Considerations on Default Logic* Computational Intelligence 4 (1988) p. 1- 16

(Mo 85) Moore, R.C.: *Semantical considerations on nonmonotonic logic* Artificial Intelligence 25 (1985) p. 75 - 94

(Rei 80) Reiter, R.: *A logic for default reasoning* Artificial Intelligence 13 (1980), p. 81 - 132

On Completeness for NP via Projection Translations

Iain A. Stewart

Computing Laboratory, Univ. Newcastle upon Tyne, Claremont Tower,
Claremont Road, Newcastle upon Tyne, NE1 7RU, England.

Abstract. We show that our logical encodings of the Satisfiability Problem and the 3-Colourability Problem are complete for NP via projection translations. However, whilst an encoding of the 3-Satisfiability Problem where all clauses have at most 3 literals is also complete for NP via projection translations, we are unable to show that a similar encoding of this decision problem where all clauses have exactly 3 literals is complete for NP via projection translations. It appears to matter how we encode decision problems (as sets of finite structures) when we deal with weak reductions such as projection translations.

1 Introduction

Constructing logics to capture complexity classes is not new. In particular, the complexity classes L, NSYMLOG (symmetric logspace), NL, P, and NP have all been characterized by extending first-order logic (with successor) with various operators (see, for example, [5-8, 13, 14]). These characterizations have yielded new results involving the completeness and equivalence of various problems via projection translations (see the afore-mentioned references), have been used to show that NL is closed under complementation [8], and have enabled new problems to be shown to be complete via logspace reductions for P and NSYMLOG, the latter being a complexity class which was previously very difficult to work with (see [15] where some problems involving free groups are considered: see [12] for more motivation concerning this logical approach).

In this paper, we investigate the logics obtained by augmenting first-order logic with operators corresponding to some problems complete for NP via logspace reductions, as was done in [13, 14]. In particular, we study logical encodings of the Satisfiability Problem (namely SAT), the 3-Satisfiability Problem (namely 3SAT), and the 3-Colourability Problem (namely 3COL) ([4]). We show that SAT and 3COL are complete for NP via projection translations (a very weak reduction between problems). One reason for the interest in these translations is that, as suggested in [2], it seems feasible to show that projection translations (of a certain arity) do not exist between certain problems, so obtaining lower bound results.

We also show that 3SAT is complete for NP via iterated projection translations (in [2] it was shown that a logical encoding of the 3-Satisfiability Problem is not complete for NP via interpretative reductions). However, it is still open as to whether 3SAT is complete for NP via projection translations. Our results also imply that there might be a difference between the two versions of the 3-Satisfiability Problem where the Boolean expressions involved are deemed to have at most and exactly 3 literals in each clause, respectively, as we show that $\leq$3SAT, a logical encoding of the 3-Satisfiability Problem where the clauses have *at most* 3 literals, *is* complete for NP via projection translations.

The logics obtained by augmenting first-order logic with operators corresponding to the problems above are shown to be (apparently) different from the logic HP*[FO_s] of [13] (or (FO+posHP) as it was called there). We show that whereas we are allowed to nest arbitrarily many positive applications of the operator HP and still obtain sentences representing problems in NP, we are only allowed one positive application of any of the above operators if we wish to remain in NP (assuming NP $\neq$ co-NP). We also answer some open questions of [14].

2 Basic Definitions

The reader is referred to [7, 13, 14] for extensive details of the concepts mentioned here and to [1] for background material.

A *vocabulary* $\tau = <\underline{R}_1,\underline{R}_2,...,\underline{R}_k,\underline{C}_1,\underline{C}_2,...,\underline{C}_m>$ is a tuple of *relation symbols* $\{\underline{R}_i : i = 1,2,..,k\}$, with $\underline{R}_i$ of arity a_i, and *constant symbols* $\{\underline{C}_i : i = 1,2,...,m\}$. A *finite structure of size n* over the vocabulary τ is a tuple $S = <\{0,1,...,n-1\}, R_1,R_2,...,R_k,C_1,C_2,...,C_m>$ consisting of a *universe* $|S| = \{0,1,...,n-1\}$, *relations* $R_1, R_2, .., R_k$ on the universe $|S|$ of arities $a_1, a_2, ..., a_k$, respectively, and *constants* $C_1, C_2, ..., C_m$ from the universe $|S|$. The size of some structure S is denoted by $|S|$. We denote the set of all finite structures over τ by STRUCT(τ) (henceforth, we do not distinguish between relations (resp. constants) and relation (resp. constant) symbols, and we assume that all structures are of size at least 2). A *problem of arity t* (≥ 0) over τ is a subset of $STRUCT_t(\tau) = \{(S,\mathbf{u}) : S \in STRUCT(\tau), \mathbf{u} \in |S|^t\}$ (we remark that we usually refer to a set of strings as a decision problem as opposed to a problem, which we reserve for a set of finite structures). If Ω is some problem then $\tau(\Omega)$ denotes its vocabulary.

The language of the *first-order logic $FO_s(\tau)$* over the vocabulary τ has as its (well-formed) formulae those formulae built, in the usual way, from the relation and constant symbols of τ, the binary relation symbols = and s, and the constant symbols 0 and max, using the logical connectives $\vee$, $\wedge$, and $\neg$, the variables $x, y, z_3, ...$ etc., and the quantifiers $\exists$ and $\forall$ (we remark that $FO_s(\tau)$ was denoted $FO_{\leq}(\tau)$ in earlier papers: we have changed the notation as the symbol $\leq$ implies the existence of a built-in linear ordering which is not the case). Any formula ϕ of $FO_s(\tau)$, with free variables those of the t-tuple $\mathbf{x}$, is interpreted in the set $STRUCT_t(\tau)$, and for each $S \in STRUCT(\tau)$ of size n and $\mathbf{u}$

$\in |S|^t$, we have that $(S,\mathbf{u}) \models \phi(\mathbf{x})$ iff $\phi^S(\mathbf{u})$ holds, where $\phi^S(\mathbf{u})$ denotes the obvious interpretation of ϕ in S, except that the binary relation symbol $=$ is always interpreted in S as equality, the binary relation symbol s is interpreted as the successor relation on $|S|$, the constant symbol 0 is interpreted as $0 \in |S|$, the constant symbol max is interpreted as $n-1 \in |S|$, and each variable of $\mathbf{x}$ is given the corresponding value from $\mathbf{u}$. (We usually write $s(x,y)$ as $y = x+1$, and $\neg(x = y)$ as $x \neq y$.) If we forbid the use of the successor relation in the logic $FO_s(\tau)$ then we denote the resulting logic by $FO(\tau)$: also, $FO_s = \cup\{FO_s(\tau) : \tau$ some vocabulary$\}$ (with FO defined similarly). The formula ϕ *describes* (or *specifies* or *represents*) the problem

$$\{(S,\mathbf{u}) : (S,\mathbf{u}) \in \mathrm{STRUCT}_t(\tau), (S,\mathbf{u}) \models \phi(\mathbf{x})\}$$

of arity t.

Having detailed how we use first-order logic to describe problems, we now illustrate how we extend first-order logic with new operatros to attain greater expressibility. Let τ_2 be the vocabulary consisting of the binary relation symbol E: so, we may clearly consider structures S over τ_2 as digraphs or graphs. Consider the problem HP of arity 2:

$$\mathrm{HP} = \{(S,u,v) \in \mathrm{STRUCT}_2(\tau_2) : \text{there is a Hamiltonian path in the digraph } S \text{ from } u \text{ to } v\}.$$

We write $(\pm\mathrm{HP})^*[FO_s]$ to denote the logic formed by allowing an unlimited number of nested applications of the operator HP, where $\mathrm{HP}[\lambda\mathbf{x}\mathbf{y}\psi^S(\mathbf{x},\mathbf{y})]$, for some formula $\psi \in (\pm\mathrm{HP})^*[FO_s]$, some k-tuples of distinct variables $\mathbf{x}$ and $\mathbf{y}$, and some relevant structure S, denotes the digraph with vertices indexed by the tuples of $|S|^k$, and where there is an edge from $\mathbf{u}$ to $\mathbf{v}$ iff there is a Hamiltonian path in the digraph described by $\psi^S(\mathbf{x},\mathbf{y})$ from $\mathbf{u}$ to $\mathbf{v}$ (this is the logic (FO + HP) of [13]). We write $(\pm\mathrm{HP})^k[FO_s]$ (resp. $\mathrm{HP}^k[FO_s]$) to denote the sub-logic of $(\pm\mathrm{HP})^*[FO_s]$ where all formulae have at most k nested applications of the operator HP (resp. where no operator appears within a negation sign): the sub-logic $\mathrm{HP}^k[FO_s]$ of $\mathrm{HP}^*[FO_s]$ is defined similarly.

In order to compare logical descriptions of decision problems, we use the notion of a logical translation (these translations play an analogous role to logspace and polynomial-time reductions between sets of strings). Let $\tau' = <R_1,R_2,\ldots,R_k,C_1,C_2,\ldots,C_m>$ be some vocabulary, where each R_i is some relation symbol of arity a_i and each C_j is some constant symbol, and let $L(\tau)$ be some logic over some vocabulary τ. Then the formulae $\Sigma = \{\phi_i(\mathbf{x}_i),\psi_j(\mathbf{y}_j) : i = 1,2,\ldots,k; j = 1,2,\ldots,m\} \subseteq L(\tau)$, where

(i) each formula ϕ_i (resp. ψ_j) is over the qa_i (resp. q) distinct variables $\mathbf{x}_i$ (resp. $\mathbf{y}_j$), for some positive integer q;

(ii) for each $j = 1, 2, \ldots, m$ and for each structure $S \in \mathrm{STRUCT}(\tau)$

$$S \models (\exists x_1)(\exists x_2)\ldots(\exists x_q)[\psi_j(x_1,x_2,\ldots,x_q) \wedge (\forall y_1)(\forall y_2)\ldots(\forall y_q)[\psi_j(y_1,y_2,\ldots,y_q) \Leftrightarrow (x_1 = y_1 \wedge x_2 = y_2 \wedge \ldots \wedge x_q = y_q)]],$$

are called τ'-*descriptive*. For each $S \in \mathrm{STRUCT}(\tau)$, the τ'-*translation of* S *with respect to* Σ is the structure $S' \in \mathrm{STRUCT}(\tau')$ with universe $|S|^q$, defined as follows: for all $i = 1, 2, \ldots, k$ and for any tuples $\{\mathbf{u}_1,\mathbf{u}_2,\ldots,\mathbf{u}_{a_i}\} \subseteq |S'| = |S|^q$

$$R_i^{S'}(\mathbf{u}_1,\mathbf{u}_2,\ldots,\mathbf{u}_{a_i}) \text{ holds iff } (S,(\mathbf{u}_1,\mathbf{u}_2,\ldots,\mathbf{u}_{a_i})) \models \phi_i(\mathbf{x}_i)$$

and for all $j = 1, 2, \ldots, m$ and for any tuple $\mathbf{u} \in |S'| = |S|^q$

$$C_j^{S'} = \mathbf{u} \text{ iff } (S,\mathbf{u}) \models \psi_j(\mathbf{y}_j)$$

(tuples are ordered lexicographically, with (0,0,..,0) < (0,0,...,1) < (0,0,...,2) < ..., and so on). Let Ω and Ω' be problems over the vocabularies τ and τ', respectively. Let Σ be a set of τ'-*descriptive* formulae from some logic $L(\tau)$, and for each $S \in \text{STRUCT}(\tau)$, let $\sigma(S) \in \text{STRUCT}(\tau')$ denote the τ'-translation of S with respect to Σ. Then Ω' is an *L-translation* of Ω iff for each $S \in \text{STRUCT}(\tau)$, $S \in \Omega$ iff $\sigma(S) \in \Omega'$.

Let $\phi \in \text{FO}_s(\tau)$, for some vocabulary τ, be of the form

$$\phi \equiv \bigvee\{\alpha_i \wedge \beta_i : i \in I\},$$

for some finite index set I, where:

(i) each α_i is a conjunction of the logical atomic relations, s, $=$, and their negations;

(ii) each β_i is atomic or negated atomic;

(iii) if $i \neq j$, then α_i and α_j are mutually exclusive.

Then ϕ is a *projective formula*. If the successor relation symbol s does not appear in ϕ (that is, $\phi \in \text{FO}(\tau)$), then ϕ is a *projective formula without successor*, and if each of the β_i (above) is atomic then ϕ is a *monotone projective formula*. Consequently, we clearly have the notions of one problem being a *first-order translation*, a *quantifier-free translation*, a *projection translation*, and a *monotone projection translation* of another, as well as all of these *without successor*. If Ω and Ω' are problems and there exist problems $\Omega_0, \Omega_1, ..., \Omega_k$, for some $k \geq 0$, such that $\Omega = \Omega_0$, Ω_{i+1} is a projection translation of Ω_i, for each $i < k$, and $\Omega_k = \Omega'$, then Ω' is an *iterated projection translation* of Ω.

First-order logic has been extended by other operators apart from HP. One such is the operator corresponding to the following problem ([14]: see [7] for other extensions not involving operators corresponding to NP-complete problems):

$$\text{3COL} = \{S \in \text{STRUCT}(\tau_2) : \text{the graph } S \text{ can be 3-coloured}\}$$

The problem HP(0,max) is defined as

$$\text{HP(0,max)} = \{S \in \text{STRUCT}(\tau_2) : \text{there is a Hamiltonian path in the digraph } S \text{ of size } n \text{ between } 0 \text{ and } n-1\}.$$

Theorem 2.1. ([13, 14])

(a) $NP = HP^1[FO_s] = HP^*[FO_s] \subseteq (\pm HP)^*[FO_s]$.

(b) $3COL^1[FO_s] \subseteq NP$.

(c) $NP \cup co\text{-}NP \subseteq 3COL^3[FO_s]$.

(d) $3COL^*[FO_s] = (\pm 3COL)^*[FO_s]$.

(e) If $3COL^1[FO_s] = 3COL^2[FO_s]$ *then* $3COL^1[FO_s] = (\pm 3COL)^*[FO_s] = NP = co\text{-}NP$.

(f) HP(0,max) is complete for NP and 3COL is complete for $3COL^1[FO_s]$ *via projection translations.* □

3 The Satisfiability Problem

In this section, we extend FO_s with an operator corresponding to the well-known NP-complete (via logspace reductions) decision problem, the Satisfiability Problem ([4]). We mention that a logical encoding of this problem has already been shown to be complete for NP via first-order translations ([10]) and interpretative reductions ([2]): interpretative reductions are similar to quantifier-free translations. Both results rely on Fagin's characterization of NP as those problems expressible in existential second-order logic ([3]). In Theorem 3.1 below, we essentially prove that SAT (defined below) is complete for NP via projection translations. Whilst Dahlhaus' proof that SAT is complete for NP via interpretative reductions can be modified to show that SAT is complete for NP via projection translations without successor, we include Theorem 3.1 as we use the actual proof later on (our proof is independent of Fagin's characterization).

The following lemma is most useful.

Lemma 3.1. *Let Ω_1, Ω_2, and Ω_3 be problems such that Ω_2 is a projection translation of Ω_1 and Ω_3 is a monotone projection translation of Ω_2. Then Ω_3 is a projection translation of Ω_1.* □

Let $\tau_{2,2} = <P,N>$, where P and N are binary relation symbols, and let $S \in \mathrm{STRUCT}(\tau_{2,2})$ be of size n. Then S can be considered as an instance of the Satisfiability Problem as follows:

$$P^S(i,j) \text{ holds iff the literal } x_i \text{ is in clause } j,$$

and

$$N^S(i,j) \text{ holds iff the literal } \neg x_i \text{ is in clause } j$$

(here, x_i is a Boolean variable). Conversely, any instance of the Satisfiability Problem can clearly be considered as a structure over $\tau_{2,2}$. Hence, we define

$$\mathrm{SAT} = \{S \in \mathrm{STRUCT}(\tau_{2,2}) : \text{the Boolean expression } S \text{ has a satisfying truth assignment}\}$$

Theorem 3.1. *There exists a monotone projection translation from HP(0,max) to SAT.*

Proof. Let $S \in \mathrm{STRUCT}(\tau_2)$ be of size n. Consider the following Boolean expressions over the Boolean variables $\{x_{ij},y_{ij},z_{ij} : i,j = 0,1,...,n-1\}$:

$\alpha_i = \bigvee\{x_{ij} : j = 0,1,...,n-1;\ i \neq j\}$;

$\beta_0 = \bigwedge\{\alpha_i : i = 0,1,...,n-2\}$;

$\beta_1 = \bigwedge\{\neg x_{n-1 i} : i = 0,1,...,n-2\}$;

$\beta_2 = \bigwedge\{\neg x_{i0} : i = 1,2,...,n-1\}$;

$\beta_3 = \bigwedge\{\neg x_{ij} \vee \neg x_{ik} : i,j,k = 0,1,...,n-1;\ i \neq j;\ i \neq k; j \neq k\}$;

$\beta_4 = \bigwedge\{\neg x_{ji} \vee \neg x_{ki} : i,j,k = 0,1,...,n-1;\ i \neq j;\ i \neq k; j \neq k\}$;

$\beta_5 = \bigwedge\{\neg x_{ij} \vee y_{ij} : i,j = 0,1,...,n-1;\ i \neq j\}$;

$\beta_6 = \bigwedge\{\neg y_{ij} \vee \neg y_{jk} \vee y_{ik} : i,j,k = 0,1,...,n-1;\ i \neq j;\ i \neq k; j \neq k\}$;

$\beta_7 = \bigwedge\{\neg y_{ij} \vee \neg y_{ji} : i,j = 0,1,...,n-1;\ i \neq j\}$;

$\beta_8 = \bigwedge\{\neg x_{ij} \vee z_{ij} : i,j = 0,1,...,n-1; i \neq j\}$;
$\beta_9 = \bigwedge\{\neg z_{ij} \vee z_{ij} : i,j = 0,1,...,n-1; i \neq j; E^S(i,j)\} \wedge$
$\bigwedge\{\neg z_{ij} : i,j = 0,1,...,n-1; i \neq j; \neg E^S(i,j)\}$
(the redundancy in β_9 will be explained soon).

Let $\gamma_0 = \beta_0 \wedge \beta_1 \wedge \beta_2 \wedge \beta_3 \wedge \beta_4$ and suppose that t is a satisfying truth assignment for γ_0. Let G be the digraph with vertices $\{0,1,...,n-1\}$ and with edges $\{(i,j) : x_{ij}$ appears in γ_0 and $t(x_{ij}) = T\}$. Then in the digraph G

(i) there are no edges entering (resp. leaving) vertex 0 (resp. $n-1$); (β_1, β_2)
(ii) there is exactly one edge leaving each vertex of $\{0,1,...,n-2\}$; (β_0, β_3)
(iii) there is at most one edge entering each vertex of $\{1,2,...,n-1\}$. (β_4)

Conversely, if t is a truth assignment for γ_0 giving rise to a digraph G satisfying the conditions (i), (ii), and (iii) above, then clearly t is a satisfying truth assignment for γ_0.

Let $\gamma_1 = \beta_5 \wedge \beta_6 \wedge \beta_7$ and suppose that t is a satisfying truth assignment for γ_1. Let H be the digraph with vertices $\{0,1,...,n-1\}$ and with edges $\{(i,j) : y_{ij}$ appears in γ_1 and $t(y_{ij}) = T\}$. Hence, we must have that

(iv) if (i,j) is an edge in G, then (i,j) is an edge in H; (β_5)
(v) if (i,j) and (j,k) are edges in H, then (i,k) is an edge in H; (β_6)
(vi) if (i,j) is an edge in H, then (j,i) is not an edge in H. (β_7)

Conversely, if t is a truth assignment for γ_1, giving rise to digraphs G and H satisfying the conditions (iv), (v), and (vi) above, then clearly t is a satisfying truth assignment for γ_1.

Let $\gamma_2 = \beta_8 \wedge \beta_9$ and suppose that t is a satisfying truth assignment for γ_2. Let K be the digraph with vertices $\{0,1,...,n-1\}$ and with edges $\{(i,j) : z_{ij}$ occurs in γ_2 and $t(z_{ij}) = T\}$. Hence, we must have that

(vii) if (i,j) is an edge in G, then (i,j) is an edge in K; (β_8)
(viii) K is a sub-digraph of the digraph S. (β_9)

Conversely, if t is a truth assignment for γ_2, giving rise to digraphs G and K satisfying the conditions (vii) and (viii) above, then clearly t is a satisfying truth assignment for γ_2.

Let $\gamma = \gamma_0 \wedge \gamma_1 \wedge \gamma_2$ and let t be a satisfying truth assignment for γ. Then by conditions (i), (ii), and (iii) above, G can be considered to be the vertex-disjoint union of a path from vertex 0 to vertex $n-1$ and a (possibly empty) collection of vertex-disjoint cycles. Suppose that there is indeed a cycle in G. Then by conditions (iv) and (v), there is a pair of edges (i,j) and (j,i) in H, thus contradicting condition (vi). Hence, the digraph G consists entirely of a Hamiltonian path from vertex 0 to vertex $n-1$. By conditions (vii) and (viii), G is a sub-digraph of the digraph S, and so the digraph S has a Hamiltonian path from vertex 0 to vertex $n-1$. By similar reasoning, it is easy to show that the digraph S has a Hamiltonian path from vertex 0 to vertex $n-1$ if and only if there is a satisfying truth assignment for γ.

To establish the result, it remains to show that there are projective formulae describing the above Boolean expression γ in terms of the digraph S (notice that γ is in conjunctive normal form). The clauses of γ (that is, the disjunctions) are considered to be of types 0, 1, ..., 9 depending on which of the expressions $\beta_0, \beta_1, ..., \beta_9$ they occur in. Four variables, u_1, u_2, u_3, and u_4 (= **u**), are used to encode the type of the clause and three variables, v_1, v_2, and v_3 (= **v**), are used to encode the index of the clause, given that it is of a particular type. For example, the clause of type 4 with index (i,j,k), that is the clause $\{\neg x_{ji} \vee \neg x_{ki}\}$, is encoded as

$$(u_1,u_2,u_3,u_4,v_1,v_2,v_3) = (0,0,\max,0,i,j,k),$$

where the type is represented in binary (with max for "1").

Similarly, the variables x_{ij}, y_{ij}, and z_{ij} are encoded using the variables p_1, p_2, q_1, and q_2. The variable x_{ij} (resp. y_{ij}, z_{ij}) is encoded as

$$(p_1,p_2,q_1,q_2,t_1,t_2,t_3) = (0,0,i,j,0,0,0) \text{ (resp. } (0,\max,i,j,0,0,0), (\max,0,i,j,0,0,0))$$

($\mathbf{t} = (t_1,t_2,t_3)$ is a tuple of dummy variables introduced to ensure that tuples encoding clauses and tuples encoding variables have equal length).

Consider the formula $\phi_0(\mathbf{p},\mathbf{q},\mathbf{t},\mathbf{u},\mathbf{v})$ defined as follows:

$$\phi_0 \equiv p_1 = p_2 = 0 \wedge \max \neq q_1 \neq q_2 \wedge \mathbf{t} = \mathbf{0} \wedge (\mathbf{u},v_2,v_3) = (\mathbf{0},0,0) \wedge v_1 = q_1$$

(our use of shorthand should be obvious). Then ϕ_0 describes whether some positive literal, given by $(\mathbf{p},\mathbf{q},\mathbf{t})$, occurs in a clause of some type, given by $\mathbf{u}$, and some index, given by $\mathbf{v}$. For example

$$\phi_0{}^S(\mathbf{p},\mathbf{q},\mathbf{t},\mathbf{u},\mathbf{v}) \text{ holds with } (\mathbf{p},\mathbf{q},\mathbf{t},\mathbf{u},\mathbf{v}) = (0,0,i,j,0,0,0,0,0,0,0,i,0,0),$$

as long as $n-1 \neq i \neq j$: this tells us that x_{ij} is a literal in the clause of type 0 and index i. Clearly, ϕ_0 describes how positive literals occur in clauses of type 0. As no negative literals occur in clauses of type 0, we need not construct a formula to describe these occurrences.

Consider the following formulae (we write, for example, "$\mathbf{u} = \mathbf{3}$" as shorthand for "$u_1 = \max \wedge u_2 = \max \wedge u_3 = 0 \wedge u_4 = 0$"):

$$\psi_1 \equiv p_1 = p_2 = 0 \wedge q_1 = \max \neq q_2 \wedge \mathbf{t} = \mathbf{0} \wedge (\mathbf{u},v_2,v_3) = (\mathbf{1},0,0) \wedge v_1 = q_2;$$

$$\psi_2 \equiv p_1 = p_2 = 0 \wedge q_2 = 0 \neq q_1 \wedge \mathbf{t} = \mathbf{0} \wedge (\mathbf{u},v_2,v_3) = (\mathbf{2},0,0) \wedge v_1 = q_1;$$

$$\psi_3 \equiv p_1 = p_2 = 0 \wedge \mathbf{t} = \mathbf{0} \wedge \mathbf{u} = \mathbf{3} \wedge q_1 = v_1 \wedge (q_2 = v_2 \vee q_2 = v_3) \wedge v_1 \neq v_2 \neq v_3 \wedge v_1 \neq v_3;$$

$$\psi_4 \equiv p_1 = p_2 = 0 \wedge \mathbf{t} = \mathbf{0} \wedge \mathbf{u} = \mathbf{4} \wedge q_2 = v_1 \wedge (q_1 = v_2 \vee q_1 = v_3) \wedge v_1 \neq v_2 \neq v_3 \wedge v_1 \neq v_3;$$

$$\phi_5 \equiv p_1 = 0 \wedge p_2 = \max \wedge \mathbf{t} = \mathbf{0} \wedge (\mathbf{u},v_3) = (\mathbf{5},0) \wedge q_1 = v_1 \wedge q_2 = v_2 \wedge v_1 \neq v_2;$$

$$\psi_5 \equiv p_1 = p_2 = 0 \wedge \mathbf{t} = \mathbf{0} \wedge (\mathbf{u},v_3) = (\mathbf{5},0) \wedge q_1 = v_1 \wedge q_2 = v_2 \wedge v_1 \neq v_2;$$

$$\phi_6 \equiv p_1 = 0 \wedge p_2 = \max \wedge \mathbf{t} = \mathbf{0} \wedge \mathbf{u} = \mathbf{6} \wedge q_1 = v_1 \wedge q_2 = v_3 \wedge v_1 \neq v_2 \neq v_3 \wedge v_1 \neq v_3;$$

$$\psi_6 \equiv p_1 = 0 \wedge p_2 = \max \wedge \mathbf{t} = \mathbf{0} \wedge \mathbf{u} = \mathbf{6} \wedge ((q_1 = v_1 \wedge q_2 = v_2) \vee (q_1 = v_2 \wedge q_2 = v_3)) \wedge v_1 \neq v_2 \neq v_3 \wedge v_1 \neq v_3;$$

$$\psi_7 \equiv p_1 = 0 \wedge p_2 = \max \wedge \mathbf{t} = \mathbf{0} \wedge (\mathbf{u},v_3) = (\mathbf{7},0) \wedge ((q_1 = v_1 \wedge q_2 = v_2) \vee (q_1 = v_2 \wedge q_2 = v_1)) \wedge v_1 \neq v_2;$$

$$\phi_8 \equiv p_1 = \max \wedge p_2 = 0 \wedge \mathbf{t} = \mathbf{0} \wedge (\mathbf{u},v_3) = (\mathbf{8},0) \wedge q_1 = v_1 \wedge q_2 = v_2 \wedge v_1 \neq v_2;$$

$$\psi_8 \equiv p_1 = p_2 = 0 \wedge \mathbf{t} = \mathbf{0} \wedge (\mathbf{u},v_3) = (\mathbf{8},0) \wedge q_1 = v_1 \wedge q_2 = v_2 \wedge v_1 \neq v_2;$$

$$\phi_9 \equiv p_1 = \max \wedge p_2 = 0 \wedge \mathbf{t} = \mathbf{0} \wedge (\mathbf{u},v_3) = (\mathbf{9},0) \wedge q_1 = v_1 \wedge q_2 = v_2 \wedge v_1 \neq v_2 \wedge E(v_1,v_2);$$

$$\psi_9 \equiv p_1 = \max \wedge p_2 = 0 \wedge \mathbf{t} = \mathbf{0} \wedge (\mathbf{u},v_3) = (\mathbf{9},0) \wedge q_1 = v_1 \wedge q_2 = v_2 \wedge v_1 \neq v_2;$$

(we do not need to specify all formulae ϕ_m and ψ_m as, for example, there are no positive literals in a clause of type 3: the redundancy in β_9 is so that we can describe the Boolean expression γ by monotone projective formulae). Set

$$\Phi \equiv \phi_0 \vee \phi_5 \vee \phi_6 \vee \phi_8 \vee \phi_9,$$

and

$$\Psi \equiv \psi_1 \vee \psi_2 \vee \psi_3 \vee \psi_4 \vee \psi_5 \vee \psi_6 \vee \psi_7 \vee \psi_8 \vee \psi_9.$$

It is easy to see that the formulae $\Phi^S(\mathbf{p},\mathbf{q},\mathbf{t},\mathbf{u},\mathbf{v})$ and $\Psi^S(\mathbf{p},\mathbf{q},\mathbf{t},\mathbf{u},\mathbf{v})$ describe the Boolean expression γ corresponding to the digraph S, and by the above reasoning we have that for any structure $S \in \mathrm{STRUCT}(\tau_2)$

$$S \models \mathrm{HP}[\lambda xyE(x,y)](0,\max) \text{ (i.e. } (S,0,\max) \in \mathrm{HP})$$

if and only if

$$S \models \mathrm{SAT}[\lambda(\mathbf{p},\mathbf{q},\mathbf{t})(\mathbf{u},\mathbf{v})\Phi(\mathbf{p},\mathbf{q},\mathbf{t},\mathbf{u},\mathbf{v}),(\mathbf{p},\mathbf{q},\mathbf{t})(\mathbf{u},\mathbf{v})\Psi(\mathbf{p},\mathbf{q},\mathbf{t},\mathbf{u},\mathbf{v})].$$

The formulae Φ and Ψ are clearly monotone projective and so the result follows. □

Corollary 3.1. *SAT is complete for NP = $SAT^1[FO_s]$ via projection translations.* □

Proposition 3.1. *$(\pm SAT)^1[FO_s] \subseteq SAT^2[FO_s]$.*

Proof. Let Ω be the sentence $\neg\mathrm{SAT}[\lambda\mathbf{xy}\phi(\mathbf{x},\mathbf{y}),\mathbf{xy}\psi(\mathbf{x},\mathbf{y})]$ of $(\pm\mathrm{SAT})^1[\mathrm{FO}_s]$, where ϕ and ψ are first-order formulae with free variables from among the k-tuples $\mathbf{x}$ and $\mathbf{y}$. Consider the sentence $\Omega' \in \mathrm{SAT}^2[\mathrm{FO}_s]$ defined as follows:

$$\Omega' \equiv \mathrm{SAT}[\lambda uv[u = 0 \wedge v = 0 \wedge \mathrm{SAT}[\lambda\mathbf{xy}\phi(\mathbf{x},\mathbf{y}),\mathbf{xy}\psi(\mathbf{x},\mathbf{y})]],$$
$$uv[u = 0 \wedge v = \max]],$$

where u and v do not occur in $\mathrm{SAT}[\lambda\mathbf{xy}\phi(\mathbf{x},\mathbf{y}),\mathbf{xy}\psi(\mathbf{x},\mathbf{y})]$. It is easy to see that Ω and Ω' are equivalent, so the result follows. □

By (the modification of) Dahlhaus' result that SAT is complete for NP via projection translations without successor, the successor relation can clearly be dispensed with in Corollary 3.1 and Proposition 3.1.

4. The 3-Satisfiability Problem

The 3-Satisfiability Problem is used as a stepping stone in showing that the 3-Colourability Problem is NP-complete via logspace reductions. As [13] has provoked our interest in the language $(\pm 3\mathrm{COL})^*[\mathrm{FO}_s]$, we now consider the 3-Satisfiability Problem from a logical point of view. There are two versions of the 3-Satisfiability Problem in the literature: the original in [9] specifies that any clause in any instance of the problem contains at most 3 literals, and the second (the usual one) specifies that any clause in any instance of the problem contains exactly 3 literals. Define

$\leq$3SAT = {$S \in \mathrm{STRUCT}(\tau_{2,2})$: the Boolean expression S has at most 3 literals in each clause and it has a satisfying truth assignment}, and

3SAT = $\{S \in \mathrm{STRUCT}(\tau_{2,2})$: the Boolean expression S has exactly 3 literals in each clause and it has a satisfying truth assignment$\}$.

Theorem 4.1. *There exists a monotone projection translation from HP(0,max) to ≤3SAT.*

Proof. Let $S \in \mathrm{STRUCT}(\tau_2)$ be of size $n \geq 4$ (the cases when $n = 2$ and $n = 3$ are trivial). Consider the following Boolean expression over the Boolean variables $\{x_{ij}, y_{ij}, z_{ij}, w_{ij} : i,j = 0,1,...,n-1\}$:

$$\alpha_i = (x_{i0} \vee x_{i1} \vee w_{i1}) \wedge \bigwedge\{\neg w_{ij\text{-}1} \vee x_{ij} \vee w_{ij} : j = 2,3,...,n-3\} \wedge (\neg w_{in\text{-}3} \vee x_{in\text{-}2} \vee x_{in\text{-}1}) \wedge \neg x_{ii}$$

together with the formulae $\beta_0, \beta_1, ..., \beta_9$ of the proof of Theorem 3.1. Let $\gamma = \bigwedge\{\beta_i : i = 0,1,...,9\}$. Then γ has at most 3 distinct literals in each clause. By reasoning as in the proof of Theorem 3.1, it is easy to see that the digraph S has a Hamiltonian path from vertex 0 to vertex $n-1$ iff there is a satisfying truth asignment for γ. There are projective formulae describing the Boolean expression γ in terms of the digraph S as in the proof of Theorem 3.1. □

Analogies of Corollary 3.1 and Proposition 3.1 hold for ≤3SAT.

Theorem 4.2. *There exists a projection translation from HP(0,max) to 3SAT.*

Proof. Let $S \in \mathrm{STRUCT}(\tau_2)$ be of size $n \geq 4$ (again, the cases when $n = 2$ and $n = 3$ are trivial). Let the formulae $\beta_0', \beta_1', ..., \beta_9'$ be analogous to the formulae $\beta_0, \beta_1, ..., \beta_9$ of the proof of Theorem 4.1 except that the literals $\neg y_{00}$ and $\neg z_{00}$ are introduced to pad out each clause so that it contains exactly three literals, and let β_{10}' and β_{11}' be defined as follows:

$$\beta_{10}' = (y_{00} \vee y_{n\text{-}1n\text{-}1} \vee z_{n\text{-}1n\text{-}1}) \wedge (y_{00} \vee \neg y_{n\text{-}1n\text{-}1} \vee z_{n\text{-}1n\text{-}1}) \wedge (y_{00} \vee y_{n\text{-}1n\text{-}1} \vee \neg z_{n\text{-}1n\text{-}1}) \wedge (y_{00} \vee \neg y_{n\text{-}1n\text{-}1} \vee \neg z_{n\text{-}1n\text{-}1});$$

$$\beta_{11}' = (z_{00} \vee y_{n\text{-}1n\text{-}1} \vee z_{n\text{-}1n\text{-}1}) \wedge (z_{00} \vee \neg y_{n\text{-}1n\text{-}1} \vee z_{n\text{-}1n\text{-}1}) \wedge (z_{00} \vee y_{n\text{-}1n\text{-}1} \vee \neg z_{n\text{-}1n\text{-}1}) \wedge (z_{00} \vee \neg y_{n\text{-}1n\text{-}1} \vee \neg z_{n\text{-}1n\text{-}1}).$$

Let $\gamma' = \bigwedge\{\beta_i' : i = 0,1,...,11\}$. Then γ' has exactly 3 distinct literals in each clause (as $n \geq 4$). By reasoning as in the proof of Theorem 3.1, it is easy to see that the digraph S has a Hamiltonian path from vertex 0 to vertex $n-1$ iff there is a satisfying truth asignment for γ'. (The formulae β_{10}' and β_{11}' ensure that for any satisfying truth assignment t for γ', we have that $t(y_{00}) = t(z_{00}) = T$ and so all occurrences in γ' of the literals $\neg y_{00}$ and $\neg z_{00}$ are "dummy occurrences" used to pad out the clauses where necessary.) There are projective formulae describing the Boolean expression γ' in terms of the digraph S (as in the proof of Theorem 3.1 except that the variable w_{ij} is encoded by setting both p_1 and p_2 at max). □

Again, analogies of Corollary 3.1 and Proposition 3.1 hold for 3SAT except that 3SAT is only complete for NP via iterated projection translations.

In Section 6 of [2], it is shown that an encoding of the 3-Satisfiability Problem (over a different vocabulary to ours), 3SAT′, is not complete for NP via interpretative reductions. The reasoning in [2] is as follows:

(i) the problem 3SAT′ can be expressed by a second-order existential sentence where the first-order part, in prenex normal form, is such that the prefix of quantifiers involves only universal quantifiers;

(ii) for any relevant structure S of size n satisfying this sentence, we may form the *sub-structure* S_i, for $1 < i < n$, over the same vocabulary as S, such that $|S_i| = \{0,1,...,i-1\}$ and the relations of S_i are those of S restricted to $\{0,1,...,i-1\}$;

(iii) clearly, S_i satisfies our sentence also (we point out that all problems in [2] are over vocabularies consisting entirely of relation symbols, the constant symbols 0 and max are not involved, and the successor relation is not involved.);

(iv) if 3SAT′ is complete for NP via interpretative reductions and Ω is some problem in NP, then we have that any sub-structure of any structure of Ω must also be in Ω; that is, Ω is closed under the operation of forming sub-structures; that is, any problem in NP is closed under the operation of forming sub-structures;

(v) the problem, over some vocabulary τ, consisting of all those structures S over τ such that $|S|$ is even is clearly in NP, and so must be closed under the operation of forming sub-structures: this yields a contradiction; thus, 3SAT′ is not complete for NP via interpretative reductions.

The reasoning above cannot be applied to show that 3SAT (resp. $\leq$3SAT) is not complete for NP via projection translations because:

(a) the existence of constant symbols spoil the formation of sub-structures, as in (ii) above;

(b) it is unknown whether our encodings of the 3-Satisfiability Problem, 3SAT and $\leq$3SAT, can be expressed in the logical form of (i) above.

We remark that it is unknown whether the problem 3SAT is complete for NP via projection translations. This discussion should convince the reader that when we are considering weak reductions between problems, such as those mentioned above, it seems to matter as to over which vocabulary we choose to encode our problems.

5. The 3-Colourability Problem

We now exhibit a projection translation from the problem 3SAT to the problem 3COL: mimicking the logspace reduction in [11] will not suffice as the resulting formula is not projective.

Theorem 5.1. *There is a projection translation from 3SAT to 3COL.*

Proof. Let $S \in \mathrm{STRUCT}(\tau_{2,2})$ be of size n such that each clause contains exactly 3 distinct literals. We define the graphs G, $\{G_j : j = 0,1,\ldots,n-1\}$, and H as follows.

The graph G has vertices $\{y_i, \neg y_i, z_{ij}, \neg z_{ij}, a, b, c : i,j = 0,1,\ldots,n-1\}$ and edges $\{\{a,y_i\},\{\neg y_i,y_i\},\{\neg y_i,a\},\{a,b\},\{b,c\},\{c,a\},\{c,z_{ij}\},\{c,\neg z_{ij}\} : i,j = 0,1,\ldots,n-1\}$.

The graph G_j corresponds to the jth clause of the Boolean expression S, and has vertices $\{w_{ij}, \neg w_{ij}, p_{ij}, \neg p_{ij}, q_{ij}, \neg q_{ij} : i = 0,1,\ldots,n-1\}$.

Among its edges are

$\{\{w_{ij},p_{ij}\},\{w_{ij},q_{ij}\},\{p_{ij},\neg w_{ij}\},\{q_{ij},\neg w_{ij}\},\{\neg w_{ij},\neg p_{ij}\},\{\neg w_{ij},\neg q_{ij}\},$
$\{\neg p_{ij},w_{i+1j}\},\{\neg q_{ij},w_{i+1j}\} : i = 0,1,\ldots,n-1$ and addition is modulo $n-1\}$.

However, G_j has other edges depending on the jth clause. For example, if the jth clause is $x_k \vee x_m \vee \neg x_r$, say, there are edges

$\{\{\{p_{ij},q_{ij}\} : i = 0,1,\ldots,n-1;\ i \neq k,m\},\{w_{kj},\neg w_{kj}\},\{w_{mj},\neg w_{mj}\},\{\{\neg p_{ij},\neg q_{ij}\} :$
$i = 0,1,\ldots,n-1;\ i \neq r\},\{\neg w_{rj},w_{r+1j}\}\}$ (with addition modulo n-1)

(G_j can be visualized as in Fig. 1). If the jth clause has a different structure

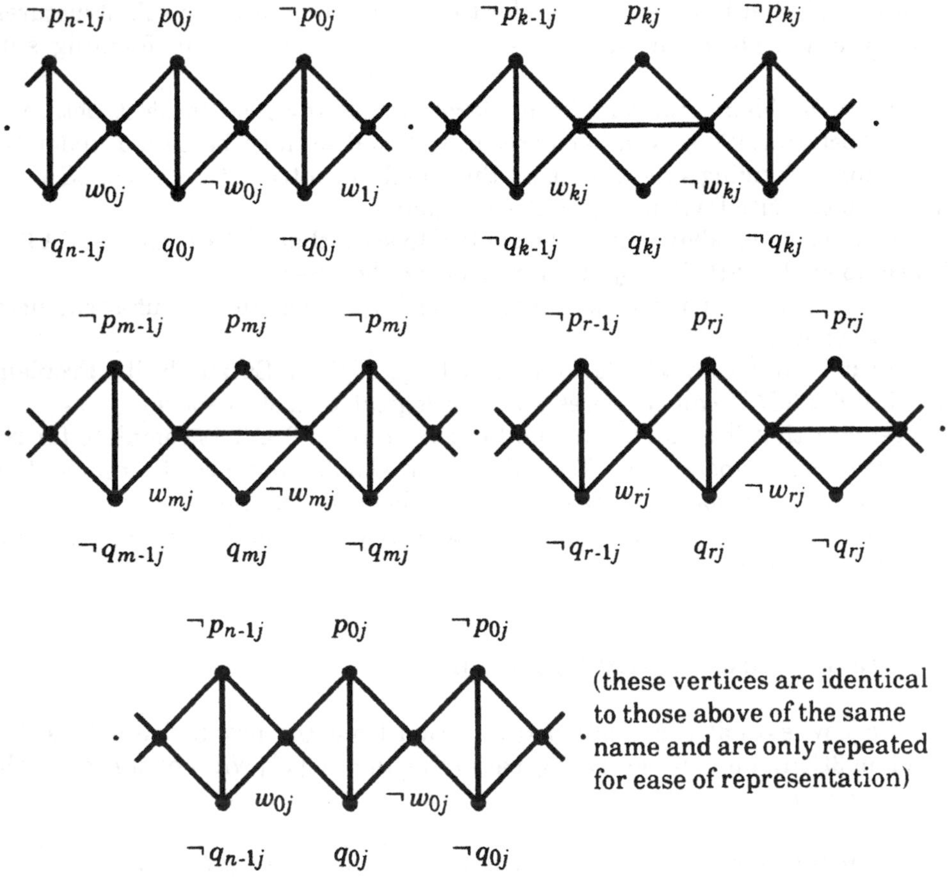

Figure 1 : the graph G_j

then the extra edges of G_j are defined in the obvious manner.

The graph H is made up from the graphs G and $\{G_j : j = 0,1,\dots,n-1\}$, and extra edges are added as follows: if the jth clause is of the form $x_k \vee x_m \vee \neg x_r$, say, then the following edges are added:

$$\{\{z_{kj},w_{kj}\},\{z_{mj},w_{mj}\},\{\neg z_{rj},\neg w_{rj}\},\{y_k,z_{kj}\},\{y_m,z_{mj}\},\{\neg y_r,\neg z_{rj}\}\}$$

(and similarly for clauses with a different structure: such an H can be visualized as in Fig. 2).

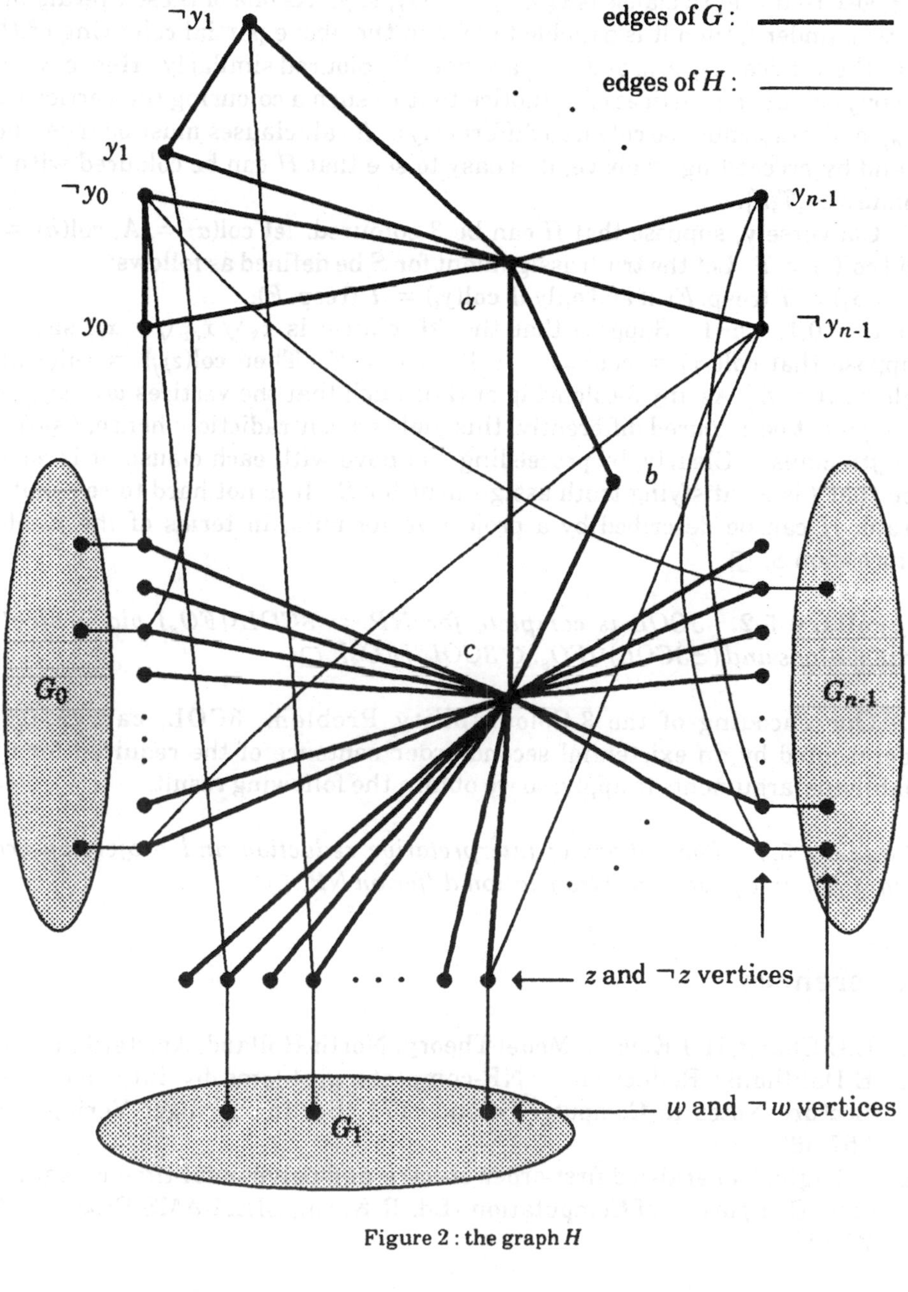

Figure 2 : the graph H

Suppose that t is a satisfying truth assignment for the Boolean expression S (over the Boolean variables $\{x_i : i = 0,1,\ldots,n-1\}$). Then we colour the vertices $\{y_i, \neg y_i, a, b, c : i = 0,1,\ldots,n-1\}$ of H with the colours $\{A,T,F\}$ as follows:

$\mathrm{col}(y_i) = T$ (resp. F) iff $t(x_i) = T$ (resp. F);

$\mathrm{col}(\neg y_i) = T$ (resp. F) iff $t(x_i) = F$ (resp. T);

$\mathrm{col}(a) = A$; $\mathrm{col}(b) = F$; $\mathrm{col}(c) = T$.

Suppose that the jth clause is $x_k \vee x_m \vee \neg x_r$, say. As one of these literals must be true under t, then it is possible to extend the above partial colouring of H so that the vertices z_{kj}, z_{mj}, and $\neg z_{rj}$ are not all coloured similarly. Hence, we can clearly colour the vertices of G_j (notice that in such a colouring the vertices w_{kj}, w_{mj}, and $\neg w_{rj}$ must be coloured differently). As all clauses must be true under t, and by proceeding as above, it is easy to see that H can be coloured with the colours $\{A,T,F\}$.

Conversely, suppose that H can be 3-coloured: let $\mathrm{col}(a) = A$, $\mathrm{col}(b) = F$, and $\mathrm{col}(c) = T$. Let the truth assignment for S be defined as follows:

$t(x_i) = T$ (resp. F) if and only if $\mathrm{col}(y_i) = T$ (resp. F),

for $i = 0,1,\ldots,n-1$. Suppose that the jth clause is $x_k \vee x_m \vee \neg x_r$, say, and suppose that $\mathrm{col}(y_k) = \mathrm{col}(y_m) = \mathrm{col}(\neg y_r) = F$. Then $\mathrm{col}(z_{kj}) = \mathrm{col}(z_{mj}) = \mathrm{col}(\neg z_{rj}) = A$. As any 3-colouring of G_j is such that the vertices w_{kj}, w_{mj}, and $\neg w_{rj}$ must be coloured differently, this yields a contradiction: hence, t satisfies the jth clause. Clearly, by proceeding as above with each clause, it is easy to see that t is a satisfying truth assignment for S. It is not hard to see that the graph H can be described by a projective formula, in terms of the Boolean expression S. □

Corollary 5.2. *3COL is complete for NP = 3COL1[FO$_s$] via projection translations and (±3COL)1[FO$_s$] ⊆ 3COL2[FO$_s$].* □

Our encoding of the 3-Colourability Problem, 3COL, can easily be represented by an existential second-order sentence of the required form for Dahlhaus' arguments to apply, so we obtain the following result.

Corollary 5.3. *The notions of interpretative reduction and projection (resp. quantifier-free first-order) translation differ on NP.* □

References

1. C.C.Chang, H.J.Keisler: Model Theory, North-Holland, Amsterdam, 1977.
2. E.Dahlhaus: Reduction to NP-complete problems by interpretations, Lecture Notes in Computer Science 171, Berlin: Springer-Verlag, 1979, 357-365.
3. R.Fagin: Generalized first-order spectra and polynomial-time recognizable sets, Complexity of Computation (Ed. R.Karp), SIAM-AMS Proc., 7, 1974, 27-41.

4. M.R.Garey and D.S.Johnson: Computers and intractability, W.H.Freeman and Co., San Francisco, 1979.
5. Y.Gurevich: Toward logic tailored for computational complexity, Computation and Proof Theory (Ed. E. Börger et al.), Lecture Notes in Mathematics 1104, Berlin: Springer-Verlag, 1984, 175-216.
6. Y.Gurevich: Logic and the challenge of computer science, in: Current Trends in Theoretical Computer Science (Ed. E. Börger), Computer Science Press, 1987, 1-57.
7. N.Immerman: Languages that capture complexity classes, SIAM J. Comput., 16, 4, 1987, 760-778.
8. N.Immerman: Nondeterministic space is closed under complementation, SIAM J. Comput., 17, 5, 1988, 935-938.
9. R.M.Karp, Reducibility among combinatorial problems, Complexity of computer computations (Ed. R.E.Miller and J.W.Thatcher), Plenum Press, New York, 1972, 85-103.
10. L.Lovasz and P.Gacs: Some remarks on generalized spectra, Zeitschr. f. math., Logik und Grundlagen d. Math., 23, 1977, 547-554.
11. U.Manber: Introduction to Algorithms: a creative approach, Addison-Wesley, Reading, Mass. 1989.
12. I.A.Stewart: The demise of the Turing machine in complexity theory, Proc. TURING 1990: A colloquium in celebration of Alan Turing, Brighton, England, April, 1990, Oxford Univ. Press, to appear.
13. I.A.Stewart: Using the Hamiltonian operator to capture NP, to appear, J. Comput. System Sci..
14. I.A.Stewart: Comparing the expressibility of languages formed using NP-complete operators, J. Logic Computat., 1, 3, 1991, 305-330.
15. I.A.Stewart: Complete problems for symmetric logspace involving free groups, Inform. Process. Lett., 40, 1991, 263-267.

Control of ω-Automata, Church's Problem, and the Emptiness Problem for Tree ω-Automata*

J.G. Thistle

Visitor
Department of Electrical Engineering
and Computer Sciences
University of California
Berkeley, CA 94720 U.S.A.

W.M. Wonham

Systems Control Group
Department of Electrical Engineering
University of Toronto
Toronto, Ontario
Canada M5S 1A4

Abstract. Church's problem and the emptiness problem for Rabin automata on infinite trees, which represent basic paradigms for program synthesis and logical decision procedures, are formulated as a control problem for automata on infinite strings. The alphabet of an automaton is interpreted not as a set of input symbols giving rise to state transitions but rather as a set of output symbols generated during spontaneous state transitions; in addition, it is assumed that automata can be "controlled" through the imposition of certain allowable restrictions on the set of symbols that may be generated at a given instant. The problems in question are then recast as that of deciding membership in a deterministic Rabin automaton's *controllability subset* – the set of states from which the automaton can be controlled to the satisfaction of its own acceptance condition. The new formulation leads to a direct, efficient and natural solution based on a fixpoint representation of the controllability subset. This approach combines advantages of earlier solutions and admits useful extensions incorporating liveness assumptions.

1 Introduction

This article summarizes recent research linking control synthesis with problems in program synthesis and logical decision procedures. In particular, it discusses a novel, control-theoretic formulation of *Church's problem* and *the emptiness problem for automata on infinite trees*.

Church's problem is a basic theoretical paradigm for program synthesis. Proposed in [Chu63] (see also [TB73, Tra86]), it concerns the construction of a string transducer whose infinite behaviour satisfies a given logical formula. Suitable logical languages include Büchi's sequential calculus (SC) [McN90] – also known as the second-order monadic theory of one successor function (S1S) – and modal program logics such as linear-time propositional temporal logic [PR89a, PR89b].

Solutions to Church's problem have arisen from the study of decision problems for the associated logical languages. Of particular importance in this regard have been

*This work was supported by the Natural Sciences and Engineering Research Council of Canada under Grant number OGP0007399, a Postgraduate Scholarship and a Postdoctoral Fellowship, and through S.T.A.R. Awards and Open Fellowships from the University of Toronto and College Fellowships from Trinity College in the University of Toronto. The report was prepared while the first author was visiting the Department of Engineering of the University of Cambridge.

decision procedures based on automata on infinite objects, also known as *ω-automata* [Tho90, Eme90]. Büchi's [Büc62, Büc90] decision procedure for the satisfiability of S1S formulas included a construction which, given any S1S formula, produced a finite ω-automaton on infinite strings that accepted exactly those strings representing models for the formula; the satisfiability problem was thus reduced to the decidable "emptiness problem" for (Büchi) automata on infinite strings. Büchi and Landweber later used Büchi's construction to reformulate Church's problem in purely automata-theoretic terms, then solved it by reduction to the equivalent problem of constructing winning strategies for zero-sum, two-player, ω-regular games of perfect information [Lan69, BL69] (see also [Dil89]). More recent approaches to Church's problem have retained the automata-theoretic formulation but have produced simpler solutions through a reduction to the emptiness problem for automata on infinite trees [Tho90].

Introduced by Rabin [Rab69, Rab70] in the course of generalizing Büchi's S1S decision procedure to S2S (the second order monadic theory of two successor functions, or the theory of infinite binary trees), automata on infinite trees generalize Büchi's automata on infinite strings. Rabin showed that given any S2S formula, an automaton could be constructed that accepted exactly those trees representing models of the formula. This reduced the satisfiability problem to the emptiness problem for automata on infinite trees, which was then shown to be decidable. Rabin also solved Church's problem by constructing an automaton that accepted exactly those trees representing suitable transducers [Rab72]. Indeed, reduction to the emptiness problem for automata on infinite trees has remained an important technique in both program synthesis [PR89a, PR89b] and logical decision procedures [Str81, Eme90].

Because of these applications, the emptiness problem has attracted considerable study [Tho90]. A key result of Rabin's [Rab69, Rab70, Rab72] and Hossley and Rackoff's [HR72] early solutions is the *finite model theorem*, which states roughly that a finite automaton accepts some infinite tree only if it accepts some infinite tree having a finite representation; moreover, if there exists an accepted tree then a finite representation of an accepted tree can be effectively constructed. Emerson showed that this led to a *small model theorem* stating that a Rabin tree ω-automaton accepts some tree only if it accepts a tree obtained by "unwinding" some graph embedded in its own transition structure [Eme85]. (For a related result pertaining to general Muller automata see [GH82].)

More recently Emerson and Jutla [EJ88] and Pnueli and Rosner [PR89a] have obtained solutions to the emptiness problem that are essentially optimal in their computational complexity. Applying the small model theorem, Emerson and Jutla expressed the acceptance condition as a μ-calculus formula and checked for the existence of a model of the formula embedded in the automaton's transition structure; Pnueli and Rosner used a refinement of the method of Hossley and Rackoff.

This article discusses a new version of the emptiness problem that arises from research on supervisory control of infinite behaviour of discrete-event systems [Thi91]. Based on an interpretation of string ω-automata as *generators* of output strings, rather than recognizers of inputs, the new problem formulation concerns the "control" of such automata through the imposition of certain allowable restrictions on the set of output symbols that may be generated at a given time. Within this framework the emptiness problem for tree ω-automata becomes that of determining whether a string ω-automaton can be controlled to the satisfaction of its own acceptance condition.

More precisely, the emptiness problem is formally equivalent to that of deciding membership in a deterministic string ω-automaton's ***controllability subset*** – the set of states from which the automaton can be controlled to the satisfaction of its acceptance condition. The key result of the report is the representation of the controllability subset as a certain fixpoint of a simple ***inverse dynamics*** operator. The fixpoint characterization allows for straightforward computation that matches the essentially optimal upper bound on computational complexity of [EJ88, PR89a]. Moreover, intermediate results of the calculation allow the construction of a ***state feedback*** control strategy – one in which the choice of control action depends only on the current state of the automaton. This provides a new version of the small model theorem that, unlike Emerson's original proof [Eme85], is directly constructive.

In its use of structural induction and a fixpoint characterization, the approach presented here synthesizes techniques of Rabin [Rab72] and Emerson and Jutla [EJ88]; the resulting solution is computationally more efficient than that of [Rab72] and mathematically more direct than that of [EJ88]. In particular, the new approach does not depend on the small model theorem but rather yields it as a corollary.

Perhaps more importantly, the control formulation is sufficiently natural that it allows easy extension. For instance, the method can be generalized to allow for liveness assumptions on the automaton's behaviour: this article outlines the extension to liveness assumptions represented by Büchi acceptance conditions; assumptions of more general forms will be dealt with in future reports. The incorporation of such liveness assumptions solves one instance of an extension of Church's problem that was posed, but not constructively solved, in [ALW89]; it also has potential applications to the satisfiability of logical formulas in, for example, fair models of concurrent computation [CVW86]. The techniques of [HR72, PR89a] do not appear to admit such assumptions.

The control-theoretic problem formulation is presented in section 2. Section 3 then discusses some fixpoint preliminaries. In section 4 we introduce operations on automata based on those of [Rab72] – these simplify induction on the complexity of the acceptance condition [EJ88]. The key ***inverse dynamics*** and ***p-reachability*** operators are introduced in section 5. In section 6 the controllability subset is characterized in terms of fixpoints of these operators. The state feedback result follows from this fixpoint representation. It is then shown in section 7 that straightforward computation of the fixpoint matches the essentially optimal upper bounds established in [EJ88, PR89a]. The results are generalized to the case of Büchi liveness assumptions in section 8. Finally, section 9 presents some conclusions. Owing to lack of space, details of proofs are omitted from the present report; these appear in the thesis [Thi91] and will be published in the control literature.

2 Controllability subsets of Rabin automata

Before defining Rabin automata and their controllability subsets we establish some notation for formal languages.

Σ^* denotes the set of all finite strings (including the *empty string*, denoted by 1, having length 0) over some finite symbol alphabet Σ; Σ^ω denotes the set of (countably) infinite strings over Σ; Σ^∞ denotes $\Sigma^* \cup \Sigma^\omega$. A *language* is any set of strings over Σ; in particular, an *ω-language* is a subset of Σ^ω.

A finite string $k \in \Sigma^*$ is a *prefix* of $v \in \Sigma^\infty$ if k is an initial substring of v – in this case we write $k \leq v$. For any string $v \in \Sigma^\infty$ we let $\text{pre}(v)$ denote its set of (finite) prefixes; that is, $\text{pre}(v) := \{k \in \Sigma^* : k \leq v\}$.

A *Rabin automaton* [McN66, Rab72] is a 5-tuple

$$\mathcal{A} = (\Sigma, X, \delta, x_0, \{(R_p, I_p) : p \in P\})$$

consisting of:

a finite *alphabet* Σ,

a finite *state set* X,

a *transition function* $\delta : \Sigma \times X \longrightarrow 2^X$,

an *initial state* x_0,

and a *family of accepting pairs* $(R_p, I_p) \in 2^X \times 2^X$ with index set P.

It is convenient to extend δ to a map $\delta : \Sigma^* \longrightarrow 2^X$ according to

$$(1, x) \overset{\delta}{\mapsto} x$$

$$(k\sigma, x) \overset{\delta}{\mapsto} \bigcup \{\delta(\sigma, x_k) : x_k \in \delta(k, x)\}, \ \forall\, k \in \Sigma^*, \ \sigma \in \Sigma.$$

A *path* on $\mathcal{A}_M$ of a string $v \in \Sigma^\infty$ is a total function $\pi : \text{pre}(v) \longrightarrow X$ such that

$$\pi(1) = x_0 \ \& \ \forall\, k \in \text{pre}(v), \ \sigma \in \Sigma \ : \ k\sigma \in \text{pre}(s) \Rightarrow \pi(k\sigma) \in \delta(\sigma, \pi(k))$$

Thus a path determines a state sequence consistent with the form of the string and the transition structure of the automaton.

We shall sometimes wish to discuss "paths" that do not begin with the initial state (i.e. for which $\pi(1) \neq x_0$). For this we define $\mathcal{A}_x$ to be the automaton obtained by designating $x \in X$ the initial state.

In keeping with our interpretation of automata as generators we say that a string $v \in \Sigma^\infty$ is *generated* by $\mathcal{A}$ if v has a path on $\mathcal{A}$.

The *recurrence set* of a path $\pi : \text{pre}(s) \longrightarrow X$ is

$$\Omega_\pi := \{x \in X : |\pi^{-1}(x)| = \omega\}$$

In other words, the recurrence set is the set of states entered infinitely often on the given path. The recurrence set Ω_π is nonempty if and only if the string s is infinite.

For our purposes it is convenient to adopt a slight modification of the standard definition of acceptance. A path π is *accepted* if

$$\exists p \in P \ : \ \Omega_\pi \cap R_p \neq \emptyset \ \& \ \Omega_\pi \subseteq I_p$$

Such paths represent infinite state sequences along which R_p is "continually recurrent" and I_p is "eventually invariant," for some $p \in P$. (Usual practice is to specify in place of I_p its complement, say $\overline{I_p}$, and require the equivalent condition that $\Omega_\pi \cap \overline{I_p} = \emptyset$ – in other words, that $\overline{I_p}$ be "finitely recurrent.")

The ω-language *accepted* by $\mathcal{A}$ is the set of all (infinite) strings over Σ that have accepted paths on $\mathcal{A}$.

We say that $\mathcal{A}$ is *deterministic* if $|\delta(\sigma, x)| \leq 1$, for all $\sigma \in \Sigma, x \in X$. In this case we represent the transition function as a partial function $\delta : \Sigma^* \times X \longrightarrow X$, writing $\delta(k, x)!$ to signify that the function is defined for the particular argument (k, x). If $\mathcal{A}$ is deterministic then every string has at most one path π on $\mathcal{A}$.

To introduce a control feature we assume that a family $\mathbf{C} \subseteq 2^\Sigma$ of *control patterns* is given, representing subsets of the alphabet to which we can restrict, at any point in the operation of the automaton, the set of symbols that it may generate. Control strategies can be represented as "feedback maps"

$$f : \Sigma^* \longrightarrow \mathbf{C}$$

which can be interpreted as associating with the sequence of all past symbols generated by the automaton a corresponding control action. In keeping with this interpretation, we say that $v \in \Sigma^\infty$ is generated by $\mathcal{A}_x$ *under* f if v is generated by $\mathcal{A}_x$ and for all $k\sigma \in \text{pre}(v)$, k is generated by $\mathcal{A}_x$, $f(k)$ is defined and $\sigma \in f(k)$.

We can now define the state subset from which the infinite behaviour of an automaton can be controlled to the satisfaction of its acceptance condition:

Definition 2.1

For any Rabin automaton $\mathcal{A} = (\Sigma, X, \delta, x_0, \{(R_p, I_p) : p \in P\})$, define $F^\mathcal{A} \subseteq X$ as the set of all $x \in X$ for which there exists a map $f : \Sigma^* \longrightarrow \mathbf{C}$ such that:

i. every $s \in \Sigma^\omega$ generated by $\mathcal{A}_x$ under f is accepted by $\mathcal{A}_x$; and

ii. for any $k \in \Sigma^*$ generated by $\mathcal{A}_x$ under f, there exists $\sigma \in \Sigma$ such that $k\sigma$ is generated by $\mathcal{A}_x$ under f.

□

Clause i. captures the notion that f should control $\mathcal{A}$ to the satisfaction of its acceptance condition. Clause ii. rules out trivial solutions by requiring that every finite string generated by $\mathcal{A}_x$ under f have proper extensions that are also generated by $\mathcal{A}_x$ under f; for deterministic $\mathcal{A}$, this means that the control strategy represented by f must prevent deadlocks. The reader may readily verify that the Rabin tree ω-automaton emptiness problem [Rab72, Tho90] reduces in polynomial time to that of deciding membership in $F^\mathcal{A}$ for a deterministic Rabin string ω-automaton $\mathcal{A}$, and vice versa.

We call $F^\mathcal{A}$ the *controllability subset* of $\mathcal{A}$. While straightforward and intelligible from an intuitive standpoint, the above definition is of limited mathematical usefulness. The main results of the article establish an alternative representation of the controllability subset of a deterministic Rabin automaton as a certain fixpoint of an operator that depends in a simple manner on the transition structure of the automaton and the family of control patterns. This new characterization allows both efficient computation of the controllability subset and effective synthesis of appropriate controls.

3 Preliminaries: Fixpoints of Monotone Operators

This section reviews basic results of fixpoint theory as they apply to the methods of the article. Our presentation of these preliminaries is adapted from that of [EL86].

A k-ary *operator* on a power set 2^X is a map $f : (2^X)^k \to 2^X$.

An operator is *monotone* if it preserves inclusion; that is,

$$X_i \subseteq X_i' \Longrightarrow f(X_1, \ldots, X_i, \ldots X_k) \subseteq f(X_1, \ldots, X_i', \ldots X_k)\ ,\ 1 \le i \le k\ .$$

An operator is $\cup$-*continuous* if, for any i, $1 \le i \le k$, and any nondecreasing sequence $X_i^0 \subseteq X_i^1 \subseteq X_i^2 \ldots$,

$$\bigcup_{j=0}^{\infty} f(X_1, \ldots, X_i^j, \ldots, X_k) = f(X_1, \ldots, \bigcup_{j=0}^{\infty} X_i^j, \ldots, X_k)$$

An operator is $\cap$-*continuous* if, for any i, $1 \le i \le k$ and any nonincreasing sequence $X_i^0 \supseteq X_i^1 \supseteq X_i^2 \ldots$,

$$\bigcap_{j=0}^{\infty} f(X_1, \ldots, X_i^j, \ldots, X_k) = f(X_1, \ldots, \bigcap_{j=0}^{\infty} X_i^j, \ldots, X_k)$$

Both $\cup$-continuity and $\cap$-continuity imply monotonicity; for operators on finite power sets the reverse implications hold.

It is convenient to introduce a special notation for extremal fixpoints of monotone operators. The expressions of this calculus are made up of monotone operators applied to subsets, and the "fixpoint quantifiers" μ and ν (see, e.g., [EL86]). Expressions of the form

$$\mu Y.\ \phi(Y) \quad (\text{resp. } \nu Y.\ \phi(Y))$$

represent the least (resp. greatest) $Y \subseteq X$ such that $Y = \phi(Y)$ – in other words the least (resp. greatest) fixpoint of the operator that maps every $Y \subseteq X$ to $\phi(Y)$. The existence of such fixpoints follows from the lattice-theoretic results of Tarski and Knaster[Tar55]:

Theorem 3.1 (Tarski-Knaster)

Let $f : 2^X \to 2^X$ be a monotone operator on X. Then f has least and greatest fixpoints; in fact,

(i) $\mu Y.\ f(Y) = \bigcap\{Y' \subseteq X :\ Y' = f(Y')\} = \bigcap\{Y' \subseteq X :\ Y' \supseteq f(Y')\}$

(i′) $\nu Y.\ f(Y) = \bigcup\{Y' \subseteq X :\ Y' = f(Y')\} = \bigcup\{Y' \subseteq X :\ Y' \subseteq f(Y')\}$

(ii) If f is $\cup$-continuous then $\mu Y.\ f(Y) = \bigcup_{i=0}^{\infty} f^i(\emptyset)$.

(ii′) If f is $\cap$-continuous then $\nu Y.\ f(Y) = \bigcap_{i=0}^{\infty} f^i(X)$.

(where f^i denotes the i-fold composition of f with itself).

□

4 Automaton structure

The recursive form of our fixpoint characterization of the controllability subset and the inductive nature of some of our proofs necessitate a precise notion of the structural complexity of automata and methods of converting complex automata to simpler ones. One useful measure of structural complexity is the number of pairs in the acceptance condition; another is the number of "live" states as defined by Rabin [Rab72].

Let $\mathcal{A} = (\Sigma, X, \delta, x_0, \{(R_p, I_p) : p \in P\})$ be a Rabin automaton. The set of *live* states of $\mathcal{A}$ is given by[1]

$$L(\mathcal{A}) := \{x \in X : (\exists \sigma \in \Sigma)\ [\delta(\sigma, x)! \ \&\ \delta(\sigma, x) \neq x]\}$$

where $\delta(\sigma, x)!$ signifies that δ is defined for (σ, x). In other words, a state is live if other states can be reached from it. An approximate opposite to liveness is "degeneracy." A state $x \in X$ is *degenerate* if there are transitions leaving x but all of them simply lead back to x; more precisely, the set of degenerate states of $\mathcal{A}$ is given by

$$D(\mathcal{A}) := \{x \in X : (\exists \sigma \in \Sigma)\ [\delta(\sigma, x)!] \ \&\ (\forall \sigma \in \Sigma)\ [\delta(\sigma, x)! \Longrightarrow \delta(\sigma, x) = x]\}$$

For any $X' \subseteq X$ and $p' \in P$ the following operations[2] on the Rabin automaton $\mathcal{A} = (\Sigma, X, \delta, x_0, \{(R_p, I_p) : p \in P\})$ potentially reduce its structural complexity as measured by $|L(\mathcal{A})|$ and $|P|$:

self-looping of a subset: $\mathcal{A}(\hookrightarrow X') := (\Sigma, X, \delta', x_0, \{(R'_p, I'_p) : p \in P\})$, where

$$\delta'(\sigma, x) = \begin{cases} x & \text{if } x \in X' \\ \delta(\sigma, x) & \text{otherwise} \end{cases} \quad (\forall \sigma \in \Sigma,\ x \in X)$$
$$\&\ R'_p = R_p \cup X' \ \&\ I'_p = I_p \cup X' \quad (\forall p \in P)$$

restriction to a subset: $\mathcal{A} \restriction X' := (\Sigma, X, \delta', x_0, \{(R'_p, I'_p) : p \in P\})$, where

$$\delta'(\sigma, x) = \begin{cases} \delta(\sigma, x) & \text{if } x \in X' \\ x & \text{otherwise} \end{cases} \quad (\forall \sigma \in \Sigma,\ x \in X)$$
$$\&\ R'_p = R_p \cap X' \ \&\ I'_p = I_p \cap X' \quad (\forall p \in P)$$

exclusion of a pair: Let $\mathcal{A} \restriction (I_{p'} \cup D(\mathcal{A})) = (\Sigma, X, \delta', x_0, \{(R'_q, I'_q) : q \in Q\})$. Then

$$\mathcal{A} \downarrow p' := \begin{cases} (\Sigma, X, \delta', x_0, \{(R'_q, I'_q) : q \in Q\}) & \text{if } |P| = 1 \\ (\Sigma, X, \delta', x_0, \{(R'_q, I'_q) : q \in Q \setminus \{p'\}\}) & \text{if } |P| > 1 \end{cases}$$

Self-looping of a subset $X' \subseteq X$ turns every $x \in X'$ into a degenerate state and ensures that the singleton $\{x\}$ satisfies the acceptance criterion. On the other hand, restriction to the subset $X' \subseteq X$ turns all other states into degenerate states that do *not* satisfy the acceptance condition. Finally, exclusion of a pair indexed by $p' \in P$ restricts the automaton to the subset $I_{p'} \cup D(\mathcal{A})$ and, provided $|P| > 1$, eliminates the pair indexed by p'. All three of these operations potentially reduce the number of live states while the third potentially reduces the number of pairs in the acceptance condition.

---

[1] Rabin excludes the initial state from the set of live states.

[2] "Self-looping" and "restriction" are based on similar operations defined in [Rab72].

5 Inverse dynamics and p-reachability operators

We shall characterize $F^{\mathcal{A}} \subseteq X$ as a certain fixpoint of the following monotone operator:

Definition 5.1
Let $\mathcal{A} = (\Sigma, X, \delta, x_0, \{(R_p, I_p) : p \in P\})$ be a deterministic Rabin automaton. Its *inverse dynamics operator* is given by

$$\theta^{\mathcal{A}} : 2^X \longrightarrow 2^X$$
$$X' \mapsto \{x \in X : (\exists \Gamma \in \mathbf{C})[(\exists \sigma \in \Gamma)[\delta(\sigma, x)!] \;\&\; (\forall \sigma \in \Gamma)[\delta(\sigma, x)! \Rightarrow \delta(\sigma, x) \in X']]\}$$

□

For any $X' \subseteq X$, $\theta^{\mathcal{A}}(X')$ is the set of all states in which the automaton can be controlled so that its next state will belong to X'. In other words, $\theta^{\mathcal{A}}(X')$ is the "weakest precondition" that ensures that $\mathcal{A}$ can be controlled to enter X'.

A number of state subsets having significance for control can be characterized as extremal fixpoints involving the inverse dynamics operator. For instance, the fixpoint $\mu X_2.\ \theta^{\mathcal{A}}(X_1 \cup X_2)$ represents the state subset from which $\mathcal{A}$ can be controlled eventually to enter $X_1 \subseteq X$ (see Theorem 3.1 (ii)); for any I_p, the fixpoint $\mu X_2.\ [\theta^{\mathcal{A}}(X_1 \cup X_2) \cap I_p]$ represents the subset from which $\mathcal{A}$ can be controlled to enter X_1 by way of a path that lies within I_p (again, see Theorem 3.1 (ii)). It is convenient to generalize this second fixpoint by defining the following operators:

Definition 5.2
Let $\mathcal{A} = (\Sigma, X, \delta, x_0, \{(R_p, I_p) : p \in P\})$ be a Rabin automaton. For any $p \in P$, the *p-reachability operator* of $\mathcal{A}$ is given by

$$\rho_p^{\mathcal{A}} : (2^X)^2 \longrightarrow 2^X$$
$$(X_1, X_2) \mapsto \mu X_3.\ [\theta^{\mathcal{A}}(X_1) \cup [\theta^{\mathcal{A}}(X_1 \cup X_2 \cup X_3) \cap I_p]]$$

□

The reader may readily verify that $\rho_p^{\mathcal{A}}(X_1, X_2)$ is the set of states from which $\mathcal{A}$ can be controlled to reach X_2 by way of a path that lies within $I_p \subseteq X$ *or*, failing that, to reach X_1 (and to do so in at most one state transition after leaving I_p).

6 Fixpoint characterization of $F^{\mathcal{A}}$

The above p-reachability operators admit a concise representation of $F^{\mathcal{A}}$:

Definition 6.1
Let $\mathcal{A} = (\Sigma, X, \delta, x_0, \{(R_p, I_p) : p \in P\})$ be a deterministic Rabin automaton. Then $C^{\mathcal{A}} \subseteq X$ is given by[3]

$$C^{\mathcal{A}} := \begin{cases} \mu X_1.\ \nu X_2.\ \rho_p^{\mathcal{A}}(X_1, X_2 \cap R_p)\ , & \text{if } |P| = 1, \\ \mu X_1.\ [\bigcup_{p \in P}\ \nu X_2.\ \rho_p^{\mathcal{A}}(X_1, C^{\mathcal{A}(\hookrightarrow X_1 \cup (X_2 \cap R_p))\downarrow p})], & \text{if } |P| > 1. \end{cases}$$

□

[3]See [Thi91] for a proof of the existence of $C^{\mathcal{A}}$.

Consider the case where $|P| = 1$. By Theorem 3.1 (ii), $C^{\mathcal{A}}$ is the least upper bound of the nondecreasing sequence of subsets $C_0 \subseteq C_1 \subseteq C_2 \subseteq \ \ldots$ given by

$$\begin{aligned} C_0 &:= \emptyset \\ C_{i+1} &:= \nu X_2.\ \rho_p^{\mathcal{A}}(C_i, X_2 \cap R) \end{aligned}$$

Intuitively, $C_1 \subseteq X$ is the largest $X_2 \subseteq X$ from which the automaton can be controlled to reach $X_2 \cap R_p$ by way of a path that lies within I_p (see Theorem 3.1 (i′)) – in other words, C_1 is the set of states from which the automaton can be controlled to remain forever within I_p and to enter R_p infinitely often. By induction, C_i represents the subset from which the automaton can be controlled so that it enters R_p infinitely often and enters $X \setminus I_p$ fewer than i times. Thus $C^{\mathcal{A}}$ is indeed the set of states from which $\mathcal{A}$ can be controlled so that along any state trajectory R_p is "continually recurrent" and I_p is "eventually invariant."

If $|P| > 1$, $C^{\mathcal{A}}$ is the least upper bound of the nondecreasing sequence $C_0 \subseteq C_1 \subseteq C_2 \subseteq \ \ldots$ given by

$$\begin{aligned} C_0 &:= \emptyset \\ C_{i+1} &:= \bigcup_{p \in P} \nu X_2.\ \rho_p^{\mathcal{A}}(C_i, C^{\mathcal{A}(\hookrightarrow C_i \cup (X_2 \cap R_p))\downarrow p}) \end{aligned}$$

Thus $C_1 \subseteq X$ is the set of states from which, for some $p \in P$, $\mathcal{A}$ can be controlled to remain forever within I_p and either enter R_p infinitely often or satisfy the acceptance condition obtained by excluding the pair (R_p, I_p) – or in other words, to remain forever within I_p and satisfy its own acceptance condition. By induction, C_i is the subset in which a sequence $p_i, p_{i-1}, p_{i-2}, \ \ldots \ , p_1$ of i elements of P can be inductively chosen in such a way that if p is the latest element to have been selected then $\mathcal{A}$ can be controlled so that it either remains within I_p forever and satisfies its acceptance condition or else, provided $i > 1$, it eventually reaches a state in which a new element in the sequence can be chosen.

As this interpretation suggests, $C^{\mathcal{A}}$ coincides with $F^{\mathcal{A}}$:

Theorem 6.2

For any deterministic Rabin automaton $\mathcal{A}$, $\quad F^{\mathcal{A}} = C^{\mathcal{A}}$.

□

The inclusion $F^{\mathcal{A}} \subseteq C^{\mathcal{A}}$ can be proved by induction on the number of live states of $\mathcal{A}$, in the manner of [Rab72]. For the reverse inclusion we consider every least fixpoint as the limit of a nondecreasing sequence of state subsets, by Theorem 3.1. Membership in these subsets is used to define a well-founded partial ordering or "ranking" of states: the earlier a state occurs in the sequence of subsets, the lower its rank. The theorem is proved through arguments concerning the effect of a suitable control pattern on various ranks of the system state; these typically show that under appropriate conditions a particular rank is either nonincreasing or strictly decreasing. The proof yields the following corollary, which constitutes a new version of the small model theorem. In control-theoretic terms, it states that the automaton can be controlled to satisfy its acceptance condition only if it can be so controlled by means of "state feedback" – that is, by means of a feedback map that factors through the transition function, so that the selection of control patterns depends only on the current state of the automaton.

Corollary 6.3 (State feedback)
Let $\mathcal{A} = (\Sigma, X, \delta, x_0, \{(R_p, I_p) : p \in P\})$ be a deterministic Rabin automaton. Then there exists a total map $\phi^{\mathcal{A}} : C^{\mathcal{A}} \longrightarrow \mathbf{C}$ such that for any $x \in C^{\mathcal{A}}$ we have

i. for any $s \in \Sigma^\omega$ and any path $\pi : \text{pre}(s) \longrightarrow X$ of s on $\mathcal{A}_x$, the condition $\forall k\sigma \in \text{pre}(s) : \sigma \in \phi^{\mathcal{A}}(\pi(k))$ implies $\exists p \in P : \Omega_\pi \cap R_p \neq \emptyset \;\&\; \Omega_\pi \subseteq I_p$, and

ii. $\exists \sigma \in \phi^{\mathcal{A}}(x) : \delta(\sigma, x)!$ and $\forall \sigma' \in \phi^{\mathcal{A}}(x) : \delta(\sigma', x)! \implies \delta(\sigma', x) \in C^{\mathcal{A}}$.

□

7 The complexity of computing controllability subsets

The controllability subset of a deterministic Rabin automaton $\mathcal{A}$ can be computed, in the manner suggested by Theorem 3.1 (ii) and (iii), in time $\bigcirc(kl(mn)^{3m})$, where k is the number of control patterns, l is the size of the alphabet, m is the number of state subset pairs in the acceptance condition and n is the number of states. This matches the best known upper bound for the Rabin tree ω-automaton emptiness problem [EJ88, PR89a].

Indeed, the problem of deciding membership in controllability subsets is polynomially equivalent to the tree automaton emptiness problem, and hence NP-complete [EJ88] (in fact, its membership in NP follows immediately from the state feedback corollary 6.3). Thus the complexity of our algorithm, singly exponential in the number of state subset pairs and polynomial in the number of states, can be regarded as essentially optimal [EJ88, PR89a].

8 Liveness assumptions

To demonstrate the extensibility of the above approach we now consider control synthesis under liveness assumptions. In particular we suppose that, regardless of control action, the behaviour of the automaton satisfies a Büchi acceptance condition [Büc62]. As was mentioned in the introduction, the resulting synthesis problem represents one instance of that proposed in [ALW89] and has applications to logical decision problems for fair models of computation [CVW86]. The methods of [HR72, PR89a] do not appear to accommodate liveness assumptions.

Formally, we define a *Rabin-Büchi automaton* to be a 6-tuple

$$(\Sigma, X, \delta, x_0, \{(R_p, I_p) : p \in P\}, R)$$

where the first five components are those of a Rabin automaton and the sixth is a state subset $R \subseteq X$ as employed in a Büchi acceptance criterion [Büc62].

Paths of strings on such automata are defined as in section 2. A string $k \in \Sigma^*$ is said to be generated by $\mathcal{A}$ if it has a path on $\mathcal{A}$; a string $s \in \Sigma^\omega$ is generated by $\mathcal{A}$ if s has a path π on $\mathcal{A}$ such that $\Omega_\pi \cap R \neq \emptyset$. This represents the liveness assumption that all infinite trajectories followed by $\mathcal{A}$ will pass through R infinitely often.

As before, a string $v \in \Sigma^\omega$ is *generated under* a feedback map $f : \Sigma^* \longrightarrow \mathbf{C}$ if $v \in \Sigma^\omega$ is generated by $\mathcal{A}$ and, for all $k\sigma \in \text{pre}(v)$, $f(k)$ is defined and $\sigma \in f(k)$.

The definition of acceptance remains the same; that is, $v \in \Sigma^\omega$ is accepted by $\mathcal{A}$ if it has a path $\pi : \text{pre}(v) \longrightarrow X$ on $\mathcal{A}$ such that

$$\exists p \in P \ : \ \Omega_\pi \cap R_p \neq \emptyset \ \& \ \Omega_\pi \subseteq I_p$$

It will be convenient to assume (without loss of generality) that for all $x \in X$, the ω-language accepted by $\mathcal{A}_x$ is contained in that generated by $\mathcal{A}_x$. When this holds we shall say that $\mathcal{A}$ is *well-posed*.

Definition 8.1

Let $\mathcal{A} = (\Sigma, X, \delta, x_0, \{(R_p, I_p) : p \in P\}, R)$ be a Rabin-Büchi automaton. Define $F^{\mathcal{A}} \subseteq X$ as the set of all $x \in X$ for which there exists a map $f : \Sigma^* \longrightarrow C$ such that:

i. every $s \in \Sigma^\omega$ generated by $\mathcal{A}_x$ under f is accepted by $\mathcal{A}_x$; and

ii. for any $k \in \Sigma^*$ generated by $\mathcal{A}_x$ under f, there exists $v \in \Sigma^\omega$ such that kv is generated by $\mathcal{A}_x$ under f.

□

Again we call $F^{\mathcal{A}}$ the *controllability subset* of $\mathcal{A}$. $F^{\mathcal{A}}$ intuitively represents the set of states of $\mathcal{A}$ from which the automaton can be controlled in a deadlock-free manner to satisfy its Rabin acceptance criterion, under the assumption that all infinite trajectories followed by $\mathcal{A}$ satisfy the Büchi acceptance criterion.

8.1 Automaton structure

To extend our structural operations on Rabin automata to Rabin-Büchi automata $\mathcal{A} = (\Sigma, X, \delta, x_0, \{(R_p, I_p) : p \in P\}, R)$ it suffices to define their effect on the last component $R \subseteq X$:

self-looping of a subset $X' \subseteq X$ replaces R with $R \cup X'$;

restriction to a subset $X' \subseteq X$ replaces R with $R \cup (X \setminus X')$; and

exclusion of a pair $p \in P$ replaces R with $R \cup (X \setminus (I_p \cup D(\mathcal{A})))$.

These definitions effectively mean that it is possible under our liveness assumption for the automaton to loop forever through any states made degenerate by the operations of self-looping and restriction.

8.2 Inverse dynamics and p-reachability operators

The main difference in the fixpoint characterization of the controllability subset under Büchi liveness assumptions lies in the definition of the "inverse dynamics operator": whereas for Rabin automata this operator maps every state subset X' to the set of states from which the automaton can be controlled to reach X' in a single step, for Rabin-Büchi automata it maps X' to the set of states from which the automaton can be controlled to reach X' immediately upon entering the subset R associated with the Büchi acceptance criterion.

To generalize our inverse dynamics operator we first extend the map $\theta^{\mathcal{A}}$ of previous sections to pairs of subsets and then define a new inverse dynamics operator $\tilde{\theta}^{\mathcal{A}}$ in terms of extremal fixpoints of $\theta^{\mathcal{A}}$.

Definition 8.2
Let $\mathcal{A} = (\Sigma, X, \delta, x_0, \{(R_p, I_p) : p \in P\}, R)$ be a deterministic Rabin-Büchi automaton. Define

$$\begin{aligned}\theta^{\mathcal{A}} : (2^X)^2 &\longrightarrow 2^X \\ (X_1, X_2) &\mapsto \{x \in X : \ (\exists \Gamma \in \mathbf{C})[\ (\exists \sigma \in \Gamma)[\delta(\sigma, x)! \ \& \ \delta(\sigma, x) \in X_1] \\ &\qquad \& (\forall \sigma \in \Gamma)[\delta(\sigma, x)! \Rightarrow \delta(\sigma, x) \in X_1 \cup X_2]\ ]\}\end{aligned}$$

□

Thus the subset $\theta^{\mathcal{A}}(X_1, X_2)$ is the set of all states from which the automaton can be controlled to enter $X_1 \cup X_2$ in a single transition without being prevented from entering X_1.

Definition 8.3
Let $\mathcal{A}$ be a deterministic Rabin-Büchi automaton. Its *inverse dynamics operator* is given by

$$\begin{aligned}\tilde{\theta}^{\mathcal{A}} : \ (2^X)^2 &\longrightarrow 2^X \\ (X_1, X_2) &\mapsto \nu X_3.\ \mu X_4.\ [\theta^{\mathcal{A}}(X_1 \cup (X_4 \setminus R), X_3 \setminus R) \cap X_2]\end{aligned}$$

□

Intuitively, $\tilde{\theta}^{\mathcal{A}}(X_1, X_2)$ is the set of all states from which $\mathcal{A}$ can be controlled to enter X_1 immediately upon entering R – and until it does so to remain within X_2. (Recall the liveness assumption that all infinite trajectories followed by $\mathcal{A}$ enter the subset R infinitely often.)

Definition 8.4
Let $\mathcal{A} = (\Sigma, X, \delta, x_0, \{(R_p, I_p) : p \in P\}, R)$ be a deterministic Rabin-Büchi automaton. For any $p \in P$, the *p-reachability operator* of $\mathcal{A}$ is given by

$$\begin{aligned}\rho_p^{\mathcal{A}} : \ (2^X)^2 &\longrightarrow 2^X \\ (X_1, X_2) &\mapsto \mu X_3.\ [\tilde{\theta}^{\mathcal{A}}(X_1, X) \cup \tilde{\theta}^{\mathcal{A}}(X_1 \cup X_2 \cup X_3, I_p)]\end{aligned}$$

□

Thus $\rho_p^{\mathcal{A}}(X_1, X_2)$ is the state subset from which, under our liveness assumption, $\mathcal{A}$ can be controlled to reach X_2 by way of a path that lies within I_p – or failing that to reach X_1 immediately upon leaving I_p and entering R.

8.3 Fixpoint characterization of the controllability subset

$F^{\mathcal{A}}$ can now be represented by an expression of the same form as in the earlier setting.

Definition 8.5
Let $\mathcal{A} = (\Sigma, X, \delta, x_0, \{(R_p, I_p) : p \in P\}, R)$ be a deterministic Rabin-Büchi automaton. Then $C^{\mathcal{A}} \subseteq X$ is given by[4]

$$C^{\mathcal{A}} := \begin{cases} \mu X_1.\ \nu X_2.\ \rho_p^{\mathcal{A}}(X_1, X_2 \cap R_p)\ , & \text{if } |P| = 1, \\ \mu X_1.\ [\ \bigcup_{p \in P}\ \nu X_2.\ \rho_p^{\mathcal{A}}(X_1, C^{\mathcal{A}(\hookrightarrow X_1 \cup (X_2 \cap R_p))\downarrow p})], & \text{if } |P| > 1. \end{cases}$$

□

The interpretation of the above definition is similar to that of definition 6.1.

Theorem 8.6
For any well-posed deterministic Rabin-Büchi automaton $\mathcal{A}$, $F^{\mathcal{A}} = C^{\mathcal{A}}$. □

Corollary 8.7 (State feedback)
Let $\mathcal{A} = (\Sigma, X, \delta, x_0, \{(R_p, I_p) : p \in P\}, R)$ be a well-posed deterministic Rabin-Büchi automaton. Then there exists a total map $\phi^{\mathcal{A}} : C^{\mathcal{A}} \longrightarrow \mathbf{C}$ such that for any $x \in C^{\mathcal{A}}$ we have

i. for any $s \in \Sigma^\omega$ and any path $\pi : \text{pre}(s) \longrightarrow X$ of s on $\mathcal{A}_x$, the conditions $\forall k\sigma \in \text{pre}(s)\ :\ \sigma \in \phi^{\mathcal{A}}(\pi(k))$ and $\Omega_\pi \cap R \neq \emptyset$ together imply $\exists p \in P\ :\ \Omega_\pi \cap R_p \neq \emptyset\ \&\ \Omega_\pi \subseteq I_p$; and

ii. there exists $t \in \Sigma^\omega$ and a path $\pi : \text{pre}(s) \longrightarrow X$ of t on $\mathcal{A}_x$ such that $\forall k\sigma \in \text{pre}(t)\ :\ \sigma \in \phi^{\mathcal{A}}(\pi(k))$ and $\Omega_\pi \cap R \neq \emptyset$.

□

8.4 Computational complexity

Calculation of the controllability subset of a deterministic Rabin-Büchi automaton in the manner suggested by Theorem 3.1 (ii) and (ii′) satisfies the same upper bound as that given in section 7. This algorithm is therefore essentially optimal.

9 Conclusion

The system-theoretic character of the present control formulation contrasts strongly with the information-processing style of the emptiness problem and its earlier, more combinatorial solutions; the directness and accessibility of the control approach suggest that such system-theoretic methods can shed useful light on dynamical aspects of computer systems and their logics.

Acknowledgments

The authors are grateful to the first author's thesis examiners, and particularly Professors Amir Pnueli, Eric Hehner and Raymond Kwong, for reading a preliminary version of these results.

[4]See [Thi91] for a proof of the existence of this fixpoint.

References

[ALW89] Martín Abadi, Leslie Lamport, and Pierre Wolper. Realizable and unrealizable specifications of reactive systems. In *Automata, Languages and Programming,* 16th International Colloquium, Stresa, Italy, July 1989, Proceedings (Lecture Notes in Computer Science no. 372), pages 1–17. Springer-Verlag, 1989.

[Büc62] J. Richard Büchi. On a decision method in restricted second order arithmetic. In *Logic, Methodology and Philosophy of Science,* Proceedings of the 1960 International Congress, pages 1–11, Stanford, Calif., 1962. Stanford University Press.

[Büc90] J. Richard Büchi. Section 5: Automata and monadic theories. In Saunders MacLane and Dirk Siefkes, editors, *The Collected Works of J. Richard Büchi.* Springer-Verlag, 1990.

[BL69] J. Richard Büchi and Lawrence H. Landweber. Solving sequential conditions by finite-state strategies. *Transactions of the American Mathematical Society,* 138:295–311, 1969.

[Chu63] Alonzo Church. Logic, arithmetic and automata. In *Proceedings of the International Congress of Mathematicians, 15-22 August, 1962,* pages 23–35, Djursholm, Sweden, 1963. Institut Mittag-Leffler.

[CVW86] Constantin Courcoubetis, Moshe Y. Vardi, and Pierre Wolper. Reasoning about fair concurrent programs (extended abstract). In *Symposium on the Theory of Computing,* pages 283–294. ACM, 1986.

[Dil89] David L. Dill. *Trace Theory for Automatic Hierarchical Verification of Speed-Independent Circuits.* ACM Distinguished Dissertations. MIT Press, 1989.

[EJ88] E. Allen Emerson and Charanjit S. Jutla. The complexity of tree automata and logics of programs (extended abstract). In *29th Annual Symposium on Foundations of Computer Science,* pages 328 – 337, 1988.

[EL86] E. Allen Emerson and Chin-Laung Lei. Efficient model checking in fragments of the propositional mu-calculus (extended abstract). In *Symposium on Logic in Computer Science,* pages 267 – 278. IEEE, June 1986.

[Eme85] E. Allen Emerson. Automata, tableaux and temporal logics. In Rohit Parikh, editor, *Logics of Programs,*(Lecture Notes in Computer Science vol. 193), pages 79 – 87. Springer-Verlag, June 1985.

[Eme90] E. Allen Emerson. Temporal and modal logic. In Jan van Leeuwen, editor, *Handbook of Theoretical Computer Science, vol. B: Formal Models and Semantics.* Elsevier, The MIT Press, 1990.

[GH82] Yuri Gurevich and Leo Harrington. Trees, automata and games. In *Symposium on the Theory of Computing,* pages 60–65. ACM, 1982.

[HR72] R. Hossley and C. Rackoff. The emptiness problem for automata on infinite trees. In *Switching and Automata Theory Symposium,* pages 121–124. IEEE, October 1972.

[Lan69] Lawrence H. Landweber. Synthesis algorithms for sequential machines. In *Information Processing 68*, pages 300–304, Amsterdam, 1969. North-Holland.

[McN66] Robert McNaughton. Testing and generating infinite sequences by a finite automaton. *Information and Control*, 9:521–530, 1966.

[McN90] Robert McNaughton. Büchi's sequential calculus. In Saunders MacLane and Dirk Siefkes, editors, *The Collected Works of J. Richard Büchi*, pages 382–397. Springer-Verlag, 1990.

[PR89a] Amir Pnueli and Roni Rosner. On the synthesis of a reactive module. In *Sixteenth Annual Symposium on Principles of Programming Languages*, pages 179–190. Association for Computing Machinery, January 1989.

[PR89b] Amir Pnueli and Roni Rosner. On the synthesis of an asynchronous reactive module. In *Automata, Languages and Programming*, 16th International Colloquium, Stresa, Italy, July 1989, Proceedings (Lecture Notes in Computer Science no. 372), pages 652–671. Association for Computing Machinery, January 1989.

[Rab69] Michael O. Rabin. Decidability of second-order theories and automata on infinite trees. *American Mathematical Society Transactions*, 141:1–35, July 1969.

[Rab70] Michael O. Rabin. Weakly definable relations and special automata. In Y. Bar-Hillel, editor, *Mathematical Logic and Foundations of Set Theory*, pages 1–23. North-Holland, 1970.

[Rab72] Michael O. Rabin. *Automata on Infinite Objects and Church's Problem*. Conference Board of the Mathematical Sciences Regional Conference Series in Mathematics No. 13. American Mathematical Society, Providence, Rhode Island, 1972. Lectures from the CBMS Regional Conference held at Morehouse College, Atlanta, Georgia, September 8 – 12, 1969.

[Str81] R.S. Streett. A propositional dynamic logic of looping and converse. Technical Report TR-263, MIT Laboratory for Computer Science, 1981.

[Tar55] Alfred Tarski. A lattice-theoretical fixpoint theorem and its applications. *Pacific Journal of Mathematics*, 5:285–309, 1955.

[TB73] B.A. Trakhtenbrot and Ya. M. Barzdin'. *Finite Automata: Behavior and Synthesis.* North-Holland, 1973.

[Thi91] John Graham Thistle. *Control of Infinite Behaviour of Discrete-Event Systems.* PhD thesis, University of Toronto, Toronto, Canada, January 1991. Available as Systems Control Group Report No. 9012, Systems Control Group, Department of Electrical Engineering, University of Toronto, January 1991.

[Tho90] Wolfgang Thomas. Automata on infinite objects. In Jan van Leeuwen, editor, *Handbook of Theoretical Computer Science, vol. B: Formal Models and Semantics*. Elsevier, The MIT Press, 1990.

[Tra86] Boris A. Trakhtenbrot. *Selected Developments in Soviet Cybernetics: Finite Automata, Combinational Complexity, Algorithmic Complexity.* Delphic Associates Inc., Falls Church, VA, USA, 1986.

Comparing the Theory of Representations and Constructive Mathematics

A.S.Troelstra*

Faculteit Wiskunde en Informatica
Universiteit van Amsterdam
Plantage Muidergracht 24
1018TV AMSTERDAM (NL)
email: anne@fwi.uva.nl

Abstract

The paper explores the analogy between reducibility statements of Weihrauch's theory of representations and theorems of constructive mathematics which can be reformulated as inclusions between sets. Kleene's function-realizability is the key to understanding of the analogy, and suggests an alternative way of looking at the theory of reducibilities.

1 Introduction

In a series of interesting papers Kreitz and Weihrauch ([KW1, KW2, KW3, W1, W2, WK]) have developed a theory of representations (henceforth "TR" for short), lifting notions of "effectiveness" on Baire space $\mathbb{B}$ to other structures, of cardinality not greater than $\mathbb{B}$, via the notion of "representation" which is analogous to Ershov's notion of numeration for countable structures. TR has been proposed as an alternative to the various forms of constructive mathematics and has been applied in complexity theory (see e.g.[M]). Just as classical recursive mathematics, TR introduces constructivity considerations in a classical setting.

Looking at examples of representations and the reducibility relations between them, one is struck by an obvious informal parallel or analogy between these results and certain well-known facts from constructive mathematics ("CM" for short) in the style of Bishop, Brouwer and Markov. (For general background on CM, and references, see [TD].) In this paper we discuss and explain to some extent this parallelism, and show by means of some examples how results from constructive mathematics may be converted into results on the reducibility of representations,

*I am indebted to K.Weihrauch for discussions and helpful comments on an earlier version of this paper.

and conversely, how many reducibility results in the theory of representations might actually have been obtained by "translating" a theorem of constructive mathematics.

Although we set in this paper some steps on the road towards a more explicit formulation of the analogies, we are far from having demonstrated "equivalence" between TR and some form of CM. In order to make the idea of "equivalence" precise, one would first have to agree on an appropriate and natural formalism for CM and on a suitable formal setting for TR. Even under the assumption that such an agreement can be reached, I do not think it likely that full equivalence results. However, even the exploration of partial and limited analogies between TR and CM can be worthwhile, e.g. because known results from CM may show us what to expect and look for in the case of TR.

In the papers by Kreitz and Weihrauch we encounter many different representations for (classically) the same set of objects. This is reminiscent of early work in intuitionistic mathematics and Markov's constructive mathematics, where many different constructive analogues of the same classical notions are studied. There later developments showed that only very few of the many alternatives were well-behaved mathematically. (Bishop's constructive mathematics seems to have skipped this "proliferation-of-notions" stage.)

Similarly, it is to be expected that on further systematic development of TR, only a few of the many possible representententations for the same set will turn out to have good properties. Here too analogies might be exploited to advantage.

NOTATION. Let α, β, γ, possibly sub-or superscripted, be used for elements of $\mathbb{B}$. We assume that our standard coding of finite sequences of $\mathbb{N}$ is *onto* $\mathbb{N}$. j_1, j_2 are the inverses to a surjective pairing $(\ ,\ ) : \mathbb{N}^2 \to \mathbb{N}$; pairing and its inverses are lifted to $\mathbb{B}$ by $(\alpha, \beta) := \lambda n.(\alpha n, \beta n)$ etc. We write $\bar{\beta}n$ for the (code of) the sequence $\langle \beta 0, \beta 1, \ldots, \beta(n-1)\rangle$, $\bar{\beta}0 \equiv \langle\ \rangle$ (empty sequence). We write $\hat{n}$ as short for $\langle n \rangle$. $*$ denotes concatenation of codes.

$$\begin{aligned} \alpha(\beta) = x \quad &:= \quad \alpha(\bar{\beta}(\min_z[\alpha(\bar{\beta}z) > 0]) = x + 1, \\ \alpha|\beta = \gamma \quad &:= \quad \forall x(\alpha(\hat{x} * \beta) = \gamma x). \end{aligned}$$

$\Phi_\alpha : \mathbb{B} \rightharpoonup \mathbb{N}$ and $\Psi_\alpha : \mathbb{B} \rightharpoonup \mathbb{B}$ are partial continuous functionals given by

$$\begin{aligned} &\Phi_\alpha(\beta) = n := \alpha(\beta) = n, \\ &\Psi_\alpha(\beta) = \gamma := \alpha|\beta = \gamma. \end{aligned}$$

Let $\langle I_n \rangle_n$ be a standard enumeration of intervals with dyadic rational endpoints, such that $I_{(j,m)} := (\nu_D(j) - 2^{-m}, \nu_D(j) + 2^{-m})$ and ν_D is a standard enumeration of the dyadic rationals $k2^{-n}, k \in \mathbb{Z}, n \in \mathbb{N}$. Let $\langle r_n \rangle_n$ be a standard enumeration of $\mathbb{Q}$

2 Representations

DEFINITION. A *representation* of a set M is a partial surjective mapping $\delta : \mathbb{B} \rightharpoonup M$. We use $\delta, \delta', \delta''$ for arbitrary representations.

If δ, δ' are two representations of M, we say that δ is *reducible* to δ' iff there is a partial continuous $\Gamma : \mathbb{B} \rightharpoonup \mathbb{B}$ such that

$$\forall \alpha \in \mathrm{dom}(\delta)(\delta(\alpha) = \delta'(\Gamma\alpha))$$

If Γ is recursive, we say that δ is *computably reducible* to δ' (c-*reducible*). We write $\delta \preceq \delta'$, $\delta \preceq_c \delta'$ for reducibility and c-reducibility respectively.

δ and δ' are *equivalent* (*c-equivalent*) iff $\delta \preceq \delta'$ and $\delta' \preceq \delta$ ($\delta \preceq_c \delta'$ and $\delta' \preceq_c \delta$). Notation: $\delta \equiv \delta'$ ($\delta \equiv_c \delta'$).

Examples of representations

(A) $\alpha \mapsto \Phi_\alpha$, $\alpha \mapsto \Psi_\alpha$ are representations of certain classes $[\mathbb{B} \rightharpoonup \mathbb{N}]$, $[\mathbb{B} \rightharpoonup \mathbb{B}]$ of partial continuous functions.

(B) Let $\mathrm{En}(\alpha) := \{n : \exists m(\alpha m = n+1)\}$, then En is a representation of $\mathrm{P}(\mathbb{N})$, the powerset of $\mathbb{N}$, by enumerations; $\mathrm{En}^c(\alpha) := \{n : \forall m(\alpha m \neq n+1)\}$.

(C) Let $\mathrm{Cf}(\alpha) := \{n : \alpha n = 0\}$, then Cf is a representation of $\mathrm{P}(\mathbb{N})$ by characteristic functions.

(D) We define a representation $\rho_<$ of $\mathbb{R}$ as follows.

$$\begin{aligned} \alpha \in \mathrm{dom}(\rho_<) &:= \exists m \in \mathbb{N} \forall n \in \mathrm{En}(\alpha)(r_n < m) \\ \rho_<(\alpha) &:= \mathrm{l.u.b.}\{r_n : n \in \mathrm{En}(\alpha)\} \end{aligned}$$

(E) Similarly for $\rho_>$, with (least) upper bound replaced by (greatest) lower bound.

(F) Let $\mathbb{Q}_n = \{k2^{-n} : k \in \mathbb{Z}\}$, $n \in \mathbb{N}$, and put

$$\begin{aligned} \alpha \in \mathrm{dom}(\rho_{\mathbb{R}}) &:= \forall n(r_{\alpha n} \in \mathbb{Q}_n \wedge |r_{\alpha n} - r_{\alpha(n+1)}| \leq 2^{-(n+1)}), \\ \rho_{\mathbb{R}}(\alpha) &:= \lim \langle r_{\alpha n} \rangle_n. \end{aligned}$$

(G) A complete, separable metric space $\mathcal{M}$ is a triple $(M, d, \langle p_n \rangle_n)$, where d is a distance function on the set M, and $\langle p_n \rangle_n$ is dense in M. $\mathcal{M}$ is in fact completely determined by the distance function on $\langle p_n \rangle_n$, that is to say $\mathcal{M}$ is determined by a function α defined on (coded) triples of natural numbers, such that

$$\forall k(|d(p_i, p_j) - r_{\alpha(i,j,k)}| < 2^{-k})$$

A representation δ_{NC} (Cf. [KW2, 4.12]) may be given for the points of $\mathcal{M}$ by

$$\begin{aligned} \mathrm{dom}(\delta_{\mathrm{NC}}) &:= \{\beta : \forall m(d(p_{\beta(m+1)}, p_{\beta m}) \leq 2^{-m}\} \\ &= \{\beta : \forall m, k(r_{\alpha(\beta(m+1), \beta m, k)} < 2^{-m} + 2^{-k})\} \\ \delta_{\mathrm{NC}}(\beta) &:= \lim \langle p_{\delta n} \rangle_n \end{aligned}$$

Less well-behaved is the representation δ_{C}, where the Cauchy modulus is left implicit:

$$\begin{aligned} \mathrm{dom}(\delta_{\mathrm{C}}) &:= \{\beta : \langle p_{\beta n} \rangle_n \text{ is a Cauchy sequence}\} \\ \delta_{\mathrm{C}}(\beta) &:= \lim \langle p_{\beta n} \rangle_n. \end{aligned}$$

(H) A slightly different representation $\delta_{\mathcal{M}}$ has as domain all of $\mathbb{B}$: let $\langle p_{\gamma(k,m,n)} \rangle_n$ be a standard enumeration (possibly with repetitions) of $\{p_i : d(p_i, p_m) < 2^{-k}\}$, and let

$$\delta_{\mathcal{M}}(\beta) := \lim \langle p_{\delta n} \rangle_n$$

where $\delta 0 = \beta 0$, $\delta(n+1) = \gamma(n, \delta n, \beta(n+1))$. (Special case: $\delta_{\mathbb{R}}$ with $\langle p_n \rangle_n$ a standard enumeration of $\mathbb{Q}$.)

(I) The (listable) open sets of $\mathbb{R}$ may be represented by ω with

$$\mathrm{dom}(\omega) := \mathbb{B},\ \omega(\alpha) := \bigcup\{I_n : n \in \mathrm{En}(\alpha)\}.$$

This suggests a complementary representation of closed sets as $\delta_{\mathrm{cl}}(\alpha) := \mathbb{R} \setminus \omega(\alpha)$.

It is not difficult to show that

(1) $$\mathrm{Cf} \preceq \mathrm{En},\ \neg\mathrm{En} \preceq \mathrm{Cf},$$

(2) $$\rho_{\mathbb{R}} \preceq \rho_<,\ \rho_{\mathbb{R}} \preceq \rho_>,\ \neg\rho_< \preceq \rho_{\mathbb{R}},\ \neg\rho_> \preceq \rho_{\mathbb{R}}, \delta_{\mathbb{R}} \equiv \rho_{\mathbb{R}}.$$

It is also easy to see that given two representations δ, δ' of a set, we can define a greatest lower bound $\delta \sqcap \delta'$ by

$$\alpha \in \mathrm{dom}(\delta \sqcap \delta') := \mathrm{j}_1\alpha \in \mathrm{dom}(\delta) \wedge \mathrm{j}_2\alpha \in \mathrm{dom}(\delta') \wedge \delta(\mathrm{j}_1\alpha) = \delta'(\mathrm{j}_2\alpha),$$
$$(\delta \sqcap \delta')(\alpha) := \delta(\mathrm{j}_1\alpha)\ (= \delta'(\mathrm{j}_2\alpha)).$$

Then

(3) $$\mathrm{En} \sqcap \mathrm{En}^c \equiv \mathrm{Cf},\ \rho_< \sqcap \rho_> \equiv \rho_{\mathbb{R}}$$

Remarks on CM versus TR. Results such as (1) and (2) invite comparison between CM and TR. In the classical setting of TR differences between representations δ, δ' of the same set X correspond to differences in the amount of information acessible (obtainable) from the representing functions (i.e. the amount of information encoded in the representing functions) concerning the elements of the set. In CM, where all objects are supposed to be given to us (are defined) by the presentation of certain bits of information, the δ, δ' correspond to different sets. Thus, in TR, Cf and En reflect different information concerning elements of $P(\mathbb{N})$; in CM these correspond to decidable and enumerable sets respectively.

Similarly $\rho_{\mathbb{R}}$, $\rho_<$, $\rho_>$ correspond in CM to respectively the reals (in the sense of one of the standard definitions), "limits" of monotone increasing sequences in $\mathbb{Q}$ with upper bound, and "limits" of monotone decreasing sequences in $\mathbb{Q}$ with lower bound (see [TD]).

The analogue of (1) in CM is : decidable implies enumerable, but not conversely; the analogue of (2) is (approximately): every real is approximated by a bounded monotone increasing sequence in $\mathbb{Q}$ with upper bound, and by a bounded monotone decreasing sequence in $\mathbb{Q}$ with lower bound, but not conversely.

Results such as (3) also have their counterpart in CM: $\mathrm{En} \sqcap \mathrm{En}^c \equiv \mathrm{Cf}$ expresses that a set which is enumerable with enumerable complement is decidable (with an appeal to Markov's principle $\forall x(A \vee \neg A) \wedge \neg\neg\exists x A \rightarrow \exists x A$), and $\rho_< \sqcap \rho_> \equiv \rho_{\mathbb{R}}$ corresponds to the fact that if we have monotone increasing $\langle r_n \rangle_n \subset \mathbb{Q}$, monotone decreasing $\langle s_n \rangle_n \subset \mathbb{Q}$, $\forall nm(r_n < s_m)$, $\forall k \exists n(s_n - r_n < 2^k)$, then $\lim\langle r_n \rangle_n = \lim\langle s_n \rangle_n$ is a real.

As to the representations δ_{NC}, δ_{C}, it is not hard to see that for the reals $\delta_{\mathrm{NC}} \preceq \delta_{\mathrm{C}}$, but $\neg\delta_{\mathrm{C}} \preceq \delta_{\mathrm{NC}}$, while $\delta_{\mathrm{NC}} \equiv \delta_{\mathbb{R}}$. $\neg\delta_{\mathrm{C}} \preceq \delta_{\mathrm{NC}}$ expresses that the Cauchy modulus of a fundamental sequence of basis points cannot be computed continuously from the sequence itself; this reflects the distinction between F-numbers and FR-numbers in Markov's constructive mathematics (see [Sh]).

3 From CM to TR

The obvious parallels between certain facts from CM on the one hand, and certain reducibilities in TR on the other hand, raises the question whether we can perhaps systematically translate (certain classes of) statements of CM into reducibility statements of TR.

What the examples of analogies suggest, is the following. Let us assume that we have settled on a common formal language for (parts of) CM and TR, containing at least variables ranging over $\mathbb{N}$ and $\mathbb{B}$.

On the side of CM, we deal with sets $\mathcal{X}$, the elements of which are regarded as given by information encodable by elements of $\mathbb{B}$. That is to say, $\mathcal{X}$ may be thought of as a pair $(X, \delta : X \twoheadrightarrow X^*)$ with $X \subset \mathbb{B}$, $\delta(\alpha)$ the element described by α, δ surjective; this is nothing but a representation of X^*. Modulo an isomorphism, we can always take δ to be a map assigning to α its equivalence class $\alpha_\sim$, i.e. $\alpha \sim \beta \leftrightarrow \delta\alpha = \delta\beta$. (Examples: Cf can be presented as $(\mathbb{B}, \sim_0)$, where $\alpha \sim_0 \beta := \forall n(\alpha n = 0 \leftrightarrow \beta n = 0)$, and $\rho_{\mathbb{R}}$ is $(\text{dom}(\rho_{\mathbb{R}}), \sim_1)$ where $\alpha {\sim_1} \beta := \forall n(|r_{\alpha k} - r_{\beta k}| \leq 2^{-k+1}$.)

Let $\mathcal{Y} = (Y, \delta' : Y \twoheadrightarrow Y^*)$. $\mathcal{X} \subset \mathcal{Y}$ is expressible in codes as $\forall \alpha \in X \exists \beta \in Y(\delta\alpha = \delta'\beta)$. Within a classical setting, a more constructive reading of this can be enforced by requiring the existence of a continuous Γ such that $\forall \alpha \in X(\delta\alpha = \delta'(\Gamma\alpha))$ (i.e. $\delta \preceq \delta'$. So $\delta \preceq \delta'$ appears as the TR-analogue of the constructive reading of $\forall \alpha \in X \exists \beta \in Y(\delta\alpha = \delta'\beta)$.

It is a wellknown fact from the metamathematics of constructive systems, that there is a wide class of sets X and constructive formal systems **S** such that $\mathbf{S} \vdash \forall \alpha \in X \exists \beta \in Y A(\alpha, \beta) \Rightarrow \mathbf{S} \vdash \exists \Gamma \forall \alpha \in X\, A(\alpha, \Gamma\alpha)$ (Γ continuous) holds (more on this in the next section).

Thus, as a rule, a *constructive* proof of an inclusion will automatically generate a (constructively proved) reducibility statement. Of course, in the setting of TR reducibility statements may also be proved by classical means. Many mathematical theorems can be reworded as inclusions (for an example, see section 5).

4 Realizability

We shall restrict attention to statements of CM and TR formalizable in the language of elementary analysis **EL**, with variables for natural numbers (x, y, z, n, m) and variables ranging over $\mathbb{B}$ (α, β, γ). For a precise description see e.g. [TD, page 144].

Although the results below do have generalizations, there are several reasons for this restriction. First of all, we want to avoid undue complications. We are in this section not aiming at maximal generality, but are satisfied with an illustration of the basic ideas; secondly, if we want to formulate analogues in TR for statements in CM, it is pretty obvious what to do with $\mathbb{N}, \mathbb{B}, \mathbb{R}$ etc., since the elements of these sets are obviously encodable by elements of $\mathbb{B}$. But for the interpretation of P($\mathbb{N}$) (and higher powersets) in the context of TR, we must choose (what from the viewpoint of CM is) a subclass encodable by a subset of $\mathbb{B}$ — such as the decidable or the enumerable sets. There is a certain arbitrariness in this choice, and it seems better to regard En, Cf as the TR-analogues of enumerable and decidable sets respectively, and not as analogues of P($\mathbb{N}$).

DEFINITION. The class of *almost negative formulas* is the least class containing formulas $t = s$, $\exists x(t = s)$, $\exists \alpha(t = s)$ and closed under $\wedge, \rightarrow$, $\forall x$, $\forall \alpha$, $\neg$. □
N.B. $\neg A$ may actually be rendered as $A \rightarrow 1 = 0$, $A \vee B$ is definable as $\exists x((x = 0 \rightarrow A) \wedge (x \neq 0 \rightarrow B))$.

Almost negative formulas do not contain "essential" existential quantifiers. That is to say, $\exists x(t = s)$ is innocent in the sense that if true, the x can be found as $\min_x[t = s]$; $\exists \alpha(t = s]$ is reduced to the preceding case since it is equivalent to $\exists x(t[\alpha/x * \lambda x.0] = s[\alpha/x * \lambda x.0])$ (all terms of **EL** are continuous in their function parameters and $\{x * \lambda x.0 : x \in \mathbb{N}\}$ ranges over a dense subset of $\mathbb{B}$). In the presence of Markov's principle $\forall x(A \vee \neg A) \wedge \neg\neg\exists x A \rightarrow \exists x A$, almost negative formulas are equivalent to negative ones not containing any existential quantifier or disjunction.

For notions of constructive mathematics definable in **EL** we can easily give a recipe for finding a corresponding representation with almost negative domain, namely Kleene's function realizability, formalized in detail in [K].

Function realizability associates to each formula A an almost negative formula with an extra function variable α (α not free in A), written $\alpha\,\mathbf{r}\,A$ ("α realizes A").

If the elements of a set X are presented as a subset of $\mathbb{B}^l \times \mathbb{N}^k$ definable by a formula $A(\vec{\beta}, \vec{x})$ in **EL**, with an appropriate equivalence relation $\sim$, the obvious choice of representation is δ_A with

$$\begin{aligned} \mathrm{dom}(\delta_A) &:= \{(\alpha, \vec{\beta}, \vec{x}) : \alpha\,\mathbf{r}\,A(\vec{\beta}, \vec{x})\} \\ \delta_A(\alpha, \vec{\beta}, \vec{x}) &:= (\vec{\beta}, \vec{x})/\sim \end{aligned}$$

We recall the definition of $\alpha\,\mathbf{r}\,A$, presented by induction on the complexity of A:

DEFINITION. ($A \vee B$ treated as defined)

$$\begin{aligned} \alpha\,\mathbf{r}\,(t = s) &:= t = s \\ \alpha\,\mathbf{r}\,(A \wedge B) &:= (\mathrm{j}_1\alpha\,\mathbf{r}\,A) \wedge (\mathrm{j}_2\alpha\,\mathbf{r}\,B) \\ \alpha\,\mathbf{r}\,(A \rightarrow B) &:= \forall \beta(\beta\,\mathbf{r}\,A \rightarrow \alpha|\beta\,\mathbf{r}\,B) \\ \alpha\,\mathbf{r}\,\forall \beta A &:= \forall \beta(\alpha|\beta\,\mathbf{r}\,A) \\ \alpha\,\mathbf{r}\,\forall x\,A &:= \forall \beta(\alpha|\beta\,\mathbf{r}\,A[x/\beta 0] \\ \alpha\,\mathbf{r}\,\exists \beta A &:= (\mathrm{j}_2\alpha)\,\mathbf{r}\,A[\beta/\mathrm{j}_1\alpha] \\ \alpha\,\mathbf{r}\,\exists x\,A &:= (\mathrm{j}_2\alpha)\,\mathbf{r}\,A[x/(\mathrm{j}_1\alpha)0] \end{aligned}$$

(Alternative clauses for $\forall x, \exists x$ leading to an equivalent notion are: $\alpha\,\mathbf{r}\,\forall x\,A := \forall x(\lambda y.\alpha(x, y)\,\mathbf{r}\,A)$, $\alpha\,\mathbf{r}\,\exists x\,A := \lambda y.\alpha(y + 1)\,\mathbf{r}\,A[x/\alpha 0])$. □

It is easy to see that $\alpha\,\mathbf{r}\,A$ is logically equivalent to an almost negative formula. For almost negative formulas B realizability is equivalent to truth: $B \leftrightarrow \exists \alpha(\alpha\,\mathbf{r}\,B)$. But in general $\exists \alpha(\alpha\,\mathbf{r}\,A)$ is not provably equivalent to A.

If we apply this recipe to the concept underlying $\rho_<$, i.e. "least-upper-bounds" of enumerable sets of rationals with an upper bound, we have α's encoding such sets and satisfying

$$A(\alpha) := \exists n \in \mathbb{Z} \forall m \in \mathrm{En}(\alpha)(r_m < n).$$

$\forall m \in \mathrm{En}(\alpha)(r_m < n)$ is almost negative, and the representation suggested by the recipe is not $\rho_<$ but the modified $\rho^*_<$ in which the existential quantifier has been made explicit:

$$\begin{array}{lll}\mathrm{dom}(\rho^*_<) & := & \{\alpha : \forall m \in \mathrm{En}(\lambda y.\alpha(y+1))(r_m < \alpha 0)\} \\ \rho^*_< & := & \text{l.u.b.}\{r_m : m \in \mathrm{En}(\alpha)\}.\end{array}$$

It is easy to see that $\rho^*_< \preceq \rho_<$, but not $\rho_< \preceq \rho^*_<$. In many other examples the predicates defining the domain of a representation δ are in fact almost negative and can be straightforwardly read as deriving from a definition of a corresponding notion in CM. For an interesting example, see also δ_{loc} in the next section.

In the case of sets in CM with a complicated definition, replacing $A(\beta)$ by $\exists\alpha(\alpha\, \mathbf{r}\, A(\beta))$ as the notion considered (which is what our proposal for a "privileged" representation amounts to) may actually involve a *change* in the notion studied, since in general $A \leftrightarrow \exists\alpha(\alpha\, \mathbf{r}\, A)$ is not provable.

On the other hand, the logical schema $A \leftrightarrow \exists\alpha(\alpha\, \mathbf{r}\, A)$ is equivalent to a generalized continuity schema:

$$\text{GC} \qquad \forall\alpha[A\alpha \to \exists\beta\, B(\alpha,\beta)] \to \exists\gamma\forall\alpha[A\alpha \to B(\alpha,\gamma|\alpha]$$

where A is almost negative, and the soundness theorem for function realizability establishes the consistency of GC relative to many constructive systems, in particular relative to $\mathbf{EL} + \mathrm{BI} + \mathrm{M}$ (containing FAN, AC_{01}, CONT_1).

So, looking at examples of concrete representations as found in the papers by Kreitz and Weihrauch, we see that in many cases the domains are definable by almost negative formulas, and hence the representations may be regarded as being derived from a CM-notion by our recipe.

But what about representations with a domain not defined by an almost negative formula, can we also think of these as derived from a suitable CM-notion? The principal difficulties we meet in the examples are of two types: (a) the definition of $\mathrm{dom}(\delta)$ contains set-quantifiers, and (b) the definition of $\mathrm{dom}(\delta)$ contains (essential) quantifiers $\exists\alpha$ or $\exists x$.

As long as we are working in the context of **EL**, the first difficulty can only be circumvented by finding an equivalent definition in the language of **EL**.

In the second case, where a definition of $\mathrm{dom}(\delta)$ in **EL** contains quantifiers $\exists\alpha$ or $\exists n$, there is a quite general solution: we may replace these quantifiers by $\neg\neg\exists\alpha, \neg\neg\exists n$, or equivalently by $\neg\forall\alpha\neg, \neg\forall n\neg$. In the classical setting of TR this is actually the same representation, but now the definition is (almost) negative. (Under this transformation, δ_{C} corresponds to the quasi-numbers of constructive recursive mathematics, and $\rho_<$ to monotone increasing sequences which are not unbounded.)

An example of the elimination of a set quantifier is the following. In [KW3, page 29] the following representation δ'_{cl} of a class of closed subsets of $\mathbb{R}$ is considered:

$$\begin{array}{ll}\alpha \in \mathrm{dom}(\delta'_{\mathrm{cl}}) & := \exists X \subset \mathbb{R}(X \text{ closed } \wedge \mathrm{En}(\alpha) = \{k : I_k \cap X \neq \emptyset\}), \\ \delta'_{\mathrm{cl}}(\alpha) & := \bigcap_m \bigcup_n \{I_{(n,m)} : (n,m) \in \mathrm{En}(\alpha)\}.\end{array}$$

We can replace the quantifier $\exists X \subset \mathbb{R}$ by an extra condition on the enumeration expressing that whenever a finite set $I_{n_1}, \ldots, I_{n_p}$ covers an I_m with $m \in \mathrm{En}(\alpha)$ then at least one $n_i \in \mathrm{En}(\alpha)$. One can then prove that $\delta'_{\mathrm{cl}}(\alpha)$ as defined above is indeed closed, and that α indeed enumerates all I_k with nonempty intersection with δ'_{cl}. In passing from an inclusion statement $\forall\alpha \in \mathrm{dom}(\delta)\exists\beta \in \mathrm{dom}(\delta')(\delta(\alpha) = \delta'(\beta))$ in CM to $\delta \preceq \delta'$ the statement is strengthened by requiring continuous dependence of β from α; but this is usually an automatic consequence of a constructive proof of the conclusion, by the following

THEOREM. *For suitable formal theories* **S** *in the language of* **EL**, *the following derived rule holds:*

$$\mathbf{S} \vdash \forall\alpha[A\alpha \to \exists\beta B(\alpha,\beta)] \Rightarrow \mathbf{S} \vdash \exists\gamma\forall\alpha[A(\alpha) \to \gamma|\alpha \downarrow \wedge B(\alpha,\gamma|\alpha)]$$

(So γ codes a partial continuous function in $[\mathbb{B} \rightharpoonup \mathbb{B}]$). Moreover, if $\forall\alpha\exists\beta A(\alpha,\beta)$ is closed, then γ may be taken to be recursive. For **S** *we can take* **EL**, *with some of the following schemata added:*

(1) bar induction BI;

(2) the fan theorem FAN, *a consequence of* BI*:*
$\forall\alpha \le \beta\exists x A(\bar{\alpha}x) \to \exists z\forall\alpha \le \beta\exists x \le zA(\bar{\alpha}x)$

(3) Markov's principle (cf. end of section 2)

(4) the countable axiom of choice AC_{01}*:* $\forall x\exists\alpha A(x,\alpha) \to \exists\beta\forall x A(x,\lambda y.\beta(x,y))$;

(5) the continuity principle CONT_1*:* $\forall\alpha\exists\beta A(\alpha,\beta) \to \exists\gamma\forall\alpha A(\alpha,\gamma|\alpha)$, *or some weaker version.*

PROOF. By **q**-realizability for functions; see [T, 3.3, 3.7.9]. □

The theorem may be applied as follows. Suppose that we can express in **EL**, for representations δ, δ', δ with an almost negative domain,

$$\forall\alpha \in \mathrm{dom}(\delta)\exists\beta(\delta(\alpha) = \delta'(\beta))$$

and this is provable in a suitable system **S** (e.g. $\mathbf{EL} + \mathrm{FAN} + \mathrm{MP} + \mathrm{AC}_{01}$), which is a subsystem of classical analysis, then by the theorem

$$\mathbf{S} \vdash \exists\gamma\forall\alpha \in \mathrm{dom}(\delta)(\gamma|\alpha \downarrow \wedge \delta(\alpha) = \delta'(\gamma|\alpha)).$$

That is to say, in cases where δ, δ' have been chosen to correspond to the constructive definitions of sets X, Y, the constructive proof of $\forall x \in X\exists y \in Y(x = y)$ automatically yields the stronger $\delta \preceq \delta'$ (established constructively as well as classically).

For an extension to stronger systems, see e.g. [Fr], where an extension of **q**-realizability to intuitionistic second-order arithmetic is given. For further generalizations to higher-order logic, see e.g. [vO].

5 An Example

How translation of CM-statements, and the use of the metatheorem works out in practice, we can see from the following more or less representative example. In [Bi, page 177] we find the following

Theorem. *Let X be an inhabited closed located set in a complete separable metric space $\mathcal{M} \equiv (M, d, \langle p_n \rangle_n)$, $x \in M$ such that $\forall y \in X(d(y,x) > 0)$, then $d(X,x) > 0$.* □

Proof. For a constructive proof, see [Bi]. A classical proof is even easier.

We give a reformulation in TR. We first define a representation of inhabited closed located sets.

Definition.

$$\begin{aligned} \alpha \in \mathrm{dom}(\delta_{\mathrm{loc}}) &:= \exists X \subset M(\exists x(x \in X) \wedge \mathrm{X\ located} \wedge \\ &\qquad \forall \beta \in \mathrm{dom}(\delta_{\mathcal{M}})(d(X, \delta_{\mathcal{M}}(\beta)) = \delta_{\mathbb{R}}(\alpha \mid \beta))) \\ \delta_{\mathrm{loc}}(\alpha) &:= \{\delta_{\mathcal{M}}(\beta) : \delta_{\mathbb{R}}(\alpha \mid \beta) = 0\}. \ \square \end{aligned}$$

So $\lambda\beta.(\alpha|\beta)$ describes the continuous distance function $d(X, \delta_{\mathcal{M}}(\beta))$ ($\delta_{\mathcal{M}}$ as defined in section 2, (H)).

Definition. For pairs (X,x), X a closed located set of $\mathcal{M}$, $x \in M$, $d(X,x) > 0$, we choose a representation ρ^* as follows. If $\alpha = (\beta, \hat{n} * \gamma)$, and $X = \delta_{\mathrm{loc}}(\beta)$, $x = \delta_{\mathcal{M}}(\gamma)$ and $d(X,x) > 2^{-n}$, then $\alpha \in \mathrm{dom}(\rho^*)$, and $\rho^*(\beta, \hat{n} * \gamma) = (\delta_{\mathrm{loc}}(\beta), \delta_{\mathcal{M}}(\gamma))$.

For pairs (X,x), X closed located, inhabited, $x \in M$ such that $\forall y \in X(d(y,x) > 0)$, we define ρ^{**} as follows. If $\alpha = (\beta, \beta', \beta'')$, $\delta_{\mathrm{loc}}(\beta) = X$, $\delta_{\mathcal{M}}(\beta') = x$, and β'' such that

$$\delta_{\mathcal{M}}(\gamma) \in X \to \beta''(\gamma)\downarrow \wedge\, d(\delta_{\mathcal{M}}(\gamma), \delta_{\mathcal{M}}(\beta')) > 2^{-\beta''(\gamma)},$$

then $\alpha \in \mathrm{dom}(\rho^{**})$, and clearly

$$\rho^{**}(\beta, \beta', \beta'') = (\delta_{\mathrm{loc}}(\beta), \delta_{\mathcal{M}}(\beta')).$$

It is not difficult to strengthen the proof of the theorem to a constructive proof of the equivalence $\rho^* \equiv \rho^{**}$. (Even if this reformulation of the theorem looks a bit complicated at first sight, it is nevertheless a straightforward spelling out of the required explicit information.)

If we want to obtain $\rho^* \equiv \rho^{**}$ via an appeal to the derived rule in the preceding section, we must put in some extra work.

Reformulation of the representation δ_{loc}

Let X be inhabited, located in a complete, separable metric space $\mathcal{M}$, and let $f(x) = d(X,x)$. the $X = \{x : f(x) = 0\}$. Our definition of δ_{loc} given above is not satisfactory, inasmuch its definition is not in the language of **EL**. We can correct this as follows. It is not difficult to see that X itself is a complete, separable metric space, and that we can construct explicitly a sequence $\langle q_n \rangle_n$ of points in X, dense in X. The distance function $\lambda x.d(X,x)$ must satisfy

$$\begin{aligned} &\forall x, y \in M(d(X,y) = 0 \to d(x,y) \geq d(x,X)), \\ &\forall x, k \exists y(d(X,y) = 0 \wedge d(x,y) < d(x,X) + 2^{-k}). \end{aligned}$$

Now an arbitrary continuous $f : M \longrightarrow \mathbb{R}$ represents the distance function of $X_f \equiv \{x : f(x) = 0\}$, i.e. $d(X_f, x) = f(x)$, if the following conditions are satisfied:

$$(*) \qquad \begin{cases} \forall x, y(f(y) = 0 \to d(x,y) \geq f(x)),\ \forall n(f(q_n) = 0) \\ \forall x, k \exists n(d(x, q_n) < f(x) + 2^{-k}) \end{cases}$$

If f is represented by a γ such that $\delta_{\mathbb{R}}(\gamma|\alpha) = f(\delta_{\mathcal{M}}(\alpha))$, and d by a γ' such that $\delta_{\mathbb{R}}(\gamma'|(\alpha,\beta)) = d(\delta_{\mathcal{M}}(\alpha), \delta_{\mathcal{M}}(\beta))$, and the sequence $\langle q_n \rangle_n$ by a γ'' such that $q_n = \delta_{\mathcal{M}}(\lambda y.\gamma''(n,y))$, it is not hard to verify that $(*)$ can be expressed by an almost negative formula.

We can take $(*)$ as the condition (on γ and γ'') determining the domain of an appropriate representation of the located, closed, inhabited sets. Now the theorem on the derived rule applies. The "balance of work" in the case of a constructive proof is more or less neutral: compared with a direct constructive proof of $\rho^* \equiv \rho^{**}$, we had to put in a bit extra work in order to formulate things in **EL**, while we saved a little by an appeal to the derived rule. In this particular case we can give a much shorter proof by reasoning classically: from the classically proven $\forall y d(\delta_{\mathrm{loc}}(\alpha), y) > 0 \to \exists k\, d(\delta_{\mathrm{loc}}(\alpha), x) > 2^{-k}$ we can find the k continuously in α.

REMARK. In adopting a coding as an appropriate rendering of a concept of constructive mathematics, one often uses a lemma: if f is a continuous map from $\mathcal{M}$ to $\mathcal{M}'$, $\mathcal{M}$ and $\mathcal{M}'$ complete separable metric spaces, then (classically, or if we assume intuitionistic continuity axioms) there is a continuous $\Gamma : \mathbb{B} \longrightarrow \mathbb{B}$ such that $f(\delta_{\mathcal{M}}(\alpha)) = \delta_{\mathcal{M}'}(\Gamma\alpha)$ (proof left to the reader).

6 Compactness of Bounded Closed Subsets

As a second illustrative example we show how conversely a reducibility result from [KW] might also have been obtained from a proof of CM with application of the derived rule.

DEFINITION. δ_{cl} represents closed sets as countable intersections of complements of basis intervals:

$$\delta_{\mathrm{cl}} := \bigcap\{\mathbb{R} \setminus I_k : k \in \mathrm{En}(\alpha)\} = \mathbb{R} \setminus \bigcup\{I_k : k \in \mathrm{En}(\alpha)\}.$$

For the corresponding notion of bounded sets we define δ_{bcl}:

$$\hat{n} * \alpha \in \mathrm{dom}(\delta_{\mathrm{bcl}}) := \delta_{\mathrm{cl}}(\alpha) \subset [-n, n]\ (n \in \mathbb{N}),$$
$$\delta_{\mathrm{bcl}}(\hat{n} * \alpha) := \delta_{\mathrm{cl}}(\alpha) \text{ whenever } \hat{n} * \alpha \in \mathrm{dom}(\delta_{\mathrm{bcl}}). \ \square$$

$\mathrm{dom}(\delta_{\mathrm{bcl}})$ is easily seen to be definable by an almost negative formula (e.g. $\hat{n} * \alpha \in \mathrm{dom}(\delta_{\mathrm{bcl}}) \leftrightarrow \forall r \in \mathbb{Q}(r < -n \vee n < r \to \exists m(\alpha m > 0 \wedge r \in I_{\alpha m - 1})))$. We want to compare δ_{bcl} with δ_{whb}, a representation of the compact subsets of $\mathbb{R}$.

Definition. Put

$$\begin{aligned} C_\alpha &:= \{I_j : j \in \mathrm{En}(\alpha)\}, \\ C_{\alpha,n} &:= \{I_j : \exists i \in D_n(\alpha i = j+1)\} \end{aligned}$$

where $\langle D_n\rangle_n$ is some standard enumeration of finite sets, e.g. $D_n = \{m_0 < \ldots < m_{p-1}\} \Leftrightarrow n = \sum_{i<p} 2^{m_i}$.

A function $\beta \in \mathbb{B}$ *witnesses the compactness* of $X \subset \mathbb{R}$ (*is a witness for compactness* of X) iff

$$\begin{aligned} &\forall\alpha[(X \subset \bigcup C_\alpha) \leftrightarrow (\alpha \in \mathrm{dom}(\Phi_\beta))] \text{ and} \\ &\forall\alpha[(X \subset \bigcup C_\alpha) \rightarrow X \subset \bigcup C_{\alpha,\beta(\alpha)}]. \end{aligned}$$

The *weak Heine-Borel representation* is then given by

$$\begin{aligned} \alpha \in \mathrm{dom}(\delta_{\mathbf{whb}}) &:= \exists X \subset \mathbb{R}(\alpha \text{ witnesses compactness of X}) \\ \delta_{\mathbf{whb}}(\alpha) &:= \bigcap\{\bigcup C_{\beta,\alpha(\beta)} : \alpha(\beta)\downarrow \wedge \beta \in \mathbb{B}\} \end{aligned}$$

Theorem. $\delta_{\mathrm{bcl}} \preceq \delta_{\mathbf{whb}}$.

Proof. We show constructively $\forall\alpha \in \mathrm{dom}(\delta_{\mathrm{bcl}})\exists\gamma \in \mathrm{dom}(\delta_{\mathbf{whb}})(\delta_{\mathrm{bcl}}(\alpha) = \delta_{\mathbf{whb}}(\gamma))$. Let $\hat{n}*\alpha \in \mathrm{dom}(\delta_{\mathrm{bcl}})$, then $\delta_{\mathrm{bcl}}(\hat{n}*\alpha) = [-n,n]\backslash\bigcup\{I_j : j \in \mathrm{En}(\alpha)\} = [-n,n]\backslash\bigcup C_\alpha$. *Assuming the compactness of* $[-n,n]$ (which is a consequence of FAN) we have

$$\delta_{\mathrm{bcl}}(\hat{n}*\alpha) \subset \bigcup C_\beta \Leftrightarrow \exists m([-n,n] \subset \bigcup C_{\alpha,m} \cup \bigcup C_{\beta,m})$$

and

$$[-n,n] \subset \bigcup C_{\alpha,n} \cup \bigcup C_{\beta,m} \Rightarrow \delta_{\mathrm{bcl}}(\hat{n}*\alpha) \subset \bigcup C_{\beta,m}.$$

Since $[-n,n] \subset \bigcup C_{\alpha,n} \cup \bigcup C_{\beta,m}$ is decidable in n, m, α, β, we can compute m for any β covering $\delta_{\mathrm{bcl}}(\hat{n}*\alpha)$. □

This proof is constructive (assuming FAN), so, with an appeal to our derived rule M may be found continuously from β, by Φ_γ say. Then γ is a witness for the compactness of $\delta_{\mathrm{bcl}}(\hat{n}*\alpha)$, and may be found (again appealing to the derived rule) continuously from α.

Remark. If we compare the proof with the similar argument in [KW], we see that we have saved little work by an appeal to the derived rule, since the continuity of the dependencies is not difficult to see. But the proof shows at least that this result fits into our "metamathematical schema".

More interesting is a proof of the converse, $\delta_{\mathbf{whb}} \preceq \delta_{\mathrm{bcl}}$. The proof in [KW3] looks (nearly) constructive, but does not bring the statement under our schema, since (a) constructively we also need to show that $\delta_{\mathbf{whb}}(\alpha)$ defines a closed set (in CM a witness of compactness does not uniquely determine the set being witnessed: $[1,2] \cup [2,3]$ and $[1,3]$ have the same witnesses of compactness, but $[1,2] \cup [2,3]$ is not closed), and (b) the domain of $\delta_{\mathbf{whb}}$ has not been defined in **EL**. So we have to prove a

LEMMA. *$\alpha \in \mathrm{dom}(\delta_{\mathbf{whb}})$ is expressible by an almost negative predicate.*
PROOF.

$$(1) \qquad \forall\gamma[(\bigcap_{\beta}\{\bigcup C_{\beta,\alpha(\beta)} : \alpha(\beta)\downarrow\} \subset \bigcup C_{\gamma}) \to \alpha(\gamma)\downarrow]$$

expresses that α witnesses compactness of $\bigcap_{\beta}\{\bigcup C_{\beta,\alpha(\beta)} : \alpha(\beta)\downarrow\}$. We can rewrite (1) as

$$(2) \qquad \forall\gamma[\forall\gamma'(\forall\beta(\alpha(\beta)\downarrow \to \delta_{\mathbb{R}}(\gamma') \in \bigcup C_{\beta,\alpha(\beta)}) \to \delta_{\mathbb{R}}(\gamma') \in \bigcup C_{\gamma}) \to \alpha(\gamma)\downarrow]$$

and this is easily verified to be equivalent to an almost negative statement.

In order to meet objection (a) above, we need a

LEMMA. *In* **EL** + **MP** *we can prove that whenever*

$$(1) \qquad \forall\gamma[(\bigcap_{\beta}\{\bigcup C_{\beta,\alpha(\beta)} : \alpha(\beta)\downarrow\} \subset \bigcup C_{\gamma}) \to \alpha(\gamma)\downarrow]$$

then $\delta_{\mathbf{whb}}(\alpha) \equiv X_\alpha$ is closed.
PROOF. Let $x \notin X_\alpha$, $X_\alpha \subset [-n, n]$. (We may assume this without loss of generality, since $C_{\lambda n.n}$ certainly covers, so $\alpha(\lambda n.n)\downarrow$, and $\bigcup C_{\lambda n.n,\alpha(\lambda n.n)}$ provides us with a bound.) Consider the following subset of $\langle I_n \rangle_n$ consisting of intervals of the forms

$$\{(r, r') : r \le -n-1 \wedge r' < x\},\ \{(r, r') : x < r \wedge n+1 \le r'\}.$$

This set covers X_α, for if $x' \in X_\alpha$, $x \neq x$, then by Markov's principle $x' \# x$, i.e. $x' < x$ or $x < x'$, hence $x' \in (-n-1, r)$ for some $r < x$ or $x' \in (r', n+1)$ for some $r' > x$. Since X_α is compact, there is a finite collection

$$\{(r_1', r_1), \ldots, (r_p', r_p)\} \cup \{(s_1, s_1'), \ldots, (s_q, s_q')\}$$

with $r_i' \le -n-1, r_i < x, x < s_j, n+1 \le s_j'$ covering X_α. Now each $x' \in X_\alpha$ has a distance to x of at least $\inf\{x - \sup\{r_1, \ldots, r_p\}, \inf\{s_1, \ldots, s_q\} - x\} > 0$. Therefore, if x is in the closure of X_α, also $\neg\neg x \in X_\alpha$. If $x \in X_\alpha$, then

$$\forall\beta(\alpha(\beta)\downarrow \ \to \exists j(x \in I_j \in C_{\beta,\alpha(\beta)})),$$

hence if $\neg\neg x \in X_\alpha$, then

$$\forall\beta(\alpha(\beta)\downarrow \to \neg\neg\exists j(x \in I_j \in C_{\beta,\alpha(\beta)})).$$

By Markov's principle, we can drop $\neg\neg$, so $x \in X$. □

By means of the preceding two lemma's, the proof in [KW] is now easily adapted to obtain

PROPOSITION. $\mathrm{dom}(\delta_{\mathbf{whb}}) \subset \mathrm{dom}(\delta_{\mathbf{bcl}})$ *is constructively provable.* □

This brings the reducibility $\delta_{\mathbf{whb}} \preceq \delta_{\mathbf{bcl}}$ under our general schema. But note that in this case the balance of work is even *negative*: in order to make the proof of the proposition constructive we needed to put in extra work.

7 Concluding Remarks

(1) Our discussion covers most, but not all specific representations and reducibilities discussed in the work of Weihrauch and Kreitz. Thus in [W2] representations for *sets* of continuous maps between separable metric spaces in which a pointset X appears as a parameter (the domain of definition of the function). To bring this under the schema, we need to extend the results of section 4 to **EL** extended with a (purely schematic) set variable. We have checked that this is possible, but we are not really satisfied with our treatment.

The formulation of an TR-analogue to a CM-inclusion or -equality statement may become awkward if the CM-statement is expressed in a language with set variables. As an example, consider the following simple theorem taken from [BB, page 37]:

THEOREM. *For inhabited sets $X \subset \mathbb{R}$ with an upper bound, X has an l.u.b. iff* $(*)$ $\forall x, y \in \mathbb{R}(x < y \to y \geq X \vee \exists x' \in X(x < x'))$, *where* $y \geq X := \forall x \in X(y \geq x)$. □

This theorem may be recast as $\delta_{1,X} \equiv \delta_{2,X}$ where $\delta_{1,X}$ and $\delta_{2,X}$ are representations depending on a parameter X. A code in $\mathrm{dom}(\delta_{1,X})$ should specify an $x_0 \in X$, an upper bound for X, and a decision function applicable to pairs x, y with $x < y$ for $(*)$, plus a function yielding the $x' \in X$ if the second alternative in $(*)$ holds. $\delta_{1,X}$ assigns to this code simply l.u.b.(X). A code for $\delta_{2,X}$ should specify an element of X, an upper bound x_1 of X, and a sequence $\langle y_n \rangle_n \subset X$ such that $y_n + 2^{-n} > x_1$.

As the example shows, the fact that $P(\mathbb{R})$ is not encodable by $\mathbb{B}$ may be circumvented by parametric representations. We suspect that this can be done quite generally (cf. the representations of the continuous partial functions between metric spaces mentioned above), but this aspect calls for further investigation.

(2) For reducibilities between representations expressible in **EL**, function realizability provides a key to understanding the analogies between CM and TR. However, in general there is no saving of labour in deriving reducibilities from CM-results. But the CM-results suggest reducibilities, and the discussion shows that attention to the logical form of the definitions of domains of representations may help to explain why certain representations are better behaved than others — a type of explanation different in spirit from the topological criteria in [KW2], hence adding a little bit of insight.

(3) We believe that comparison between CM and TR may be useful in selecting the mathematically best-behaved representations.

References

[B] M.J.Beeson, *Foundations of Constructive Mathematics*, Springer-Verlag, Berlin 1985.

[BB] E. Bishop, D.S. Bridges, *Constructive Analysis*, Springer-Verlag, Berlin 1985.

[Bi] E.A. Bishop, *Foundations of Constructive Analysis* (1967), McGrawHill, New York.

[Fr] H.M. Friedman, On the derivability of instantiation properties, *The Journal of Symbolic Logic 42* (1977), 506–514.

[K] S.C. Kleene, Formalized recursive functionals and formalized realizability, *memoirs of the American mathematical Society 89* (1969).

[KW1] C. Kreitz, K. Weihrauch, A unified approach to constructive and recursive analysis, in: M.M. Richter et al. (eds.), *Computation and Proof theory*, Springer-Verlag, Berlin 1984, 259-278.

[KW2] C. Kreitz, K. Weihrauch, Theory of representations, *Theoretical Computer Science 38* (1985), 35–53.

[KW3] C. Kreitz, K. Weihrauch, Compactness in constructive analysis revisited, *Annals of Pure and Applied Logic 36* (1987), 29–38.

[M] N. Th. Müller, Computational complexity of real functions and real numbers. Informatik Berichte 59 (1986), Fern-Universität Hagen, BRD.

[ML] P. Martin-Löf, *Intuitionistic Type Theory*, Bibliopolis, Napoli.

[Sh] N.A. Shanin, *Constructive Real Numbers and Function Spaces*, American Mathematical Society, providence (RI), 1968 (translation of the russian original).

[T] A.S. Troelstra (ed.), *Metamathematical investigation of Intuitionistic Arithmetic and Analysis*, Springer verlag, berlin 1973.

[TD] A.S. Troelstra, D. van Dalen, *Constructivism in Mathematics* (1988), North-Holland Publ. Co., Amsterdam

[vO] J. van Oosten, Exercises in Realizability. Ph.D. thesis, Universiteit van Amsterdam, 1990.

[W1] K. Weihrauch, Type 2 recursion theory, *Theoretical Computer Science 38* (1985), 17–33.

[W2] K. Weihrauch, Computability on computable metric spaces, *Theoretical Computer Science*, to appear.

[WK] K. Weihrauch, C. Kreitz, Representations of the real numbers and of the open subsets of the set of real numbers, *Annals of Pure and Applied Logic 35* (1987), 247–260.

Infinitary Queries and Their Asymptotic Probabilities I: Properties Definable in Transitive Closure Logic

Jerzy Tyszkiewicz*

Institute of Informatics, University of Warsaw, ul. Banacha 2, 02-097 Warszawa, Poland. e-mail jurekty@plearn.bitnet

Abstract

We present new general method for proving that for certain classes of finite structures the limit law fails for properties expressible in transitive closure logic. In all such cases also all associated asymptotic problems are undecidable.

1 Introduction

The problems considered in this paper belong to the research area called *random structure theory,* and, more specifically, to its logical aspect.
To explain (very imprecisely and incompletely) what does it mean, let us imagine that we have a class of some structures (say, finite ones over some fixed signature), equipped with a probability space structure (this probability is usually assumed to be only *finitely* additive). Then we draw one structure at random and ask:
- what does the drawn structure look like?
- does the drawn structure have a given property?

Those questions are typical in random structure theory. To turn to the logical part of it, look at the drawn structure through logical glasses: we can only notice properties definable in some particular logic. Then new questions become natural:
- does every property we can observe have a probability (is it measurable)?
- if so, what is this probability?
- can we compute this probability, and, eventually, how difficult is it?

It becomes clear from the above that the random structure theory is tightly connected with combinatorics, finite model theory, mathematical logic, and, last but not least, computer science. A much more complete exposition of the logical part of the random structure theory may be found in a nice survey [2].
The present paper is devoted to study of asymptotic probabilities of these properties of finite structures, which are definable in *transitive closure logic* (denoted *TC* in the sequel). The results we obtain provide a contribution to problem 8.2 in the paper [2]:

> *Develop techniques for showing the existence of asymptotic probabilities of* least fixed point logic[1] *sentences in slow growing classes.*

---

*This research was completed in the Institute of Mathematics, University of Warsaw.

[1] Transitive closure logic is a sublogic of the least fixed point logic (J.T.).

As a matter of fact, we actually deal with the other side of the coin, and present general conditions under which asymptotic probabilities of TC sentences fail to exist. In a sequel to the present paper [14] we prove results concerning similarly stated problem for properties definable in least fixed point logic. In [11] authors prove similar results for the infinitary logic $L^{\omega}_{\infty\omega}$.

2 Probability, Logic and Semi–Forests

2.1 Probability

Let $\mathcal{A}$ be an arbitrary class of finite structures over (fixed) signature τ with equality. Let $\mathcal{A}(n) \subseteq \mathcal{A}$ be a subclass of $\mathcal{A}$ containing all structures with carrier set $n = \{0, \ldots, n-1\}$, for $n \in \omega$. Let $\mu_n$be a probability distribution on $\mathcal{A}(n)$. Then for any property φ of some structures in $\mathcal{A}$ (often defined by (and identified with) a sentence of some logic over τ) we can define

$$\mu_n(\varphi) = \mu_n(\{\mathcal{A} \in \mathcal{A}(n) \mid \mathcal{A} \models \varphi\}).$$

Often considered examples of probabilities are the *uniform labelled* probabilities (i.e. each structure in $\mathcal{A}(n)$ is weighted equally), and *uniform unlabelled* probabilities (i.e. each isomorphism class in $\mathcal{A}(n)$ is weighted equally).
Writing μ for $\{\mu_n\}_{n\in\omega}$ we may consider $\langle\mathcal{A}, \mu\rangle$ as *randomized class of finite structures* and make it an object of our study.
We now fix a randomized class $\langle\mathcal{A}, \mu\rangle$. All notions we define should be understood to refer to this fixed $\langle\mathcal{A}, \mu\rangle$. Unless otherwise stated, all results in this paper hold for any such randomized class.
We are interested in asymptotic properties of $\mu_n(\varphi)$, and especially whether the limit $\mu(\varphi) = \lim_{n\to\infty} \mu_n(\varphi)$ exists, for φ being a sentence of the logic under consideration. If it exists, we call it an *asymptotic probability of* φ. If this is the case for every sentence of the logic L, we say that the *convergence law* holds (for L and $\langle\mathcal{A}, \mu\rangle$). If, in addition, every sentence has probability either 0 or 1, we say that the *0–1 law* holds.
By an *asymptotic problem* (for L and $\langle\mathcal{A}, \mu\rangle$) we mean any set of sentences of L of the form $\{\varphi \in L \mid \mu(\varphi) \geq r\}$, where r is rational, $0 < r \leq 1$. We are interested in complexities of asymptotic problems, and especially in the complexity of the *almost sure theory,* i.e. the set of all L sentences of asymptotic probability one. In this paper we consider only the decidable/undecidable distinction between complexities.
Sometimes, if $\mu(\varphi) = 1$, we say that φ *holds almost surely* (we often use this convention when φ is described in an informal way).
We also say that sequence $\{a_n\}_{n\in\omega}$ *converges recursively* to limit $0 < a < \infty$ iff there exists a recursive function $g : \omega \to \omega$ such that for every positive $k \in \omega$, and every $n > g(k)$ we have $|a_n - a| < 1/k$. If $a = \infty$, we change the last condition to: for every positive $k \in \omega$, and for every $n > g(k)$ we have $a_n > k$.

2.2 Logic

We replace TC with a logical formalism of the same expressive power, but much easier to deal with. This formalism is a *Dynamic Logic of While Programs with Rich*

Tests and Nondeterministic Assignments [12]. A result stating equivalence of these formalisms is proved in [7]. Our choice of the formalism is motivated by simpler verification of *dynamic* formulas, which results in argumentation which we believe to be more clear and easier to pursue.
We simultaneously define the sets of *programs* and *formulae*. For the set of formulae we use symbol DL.
First we define the class of *instructions* as the least class of expressions containing

nondeterministic assignment

$$x_i := y \texttt{ such that } \varphi(y, \vec{x})$$

for all φ in DL; intended meaning of such an assignment is that just after it is executed, register x_i contains nondeterministically chosen element of the universe satisfying φ with parameters – the values of registers appearing in $\vec{x}$. If there is no such an element, then this instruction loops forever. Variable y is *bounded* in this instruction.

halt instruction *halt*

abort instruction *abort*

and closed under the following formation rules (we assume S_1, S_2 to be instructions and φ to be a formula in DL, braces are *not* parts of the compound instructions):

conditional instruction
```
if φ(x⃗)
then
S1
else
S2
fi
```

while loop
```
while φ(x⃗)
do
S1
od
```

sequential composition
```
S1
S2
```

Program is just an instruction with the last line *halt* and with all lines numbered with consecutive natural numbers, starting at the top. We define in standard manner free variables of a program.
The class of formulae is the least class of expressions containing all first order formulae over τ, *halting formulae* $\langle Q\rangle true$ for all programs Q (intended meaning of such formula is that *there exists a terminating computation of* Q*, starting from actual valuation,*) closed under classical propositional connectives ($\neg$, $\wedge$, $\vee$, $\Rightarrow$, $\Leftrightarrow$), and first order quantification ($\forall$, $\exists$). Free variables of a formula are defined in standard way, except that free variables of a formula $\langle Q\rangle true$ are all free variables of Q.

Semantics of programs and formulae in a τ-structure $\mathcal{A}$ is intuitively clear, and thus is moved to appendix A. The only thing we mention here is that one program may have many different computations in $\mathcal{A}, \vec{x}:\vec{a}$.
Definition of syntax of programs is oriented towards formal semantics definition. In the sequel we often use an informal notation for programs, e.g. we skip line numbering, omit *halt* instructions, etc. Sometimes we neglect initial values of these variables which are irrelevant to computation of a program; e.g. initial value of z in program from proof of lemma 3 does not play any role in the computation. We adopt also one syntactic abbreviation concerning programs: $x_i := x_j$ is to be understood as $x_i := y$ **such that** $y = x_j$.

2.3 Semi–Forests

Let τ_0 be a signature consisting exactly of one binary relation symbol r and equality sign $=$. In what follows we denote by DL_0 the logic $DL(\tau_0)$.
Let $A \subseteq \omega$ be a finite set and let $R \subseteq A \times A$ be a binary relation. Of course $\mathcal{A} = \langle A, R\rangle$ may be considered as a τ_0-structure (directed graph, in other words).
We call such a τ_0-structure $\mathcal{A}$ a *semi-forest* if for every $a, b \in A$ there is at most one R-path (R-cycle if $a = b$, respectively) with no repeating vertices, between a and b. Natural examples of semi–forests are forests themselves (in directed version), as well as unary functions, permutations and many others.
Now let $\mathcal{A} = \langle A, R\rangle$ be a semi–forest, and let $a, b \in A$. Then we can naturally define an *ordered interval* $[\![a, b]\!] \subseteq A$, consisting of all elements subsequently appearing on the unique path from a to b. (It may happen that $[\![a, b]\!] = \emptyset$ or $[\![a, a]\!] \supsetneq \{a\}$.)
The *height* $h(\mathcal{A})$ of a semi–forest $\mathcal{A}$ is a maximal length of such an interval.
The following two technical lemmas will be the only facts about semi–forests we will need in further investigations:

Lemma 1 *For any two semi–forests $\mathcal{A}_1$ and $\mathcal{A}_2$ their disjoint union $\mathcal{A}_1 \sqcup \mathcal{A}_2$ is a semi–forest, too, and $h(\mathcal{A}_1 \sqcup \mathcal{A}_2) = \max\{h(\mathcal{A}_1), h(\mathcal{A}_2)\}$.* □

Lemma 2 (due to anonymous referee of the paper) *Let $\mathcal{A} = \langle A, R\rangle$ be any semi–forest. Let $a, b \in A$, be such that $[\![a, b]\!] \neq \emptyset$. Then $c \in [\![a, b]\!]$ iff there is no R-path from a to b on which c does not appear.*

Proof: Suppose that $c \in [\![a, b]\!]$, and let p be an arbitrary R-path from a to b in a semi–forest $\mathcal{A}$. Then after cutting off all repetitions, p becomes $[\![a, b]\!]$, so c was present on p.
To prove the converse observe that $[\![a, b]\!]$ is an R-path from a to b, so c appears on it. □

3 Expressive Power of DL_0 Over Semi–Forests

Lemma 3 *There exists a DL formula $Path(x, y)$ such that for any τ-structure $\mathcal{A}$*

$$\mathcal{A}, x:a, y:b \models Path \qquad \text{iff} \qquad \text{there is an } R\text{-path from } a \text{ to } b.$$

Proof: Take

$$Path \equiv \langle z := x \texttt{ while } z \neq y \texttt{ do } z := w \texttt{ such that } r(z,w) \texttt{ od}\rangle true.$$

□

Lemma 4 *There is DL_0 formula $Memb(x;t,u)$ such that for any semi-forest $\mathcal{A}$:*

$$\mathcal{A}, x:c, t:d, u:e \models Memb \qquad \text{iff} \qquad c \in [\![d,e]\!].$$

Proof: Let $Q_1(y;t,u)$ be the following program:

$$v := t;\ \texttt{while } v \neq u \texttt{ do } v := w \texttt{ such that } (R(v,w) \wedge w \neq y) \texttt{ od}.$$

Then let $Memb(x;t,u)$ be $\langle Q\rangle true$ with

$$Q \equiv v := t;\ \texttt{while } v \neq u \texttt{ do } v := y \texttt{ such that } (R(v,y) \wedge \neg\langle Q_1\rangle true) \texttt{ od}.$$

Verification that this is a proper choice is based on lemma 2, and is left for the reader. □

We immediately infer the following:

Lemma 5 *There is a DL_0 formula $Succ(x,y;t,u)$ such that for any semi-forest $\mathcal{A}$*

$\mathcal{A}, x:a, y:b, t:d, u:e \models Succ$ iff $a \in [\![d,e]\!]$ is an immediate successor of $a \in [\![d,e]\!]$.

Lemma 6 *There exist DL_0 formulae $Nat(x;t,u)$ and $Zero(x;t,u)$ such that for any semi-forest $\mathcal{A}$ and any $d,e \in A$ the set $I_{de} = \{a \in M \mid \mathcal{A}, x:a, t:d, u:e \models Nat\}$ is either empty or is an interval $[\![d,e]\!]$ of length $h = h(\mathcal{A})$, and, in the latter case,*

$$\langle I_{de},\ Zero(x;d,e), Succ(x,y;d,e)\rangle \simeq \langle [0,h-1]\ ,0\ ,succ\rangle.$$

Proof: Consider a program $Q(p,q,t,u)$

```
if ¬(∃z Memb(z;t,u) ∧ ∃z Memb(z;p,q))
then
    abort
else
    w := p; s := t
    while w ≠ q ∧ s ≠ u
    do
        w := v such that Succ(v,w;p,q);
        s := v such that Succ(v,s;t,u)
    od
    if s ≠ u then abort else halt fi
fi
```

It is not hard to see that Q has a terminating computation in $\mathcal{A}, p:a, q:b, t:d, u:e$ iff the length of $[\![d,e]\!]$ is no less than the length of $[\![a,b]\!]$, and both intervals are nonempty. It follows that we can take

$$Nat(x;t,u) \equiv (\forall pq\ (\neg\exists z\ Memb(z;p,q) \vee \langle Q\rangle true)) \wedge Memb(x;t,u)$$
$$Zero(x;t,u) \equiv Nat(x;t,u) \wedge x = t.$$

□

The last tool we may (eventually) need is the following proposition, useful when semi-forest structure is defined by some formula, possibly with parameters.

Proposition 1 *Suppose that for a finite τ-structure $\mathcal{A} = \langle n, \vec{R}\rangle$ and a DL formula $\varphi(\vec{x}, \vec{y}; \vec{z})$ with $\vec{x}$ and $\vec{y}$ of length k and $\vec{z}$ of length m, for every choice of $\vec{c} \in n^m$ to be the value of $\vec{z}$, the structure*

$$\mathcal{A}_{\varphi,\vec{z}:\vec{c}} = \langle n^k, \varphi^{\mathcal{A}}(\vec{x}, \vec{y}; \vec{c})\rangle = \langle n^k, \{(\vec{a}, \vec{b}) \in n^k \times n^k \mid \mathcal{A}, \vec{x}:\vec{a}, \vec{y}:\vec{b}, \vec{z}:\vec{c} \models \varphi\}\rangle$$

is a semi-forest of height $h(\mathcal{A}_{\varphi,\vec{z}:\vec{c}})$. Then there is a formula $\widehat{\varphi}(\vec{x}, \vec{z}; \vec{y}, \vec{z}')$ in DL which defines a semi-forest $\mathcal{A}_{\widehat{\varphi}}$ of height $h(\mathcal{A}_{\widehat{\varphi}}) = \max_{\vec{c} \in n^m} h(\mathcal{A}_{\varphi,\vec{z}:\vec{c}})$.

Proof: Taking $\widehat{\varphi}(\vec{x}, \vec{z}; \vec{y}, \vec{z}') \equiv \varphi(\vec{x}, \vec{y}, \vec{z}) \wedge \vec{z} = \vec{z}'$ we get structure

$$\mathcal{A}_{\widehat{\varphi}} = \langle n^{k+m}, \{((\vec{a}, \vec{c}), (\vec{b}, \vec{c})) \in n^{k+m} \times n^{k+m} \mid \mathcal{A}, \vec{x}:\vec{a}, \vec{y}:\vec{b}, \vec{z}:\vec{c} \models \varphi\}\rangle,$$

(isomorphic to) disjoint union of all $\mathcal{A}_{\varphi,\vec{z}:\vec{c}}$, so the claim follows by lemma 1. □

4 Tools: Expressibility Over Structures with Successor

Now we turn to the main technical tools we will use to prove our results.
But first we need some auxiliary terminology. Let us consider the class $\mathcal{W}$ of all finite models of the form $\mathcal{W}(n) \ni W = \langle n, Succ, U\rangle$, where $Succ$ is a standard successor relation restricted to n, and U is an unary predicate. Then every element $W \in \mathcal{W}(n)$ may be treated as a binary string of length n. We say that the logic L over $\langle \leq, u, =\rangle$ *captures complexity class* C iff every subset of $\mathcal{W}$ that belongs to C as set of words is definable by some sentence of L. In our situation unary predicate is usually absent, so we have only structures which correspond to strings over one letter alphabet.
Let us also adopt that a recursive function $G : \omega \to \omega$ is called *space constructible*[2] iff there exists an on-line Turing Machine M which computes G, and for some constant c_G and all $n \in \omega$, machine M uses at most $c_G \cdot length(G(n))$ tape cells during computation with input n (input and output are represented as binary strings).
Let $\widetilde{G}$ be a function from ω into ω defined by:

- $\widetilde{G}(0) = 1$;
- $\widetilde{G}(m) = GG(\widetilde{G}(m-1) + 1)$ for $m > 0$.

It is easy to see that if G is space constructible and strictly growing, then so is $\widetilde{G}$. If $G : \omega \to \omega$ is nondecreasing and unbounded, then for such a function we denote by $\overline{G}^{-1}$ an *upper converse of* G, i.e., a function $\omega \to \omega$ which maps n onto the least natural number m such that $G(m) \geq n$. Similarly we define *lower converse* $\underline{G}^{-1}$ which maps n onto the greatest natural number m such that $G(m) \leq n$.

Lemma 7 *For every space constructible, strictly growing function $G : \omega \to \omega$ and for every logic L which captures DLOGSPACE over structures with successor relation, there is a sentence φ in L such that for any finite structure $\mathcal{A}$ with successor, if n is the size of $\mathcal{A}$, one has*

$$\mathcal{A} \models \varphi \quad \text{iff} \quad \text{there is a value of } \widetilde{G} \text{ in the integer interval } [\underline{G}^{-1}(n), n].$$

[2] Note that our definition slightly differs from the usual one.

Proof: It suffices to show that the property "there is a value of $\widetilde{G}$ in the integer interval $[\underline{G}^{-1}(n), n]$" is in DLOGSPACE (recall that in our situation input value n is given as *unary* string).
The machine we need cycles through all input numbers m of at most $c_{\widetilde{G}} \cdot \log n$ bits and for each of them
1. it tries to compute $r = \widetilde{G}(m)$ in $c_{\widetilde{G}} \cdot \log n$ tape cells to check whether $r \leq n$;
2. it tries to compute $G(r)$ (and eventually $G(r+1)$, if $G(r) = n$) in $c_G \cdot \log n$ tape cells to check whether $r \geq \underline{G}^{-1}(n)$;
3. if $\underline{G}^{-1}(n) \leq r \leq n$, it accepts;
4. if it does not accept for any input of at most $c_{\widetilde{G}} \cdot \log n$ bits, it rejects. □

Lemma 8 *With assumptions as above there exist L formulae Zero(x), Max(x), Add(x, y, z) and Mult(x, y, z) such that for any finite A with successor and of size n one has*

$$\langle M, \preceq, Zero, Max, Add, Mult\rangle \simeq \langle n, \leq, 0, n-1, +, \times\rangle.$$

Proof: Addition and multiplication of numbers bound by n are in DLOGSPACE.□

5 Main Theorem

The main theorem is formulated and proved in version essentially weaker than it could be. We decided to do that for three major reasons: eventual stronger formulations are essentially more complicated (e.g. existence of asymptotic probabilities and complexity of the almost sure theory should be treated separately); the current one is still powerful enough to be applied in many interesting situations; finally, similar theorem is proved in the strongest known to author version in the sequel to the current paper.

Theorem 1 *Suppose that for a class $\mathcal{A}$ of finite structures and a sequence μ of probabilities there exists a DL formula $\varphi(\vec{x}, \vec{y})$ with $\vec{x}$ and $\vec{y}$ of length k, such that for every $A \in \mathcal{A}(n)$ the structure*

$$A_\varphi = \langle n^k, \varphi^A(\vec{x}, \vec{y})\rangle = \langle n^k, \{(\vec{a}, \vec{b}) \in n^k \times n^k \mid A, \vec{x}:\vec{a}, \vec{y}:\vec{b} \models \varphi\}\rangle$$

is a semi-forest and for some nondecreasing, unbounded and recursive function $g: \omega \longrightarrow \omega$ one has

$$\lim_{n\to\infty} \mu_n(\{A \in \mathcal{A}(n) \mid g(n) \leq h(A_\varphi)\}) = 1. \tag{1}$$

Then the limit law fails for DL and $\langle \mathcal{A}, \mu\rangle$ and no associated asymptotic problem is decidable; moreover, there exists a sentence ψ with $\liminf_{n\to\infty} \mu_n(\psi) = 0$ and $\limsup_{n\to\infty} \mu_n(\psi) = 1$.

Proof: We assume $k = 1$. Generalization to higher arities causes no problems. In the examples of applications k is always 1.
Let us note that w.l.o.g. we can assume g to be $\overline{G}^{-1}$ for some space constructible strictly growing function G. Indeed, g is nondecreasing and unbounded, so there exists recursive $\overline{g}^{-1}$, and then there is a space constructible strictly growing G such that $G \geq \overline{g}^{-1}$. Then $\overline{G}^{-1} \leq g$, so $\overline{G}^{-1}$ satisfies (1) as well.

Lemma 6 guarantees that we can define a structure with successor of size $h(\mathcal{A}_\varphi)$ in every $\mathcal{A}$, by means of DL formulae. By a famous theorem in [9] DL captures NLOGSPACE over structures with successor, so there exists, by lemma 7, a DL sentence expressing property "there is a value of $\widetilde{G}$ in the integer interval $[\underline{G}^{-1}(h(\mathcal{A}_\varphi)), h(\mathcal{A}_\varphi)]$". Let us call this sentence Θ. Below we use h to denote $h(\mathcal{A}_\varphi)$. Now let $\varepsilon > 0$ be arbitrary.
Choose n_0 large enough to have for every $n > n_0$:

$$\mu_n(\{\mathcal{A} \in \mathbf{A}(n) \mid g(n) \leq h\}) \geq 1 - \varepsilon \tag{2}$$

- For all $n > n_0$ such that $n = G\widetilde{G}(p)$ for some p (there are infinitely many such n), we have $g(n) = \overline{G}^{-1}(n) = \underline{G}^{-1}(n) = \widetilde{G}(p)$ since G is strictly growing, and therefore

$$\mu_n(\Theta) \geq \mu_n(\{g(n)\} \subseteq [\underline{G}^{-1}(h), h]) = \mu_n(g(n) \leq h \leq n) \geq 1 - \varepsilon,$$

by (2).

- For all $n > n_0$ such that $n = \widetilde{G}(p) - 1$ for some p (there are infinitely many such n), we have no value of $\widetilde{G}$ in $[gg(n), n]$ since $\widetilde{G}(p) > n$ and $\widetilde{G}(p-1) < gg(\widetilde{G}(p) - 1) - 1 < gg(n)$ (recall that $\widetilde{G}$ is strictly growing, too) and thus

$$\mu_n(\Theta) \leq 1 - \mu_n([\underline{G}^{-1}(h), h] \subseteq [gg(n), n]) = 1 - \mu_n(g(n) \leq h \leq n) \leq \varepsilon,$$

by (2).

Therefore, as ε is arbitrary, we get immediately that $\liminf_{n\to\infty} \mu_n(\Theta) = 0$ and $\limsup_{n\to\infty} \mu_n(\Theta) = 1$. In particular $\mu_n(\Theta)$ does not converge.
Now we turn to the undecidability result.
In the paper [3] one may find proof of the following fact, which combined with lemmas 6 and 8 immediately gives the desired consequence:

Proposition 2 *If arbitrarily large initial segments of arithmetics are almost surely definable by means of formulae of logic L closed under classical propositional connectives and first order quantification, then sets $\{\varphi \in L \mid \mu(\varphi) = 1\}$ and $\{\varphi \in L \mid \mu(\varphi) = 0\}$ are recursively inseparable, and hence no asymptotic problem for L is decidable.* □

6 Examples of Applications

The interest of the main theorem is that the semi–forest structure is relatively easy to define, and the function g which should approximate from below the height of this semi–forest may grow very slowly. This allows us to find many interesting situations where we may apply the main theorem to get interesting, new results.
The examples we have chosen are intended to show how we can define the semi–forest structure, and how easy is to find a nondecreasing, unbounded and recursive g to approximate its height. However, we believe that these examples are interesting for their own.

6.1 Slow Growing Classes

We deal with randomized classes considered in the paper [1], to which we refer the reader for necessary definitions.
Suppose that a recursive class $\mathcal{A}$ of semi-forests is closed under disjoint unions and components, and that μ is a sequence of uniform labelled probabilities, and ν is a sequence of uniform unlabelled probabilities. Let $\sum_{n=0}^{\infty}(a_n/n!)x^n$ be the exponential generating series for $\mathcal{A}$ with radius of convergence $0 < R \leq \infty$, and let $\sum_{n=0}^{\infty} b_n x^n$ be the ordinary generating series for $\mathcal{A}$ with radius of convergence $0 < S \leq 1$.
Suppose that the situation is nontrivial, i.e. asymptotic uniform labelled (unlabelled) probabilities of first order sentences do exist. Then, as it has been proved in [1], the following equalities hold:

labelled case

$$\lim_{n\to\infty} \frac{a_{n-m}/(n-m)!}{a_n/n!} = R^m \tag{3}$$

unlabelled case

$$\lim_{n\to\infty} \frac{b_{n-m}}{b_n} = S^m \tag{4}$$

for every m being a cardinality of some connected structure in $\mathcal{A}$.

Theorem 2 *Suppose that the convergence in the above is recursive. Then any of the following conditions is sufficient for the DL limit law to fail, as well as for undecidability of all associated asymptotic problems (below $\mathcal{H}^d$ denotes the subclass of all connected structures of height at least d, symbol $|\mathcal{A}|$ denotes the cardinality of $\mathcal{A}$ and $\sigma(\mathcal{A})$ – the number of automorphisms of $\mathcal{A}$):*

labelled case *1. $R = \infty$ and there are connected elements of $\mathcal{A}$ of arbitrary large height*

2. $0 < R < \infty$ and for every natural d

$$\sum_{\mathcal{A}\in\mathcal{H}^d} \frac{R^{|\mathcal{A}|}}{\sigma(\mathcal{A})} = \infty,$$

unlabeled case *1. $S = 1$ and there are connected elements of $\mathcal{A}$ of arbitrary large height*

2. $0 < S < 1$ and for every natural d

$$\prod_{\mathcal{A}\in\mathcal{H}^d} (1 - S^{|\mathcal{A}|}) = 0.$$

Proof: We defer it to the extended version of the paper.

Example 1 (permutations) Let $\mathcal{P}$ be a class of finite permutations, and let μ be uniform labelled probabilities, and ν uniform unlabelled probabilities.
In that case convergence in (3) and (4) is clearly recursive, $R = 1$, $S = 1$, so we would like to apply labelled 2 and unlabelled 1. Indeed, $\mathcal{H}^d$ contains at least one

cycle of each cardinality $i \geq d$, which is a structure with $i-1$ automorphisms, and thus

$$\sum_{\mathcal{P} \in \mathcal{H}^d} \frac{R^{|\mathcal{P}|}}{\sigma(\mathcal{P})} \leq \sum_{i \geq d} 1/(i-1) = \infty,$$

so there are DL sentences with no uniform labelled asymptotic probability with respect to $\mathcal{P}$, as well as no associated asymptotic problem is decidable.
In the unlabelled case we get the same conclusion even without any calculations. This sharpens result of [10], where it is proved that in the unlabelled case the 0–1 law does not hold, for least fixed point logic. □

6.2 Trees and forests

Example 2 (trees) *Let $\mathcal{T}$ be* any *class of undirected (or directed) trees with degree of vertices almost surely bounded by m. It doesn't matter whether $\mathcal{T}$ contains all such trees or not. The only important assumption is that $\mathcal{T}(n) \neq \emptyset$ for all n. Let μ be* any *sequence of probabilities. Then the limit law for DL fails and no associated asymptotic problem is decidable.*

Proof: Directed trees are themselves semi–forests, and their height almost surely satisfies $\log_m n \leq h(T)$, for $T \in \mathcal{T}(n)$. Now the thesis immediately follows by theorem 1.
In the undirected case we have more problems. Undirected trees are not semi–forests in general, so we need a DL formula which will define semi–forest structure over them, so that its height is equal to the maximal path length in the undirected tree. We leave this for the reader. □

Remark 1 The above example may be, without any additional effort, extended to the case when almost surely all vertices of a random $T \in \mathcal{T}(n)$ have degree not greater than $g(n)$, with g satisfying:
for every rational $\varepsilon > 0$ the sequence $g(n)/n^{\varepsilon}$ recursively (with respect to both n and ε) converges to 0. □

Example 3 (oriented (rooted) forests) We consider the class $\langle \mathcal{L}, \mu \rangle$ of *oriented undirected forests* with labelled, uniform probabilities, as they are described in example 7.8 in [1]. We prove that DL limit law fails in this case, and no associated asymptotic problem is decidable.
This is not hard to see that, similarly as above, we are able to define by DL means a semi–forest from undirected forest, and the height of the resulting semi–forest will be no less than the maximal path length in the source structure. Therefore, by theorem 2, in order to establish our result it suffices to show that

$$\sum_{\mathcal{L} \in \mathcal{H}^d} R^{|\mathcal{L}|}/\sigma(\mathcal{L}) = \infty$$

(note that the suitable sequences converge recursively to $R = 1/e$). Consider the following elements in $\mathcal{H}^d$: for every $i \geq d$ the set of $i!$ chains with root in one of the ends. Each of them has only trivial automorphism. Then have indeed

$$\sum_{\mathcal{L} \in \mathcal{H}^d} R^{|\mathcal{L}|}/\sigma(\mathcal{L}) \geq \sum_{i \geq d} i!(1/e)^i = \infty.$$

6.3 Random graphs

Here we investigate the random graph construction, initiated in [6]. Let $p = p(n)$ be a function from ω into the real interval [0,1]. Then we define the probability model $\mathcal{G}(n,p) = \langle \mathcal{G}, \mu^p \rangle$ with $\mathcal{G}$ the class of all labelled, undirected finite graphs, and $\mu_n^p(\{G\}) = p^m(1-p)^{\binom{n}{2}-m}$, where m is the number of edges in G. Those graphs we treat as structures over the signature $\langle \sim, = \rangle$, following tradition.
Our aim is to prove the following result, which extends knowledge about limit laws for various logics with respect to models $\mathcal{G}(n,p)$.

Example 4 *If $p = p(n)$ satisfies*

1. *$p(n)/n^{-1}\log n$ converges recursively to 0;*

2. *for every positive rational ε the sequence $n^{-1-\varepsilon}/p(n)$ converges to 0, recursively with respect to both n and ε.*

then the limit law fails for DL and $\mathcal{G}(n,p)$, and no associated asymptotic problem is decidable.

Proof: Of course we want to apply theorem 1. In order to do that we need two things: a formula which would define the semi–forest structure over almost every G and a proof that the height of this semi–forest is bounded from the below by some nondecreasing, unbounded and recursive function of n. The idea of finding them comes from the following result (and its proof):

Proposition 3 ([6]) *If $\lim_{n\to\infty} p(n)/n^{-1}\log n = 0$ and $\lim_{n\to\infty} n^{-1-\varepsilon}/p(n) = 0$ for every rational $\varepsilon > 0$, then each finite tree almost surely appears as a component of a random $G \in \mathcal{G}(n)$.* □

Following the proof of this proposition and using information about recursive convergence of both sequences, we can show in a simple way that there is nondecreasing, unbounded and recursive function $g = g(n)$ such that $G \in \mathcal{G}(n)$ contains almost surely each finite tree of size $g(n)$ as a component.
Now we need a method to separate elements that belong to components that are trees to be able to define a semi–forest from such structure, as in previous example. Observe that the component of a is a tree iff for every b in the same component as a, $b \not\sim b$ and, moreover, all cycles $b \sim \ldots \sim b$ have the same the second and the last but one element. This allows us to write easily needed formula. □

7 Deterministic While–Programs

In this section we will work with one fixed class: the class of finite unary functions with uniform labelled probabilities: $\langle \mathcal{F}, \mu \rangle$, over signature $\tau = \langle f, = \rangle$. Our aim is to improve the proof technique in order to get similar result as in §5, but working in a formalism of smaller expressive power.
To describe this formalism let us define the set of *deterministic while–programs* (*DWP*) to be the set of these programs in which the only assignments are of form $x_i := x_j$ and $x_i := f(x_j)$ (which is intended to be a shorthand for the assignment

$x_i := y$ **such that** $y = f(x_j))$, and in which formulae in conditional instructions and while loops are restricted to be only $x_i = x_j$ and $x_i \neq x_j$. Then we are interested in formulae of the form $\langle Q\rangle true$ for deterministic while–programs Q. They play very important role in investigation of properties of programs, especially in the theory of program correctness. The reader may find more complete information about this formalism in [5, 12]. Therefore the aim of this section is to remove nondeterministic assignments, rich tests and as many quantifiers as possible from formulae resulting in our proofs, keeping them working for $\langle \mathcal{F}, \mu\rangle$.
First we cite a result proved in [13] (but in notation from [12]):

Proposition 4 *For a language with one unary function symbol and one unary predicate symbol one has*

$$SP(DWP) \approx DLOGSPACE.$$

□

We do not attempt to explain the sense of this proposition in details here.
We present a result being a corollary to the above and expressed in our terminology, instead:

Corollary 1 1. *Over finite structures with total successor function*[3] *and no other functions, formulae of the form* $\langle \vec{x} := \vec{0};\ Q\rangle true$ *with* $Q \in DWP$ *can express every DLOGSPACE–computable property.*

2. *There is a program* $TSucc(x,y) \in DWP$ *which in every* $\mathcal{F} = \langle n, F\rangle \in \mathcal{F}(n)$ *and for every* $a \in n$, *computes (defines) some total successor function on* $C_a = \{F^n(a) \mid n \in \omega\}$, *with* a *corresponding to 0.*

□

Now we can state our theorem:

Theorem 3 *In the case of* $\langle \mathcal{F}, \mu\rangle$ *the limit law fails for formulae of the form* $\exists x \forall \vec{y} \langle Q\rangle true$ *with* $Q \in DWP$, *as well as no associated asymptotic problem is decidable. Moreover, there is a program* $Q \in DWP$ *with*

$$\liminf_{n\to\infty} \mu_n(\exists x \forall \vec{y} \langle Q\rangle true) = 0$$

and

$$\limsup_{n\to\infty} \mu_n(\exists x \forall \vec{y} \langle Q\rangle true) = 1.$$

Proof (sketch): For fixed strictly growing space constructible $G : \omega \to \omega$ we have a program $Q_G(x, \vec{z}) \in DWP$ which run in $\mathcal{F}, x : a$ checks whether there is a value of $\widetilde{G}$ in the integer interval $[\underline{G}^{-1}(|C_a|), |C_a|]$; if so it stops and otherwise loops forever. Its existence is assured by proposition 1 and lemma 7.
Now we construct a program $Q_1(x, y, \vec{z}) \in DWP$ which run in $\mathcal{F}, x : a, y : b$ checks whether $|C_a| \geq |C_b|$; if so it stops, otherwise loops forever. Additionally, it should not change the value of x. It is very easy when we use program $TSucc$ as a subroutine to compute total successors on C_a and C_b.

---

[3]By a total successor function on $\{0, \ldots, n-1\}$ we mean the following total function: $\lambda m.$ **if** $m \neq n-1$ **then** $m+1$ **else** m **fi**.

Let a program Q_{1G} to be sequential composition of Q_1 and Q_G. Thus the sentence $\theta \equiv \exists x \forall y \forall \vec{z} \langle Q_{1G} \rangle true$ expresses:
"the cardinality of $|C_a|$ is maximal possible and there is a value of $\widetilde{G}$ in the integer interval $[\underline{G}^{-1}(|C_a|), |C_a|]$".
A result in [8] states that in random $\mathcal{F}$ almost surely $|C_0| > \sqrt[3]{n}$.
From now on we can follow the proof of theorem 1, taking θ as Θ, to get the thesis. The only subtle moment is that the set of formulae we are talking about is not closed under propositional connectives and quantification, which is necessary to apply proposition 2. Fortunately, they are expressible by the program itself (one may, roughly speaking, put a "model checker" subroutine inside the program). □
The last result suffices to derive the following corollary which, roughly speaking, asserts that *monadic second order logic MSO* is unable, in a very strong way, to express properties of programs.

Corollary 2 *There exists a program $Q \in DWP$ so that for every monadic second order formula φ (both without constant for 0)*

$$\liminf_{n\to\infty} \mu_n(\varphi \Leftrightarrow \langle Q \rangle true) \leq 1/2.$$

In particular, MSO $\not\geq \{\langle Q \rangle true \mid Q \in DWP\}$.

Proof: Take the program Q with

$$\liminf_{n\to\infty} \mu_n(\exists x \forall \vec{y} \langle Q \rangle true) = 0$$

and

$$\limsup_{n\to\infty} \mu_n(\exists x \forall \vec{y} \langle Q \rangle true) = 1.$$

Then let φ be arbitrary formula in monadic second order logic. It is easy to see that

$$\mu_n(\varphi \Leftrightarrow \langle Q \rangle true) \leq \mu_n(\exists x \forall \vec{y} \varphi \Leftrightarrow \exists x \forall \vec{y} \langle Q \rangle true) \leq$$
$$\leq 1 - |\mu_n(\exists x \forall \vec{y} \varphi) - \mu_n(\exists x \forall \vec{y} \langle Q \rangle true)|.$$

By a theorem proved in [4] $\lim_{n\to\infty} \mu_n(\exists x \forall \vec{y} \varphi)$ exists, so lim inf of the last number above is no greater than 1/2, which immediately implies the claim. □
Note the strength of this result. It is more than just an assertion that monadic second order logic is unable to express halting properties of programs: our result is not only qualitative, but quantitative, as well, and says that there are many counterexamples. In particular it forces that, although finite models are expressive for *DWP* (i.e. for every $Q \in DWP$ and every $\mathcal{F} \in \mathcal{F}$ there is first order formula φ such that $\mathcal{F} \models \varphi \Leftrightarrow \langle Q \rangle true$), they are not uniformly expressive (i.e. the φ above has to depend on $\mathcal{F}$), even if we allow φ to be in MSO. For discussion of expressiveness issues see [5].

Acknowledgment I wish to thank Damian Niwiński for helpful discussions and improving my poor English. Thanks are also due to an anonymous referee, whose suggestions helped me very much.

References

[1] K.J. Compton. A logical approach to asymptotic combinatorics I. First order properties, *Advances in Mathematics* **65**(1987), 65-96.

[2] K.J. Compton. 0–1 laws in logic and combinatorics, *Proc. NATO Advanced Study Institute on Algorithms and Order* (I. Rival, ed.), Reidel, Dordrecht (1988).

[3] K.J. Compton, C.W. Henson, and S. Shelah. Nonconvergence, undecidability and intractability in asymptotic problems, *Annals of Pure and Applied Logic*, **36**(1987), 207-224.

[4] K.J. Compton, and S. Shelah. A convergence theorem for unary functions, *in preparation.*

[5] P. Cousot. Methods and logics for proving programs, *in: J. van Leeuven (ed.) Handbook of theoretical computer science*, North Holland, Amsterdam, 1990.

[6] P. Erdős, and A. Rényi. On the evolution of random graphs, *Magyar Tud. Akad. Mat. Kutató Int. Kőzl.* **5**(1960), 17-61.

[7] D. Harel, and D. Peleg. On static logics, dynamic logics, and complexity classes, *Infor. and Control* **60**(1984), 86-102.

[8] B. Harris. Probability distributions related to random mappings, *The Annals of Mathematical Statistics* **31**(1960), 1045-1062.

[9] N. Immerman. Languages that capture complexity classes, *SIAM Journal of Computing* **16**, No. 4(1987).

[10] Ph. Kolaitis. On asymptotic probabilities of inductive queries and their decision problem, *in: Logics of Programs, Brooklyn, June 1985,* Lecture Notes in Comp. Sci. 193, Springer Verlag, 1985, 153-166.

[11] Ph. Kolaitis, and M.Y. Vardi. Zero–one laws for infinitary logics, *Proc. 5th IEEE Symp. on Logic in Computer Science*, 1990, 156-167.

[12] D. Kozen, and J. Tiuryn. Logics of programs, *in: J. van Leeuven (ed.) Handbook of theoretical computer science*, North Holland, Amsterdam, 1990.

[13] J. Tiuryn, and P. Urzyczyn. Some relationships between logics of programs and complexity theory, *Theoretical Computer Science*, **60**(1988), 83-108.

[14] J. Tyszkiewicz. Infinitary Queries and Their Asymptotic Probabilities II: Properties Definable in Least Fixed Point Logic, *to appear in Proc. Random Graphs'91, Poznań, Poland, August 1991.*

A Semantics of programs and *DL* formulae

A.1 Semantics of programs

Let $\mathcal{A} = \langle A, \vec{R} \rangle$ be any structure, and let $\vec{a} \in A^k$. Then a computation of a program $Q(\vec{x})$ (with k registers) in $\mathcal{A}, \vec{x} : \vec{a}$ is any (note that one program may have many different computations in $\mathcal{A}, \vec{x} : \vec{a}$), finite or infinite, sequence

$$Comp(Q, \mathcal{A}, \vec{a}) = (\vec{a}_0, l_0)(\vec{a}_1, l_1) \ldots$$

of *states* $(\vec{a}_i, l_i) \in A^k \times \omega$ such that:

1. $\vec{a}_0 = \vec{a}$, $l_0 = 0$
2. $(\vec{a}_{m+1}, l_{m+1})$ satisfies the following conditions relatively to $(\vec{a}_m, l_m)$
   (a) if l_m is a number of line with *abort*, then $(\vec{a}_{m+1}, l_{m+1}) = (\vec{a}_m, l_m)$
   (b) if l_m is a number of line with `do` or `fi` then $(\vec{a}_{m+1}, l_{m+1}) = (\vec{a}_m, l_m + 1)$
   (c) if l_m is a number of line with `if` $\varphi(\vec{x})$ then $\vec{a}_{m+1} = \vec{a}_m$ and if $\mathcal{A}, \vec{x} : \vec{a}_m \models \varphi(\vec{x})$ then l_{m+1} is the number of line just after the corresponding `then` line, and otherwise the number of line just after corresponding `else` line
   (d) if l_m is a number of line with `else` , then $\vec{a}_{m+1} = \vec{a}_m$ and l_{m+1} is the number of corresponding `fi` line
   (e) if l_m is a number of line with `while` $\varphi(\vec{x})$, then $\vec{a}_{m+1} = \vec{a}_m$ and if $\mathcal{A}, \vec{x} : \vec{a}_m \models \varphi(\vec{x})$ then $l_{m+1} = l_m + 1$, and otherwise it is the number of the line just after the corresponding odline
   (f) if l_m is a number of line with od then $\vec{a}_{m+1} = \vec{a}_m$ and l_{m+1} is the number of corresponding `while` line
   (g) if l_m is a number of line with $x_i := y$ `such that` $\varphi(y, \vec{x})$ then $\vec{a}_{m+1}$ is identical with $\vec{a}_m$ except i-th coordinate, which is now equal to some element $b \in A$ such that $\mathcal{A}, \vec{x} : \vec{a}_m, y : b \models \varphi$ and $l_{m+1} = l_m + 1$, if such b exists, and otherwise $(\vec{a}_{m+1}, l_{m+1}) = (\vec{a}_m, l_m)$
3. if l_m is a number of line with *halt*, then it is the last element in $Comp(Q, \mathcal{A}, \vec{a})$.

A.2 Semantics of *DL* Formulae

Semantics of *DL* formulae is standard, except the case of the formula $\langle Q \rangle true$. In this case we define

$\mathcal{A}, \vec{x} : \vec{a} \models \langle Q \rangle true$ iff there exists finite computation $Comp(Q, \mathcal{A}, \vec{a})$ of Q in $\mathcal{A}, \vec{x} : \vec{a}$.

On Completeness of Program Synthesis Systems

Andrei Voronkov[1]

ECRC, Arabellastr.17, 8000 Munich 81, Germany (voronkov@ecrc.de)*

Abstract. This paper addresses a question of completeness of systems extracting programs from proofs. We consequently consider specifications of total functions, partial functions, predicates and partial predicates. Our aim is to investigate the problem of existence of a logical system which allows to extract all total recursive function (all partial recursive functions, all decidable predicates) from proofs in this logical system. Another natural question is whether such systems exist among generally known systems. We prove that it is impossible to extract programs for all total recursive functions from the proofs of specifications of the form $\forall x \exists y \mathrm{Out}(x, y)$ and it is impossible to extract algorithms deciding all decidable predicates from the proofs of the specifications of the form $\forall x(P(x) \vee \neg P(x))$. Concerning specifications of the partial functions and the partial predicates we prove that in the system **HA+MP** consisting from the intuitionistic arithmetic with the Markov's principle it is possible to extract all partial recursive functions from the specifications of the form $\forall x(\mathrm{In}(x) \supset \exists y \mathrm{Out}(x, y))$ and algorithms deciding all decidable predicates from the specifications of the form $\forall x(\mathrm{In}(x) \supset (P(x) \vee \neg P(x)))$.

1 Introduction

There are many approaches to extracting programs from constructive proofs. Many papers on programming in logic address the question about the logical power of the proof systems intended for extracting programs from proofs. It means provability (or non-provability) of some principles of constructive mathematics. Another interesting question is to investigate the intensional behavior of the proof systems, for example answering the question on the relationship between the complexity of some algorithm and the complexity of programs which can be extracted from proofs and which compute the same function.

Our paper has a more modest aim - to investigate the extensional behavior of proof systems for program extraction: if f is a total recursive function (a partial recursive function, a decidable predicate), is it possible to extract a program for computing f from some proof. Of course the answer in the case of one particular function or predicate is "yes" - it is possible to extract it from a proof in some suitable calculus. But it is interesting to find logic in which we can extract all possible programs (at least extensionally) from proofs. In other words, our aim is to investigate cases when there is a logical system which allows to extract all total recursive functions (all partial recursive functions, all decidable predicates) from proofs in this

* On leave from the International Laboratory of Intelligent Systems (SINTEL), 630090, Universitetski Prospect 4, Novosibirsk 90, Russia.

logical system. Another interesting question is whether such systems exist among some known systems.

It is not easy to answer these questions in the general case because the notion of an algorithm or of a program is rather vague. So we shall concentrate on programs over the domain for which such a notion exists and had been intensively investigated - the domain of the natural numbers. From now on we restrict our considerations on the natural numbers and assume that all logics that we discuss use the first order arithmetical language $\langle 0, s, +, \times, = \rangle$. Nevertheless the ideas of the proofs from this paper are very general and applicable to other theories in different first order languages or even to type theories. We shall use **N** to denote the set of natural numbers.

We shall also freely use the overbar notation: for example, $\bar{e}$ will stand for the vector $e_1, \ldots, e_n$ and if we use $\bar{e}$ then e_i is the ith component of $\bar{e}$.

An Inference rule is any expression of the form $\varphi_1, \ldots, \varphi_n / \varphi$, where $\varphi_1, \ldots, \varphi_n, \varphi$ are formulae. Axioms are inference rules with $n = 0$. In what follows logic will state for any decidable set of inference rules containing all rules of the intuitionistic predicate calculus. (It means that there exists an algorithm which recognizes whether a given inference rule belongs to the logic or not). Provability in logic $\mathfrak{L}$ will be denoted by $\vdash_{\mathfrak{L}}$ or $\mathfrak{L}\vdash$.

A formula φ is closed iff it has no free variables.

Let m be a natural number. Then $\underline{m}$ will state for the numeral $s^n(0)$ representing this number in arithmetic. For any particular logic $\mathfrak{L}$, let $\nu_{\mathfrak{L}}$ be some fixed Godel enumeration of the set of all proofs in $\mathfrak{L}$. If $\mathfrak{L}$ is logic and φ is a formula then $\mathfrak{L} + \varphi$ stays for the logic obtained from $\mathfrak{L}$ by adding φ as an axiom.

The set Har of Harrop formulae[1] is defined as follows.

1. Any atomic formula is a Harrop formula;
2. If φ, ψ are Harrop formulae and χ is an arbitrary formula then the formulae $\varphi \wedge \psi$, $\forall x \varphi$, $\chi \supset \psi$, $\neg \chi$ are Harrop formulae.

Let f be a total function on natural numbers. The formula $\varphi(\bar{x}, y)$ specifies f iff for any tuple $\bar{m}, n$ of natural numbers $f(\bar{m}) = n$ iff $\varphi(\underline{\bar{m}}, \underline{n})$.

Let f be a partial function on natural numbers. A pair of formulae $\psi(\bar{x})$, $\varphi(\bar{x}, y)$ specifies f iff for any tuple $\bar{m}, n$ of natural numbers we have

1. If $f(\underline{\bar{m}})$ is defined then $\psi(\underline{\bar{m}})$;
2. If $\psi(\underline{\bar{m}})$ then $f(\bar{m}) = n$ iff $\varphi(\underline{\bar{m}}, \bar{n})$.

Let P be a relation on natural numbers. A formula $\varphi(\bar{x})$ specifies P iff for any tuple $\bar{m}$ of natural numbers $P(\bar{m})$ iff $\varphi(\underline{\bar{m}})$.

Via **N** we denote the set of all natural numbers. Let P be a relation on natural numbers and $S \subseteq \mathbf{N}$. A pair of formulae $\psi(\bar{x})$, $\varphi(\bar{x})$ specifies P on the set S iff for any tuple m of natural numbers

1. if $\bar{m} \in S$ then $\psi(\underline{\bar{m}})$;
2. if $\psi(\underline{\bar{m}})$ then $P(\bar{m})$ iff $\varphi(\underline{\bar{m}})$.

2 Total functions

We say that a logic $\mathfrak{L}$ has the PAP_t-property (the proofs-as-programs property for total functions) iff there exists a total recursive function ε such that if n is Gödel number of a proof Π in $\mathfrak{L}$ of a closed formula $\forall\bar{x}\exists y\text{Spec}(\bar{x},y)$ then $\varepsilon(n)$ is a number of a total recursive function f such that for any tuple $\bar{m}$ of natural numbers $\text{Spec}(\underline{\bar{m}}, f(\underline{\bar{m}}))$.

Such logic is t-complete iff for every total recursive function f there exists a formula $\text{Spec}(\bar{x},y)$ which specifies this function and a proof Π of the formula $\forall x\exists y\text{Spec}(\bar{x},y)$ in $\mathfrak{L}$.

The PAP_t-property property means that one can extract total recursive functions from proofs. If logic $\mathfrak{L}$ has PAP_t-property then ε is an algorithm which extracts programs from proofs (a program extractor). t-completeness means that we can extract *all* total recursive functions using the extracting algorithm ε.

Theorem 1. *There is no t-complete logic.*

Proof. Assume that such logic $\mathfrak{L}$ exists. Then using $\nu_{\mathfrak{L}}$ and the extracting algorithm ε one can construct an effective enumeration of the set of all total recursive functions. But it is generally known that such an enumeration does not exist.

From results of the next section follows that if we change definitions then it becomes possible to extract all total recursive functions in the following sense:

> There exists a logic $\mathfrak{L}$ and an algorithm ε such that for any total recursive function f there exists a Harrop formula φ_f which is true on natural numbers such that
> 1. ε is an extracting algorithm for $\mathfrak{L}+\varphi_f$;
> 2. There exists a formula $\text{Spec}(\bar{x},y)$ specifying f such that $\forall x\exists y\text{Spec}(\bar{x},y)$ is provable in $\mathfrak{L}+\varphi_f$.

3 Partial functions

We say that logic $\mathfrak{L}$ has the PAP_p-property (proofs-as-programs property for partial functions) iff there exists a total recursive function ε such that if n is a number of a proof Π in $\mathfrak{L}$ of a closed formula $\forall\bar{x}(\text{In}(\bar{x})\supset\text{Out}(\bar{x},y))$, where $\text{In}(\bar{x})$ is a Harrop formula, then $\varepsilon(n)$ is equal to a Kleene number of a partial recursive function f such that for any tuple $\bar{m}$ of natural numbers if $\text{In}(\underline{\bar{m}})$ then $f(\bar{m})$ is defined and $\text{Out}(\underline{\bar{m}}, f(\underline{\bar{m}}))$.

Such logic is p-complete iff for every partial recursive function f there exists a pair of formulae $\text{In}(\bar{x})$, $\text{Out}(\bar{x},y)$, where $In(\bar{x})$ is a Harrop formula, which specifies this function and a proof Π of the formula $\forall\bar{x}(\text{In}(\bar{x})\supset\exists y\text{Out}(\bar{x},y))$ in $\mathfrak{L}$.

The Markov's principle for a formula φ is the formula $\forall x(\varphi(x)\vee\neg\varphi(x))\supset(\neg\neg\exists x\varphi(x)\supset\exists x\varphi(x))$. Let **HA** be the intuitionistic (Heyting) arithmetic. **HA+MP** will stay for the calculus obtained from the intuitionistic arithmetic by adding all instances of the Markov's principle as axioms.

We say that logic $\mathfrak{L}$ has the ∃-property[2] iff for any closed formula $\exists x\varphi(x)$ from $\vdash_{\mathfrak{L}}\exists x\varphi(x)$ follows that $\vdash_{\mathfrak{L}}\varphi(\underline{n})$ for some natural number n. Logic $\mathfrak{L}$ has the disjunction property iff for closed formulas φ, ψ from $\vdash_{\mathfrak{L}}\varphi\vee\psi$ follows $\vdash_{\mathfrak{L}}\varphi$ or $\vdash_{\mathfrak{L}}\psi$.

Lemma 2. ***If χ is a true Harrop formula then* HA + MP + χ *has the disjunction property and the ∃-property.***

Proof. Let $\mathfrak{L} = \mathbf{HA} + \mathbf{MP} + \chi$. The proof is based on Kleene's characterization of logics possessing these properties [2].

First we introduce some definitions similar to those of [2]. We introduce a new relation | between sets of formulae and formulae in the following way. During the definition we assume that $T\|\!\vdash\varphi$ means $T|\varphi$ and $T\vdash\varphi$, where $\vdash$ stands for the provability in the intuitionistic logic.

1. For atomic formulae φ, $T|\varphi$ iff $T\vdash\varphi$;
2. $T|\varphi\wedge\psi$ iff $T|\varphi$ and $T|\psi$;
3. $T|\varphi\vee\psi$ iff $T\|\!\vdash\varphi$ or $T\|\!\vdash\psi$;
4. $T|\varphi\supset\psi$ iff from $T\|\!\vdash\varphi$ follows $T|\psi$;
5. $T|\neg\varphi$ iff not $T\|\!\vdash\varphi$;
6. $T|\forall x\varphi(x)$ iff for every natural number n, $T|\varphi(\underline{n})$;
7. $T|\exists x\varphi(x)$ iff for some natural number n, $T\|\!\vdash\varphi(\underline{n})$.

The relation $\|\!\vdash$ is identical to Kleene's slash from [2] except for minor changes for the quantifier cases.

As in [2] it is possible to prove that logic obtained from intuitionistic logic by adding a set of formulae S as axioms has the disjunction property and the ∃-property iff for any $\varphi \in S$ we have $S|\varphi$. Let $\mathfrak{L}$ be logic obtained from the intuitionistic predicate calculus by adding as axioms all instances of the induction scheme, the Markov's principle, and some number of true Harrop formulae. We prove that $\mathfrak{L}$ satisfies this property.

1. $\mathfrak{L}|\varphi(0)\supset(\forall x(\varphi(x)\supset\varphi(s(x)))\supset\forall x\varphi(x))$. To prove it we assume $\mathfrak{L}\|\!\vdash\varphi(0)$ and $\mathfrak{L}\|\!\vdash\forall x(\varphi(x)\supset\varphi(s(x)))$. Then trivially for any $n \in \mathbf{N}$ we have $\mathfrak{L}\|\!\vdash\varphi(\underline{n})$ and hence $\mathfrak{L}|\forall x\varphi(x)$.
2. $\mathfrak{L}|\forall x(\varphi(x)\vee\neg\varphi(x))\supset(\neg\neg\exists x\varphi(x)\supset\exists x\varphi(x))$. To prove it we assume $\mathfrak{L}\|\!\vdash\forall x(\varphi(x)\vee\neg\varphi(x))$ and $\mathfrak{L}\|\!\vdash\neg\neg\exists x\varphi(x)$. From $\mathfrak{L}\|\!\vdash\neg\neg\exists x\varphi(x)$ follows $\mathfrak{L}\vdash\neg\neg\exists x\varphi(x)$. Since all formulae provable in $\mathfrak{L}$ are true in the standard model of arithmetic, then for some $n \in \mathbf{N}$ we have $\varphi(\underline{n})$. From $\mathfrak{L}\|\!\vdash\forall x(\varphi(x)\vee\neg\varphi(x))$ we have that for every $n \in \mathbf{N}$ $\mathfrak{L}\|\!\vdash\varphi(\underline{n})\vee\neg\varphi(\underline{n})$. Hence for every $n \in \mathbf{N}$ we have $\mathfrak{L}\|\!\vdash\varphi(\underline{n})$ or $\mathfrak{L}\|\!\vdash\neg\varphi(\underline{n})$. Let $m \in \mathbf{N}$ be such that $\varphi(\underline{m})$. Then $\mathfrak{L}\|\!\vdash\neg\varphi(\underline{m})$ is impossible since only true formulae are provable in $\mathfrak{L}$. So we have $\mathfrak{L}\|\!\vdash\varphi(\underline{m})$ and hence $\mathfrak{L}|\exists x\varphi(x)$.
3. $\mathfrak{L}|\varphi$ where φ is a true Harrop formula. As in [2] it is easy to prove that for any Harrop formulae ψ from $\mathfrak{L}\vdash\psi$ follows $\mathfrak{L}|\psi$.

Theorem 3. **HA+MP** *has the* $\mathrm{PAP_p}$*-property.*

[2] In the language of arithmetic it is usually called *the numerical existence property*

Proof. Let Π be a proof in $\mathfrak{L}$ of a closed formula of the form $\forall \bar{x}(\mathrm{In}(\bar{x}) \supset \exists y \mathrm{Out}(\bar{x}, y))$, where $\mathrm{In}(\bar{x})$ is a Harrop formula. It is sufficient to show how to effectively construct by this proof a partial recursive function f meeting this specification. Let $\bar{m}$ be a tuple of natural numbers. We can effectively enumerate all proofs in the logic $\mathbf{HA} + \mathbf{MP} + \mathrm{In}(\underline{\bar{m}})$. If there exists a proof in $\mathbf{HA} + \mathbf{MP} + \mathrm{In}(\underline{\bar{m}})$ of a formula of the form $\mathrm{Out}(\underline{\bar{m}}, \underline{n})$ then we let $f(\bar{m}) = n$, where n is taken from the first such proof, otherwise $f(\bar{m})$ is not defined.

It is easy to see that f is effectively constructed and hence partial recursive. To prove that it meets the specification assume $\mathrm{In}(\underline{\bar{m}})$. Then $\mathbf{HA} + \mathbf{MP} + \mathrm{In}(\underline{\bar{m}}) \vdash \exists x \mathrm{Out}(\underline{\bar{m}}, x)$. Applying Lemma 2 we obtain that $f(\bar{m})$ is defined. By definition of f we have $\mathbf{HA} + \mathbf{MP} + \mathrm{In}(\underline{\bar{m}}) \vdash \mathrm{Out}(\underline{\bar{m}}, \underline{f(\bar{m})})$. Finally note that in $\mathbf{HA} + \mathbf{MP} + \mathrm{In}(\underline{\bar{m}})$ only true formulae are provable.

Before proving completeness of **HA+MP** we introduce some definitions and lemmata.

Let $C_n(x_1, \ldots, x_{n+1})$, $n = 2, 3 \ldots$ be the formulae representing functions for enumeration of the tuples of natural numbers defined in the following way.

$$C_2(x, y, z) \Leftrightarrow 2 \times z = (x + y) \times (x + y) + 3 \times x + y;$$

$$C_{n+1}(\bar{x}, y, z) \Leftrightarrow \exists u (C_n(\bar{x}, u) \wedge C_2(u, y, z)) \text{ for } n \geq 2.$$

Lemma 4. *The following formulas are provable in* **HA**:

1. $\forall \bar{x} y z (C_n(\bar{x}, y) \wedge C_n(\bar{x}, z) \supset y = z)$;
2. $\forall \bar{x} \bar{y} z (C_n(\bar{x}, z) \wedge C_n(\bar{y}, z) \supset \bar{x} = \bar{y})$;
3. $\forall \bar{x} \exists y C_n(\bar{x}, y)$;
4. $\forall y \exists \bar{x} C_n(\bar{x}, y)$.

Proof. Immediate by induction.

Lemma 5. *If φ is a quantifier-free formula of arithmetic and $\bar{x}$ are all variables of φ then the formula $\forall x(\varphi(\bar{x}) \vee \neg\varphi(\bar{x}))$ is provable in* **HA**.

Proof. Trivial.

Theorem 6. HA+MP *is p-complete.*

Proof. Let $f(\bar{x})$ be a partial recursive function. By the Matijasevič's theorem [3] there exists a quantifier-free formula $F(\bar{x}, \bar{y}, z)$ of the arithmetical language such that for any tuple $\bar{m}, n$ of natural numbers $f(\bar{m}) = n$ iff $\exists \bar{y}(F(\underline{\bar{m}}, \bar{y}, \underline{n}))$. Let $G(\bar{x}, u) \Leftrightarrow \exists \bar{y} z (F(\bar{x}, \bar{y}, z) \wedge C_n(\bar{y}, z, u))$ and $\mathrm{In}(\bar{x}) \Leftrightarrow \neg\neg\exists u G(\bar{x}, u)$. Let $\mathrm{Out}(\bar{x}, z) \Leftrightarrow \exists \bar{y} F(\bar{x}, \bar{y}, z)$. We prove that the pair $\mathrm{In}(\bar{x}), \mathrm{Out}(\bar{x}, z)$ satisfies the conditions of the definition of p-completeness.

From the following equivalencies we see that the pair $\mathrm{In}(\bar{x}), \mathrm{Out}(\bar{x}, z)$ specifies the function f:

$$\begin{aligned}
&\mathrm{In}(\bar{x}) \Leftrightarrow \\
&\exists u G(\bar{x}, u) \Leftrightarrow \\
&\exists \bar{y} z u (F(\bar{x}, \bar{y}, z) \wedge C_n(\bar{y}, z, u)) \Leftrightarrow \\
&\exists \bar{y} z F(\bar{x}, \bar{y}, z) \Leftrightarrow \\
&\exists v (f(\bar{x}) = v).
\end{aligned}$$

It remains to prove that $\mathbf{HA}+\mathbf{MP}\vdash\forall\bar{x}(\mathrm{In}(\bar{x})\supset\exists y\mathrm{Out}(\bar{x},y))$. Using Lemmata 4,5 it is possible to prove that $\mathbf{HA}\vdash\forall\bar{x}u(G(\bar{x},u)\vee\neg G(\bar{x},u))$. Applying the Markov's principle we have $\mathbf{HA}+\mathbf{MP}+\neg\neg\exists uG(\bar{x},u)\vdash\exists uG(\bar{x},u)$ or equivalently $\mathbf{HA}+\mathbf{MP}+\mathrm{In}(\bar{x})\vdash\exists uG(\bar{x},u)$. From this immediately follows $\mathbf{HA}+\mathbf{MP}+\mathrm{In}(\bar{x})\vdash\exists yG(\bar{x},y)$ or $\mathbf{HA}+\mathbf{MP}\vdash\forall\bar{x}(\mathrm{In}(\bar{x})\supset\exists y\mathrm{Out}(\bar{x},y))$.

The next theorem shows that the use of the Markov's principle is essential.

Theorem 7. ***HA** is not p-complete.*

Proof. Assume that **HA** is p-complete. Let f be any partial recursive function which can not be extended to a total recursive function. Let $\mathrm{In}(\bar{x}),\mathrm{Out}(\bar{x},y)$ specifies f, $\mathrm{In}(\bar{x})$ be a Harrop formula, and $\mathbf{HA}\vdash\forall\bar{x}(\mathrm{In}(\bar{x})\supset\exists y\mathrm{Out}(\bar{x},y))$. We define a general recursive function g in the following way. Let $\bar{m}$ be a tuple of natural numbers. Then $\mathbf{HA}+\mathrm{In}(\underline{\bar{m}})\vdash\exists y\mathrm{Out}(\underline{\bar{m}},y)$. Since $\mathrm{In}(\bar{x})$ is a Harrop formula, $\mathbf{HA}+\mathrm{In}(\underline{\bar{m}})$ has the $\exists$-property. So there is some $n\in\mathbf{N}$ such that $\mathbf{HA}+\mathrm{In}(\underline{\bar{m}})\vdash\mathrm{Out}(\underline{\bar{m}},\underline{n})$. We let $g(\bar{m})=n$ for the first such n (in the enumeration of all proofs in $\mathbf{HA}+\mathrm{In}(\underline{\bar{m}})$). By the construction the function g is a total recursive function. Assume that $f(\bar{m})$ is defined. Then $\mathrm{In}(\underline{\bar{m}})$ and hence $\mathrm{Out}(\underline{\bar{m}},\underline{g(\bar{m})})$. From this follows that $f(\bar{m})=g(\bar{m})$. This contradicts to the assumption that f can not be extended to a total recursive function.

4 Predicates

With minor modifications the results of the two previous sections can be generalized to the predicate case.

We say that logic $\mathfrak{L}$ has the PAP$_r$-property (proofs-as-programs property for relations) iff there exists a total recursive function ε such that if n is a number of a proof Π in $\mathfrak{L}$ of a closed formula of the form $\forall\bar{x}(\varphi(\bar{x})\vee\neg\varphi(\bar{x}))$ then $\varepsilon(n)$ is a number of a general recursive function f such that for any tuple $\bar{n}$ of natural numbers $f(\bar{n})=0$ iff $\varphi(\underline{\bar{n}})$.

Such logic is r-complete iff for every decidable predicate P there exists a formula $\varphi(\bar{x})$ which specifies this predicate and a proof Π of the formula $\forall\bar{x}(\varphi(\bar{x})\vee\neg\varphi(\bar{x}))$ in $\mathfrak{L}$.

As in Section 2 we can prove the following statement

Theorem 8. *There is no r-complete logic.*

Proof. The same as for Theorem 1. We note that if such a logic exists we can effectively construct a computable enumeration of the class of all total recursive functions with values from the set $\{0,1\}$, which is impossible.

r-completeness means that it is possible to extract programs deciding all decidable predicates from proofs of specifications of the form $\forall\bar{x}(P(\bar{x})\vee\neg P(\bar{x}))$ in one logic. Using the notion of the specification on set we can extract all decidable predicates from the specifications of the form $\forall\bar{x}(\mathrm{In}(\bar{x})\supset(P(\bar{x})\vee\neg P(\bar{x})))$.

Theorem 9. *Let* $\mathrm{In}(\bar{x})$ *be a Harrop formula of the arithmetical language and* $\mathbf{HA}+\mathbf{MP}\vdash\forall\bar{x}(\mathrm{In}(\bar{x})\supset(P(\bar{x})\vee\neg P(\bar{x})))$. *Then using a proof of this formula in* $\mathbf{HA}+\mathbf{MP}$ *one can effectively find a partial recursive function* f *such that if* $\mathrm{In}(\bar{x})$ *then* $f(\bar{x})$ *is defined and* $f(\bar{x})=0$ *iff* $P(\bar{x})$.

Proof. The same as for Theorem 3.

Let

$$y<x \Leftrightarrow \exists z(\neg z=0\wedge z+y=x);$$

$$\forall y<x\varphi \Leftrightarrow \forall y(y<x\supset\varphi).$$

Lemma 10. *Let* $\mathbf{HA}\vdash\forall x(\varphi(x)\vee\neg\varphi(x))$. *Then* $\mathbf{HA}\vdash\forall x(\forall y<x\varphi(y)\vee\neg\forall y<x\varphi(y))$.

Proof. Immediate.

Theorem 11. *Let* R *be a decidable set of natural numbers. Then there exists a pair of formulae* $\mathrm{In}(\bar{x}), P(\bar{x})$ *such that* $\mathrm{In}(\bar{x})$ *is a Harrop formula,* $\mathrm{In}(\bar{x}), P(\bar{x})$ *specify* R *on* $\mathbf{N}$ *and* $\mathbf{HA}+\mathbf{MP}\vdash\forall\bar{x}(\mathrm{In}(\bar{x})\supset(P(\bar{x})\vee\neg P(\bar{x})))$.

Proof. Let $f(\bar{x})$ be the total recursive function such that $f(\bar{x})=0$ iff $\bar{x}\in R$. Let $F(\bar{x},\bar{y},z)$ be a quantifier-free formula of the arithmetical language such that for any tuple $\bar{m},n$ of natural numbers $f(\bar{m})=n$ iff $\exists\bar{y}(F(\underline{\bar{m}},\bar{y},\underline{n})$. Let $G(\bar{x},u)\Leftrightarrow\exists\bar{y}z(F(\bar{x},\bar{y},z)\wedge C_n(\bar{y},z,u))$ and $H(\bar{x},u)\Leftrightarrow G(\bar{x},u)\wedge(\forall v<u)\neg G(\bar{x},u)$. In Theorem 6 we proved that $\mathbf{HA}\vdash\forall\bar{x}u(G(\bar{x},u)\vee\neg G(\bar{x},u))$. From this using Lemma 10 we obtain $\mathbf{HA}\vdash\forall\bar{x}u(H(\bar{x},u)\vee\neg H(\bar{x},u))$. We let $\mathrm{In}(\bar{x})\Leftrightarrow\neg\neg\exists uH(\bar{x},u)$. Since f is totally defined, $\forall\bar{x}\mathrm{In}(\bar{x})$ is true. Applying the Markov's principle we have $\mathbf{HA}+\mathbf{MP}+\mathrm{In}(\bar{x})\vdash\exists uH(\bar{x},u)$. Let $P(\bar{x})\Leftrightarrow\exists u\bar{y}(H(\bar{x},u)\wedge C_n(\bar{y},0,u))$. First we prove that the pair $\mathrm{In}(\bar{x}), P(\bar{x})$ specifies R on $\mathbf{N}$. It follows from the fact that $\forall\bar{x}\mathrm{In}(\bar{x})$ is true and from the following equivalencies:

$$\begin{aligned}&\bar{x}\in R\Leftrightarrow\\&f(\bar{x})=0\Leftrightarrow\\&\exists\bar{y}(F(\bar{x},\bar{y},0)\Leftrightarrow\\&\exists u\bar{y}(H(\bar{x},u)\wedge C_n(\bar{y},0,u))\Leftrightarrow\\&P(\bar{x}).\end{aligned}$$

From the definition of H it is possible to prove that $\mathbf{HA}\vdash\forall\bar{x}u_1u_2(H(\bar{x},u_1)\wedge H(\bar{x},u_2)\supset u_1=u_2)$. Using it we obtain $\mathbf{HA}\vdash\forall\bar{x}\bar{y}uv(H(\bar{x},u)\wedge C_n(\bar{y},v,u)\wedge\neg v=0\supset\neg P(\bar{x}))$. Finally, using it, $\mathbf{HA}+\mathbf{MP}+\mathrm{In}(\bar{x})\vdash\exists uH(\bar{x},u)$, properties of C_n and $\mathbf{HA}\vdash\forall v(v=0\vee\neg v=0)$ we can obtain $\mathbf{HA}+\mathbf{MP}+\mathrm{In}(\bar{x})\vdash P(\bar{x})\vee\neg P(\bar{x})$ or $\mathbf{HA}+\mathbf{MP}\vdash\forall\bar{x}(\mathrm{In}(\bar{x})\supset(P(\bar{x})\vee\neg P(\bar{x})))$.

The last two theorems essentially mean that we can extract all decidable predicates from proofs in $\mathbf{HA}+\mathbf{MP}$ but in general it is impossible to prove that such a predicate is defined on the set of all natural numbers.

References

1. R Harrop. Concerning formulas of the types $A \rightarrow B \wedge C$, $A \rightarrow (Ex)B(x)$ in intuitionistic formal system. *J. of Symb. Logic*, 17:27–32, 1960.
2. S.C. Kleene. Disjunction and existence under implication in elementary intuitionistic formalism. *Journal of Symbolic Logic*, 27(1), 1962.
3. Matiyasevič. The diophantiness of recursively enumerable sets (in Russian). *Soviet Mathematical Doklady*, pages 279–282, 1970.

Proving Termination for Term Rewriting Systems

Andreas Weiermann

Institut für Mathematische Logik und Grundlagenforschung,
Einsteinstr. 62, W-4400 Münster, Federal Republic of Germany

Abstract

In the first part this paper gives an order-theoretic analysis of the multiset ordering, the recursive path ordering and the lexicographic path ordering with respect to order types and maximal order types. In the second part *relativized ordinal notation systems*, i. e. "ordinary" ordinal notation systems relativized to a given partial order, are introduced and investigated for the general study of precedence-based termination orderings. It is indicated that (at least) the reduction orderings mentioned above are special cases of this construction.

1 Introduction

Current mathematical applications of term rewriting systems presume termination. A simple and common method for proving termination for such a system is to show that the rewrite relation is included in a well-founded relation on the set of terms under consideration. Such a larger well-founded relation is called a *termination ordering* for the rewriting system in question. We give some applications of tools from mathematical logic – and in particular from proof theory – to the study of these orderings.

In Section 2 we give an order-theoretic (or ordinal) analysis of the multiset ordering, the recursive path ordering and the lexicographic path ordering. These precedence-based orderings are studied with respect to well-foundedness and well-partial orderedness. Upper ordinal bounds for the corresponding order types (resp. maximal order types) are computed. We believe that these order types will be the appropriate measures for comparising termination orderings with respect to their range of applicability. Some applications of such an ordinal analysis have been indicated by Cichon in [1]. (Our computations of the maximal order types of the termination orderings mentioned above also yield "constructive" well-partial orderedness proofs in the spirit of [11] for these orderings. These proofs do not require an impredicative minimal bad sequence argument.)

To obtain strong generalizations of precedence-based termination orderings we associate in Section 3 to any (existing) ordinal notation system T and any partial ordering P a (not necessarily well-founded) notation system T_P, which is constructed out of elements of the domain of P in the same way T is constructed out of its single "base" element. For certain particular notation systems T we show that essential order-theoretic properties of the partial ordering (for example linearity, well-foundedness and well-partial orderedness) carry over to the notation system T_P. We conjecture that these lemmata are special cases of a general transfer principle relating notation systems T_P to partial orderings P (which if true should allow a number of new applications to proofs of termination).

2 Ordinal Analysis of Termination Orderings

2.1 Well-Partial Orderings: Summary of Definitions and Results

We summarize some definitions and results of [10]. A *quasi-order* is an ordered pair $\langle X, \leq \rangle$, where X is a set and $\leq$ is a transitive and reflexive binary relation on X. A *partial order* is a quasi-order in which $\leq$ is also anti-symmetric. A *linear order* is a partial order $\langle X, \leq \rangle$ in which any two elements of X are $\leq$-comparable. A partial order $\langle X, \leq \rangle$ is a *well-founded order* if and only if every nonempty subset Y of X contains at least one minimal element (with respect to $\leq$). A partial order $\langle X, \leq \rangle$ is a *well-order* if and only if every nonempty subset Y of X contains exactly one minimal element (with respect to $\leq$). (Every well-order is a linear order.) A *well-quasi-order* is a quasi-order $\langle X, \leq \rangle$ such that there is no sequence $\langle x_i : i < \omega \rangle$ of elements of X satisfying $x_i \not\leq x_j$ for all $i < j$. A *well-partial order* (wpo) is a well-quasi-order which is also a partial order. Note that if $\langle X, \leq \rangle$ is a wpo, $\leq \subseteq \leq^+ \subseteq X \times X$, and $\langle X, \leq^+ \rangle$ is a linear order, then $\langle X, \leq^+ \rangle$ is a well-order. Let $\langle X, \leq \rangle$ be a wpo. We define the *maximal order type* of $\langle X, \leq \rangle$ by $o(\langle X, \leq \rangle) := \sup\{\alpha \colon \alpha$ is the order type of $\langle X, \leq^+ \rangle$ for some extension $\leq^+$ of $\leq$ to a linear order$\}$. For $x \in X$ we define: $L_{\langle X, \leq \rangle}(x) := \{y \in X : x \not\leq y\}$ and $l_{\langle X, \leq \rangle}(x) := o(\langle L_{\langle X, \leq \rangle}(x), \leq \restriction L_{\langle X, \leq \rangle}(x) \rangle)$.

Theorem 2.1 1. *If $\langle X, \leq \rangle$ is a wpo, then there is an extension $\leq^+$ of $\leq$ to a linear order on X such that $\langle X, \leq^+ \rangle$ has order type $o(\langle X, \leq \rangle)$.*

2. *If $\langle X, \leq \rangle$ is a wpo and $Y \subset X$, then $\langle Y, \leq \restriction Y \rangle$ is a wpo and $o(\langle X, \leq \rangle) \leq o(\langle Y, \leq \restriction Y \rangle)$.*

3. *If $\langle X, \leq \rangle$ is a wpo and $x \in X$, then $l_{\langle X, \leq \rangle}(x) < o(\langle X, \leq \rangle)$.*

4. *If $\langle L_{\langle X, \leq \rangle}(x), \leq \restriction L_{\langle X, \leq \rangle}(x) \rangle$ is a wpo for all $x \in X$, then $\langle X, \leq \rangle$ is a wpo.*

5. *If $\langle X, \leq \rangle$ is a wpo, then $o(\langle X, \leq \rangle) = sup\{l_{\langle X, \leq \rangle}(x) + 1 \colon x \in X\}$.*

6. *If $\langle X, \leq_X \rangle$ and $\langle Y, \leq_Y \rangle$ are wpo's, then $\langle X \uplus Y, \leq_X \uplus \leq_Y \rangle$ is also a wpo and $o(\langle X \uplus Y, \leq_X \uplus \leq_Y \rangle) = o(\langle X, \leq_X \rangle) \oplus o(\langle Y, \leq_Y \rangle)$, where $\oplus$ denotes the (commutative) natural sum of ordinals and $\uplus$ disjoint union (of sets).*

7. *If $\langle X, \leq_X \rangle$ and $\langle Y, \leq_Y \rangle$ are wpo's, then $\langle X \times Y, \leq_X \times \leq_Y \rangle$ is also a wpo and $o(\langle X \times Y, \leq_X \times \leq_Y \rangle) = o(\langle X, \leq_X \rangle) \otimes o(\langle Y, \leq_Y \rangle)$, where $\otimes$ denotes the (commutative) natural product of ordinals and $\leq_X \times \leq_y$ is defined as follows:*
$\langle x, y \rangle \leq_X \times \leq_Y \langle x', y' \rangle \;:\Longleftrightarrow\; x \leq_X x' \wedge y \leq_Y y' \quad (x, x' \in X \wedge y, y' \in Y)$.

Let $\langle X, \leq \rangle$ be a partial order (well-partial order, well-founded order, well-order). Define $<$ on X by: $x < y :\Longleftrightarrow x \leq y$ and not $y \leq x$. By abuse of terminology we say that $\langle X, < \rangle$ is a partial order (well-partial order, well-founded order, well-order). (This is justified by the fact that for any $x, y \in X$: $x \leq y \Longleftrightarrow x < y$ or $x = y$.)

Let $\langle X, \leq\rangle$ be a well-founded order. We define for $x \in X$ $otp_<(x) := \sup\{otp_<(y) + 1 \colon y < x\}$. Finally we set $otp_<(X) := \sup\{otp_<(y) + 1 \colon y \in X\}$.

2.2 The Multiset Ordering

As usual small greek letters range over ordinals. Let $AP := \{\alpha \colon (\exists\beta)[\alpha = \omega^\beta]\}$ be the class of the *additive principal numbers* and $E := \{\alpha \colon \alpha = \omega^\alpha\}$ be the class of the *ε-numbers.* We start with some simple observations concerning the multiset ordering. Let $\langle X, <\rangle$ be a partial order. A (finite) *multiset* over X is given by a finite unordered sequence of elements in X. Let $M(X)$ be the set consisting of finite multisets over X. Multisets resemble sets, but allow multiple occurences of identical elements. If $s_1, \ldots, s_m$ are (not necessary distinct) elements of X we denote the multiset consisting of $s_1, \ldots, s_m$ by $[s_1, \ldots, s_m]$. The empty multiset is denoted by $[\,]$. Let $s = [s_1, \ldots, s_m, t_1, \ldots, t_n]$ and $t = [s_1, \ldots, s_m, u_1, \ldots, u_l]$ be multisets over X where $\{t_1, \ldots, t_n\} \cap \{u_1, \ldots, u_l\} = \emptyset$. Such a representation of s and t can always be assumed. Then $s \cap t := [s_1, \ldots, s_m]$, $s \setminus t := [t_1, \ldots, t_n]$ and $t \setminus s := [u_1, \ldots, u_l]$. We define $s \subset t \;:\Longleftrightarrow\; t \setminus s = [\,]$ and put $s = t \;:\Longleftrightarrow\; s \subset t \wedge t \subset s$. We say $x \in s \;:\Longleftrightarrow\; x \in \{s_1, \ldots, s_m\}$. The *multiset ordering* $<<$ is defined as follows:

$$s << t \quad :\Longleftrightarrow \quad s \neq t \wedge (\forall x \in s \setminus t)(\exists y \in t \setminus s)[x < y].$$

Lemma 2.1 *1. If $\langle X, \leq\rangle$ is a linear order, then $\langle M(X), <<\rangle$ is a linear order.*

2. If $\langle X, <\rangle$ is a well-founded order, then $\langle M(X), <<\rangle$ is also a well-founded order and $otp_{<<}(M(X)) \leq \omega^{otp_<(X)}$.

3. If $\langle X, \leq\rangle$ is a well-partial order, then $\langle M(X), \underline{<<}\rangle$ is a well-partial order and $o(\langle M(X), \underline{<<}\rangle) \leq \omega^{o(\langle X, \leq\rangle)}$.

Proof. 1) and 2) are obvious. (cf. [6]).
3) Higman's Lemma (see, for example, [16] Theorem 1.9 for a proof) yields that $\langle M(X), \underline{<<}\rangle$ is a wpo. We prove (with the help of Theorem 2.1 5)) $o(\langle M(X), \underline{<<}\rangle) \leq \omega^{o(\langle X, \leq\rangle)}$ by transfinite induction on $o(\langle X, \leq\rangle)$. Let $t = [t_1, \ldots, t_n] \in L_{\langle M(X), \underline{<<}\rangle}(s)$, where $s = [s_1, \ldots, s_m] \in M(X)$. Let $p := s \cap t$. Since $s \underline{<<} t$ does not hold there exists an $x \in s \setminus p$, where $x = s_{m'}$ and $m' \in \{1, \ldots, m\}$ is minimal chosen, such that for all $y \in t \setminus p$ $x \leq y$ does not hold. We regard t as $\langle p, t \setminus p\rangle$ which is an element of

$$Z := \biguplus \{\{p\} \times M(L_{\langle X, \leq\rangle}(x)) \colon p \subset s \wedge x \in s \setminus p\}.$$

So $L_{\langle M(X), \underline{<<}\rangle}(s)$ is contained in Z and the induced ordering $\leq_Z$ on Z is a subrelation of $\leq$ on $L_{\langle M(X), \underline{<<}\rangle}(s)$. So we see by i. h. and Theorem 2.1 2), 6), 7)

$$l_{\langle M(X), \underline{<<}\rangle}(s) \leq o(\langle Z, \leq_Z\rangle) \leq \bigoplus \{1 \otimes \omega^{o(L_{\langle X, \leq\rangle}(x))} \colon p \subset s \wedge x \in s \setminus p\} < \omega^{o(\langle X, \leq\rangle)}.$$

□

Note that (with the use of Theorem 2.1 4)) the proof of Lemma 2.1 3) also yields a "constructive" well-partial orderedness proof of the multiset ordering which requires only a simple transfinite induction instead of an impredicative minimal bad sequence argument.

Remark. Similar results are true for nested multisets. This follows from the results of the next Subsection.

2.3 The Recursive Path Ordering

In this Subsection we compute upper bounds for the order types of the recursive path ordering (see, for example, [3] for a definition). We define by transfinite recursion on α and ξ (cf.[14]): $\varphi\alpha\xi := \min\{\zeta \in AP\colon (\forall\eta < \alpha)[\varphi\eta\zeta = \zeta] \wedge (\forall\rho < \xi)[\varphi\alpha\rho < \zeta]\}$. Let $\overline{\varphi}\alpha\beta := \varphi\alpha(\beta+1)$, if there are β_0 and $m < \omega$ such that $\varphi\alpha\beta_0 = \beta_0$ and $\beta = \beta_0 + m$, or if $\varphi\alpha 0 = \alpha$ and $\beta < \omega$. Otherwise let $\overline{\varphi}\alpha\beta := \varphi\alpha\beta$. Then $\alpha, \beta < \overline{\varphi}\alpha\beta$ holds for all ordinals α and β.

Lemma 2.2 *Let $\delta < \varphi\gamma\delta$. Then $\varphi\alpha\beta < \varphi\gamma\delta$ holds if and only if exactly one of the following conditions is satisfied:*

1. $\alpha < \gamma \wedge \beta < \varphi\gamma\delta$,
2. $\alpha = \gamma \wedge \beta < \delta$,
3. $\gamma < \alpha \wedge \varphi\alpha\beta \leq \delta$.

Definition 2.1 *Let $\prec$ be a partial order on a (nonvoid) set F of (varyadic) function symbols.* The recursive path ordering *$\prec^*$ on the set of terms over F is defined recursively as follows:*

$$\begin{aligned}
&s = f(s_1,\ldots,s_m) \prec^* g(t_1,\ldots,t_n) = t\\
:\Longleftrightarrow\ &[f = g \wedge [s_1,\ldots,s_m] \prec^*\!\prec^* [t_1,\ldots,t_n]\,]\vee\\
&[f \prec g \wedge [s_1,\ldots,s_m] \prec^*\!\prec^* [t]\,]\vee\\
&[\neg f \prec g \wedge [s] \underline{\preceq^*\!\prec^*} [t_1,\ldots,t_n]\,].
\end{aligned}$$

The set of (closed) terms is denoted by T(F).

Lemma 2.3 *1. If $\prec$ is a well-founded ordering on F, then $\prec^*$ is a well-founded ordering on $T(F)$ and $otp_{\prec^*}(T(F)) \leq \varphi(otp_{\prec}(F))0$.*

2. If $\preceq$ is a well-partial ordering on F, then $\preceq^*$ is a well-partial ordering on $T(F)$ and $o(\langle T(F), \preceq^*\rangle) \leq \overline{\varphi}(o(\langle F, \preceq\rangle))0$.

Proof. 1) This is standard (cf. [4]).
2) Kruskal's theorem yields that $\preceq^*$ is a wpo. The assertion about the maximal order type follows from Theorem 2.2 of [16] by some additional considerations. Here we give a direct proof (which is an adaptation of Schmidt's proof-technique to the present situation). Let $\langle C, \leq\rangle$ be a well-partial order where C is a set of new constants. $T(F,C)$ denotes the set of closed terms which can be built up by F and C. We enlarge the recursive path ordering in the natural way to this set, denote the new ordering also by $\preceq^*$ and show

$$o(\langle T(F,C), \preceq^*\rangle) \leq \overline{\varphi}o(\langle F, \preceq\rangle)o(\langle C, \leq\rangle).$$

The assertion follows for $C = \emptyset$. Since $o(\langle T(F,C), \preceq^*\rangle)$ is a limit ordinal it suffices to show by Theorem 2.1 5):

$$\begin{aligned}
&l_{\langle T(F,C),\preceq^*\rangle}(s_1),\ldots,l_{\langle T(F,C),\preceq^*\rangle}(s_m) < \overline{\varphi}o(\langle F, \preceq\rangle)o(\langle C, \leq\rangle)\\
\Longrightarrow\quad &l_{\langle T(F,C),\preceq^*\rangle}(f(s_1,\ldots,s_m)) < \overline{\varphi}o(\langle F, \preceq\rangle)o(\langle C, \leq\rangle).
\end{aligned}$$

Let $g(t_1,\ldots,t_n) \in L_{\langle T(F,C),\preceq^*\rangle}(s)$ where $s := f(s_1,\ldots,s_m)$.
Case 1: $g \in L_{\langle F,\preceq\rangle}(f)$.

Case 2: $f \preceq g$.
Let $s' := [s_1, \ldots, s_m]$, $t' := [t_1, \ldots, t_n]$ and $p := s' \cap t'$. Then $s' \underline{\prec^* \prec^*} t'$ does not hold. So there exists an $x \in s' \setminus p$ (, where $x = s_{m'}$ and $m' \in \{1, \ldots, m\}$ is minimal chosen) such that for any $y \in t' \setminus p$ $x \preceq^* y$ does not hold. Therefore we can regard $g(t_1, \ldots, t_n)$ as $\langle g, p, t' \setminus p \rangle$ which is an element of

$$Y := \biguplus \{F \times \{p\} \times M(L_{\langle T(F,C), \preceq^* \rangle}(x)) : p \subset s' \wedge x \in s' \setminus p\}.$$

So assuming recursively that $t_1, \ldots, t_n$ are elements of $Z := T(L_{\langle F, \preceq \rangle}(f), C \uplus Y)$ we can regard $g(t_1, \ldots, t_n)$ as an element of Z. The induced ordering on this set is a subrelation of $\preceq^*$ on $L_{\langle T(F,C), \preceq^* \rangle}(s)$. So we conclude by using the induction hypothesis
$o(L_{\langle T(F,C), \preceq^* \rangle}(s)) \leq o(T(L_{\langle F, \preceq \rangle}(f), C \uplus Y)) \leq \overline{\varphi} l(f)(o(C) \oplus o(Y)) < \overline{\varphi} o(F) o(C)$
since $l(f) < o(F)$ and $o(Y) \leq \bigoplus \{o(F) \otimes 1 \otimes \omega^{o(L(x))} : p \subset s \wedge x \in s \setminus p\} < \overline{\varphi} o(F) o(C)$.
(At this place we have dropped some subscripts for notational convenience.) □

2.4 The Howard-Bachmann Hierarchy

To analyze the lexicographic path ordering we introduce at this place the Howard-Bachmann hierarchy of ordinals. Let Ω be a regular cardinal (usually chosen as ω_1, the first uncountable cardinal) and $\varepsilon_{\Omega+1} = min\{\xi \in E : \xi > \Omega\}$. The set of ε-numbers below Ω which are needed for the unique representation of $\alpha < \varepsilon_{\Omega+1}$ in Cantor normal form can be defined by transfinite recursion on α as follows:

1. $E_\Omega 0 := E_\Omega \Omega := \emptyset$.
2. $\alpha = \alpha_1 + \cdots + \alpha_n \wedge n > 1 \wedge \alpha_1, \ldots, \alpha_n \in AP \wedge \alpha_1 \geq \ldots \geq \alpha_n$
$\implies E_\Omega \alpha := E_\Omega \alpha_1 \cup \ldots \cup E_\Omega \alpha_n$.
3. $\alpha = \omega^\beta \wedge \beta < \alpha \implies E_\Omega \alpha := E_\Omega \beta$.
4. $\alpha \in E \wedge \alpha < \Omega \implies E_\Omega \alpha = \{\alpha\}$.

Let $\alpha^* := \max(E_\Omega \alpha \cup \{0\})$. By main recursion on $\alpha < \varepsilon_{\Omega+1}$ and side induction on $n < \omega$ we define for $\beta < \Omega$ sets $C(\alpha, \beta)$ and ordinals $\vartheta_\Omega \alpha$ recursively as follows:

1. $\{0, \Omega\} \cup \beta \subseteq C_n(\alpha, \beta)$.
2. $\gamma, \delta \in C_n(\alpha, \beta) \implies \omega^\gamma + \delta \in C_{n+1}(\alpha, \beta)$.
3. $\delta \in C_n(\alpha, \beta) \cap \alpha \implies \vartheta_\Omega \delta \in C_{n+1}(\alpha, \beta)$.
4. $C(\alpha, \beta) := \bigcup \{C_n(\alpha, \beta) : n < \omega\}$.
5. $\vartheta_\Omega \alpha := \min\{\beta < \Omega : C(\alpha, \beta) \cap \Omega \subseteq \beta \wedge \alpha \in C(\alpha, \beta)\}$.

Lemma 2.4 *$\vartheta_\Omega \alpha$ is always defined.*

Proof. Let $\beta_0 := \alpha^* + 1$. Then $\alpha \in C(\alpha, \beta_0)$. Since the cardinality of $C(\alpha, \beta_0)$ is less than Ω there exists an ordinal $\beta_1 < \Omega$ such that $C(\alpha, \beta_0) \cap \Omega \subseteq \beta_1$. Similarly we can pick inductively for each natural number $n > 1$ an ordinal $\beta_n < \Omega$ such that $C(\alpha, \beta_{n-1}) \cap \Omega \subseteq \beta_n$. Let $\beta := \sup\{\beta_n : n < \omega\}$. Then $\alpha \in C(\alpha, \beta)$, and $C(\alpha, \beta) \cap \Omega \subseteq \beta < \Omega$ by the regularity of Ω. Thus $\vartheta_\Omega \alpha \leq \beta < \Omega$. □

The following lemma is now immediate.

Lemma 2.5 *1.* $\vartheta_\Omega \alpha \notin C(\alpha, \vartheta_\Omega \alpha)$.

2. $\alpha^*, \alpha \in C(\alpha, \vartheta_\Omega \alpha)$.

3. $\vartheta_\Omega \alpha = C(\alpha, \vartheta_\Omega \alpha) \cap \Omega$.

4. $\vartheta_\Omega \alpha = \vartheta_\Omega \beta \implies \alpha = \beta$.

5. $\vartheta_\Omega \alpha < \vartheta_\Omega \beta \iff [(\alpha < \beta \wedge \alpha^* < \vartheta_\Omega \beta) \vee (\beta < \alpha \wedge \vartheta_\Omega \alpha \leq \beta^*)]$.

6. $\omega^\beta < \vartheta_\Omega \alpha \iff \beta < \vartheta_\Omega \alpha$.

The Feferman-Schütte ordinal $\Gamma_0 := \min\{\alpha : \alpha = \varphi\alpha 0\}$, the Ackermann ordinal (see, for example, [4] for a definition) and the Howard-Bachmann ordinal, are in this context denoted by $\vartheta_\Omega \Omega^2$, $\vartheta_\Omega \Omega^\omega$ and $\vartheta_\Omega \varepsilon_{\Omega+1} := \sup\{\vartheta_\Omega \alpha : \alpha < \varepsilon_{\Omega+1}\}$, where $\Omega := \omega_1$.

2.5 The Lexicographic Path Ordering

In this Subsection we compute upper bounds for the order types of the lexicographic path ordering (see, for example, [3] for a definition) by an embedding into the Howard-Bachmann hierarchy.

Definition 2.2 *Let* $\prec$ *be a partial order on a set* F *of (varyadic) function symbols. The* lexicographic path ordering $\prec^l$ *on the set of terms over* F *is defined recursively as follows:*

$$
\begin{array}{ll}
 & s = f(s_1, \ldots, s_m) \prec^l g(t_1, \ldots, t_n) = t \\
:\iff & [f \prec g \ \wedge\ (\forall i \leq m)[s_i \ \prec^l t] \vee \\
 & [f = g \ \wedge\ m \leq n \ \wedge\ (\exists i \leq n)(\forall j < i)[s_j = t_j \wedge s_i \prec^l t_i \wedge \\
 & (\forall k \in \{i+1, \ldots, n\})[s_k \prec^l t]] \vee \\
 & [\neg f \prec g \ \wedge\ (\exists i \leq n)[s \preceq^l t_i]].
\end{array}
$$

Lemma 2.6 *1. If* $\prec$ *is a well-founded ordering on* F*, then* $\prec^l$ *is a well-founded ordering on* $T(F)$ *and* $otp_{\prec^l}(T(F)) \leq \vartheta_\Omega(\Omega^\omega \cdot otp_\prec(F))$*, where* $\Omega :=$ *the least regular cardinal greater than* $otp_\prec(F)$*.*

2. If $\preceq$ *is a well-partial ordering on* F*, then* $\preceq^l$ *is a well-partial ordering on* $T(F)$ *and* $o(\langle T(F), \preceq^l \rangle) \leq \vartheta_\Omega(\Omega^\omega \cdot o(\langle F, \preceq \rangle))$*, where* $\Omega :=$ *the least regular cardinal greater than* $o(\langle F, \prec \rangle)$*.*

Proof. 1) The proof is standard (cf. [4]). Define $o : T(F) \to On$ recursively as follows: $o(f(t_1, \ldots, t_n)) := \vartheta_\Omega(\Omega^\omega \cdot otp_<(f) + \Omega^n \cdot o(t_1) + \cdots + \Omega^1 \cdot o(t_n))$. By a straightforward induction on the complexity of the subterms under consideration one can verify: $s \prec^l t \Longrightarrow o(s) < o(t)$.
2) The proof is similar to the proof of Lemma 2.3 2) but, in fact, the assertion is an

immediate corollary of Theorem 2.4 of [16] which states that the maximal order type of any simplification ordering which is based on a well-partial ordered precedence $\preceq$ is bounded by $\vartheta_\Omega(\Omega^\omega \cdot o(\langle F, \preceq\rangle))$. □

3 Relativized Ordinal Notation Systems

The ordinal analysis of the termination orderings given in Section 2 indicates that many important order-theoretic properties of the precedence ordering carry over uniformly to the induced termination ordering. Here we will try to give an analysis of this phenomenon. We define generalized precedence based termination orderings via *relativized ordinal notation systems*. (These systems are similar to the denotation systems which are defined in [7].) A typical (impredicative) ordinal notation system $\langle T, <_T\rangle$ (for example, a notation system based on ordinal diagrams, a notation systems based on Gordeev's projective axioms or a Buchholz-Rathjen-style notation system, see [9, 12, 13] for details) is given by a primitive recursively defined set T of terms and a primitive recursively defined well ordering $<_T$ on T. Unfortunately the "correct" definition of a natural ordinal notation system is one of the major open problems in infinitary proof-theory (cf. [2] or [5]). Therefore we can only give a temporary definition of an ordinal notation system which is sufficient for many applications.

Definition 3.1 *A primitive recursive set (of terms) T together with a primitive recursive binary relation $<_T$ on T is an* ordinal notation system, *if there is a primitive recursive order-isomorphism of $\langle T, <_T\rangle$ into an initial segment of $\langle T(M), <\rangle$. The system $\langle T(M), <\rangle$ is defined in full detail in [13] on pp. 383-384. $\langle T(M), <\rangle$ is by definition an* initial segment *of itself. The other* initial segments *of $\langle T(M), <\rangle$ have the form $\langle T(M) \restriction t, < \restriction t\rangle$ where $T(M) \restriction t := \{s \in T(M) : s < t\}$ and $< \restriction t := < \restriction (T(M) \restriction t)$ for some $t \in T(M)$.*

$\langle T(M), <\rangle$ is at the moment the farest-reaching ordinal notation system. We will identify ordinal notation systems with initial segments of $\langle T(M), <\rangle$. (It can be shown that the approaches via ordinal diagrams or the Gordeev axioms yield systems of the same strength.) Let $\langle X, \leq_X\rangle$ be a given partial order and let $\hat{X} = \{c_x : x \in X\}$ be a set of (new) constants for the elements of X. To define the relativization of $\langle T, <\rangle$ to $\langle X, \leq_X\rangle$ we replace the rule $0 \in T$ [cf. [13] p. 383 2.3.1.] (– such a rule can be found in any definition of a notation system –) by $\hat{X} \cup \{0\} \subset T_X$. The construction of the other terms of T_X proceeds exactly in the same way as before. The partial order $\langle X, \leq_X\rangle$ induces a partial order $\preceq$ on T_X in the natural way. We demand in addition to the rules of [13] p. 383 2.3: $x \leq x' \implies c_x \preceq c_{x'}$. The defining rules for $<$ on T carry over to defining rules for $\prec$ on T_X. To force that linearity of $<$ carries over to linearity of $\prec$ on T_X we demand furthermore the following "urelement" condition: $c_x \prec t$ for any $t \in T_X \setminus \hat{X}$. (It is also possible to demand other reasonable conditions here.)

Definition 3.2 *An ordinal notation system $\langle T, <_T\rangle$ satisfies the* transfer principle *if the following holds: If the partial order $\langle X, \leq\rangle$ is a linear order, a well-founded orderor a well-partial order, then $\langle T_X, \prec\rangle$ is it, too.*

We conjecture that any notation system $\langle T, <_T\rangle$ satisfies this transfer principle and we conjecture furthermore that the order type (resp. the maximal order type) of $\langle T_X, \prec\rangle$ can be computed "constructively" from the corresponding order type of $\langle X, \leq\rangle$. The preservation of linearity and well-foundedness causes usually no deep problems. Note that the usual well-ordering-proof-techniques for a notation system can be relativized to a given well-order (cf. [15]). Some results concerning the preservation of well partial orderedness follow from [4] and [8]. In this Section we give only some very simple examples of the transfer principle.

3.1 The Additive Closure of a Partial Order

First we reformulate the results of Subsection 2.2 in an abstract setting and construct the *additive closure* $\langle \omega^X, \prec\rangle$ of a partial order $\langle X, <\rangle$. Our terminology emphasizes *the analogy to the theory of ordinals.* To be formal, let $\hat{X} := \{c_x : x \in X\}$ be a set of constants for X and $\hat{<}$ be a binary relation symbol for $<$. In the following we should work with these syntactical counterparts of X and $<$ in a formal way. The interpretations of $\hat{X}$ and $\hat{<}$ are obvious. To simplify notation, we identify X with $\hat{X}$ and $<$ with $\hat{<}$. The context will make clear what is going on.

Definition 3.3 *Let $\langle X, <\rangle$ be a partial order. We define a set of terms ω^X as follows: $x_1, \ldots, x_n \in X \wedge x_1 \geq \ldots \geq x_n \implies \omega^{x_1} + \cdots + \omega^{x_n} \in \omega^X$.*
For $m_1 = \omega^{x_1} + \cdots + \omega^{x_m}$ and $m_2 = \omega^{y_1} + \cdots + \omega^{y_n}$ we define: $m_1 \prec m_2 \iff [m < n \wedge (\forall i \leq m) x_i = y_i] \vee [(\exists k \leq m, n)[x_k < y_k \wedge (\forall i < k) x_i = y_i]]$.
We define the formal natural sum *on ω^X as follows: $m_1 + m_2 := \omega^{z_1} + \cdots + \omega^{z_{n+m}}$ if there is a permutation $\langle z_1, \ldots, z_{n+m}\rangle$ of $[x_1, \ldots, x_n, y_1, \ldots, y_m]$ such that $z_1 \geq \ldots \geq z_{n+m}$.*

Lemma 3.1 *1. $\langle \omega^X, \prec\rangle$ is a partial order.*

2. *If $\langle X, <\rangle$ is a linear order, then $\langle \omega^X, \prec\rangle$ is a linear order.*
3. *$\langle X, <\rangle$ is a linear order, then $+$ is a total function on $\langle \omega^X, \prec\rangle$.*
4. *If $\langle X, <\rangle$ is a well-founded order, then $\langle \omega^X, \prec\rangle$ is a well-founded order and $otp_\prec(\omega^X) \leq \omega^{otp_<(X)}$.*
5. *If $\langle X, \leq\rangle$ is a well-partial order, then $\langle \omega^X, \preceq\rangle$ is also a well-partial order and $o(\langle \omega^X, \preceq\rangle) \leq \omega^{o(\langle X, \leq\rangle)}$.*
6. *If $\langle X, \leq\rangle$ is a well-partial order, then $l_{\langle \omega^X, \preceq\rangle}(\omega^{x_1} + \cdots + \omega^{x_n}) = \omega^{l_{\langle X, \leq\rangle}(x_1)} + \cdots + \omega^{l_{\langle X, \leq\rangle}(x_{n-1})} + \omega^{l_{\langle X, \leq\rangle}(x_n)}$.*

Proof. The proof is similar to the proof of Lemma 2.1. □

It is now a simple exercise to define a relativization of the ordinal notation system for ε_0 to $\langle X, \leq\rangle$.

3.2 The Predicative Closure of a Partial Order

Let $\langle X, <\rangle$ be a partial (and without loss of generality an additively closed partial) order. We now define the relativization $\langle X^\Gamma, \prec\rangle$ of the well-known notation system for Γ_0 to $\langle X, \leq\rangle$. (The notation system for Γ_0 is used in [14] for the proof theory of predicative analysis.)

Definition 3.4 *1.* $x \in X \implies x \in X^\Gamma$.

2. $t = t_1 + \cdots + t_n + t_{n+1}, t_1, \ldots, t_n \in X^\Gamma \setminus X, t_{n+1} \in X^\Gamma$, $t_1 \geq \ldots \geq t_{n+1}$, *and* $(\forall i \leq n)$ t_i *not a "sum"* $\implies t \in X^\Gamma$.

3. $t = \varphi t_1 t_2$ *and* $t_1, t_2 \in X^\Gamma \implies t \in X^\Gamma$.

For $m_1, m_2 \in X^\Gamma$ *we define:*

$$
\begin{aligned}
m_1 \prec m_2 \iff & [m_1 \in X \ \wedge \ m_2 \in X \ \wedge \ m_1 < m_2] \vee \\
& [m_1 \in X \ \vee \ m_2 \in X^\Gamma \setminus X] \vee \\
& [m_1 = s_1 + \cdots + s_{m+1} \wedge \\
& m_2 = t_1 + \cdots + t_{n+1} \wedge \\
& [s_1, \ldots, s_{m+1}] << [t_1, \ldots, t_{n+1}] \text{ as multisets }] \vee \\
& [m_1 = s_1 + \cdots + s_{m+1} \wedge \\
& m_2 = \varphi t_1 t_2 \ \wedge \ (\forall i \leq m+1) \ [s_i \prec m_2]] \\
& [m_1 = \varphi s_1 s_2 \ \wedge \ m_2 = t_1 + \cdots + t_{n+1} \\
& \wedge \, m_1 \leq t_1] \vee \\
& [m_1 = \varphi s_1 s_2 \ \wedge \ m_2 = \varphi t_1 t_2 \ \wedge \\
& [(s_1 \prec t_1 \ \wedge \ s_2 \prec m_2) \ \vee \ (s_1 = t_1 \ \wedge \ s_2 \prec t_2) \\
& \vee \, (t_1 \prec s_1 \ \wedge \ m_1 \preceq t_2)]].
\end{aligned}
$$

For $\delta \in On$ *let* $\varphi^\delta_\alpha \beta := \min\{\xi \in AP : \xi \geq \delta \ \wedge \ (\forall \eta < \alpha)[\varphi^\delta_\eta \xi = \xi] \wedge (\forall \sigma < \alpha)[\varphi \alpha \sigma < \xi]\}$.

Lemma 3.2 *1.* $\langle X^\Gamma, \prec \rangle$ *is a partial order.*

2. If $\langle X, < \rangle$ *is a linear order, then* $\langle X^\Gamma, \prec \rangle$ *is a linear order.*

3. If $\langle X, < \rangle$ *is a well-founded order, then* $\langle X^\Gamma, \prec \rangle$ *is a well-founded order and* $otp_\prec(X^\Gamma) \leq \min\{\xi : \xi = \varphi_\xi^{otp_<(X)} 0\}$.

4. If $\langle X, < \rangle$ *is a well-partial order, then* $\langle X^\Gamma, \prec \rangle$ *is a well-partial order and* $o(\langle (X^\Gamma, \prec \rangle) \leq \min\{\xi : \xi = \varphi_\xi^{o(\langle X, \leq \rangle)} 0\}$.

Proof. The proof is left to the reader. □

Remarks. The recursive path ordering can be embedded into $\langle (X^\Gamma, \prec \rangle$ in the obvious way. In the same way one can relativize the notation system for the Ackermann ordinal (cf. [4]) to get a "super-structure" for the lexicographic path-ordering. It is also possible to define an abstract Howard-Bachmann structure on $\langle X, \leq \rangle$ (this will be definitely no longer a simplification ordering).

References

[1] E. A. Cichon: *Bounds on derivation lengths from termination proofs.* Technical Report CSD-TR-622, Department of Computer Science, University of London, Surrey, England, June 1990.

[2] J. N. Crossley, J. B. Kister: *Natural well-orderings.* Archiv für Mathematische Logik und Grundlagenforschung 26 (1986/87), pp. 57-76.

[3] N. Dershowitz, J. P. Jouannaud: *Rewrite systems.* In: Handbook of Theoretical Computer Science, Part B, Elsevier 1990, pp. 243-320.

[4] N. Dershowitz, M. Okada: *Proof theoretic techniques for term rewriting theory.* Proceedings of the Third Annual Symposium on Logic in Computer Science, Edinburgh, July 1988, pp. 104-111.

[5] S. Feferman: *Proof theory: A personal report.* In the appendix of: G. Takeuti: Proof Theory. North-Holland 1987, pp. 447-485.

[6] J. H. Gallier: *What's so special about Kruskal's theorem and the ordinal* Γ_0*? A survey of some results in proof theory.* Annals of Pure and Applied Logic 53 (1991), pp. 199-260.

[7] J. Y. Girard: *Introduction to* Π^1_2*-logic.* Synthese 62 (1985), pp. 191-216.

[8] L. Gordeev: *Generalizations of the Kruskal-Friedman theorems.* Journal of Symbolic Logic 55 (1990), pp. 157-181.

[9] L. Gordeev: *Systems of iterated projective ordinal notations and combinatorial statements about binary labeled trees.* Archive for Mathematical Logic 29 (1989), pp. 29-46.

[10] D. H. J. de Jongh, R. Parikh: *Well-partial orderings and hierarchies.* Indagationes Math. 39 (1977), pp. 195-207.

[11] C. R. Murthy, J. R. Russell: *A constructive proof of Higman's lemma.* Proceedings of the Fifth Annual Symposium on Logic and Computer Science, Philadelphia, PA, June 1990, pp. 257-267.

[12] M. Okada, G. Takeuti: *On the theory of quasi-ordinal diagrams.* Contemporary Mathematics 65, Logic and Combinatorics, Proceedings of the AMS (1987), pp. 295-308.

[13] M. Rathjen: *Proof-theoretic analysis of* KPM. Archive for Mathematical Logic 30 (1990), pp. 377-403.

[14] K. Schütte: *Proof Theory.* Springer, 1977.

[15] K. Schütte: *Ein Wohlordnungsbeweis für das Ordinalzahlensystem* $T(J)$. Archive for Mathematical Logic 27 (1988), pp. 5-20.

[16] D. Schmidt: *Well-Partial Orderings and Their Maximal Order Types.* Habilitationsschrift, Heidelberg 1979.

Printing: Druckhaus Beltz, Hemsbach
Binding: Buchbinderei Schäffer, Grünstadt

Lecture Notes in Computer Science

For information about Vols. 1–535
please contact your bookseller or Springer-Verlag

Vol. 536: J. E. Tomayko, Software Engineering Education. Proceedings, 1991. VIII, 296 pages. 1991.

Vol. 537: A. J. Menezes, S. A. Vanstone (Eds.), Advances in Cryptology – CRYPTO '90. Proceedings. XIII, 644 pages. 1991.

Vol. 538: M. Kojima, N. Megiddo, T. Noma, A. Yoshise, A Unified Approach to Interior Point Algorithms for Linear Complementarity Problems. VIII, 108 pages. 1991.

Vol. 539: H. F. Mattson, T. Mora, T. R. N. Rao (Eds.), Applied Algebra, Algebraic Algorithms and Error-Correcting Codes. Proceedings, 1991. XI, 489 pages. 1991.

Vol. 540: A. Prieto (Ed.), Artificial Neural Networks. Proceedings, 1991. XIII, 476 pages. 1991.

Vol. 541: P. Barahona, L. Moniz Pereira, A. Porto (Eds.), EPIA '91. Proceedings, 1991. VIII, 292 pages. 1991. (Subseries LNAI).

Vol. 542: Z. W. Ras, M. Zemankova (Eds.), Methodologies for Intelligent Systems. Proceedings, 1991. X, 644 pages. 1991. (Subseries LNAI).

Vol. 543: J. Dix, K. P. Jantke, P. H. Schmitt (Eds.), Nonmonotonic and Inductive Logic. Proceedings, 1990. X, 243 pages. 1991. (Subseries LNAI).

Vol. 544: M. Broy, M. Wirsing (Eds.), Methods of Programming. XII, 268 pages. 1991.

Vol. 545: H. Alblas, B. Melichar (Eds.), Attribute Grammars , Applications and Systems. Proceedings, 1991. IX, 513 pages. 1991.

Vol. 546: O. Herzog, C.-R. Rollinger (Eds.), Text Understanding in LILOG. XI, 738 pages. 1991. (Subseries LNAI).

Vol. 547: D. W. Davies (Ed.), Advances in Cryptology – EUROCRYPT '91. Proceedings, 1991. XII, 556 pages. 1991.

Vol. 548: R. Kruse, P. Siegel (Eds.), Symbolic and Quantitative Approaches to Uncertainty. Proceedings, 1991. XI, 362 pages. 1991.

Vol. 549: E. Ardizzone, S. Gaglio, F. Sorbello (Eds.), Trends in Artificial Intelligence. Proceedings, 1991. XIV, 479 pages. 1991. (Subseries LNAI).

Vol. 550: A. van Lamsweerde, A. Fugetta (Eds.), ESEC '91. Proceedings, 1991. XII, 515 pages. 1991.

Vol. 551:S. Prehn, W. J. Toetenel (Eds.), VDM '91. Formal Software Development Methods. Volume 1. Proceedings, 1991. XIII, 699 pages. 1991.

Vol. 552: S. Prehn, W. J. Toetenel (Eds.), VDM '91. Formal Software Development Methods. Volume 2. Proceedings, 1991. XIV, 430 pages. 1991.

Vol. 553: H. Bieri, H. Noltemeier (Eds.), Computational Geometry - Methods, Algorithms and Applications '91. Proceedings, 1991. VIII, 320 pages. 1991.

Vol. 554: G. Grahne, The Problem of Incomplete Information in Relational Databases. VIII, 156 pages. 1991.

Vol. 555: H. Maurer (Ed.), New Results and New Trends in Computer Science. Proceedings, 1991. VIII, 403 pages. 1991.

Vol. 556: J.-M. Jacquet, Conclog: A Methodological Approach to Concurrent Logic Programming. XII, 781 pages. 1991.

Vol. 557: W. L. Hsu, R. C. T. Lee (Eds.), ISA '91 Algorithms. Proceedings, 1991. X, 396 pages. 1991.

Vol. 558: J. Hooman, Specification and Compositional Verification of Real-Time Systems. VIII, 235 pages. 1991.

Vol. 559: G. Butler, Fundamental Algorithms for Permutation Groups. XII, 238 pages. 1991.

Vol. 560: S. Biswas, K. V. Nori (Eds.), Foundations of Software Technology and Theoretical Computer Science. Proceedings, 1991. X, 420 pages. 1991.

Vol. 561: C. Ding, G. Xiao, W. Shan, The Stability Theory of Stream Ciphers. IX, 187 pages. 1991.

Vol. 562: R. Breu, Algebraic Specification Techniques in Object Oriented Programming Environments. XI, 228 pages. 1991.

Vol. 563: A. Karshmer, J. Nehmer (Eds.), Operating Systems of the 90s and Beyond. Proceedings, 1991. X, 285 pages. 1991.

Vol. 564: I. Herman, The Use of Projective Geometry in Computer Graphics. VIII, 146 pages. 1992.

Vol. 565: J. D. Becker, I. Eisele, F. W. Mündemann (Eds.), Parallelism, Learning, Evolution. Proceedings, 1989. VIII, 525 pages. 1991. (Subseries LNAI).

Vol. 566: C. Delobel, M. Kifer, Y. Masunaga (Eds.), Deductive and Object-Oriented Databases. Proceedings, 1991. XV, 581 pages. 1991.

Vol. 567: H. Boley, M. M. Richter (Eds.), Processing Declarative Kowledge. Proceedings, 1991. XII, 427 pages. 1991. (Subseries LNAI).

Vol. 568: H.-J. Bürckert, A Resolution Principle for a Logic with Restricted Quantifiers. X, 116 pages. 1991. (Subseries LNAI).

Vol. 569: A. Beaumont, G. Gupta (Eds.), Parallel Execution of Logic Programs. Proceedings, 1991. VII, 195 pages. 1991.

Vol. 570: R. Berghammer, G. Schmidt (Eds.), Graph-Theoretic Concepts in Computer Science. Proceedings, 1991. VIII, 253 pages. 1992.

Vol. 571: J. Vytopil (Ed.), Formal Techniques in Real-Time and Fault-Tolerant Systems. Proceedings, 1992. IX, 620 pages. 1991.

Vol. 572: K. U. Schulz (Ed.), Word Equations and Related Topics. Proceedings, 1990. VII, 256 pages. 1992.

Vol. 573: G. Cohen, S. N. Litsyn, A. Lobstein, G. Zémor (Eds.), Algebraic Coding. Proceedings, 1991. X, 158 pages. 1992.

Vol. 574: J. P. Banâtre, D. Le Métayer (Eds.), Research Directions in High-Level Parallel Programming Languages. Proceedings, 1991. VIII, 387 pages. 1992.

Vol. 575: K. G. Larsen, A. Skou (Eds.), Computer Aided Verification. Proceedings, 1991. X, 487 pages. 1992.

Vol. 576: J. Feigenbaum (Ed.), Advances in Cryptology - CRYPTO '91. Proceedings. X, 485 pages. 1992.

Vol. 577: A. Finkel, M. Jantzen (Eds.), STACS 92. Proceedings, 1992. XIV, 621 pages. 1992.

Vol. 578: Th. Beth, M. Frisch, G. J. Simmons (Eds.), Public-Key Cryptography: State of the Art and Future Directions. XI, 97 pages. 1992.

Vol. 579: S. Toueg, P. G. Spirakis, L. Kirousis (Eds.), Distributed Algorithms. Proceedings, 1991. X, 319 pages. 1992.

Vol. 580: A. Pirotte, C. Delobel, G. Gottlob (Eds.), Advances in Database Technology – EDBT '92. Proceedings. XII, 551 pages. 1992.

Vol. 581: J.-C. Raoult (Ed.), CAAP '92. Proceedings. VIII, 361 pages. 1992.

Vol. 582: B. Krieg-Brückner (Ed.), ESOP '92. Proceedings. VIII, 491 pages. 1992.

Vol. 583: I. Simon (Ed.), LATIN '92. Proceedings. IX, 545 pages. 1992.

Vol. 584: R. E. Zippel (Ed.), Computer Algebra and Parallelism. Proceedings, 1990. IX, 114 pages. 1992.

Vol. 585: F. Pichler, R. Moreno Díaz (Eds.), Computer Aided System Theory – EUROCAST '91. Proceedings. X, 761 pages. 1992.

Vol. 586: A. Cheese, Parallel Execution of Parlog. IX, 184 pages. 1992.

Vol. 587: R. Dale, E. Hovy, D. Rösner, O. Stock (Eds.), Aspects of Automated Natural Language Generation. Proceedings, 1992. VIII, 311 pages. 1992. (Subseries LNAI).

Vol. 588: G. Sandini (Ed.), Computer Vision – ECCV '92. Proceedings. XV, 909 pages. 1992.

Vol. 589: U. Banerjee, D. Gelernter, A. Nicolau, D. Padua (Eds.), Languages and Compilers for Parallel Computing. Proceedings, 1991. IX, 419 pages. 1992.

Vol. 590: B. Fronhöfer, G. Wrightson (Eds.), Parallelization in Inference Systems. Proceedings, 1990. VIII, 372 pages. 1992. (Subseries LNAI).

Vol. 591: H. P. Zima (Ed.), Parallel Computation. Proceedings, 1991. IX, 451 pages. 1992.

Vol. 592: A. Voronkov (Ed.), Logic Programming. Proceedings, 1991. IX, 514 pages. 1992. (Subseries LNAI).

Vol. 593: P. Loucopoulos (Ed.), Advanced Information Systems Engineering. Proceedings. XI, 650 pages. 1992.

Vol. 594: B. Monien, Th. Ottmann (Eds.), Data Structures and Efficient Algorithms. VIII, 389 pages. 1992.

Vol. 595: M. Levene, The Nested Universal Relation Database Model. X, 177 pages. 1992.

Vol. 596: L.-H. Eriksson, L. Hallnäs, P. Schroeder-Heister (Eds.), Extensions of Logic Programming. Proceedings, 1991. VII, 369 pages. 1992. (Subseries LNAI).

Vol. 597: H. W. Guesgen, J. Hertzberg, A Perspective of Constraint-Based Reasoning. VIII, 123 pages. 1992. (Subseries LNAI).

Vol. 598: S. Brookes, M. Main, A. Melton, M. Mislove, D. Schmidt (Eds.), Mathematical Foundations of Programming Semantics. Proceedings, 1991. VIII, 506 pages. 1992.

Vol. 599: Th. Wetter, K.-D. Althoff, J. Boose, B. R. Gaines, M. Linster, F. Schmalhofer (Eds.), Current Developments in Knowledge Acquisition - EKAW '92. Proceedings. XIII, 444 pages. 1992. (Subseries LNAI).

Vol. 600: J. W. de Bakker, C. Huizing, W. P. de Roever, G. Rozenberg (Eds.), Real-Time: Theory in Practice. Proceedings, 1991. VIII, 723 pages. 1992.

Vol. 601: D. Dolev, Z. Galil, M. Rodeh (Eds.), Theory of Computing and Systems. Proceedings, 1992. VIII, 220 pages. 1992.

Vol. 602: I. Tomek (Ed.), Computer Assisted Learning. Proceedings, 1992. X, 615 pages. 1992.

Vol. 603: J. van Katwijk (Ed.), Ada: Moving Towards 2000. Proceedings, 1992. VIII, 324 pages. 1992.

Vol. 604: F. Belli, F.-J. Radermacher (Eds.), Industrial and Engineering Applications of Artificial Intelligence and Expert Systems. Proceedings, 1992. XV, 702 pages. 1992. (Subseries LNAI).

Vol. 605: D. Etiemble, J.-C. Syre (Eds.), PARLE '92. Parallel Architectures and Languages Europe. Proceedings, 1992. XVII, 984 pages. 1992.

Vol. 606: D. E. Knuth, Axioms and Hulls. IX, 109 pages. 1992.

Vol. 607: D. Kapur (Ed.), Automated Deduction – CADE-11. Proceedings, 1992. XV, 793 pages. 1992. (Subseries LNAI).

Vol. 608: C. Frasson, G. Gauthier, G. I. McCalla (Eds.), Intelligent Tutoring Systems. Proceedings, 1992. XIV, 686 pages. 1992.

Vol. 609: G. Rozenberg (Ed.), Advances in Petri Nets 1992. VIII, 472 pages. 1992.

Vol. 610: F. von Martial, Coordinating Plans of Autonomous Agents. XII, 246 pages. 1992. (Subseries LNAI).

Vol. 611: M. P. Papazoglou, J. Zeleznikow (Eds.), The Next Generation of Information Systems: From Data to Knowledge. VIII, 310 pages. 1992. (Subseries LNAI).

Vol. 612: M. Tokoro, O. Nierstrasz, P. Wegner (Eds.), Object-Based Concurrent Computing. Proceedings, 1991. X, 265 pages. 1992.

Vol. 613: J. P. Myers, Jr., M. J. O'Donnell (Eds.), Constructivity in Computer Science. Proceedings, 1991. X, 247 pages. 1992.

Vol. 614: R. G. Herrtwich (Ed.), Network and Operating System Support for Digital Audio and Video. Proceedings, 1991. XII, 403 pages. 1992.

Vol. 615: O. Lehrmann Madsen (Ed.), ECOOP '92. European Conference on Object Oriented Programming. Proceedings. X, 426 pages. 1992.

Vol. 616: K. Jensen (Ed.), Application and Theory of Petri Nets 1992. Proceedings, 1992. VIII, 398 pages. 1992.

Vol. 617: V. Mařík, O. Štěpánková, R. Trappl (Eds.), Advanced Topics in Artificial Intelligence. Proceedings, 1992. IX, 484 pages. 1992. (Subseries LNAI).

Vol. 618: P. M. D. Gray, R. J. Lucas (Eds.), Advanced Database Systems. Proceedings, 1992. X, 260 pages. 1992.

Vol. 619: D. Pearce, H. Wansing (Eds.), Nonclassical Logics and Information Proceedings. Proceedings, 1990. VII, 171 pages. 1992. (Subseries LNAI).

Vol. 620: A. Nerode, M. Taitslin (Eds.), Logical Foundations of Computer Science – Tver '92. Proceedings. IX, 514 pages. 1992.

Vol. 621: O. Nurmi, E. Ukkonen (Eds.), Algorithm Theory – SWAT '92. Proceedings. VIII, 434 pages. 1992.

Vol. 623: W. Kuich (Ed.), Automata, Languages and Programming. Proceedings, 1992. XII, 721 pages. 1992.

Vol. 624: A. Voronkov (Ed.), Logic Programming and Automated Reasoning. Proceedings, 1992. XIV, 509 pages. 1992. (Subseries LNAI).

Vol. 625: W. Vogler, Modular Construction and Partial Order Semantics of Petri Nets. IX, 252 pages. 1992.

Vol. 626: E. Börger, G. Jäger, H. Kleine Büning, M. M . Richter (Eds.), Computer Science Logic. Proceedings, 1991. VIII, 428 pages. 1992.